# EMERGING WIRELESS MULTIMEDIA

# EMERGING WIRELESS MULTIMEDIA SERVICES AND TECHNOLOGIES

Edited by

**Apostolis K. Salkintzis**
*Motorola, Greece*

**Nikos Passas**
*University of Athens, Greece*

John Wiley & Sons, Ltd

*Other Wiley Editorial Offices*

John Wiley & Sons Inc., 111 River Street, Hoboken, NJ 07030, USA

Jossey-Bass, 989 Market Street, San Francisco, CA 94103-1741, USA

Wiley-VCH Verlag GmbH, Boschstr. 12, D-69469 Weinheim, Germany

John Wiley & Sons Australia Ltd, 42 McDougall Street, Milton, Queensland 4064, Australia

John Wiley & Sons (Asia) Pte Ltd, 2 Clementi Loop #02-01, Jin Xing Distripark, Singapore 129809

John Wiley & Sons Canada Ltd, 22 Worcester Road, Etobicoke, Ontario, Canada M9W 1L1

Wiley also publishes its books in a variety of electronic formats. Some content that appears in print may not be
availabe in electronic books.

*Library of Congress Cataloging-in-Publication Data*


Emerging wireless multimedia services and technologies/edited by A. Salkintzis, N. Passas.
    p. cm.
  Includes bibliographical references and index.
  ISBN 0-470-02149-7 (alk.paper)


   1. Wireless communication systems. 2. Multimedia systems. I. Salkintzis, Apostolis K. II. Passas, N. (Nikos), 1970
TK5103.2.E5157 2005
384.5– dc22                                                                              2005012179

*British Library Cataloguing in Publication Data*

A catalogue record for this book is available from the British Library


ISBN-13 978-0-470-02149-1 (HB)
ISBN-10 0-470-02149-7 (HB)

Typeset in 9/11pt Times by Thomson Press (India) Limited, New Delhi.
Printed and bound in Great Britain by Antony Rowe Ltd, Chippenham, Wiltshire.
This book is printed on acid-free paper responsibly manufactured from sustainable forestry
in which at least two trees are planted for each one used for paper production.

*To my wife Sophie, daughter Rania and son Constantine,*
*for bearing with me so many years and supporting my work*
*with endless understanding, love and patience.*

Apostolis K. Salkintzis

*To my wife Kitty, for her unconditional love, and to my daughter Dimitra,*
*for coming into my life.*

Nikos Passas

# Contents

List of Contributors     xvii

1   **Introduction**     1
*Apostolis K. Salkintzis and Nikos Passas*

1.1   Evolving Towards Wireless Multimedia Networks     1
    *1.1.1 Key Aspects of the Evolution*     3
1.2   Multimedia Over Wireless     4
    *1.2.1 IP over Wireless Networks*     4
1.3   Multimedia Services in WLANs     6
1.4   Multimedia Services in WPANs     7
1.5   Multimedia Services in 3G Networks     7
    *1.5.1 Multimedia Messaging*     8
1.6   Multimedia Services for the Enterprise     9
1.7   Hybrid Multimedia Networks and Seamless Mobility     10
1.8   Book Contents     12
References     13

**Part One    Multimedia Enabling Technologies**

2   **Multimedia Coding Techniques for Wireless Networks**     17
*Anastasios Delopoulos*

2.1   Introduction     17
    *2.1.1 Digital Multimedia and the Need for Compression*     17
    *2.1.2 Standardization Activities*     18
    *2.1.3 Structure of the Chapter*     19
2.2   Basics of Compression     19
    *2.2.1 Entropy, Entropy Reduction and Entropy Coding*     19
    *2.2.2 A General Compression Scheme*     21
2.3   Understanding Speech Characteristics     21
    *2.3.1 Speech Generation and Perception*     21
    *2.3.2 Digital Speech*     22
    *2.3.3 Speech Modeling and Linear Prediction*     23
    *2.3.4 General Aspects of Speech Compression*     24
2.4   Three Types of Speech Compressors     25
    *2.4.1 Waveform Compression*     25
    *2.4.2 Open-Loop Vocoders: Analysis – Synthesis Coders*     29
    *2.4.3 Closed Loop Coders: Analysis by Synthesis Coding*     30
2.5   Speech Coding Standards     35
2.6   Understanding Video Characteristics     36
    *2.6.1 Video Perception*     36

2.6.2  Discrete Representation of Video – Digital Video              37
2.6.3  Basic Video Compression Ideas                                 38
2.7  Video Compression Standards                                     42
2.7.1  H.261                                                         42
2.7.2  H.263                                                         42
2.7.3  MPEG-1                                                        43
2.7.4  MPEG-2                                                        44
2.7.5  MPEG-4                                                        44
2.7.6  H.264                                                         45
References                                                           46

3    Multimedia Transport Protocols for Wireless Networks          49
     Pantelis Balaouras and Ioannis Stavrakakis

3.1  Introduction                                                    94
3.2  Networked Multimedia-based Services                             50
3.2.1  Time Relations in Multimedia                                  50
3.2.2  Non-Real-time and Real-time Multimedia Services              50
3.2.3  CBR vs. VBR Encoding for Video                                52
3.2.4  Transmission of VBR Content Over Constant
       Rate Channels                                                 52
3.3  Classification of Real-time Services                            53
3.3.1  One-Way Streaming                                             53
3.3.2  Media on Demand (MoD) Delivery                                54
3.3.3  Conversational Communication                                  55
3.4  Adaptation at the Video Encoding Level                          56
3.4.1  Non-adaptive Encoding                                         56
3.4.2  Adaptive Encoding                                             56
3.4.3  Scalable/Layered Encoding                                     57
3.5  Quality of Service Issues for Real-time Multimedia Services    57
3.5.1  Bandwidth Availability                                        57
3.5.2  Delay and Jitter                                              58
3.5.3  Recovering Losses                                             58
3.6  Protocols for Multimedia-based Communication Over
     the Wireless Internet                                           61
3.6.1  Why TCP is not Suitable for Real-time Services               61
3.6.2  RTP/UDP/IP                                                    62
3.7  Real-time Transport Protocol (RTP)                              62
3.7.1  Multimedia Session with RTP or how RTP is Used              63
3.7.2  RTP Fixed Head Fields                                         64
3.7.3  RTCP Packet Format                                            65
3.7.4  How Intra-media and Inter-media Synchronization
       is Achieved                                                   70
3.7.5  Monitoring RTT, Jitter and Packet Loss Rate                 70
3.7.6  RTP and Loss Repairing Techniques                            71
3.8  RTP Payload Types                                               72
3.8.1  RTP Profiles for Audio and Video Conferences (RFC3551)      72
3.9  RTP in 3G                                                       77
3.9.1  Supported Media Types in 3GPP                                 78
3.9.2  RTP Implementation Issues For 3G                             79
References                                                           81

**4    Multimedia Control Protocols for Wireless Networks                        83**

*Pedro M. Ruiz, Eduardo Martínez, Juan A. Sánchez and Antonio*
*F. Gómez-Skarmeta*

4.1  Introduction                                                                      83
4.2  A Premier on the Control Plane of Existing Multimedia Standards                    84
     *4.2.1 ITU Protocols for Videoconferencing on Packet-switched Networks*            84
     *4.2.2 IETF Multimedia Internetworking Protocols*                                  87
     *4.2.3 Control Protocols for Wireless Networks*                                    89
4.3  Protocol for Describing Multimedia Sessions: SDP                                   90
     *4.3.1 The Syntax of SDP Messages*                                                 90
     *4.3.2 SDP Examples*                                                               93
4.4  Control Protocols for Media Streaming                                              94
     *4.4.1 RSTP Operation*                                                             94
     *4.4.2 RTSP Messages*                                                              96
     *4.4.3 RTSP Methods*                                                               97
4.5  Session Setup: The Session Initiation Protocol (SIP)                              103
     *4.5.1 Components*                                                                103
     *4.5.2 SIP Messages*                                                              104
     *4.5.3 Addresses*                                                                 106
     *4.5.4 Address Resolution*                                                        107
     *4.5.5 Session Setup*                                                             107
     *4.5.6 Session Termination and Cancellation*                                      108
4.6  Advanced SIP Features for Wireless Networks                                       109
     *4.6.1 Support of User Mobility*                                                  109
     *4.6.2 Personal Mobility*                                                         109
     *4.6.3 Session Modification*                                                      110
     *4.6.4 Session Mobility*                                                          110
4.7  Multimedia Control Panel in UMTS: IMS                                             111
     *4.7.1 IMS Architecture*                                                          111
     *4.7.2 Session Establishment in IMS*                                              113
     *4.7.3 Streaming Services in UMTS*                                                116
4.8  Research Challenges and Opportunities                                             118
Acknowledgement                                                                        119
References                                                                             119

**5    Multimedia Wireless Local Area Networks                                   121**

*Sai Shankar N*

5.1  Introduction                                                                     121
     *5.1.1 ETSI's HiperLAN*                                                          121
     *5.1.2 IEEE 802.11*                                                              123
5.2  Overview of Physical Layers of HiperLAN/2 and IEEE 802.11a                       124
5.3  Overview of HiperLAN/1                                                           125
     *5.3.1 MAC Protocol of HiperLAN/1*                                               125
5.4  Overview of HiperLAN/2                                                           127
     *5.4.1 Data Link Layer*                                                          128
     *5.4.2 MAC Protocol*                                                             128
     *5.4.3 Error Control Protocol*                                                   130
     *5.4.4 Association Control Function (ACF)*                                        130
     *5.4.5 Signaling and Radio Resource Management*                                   131

      *5.4.6 Convergence Layer*         132
      *5.4.7 Throughput Performance of HiperLAN/2*         132
   5.5 IEEE 802.11 MAC         133
      *5.5.1 Distributed Coordination Function*         133
      *5.5.2 Point Coordination Function*         135
   5.6 Overview of IEEE 802.11 Standardization         136
   5.7 IEEE 802.11e HCF         137
      *5.7.1 EDCA*         137
      *5.7.2 HCCA*         139
      *5.7.3 Support For Parameterized Traffic*         140
      *5.7.4 Simple Scheduler*         141
      *5.7.5 Admission Control at HC*         143
      *5.7.6 Power Management*         143
      *5.7.7 ACK Policies*         144
      *5.7.8 Direct Link Protocol (DLP)*         146
   5.8 Simulation Performance of IEEE 802.11         146
      *5.8.1 Throughput vs. Frame Size*         146
      *5.8.2 Throughput vs. Number of Stations*         148
      *5.8.3 EDCA ACs*         149
      *5.8.4 Effect of Bad Link*         149
   5.9 Support for VoIP in IEE 802.11e         150
      *5.9.1 Comparison of Simulation and Analysis for VoIP Traffic*         151
   5.10 Video Transmission Over IEEE 802.11E         152
      *5.10.1 Scenario 1: System Efficiency*         152
      *5.10.2 Scenario 2: TXOP Limit vs. Medium Accessing*
          *Frequency*         154
   5.11 Comparison of HiperLAN/2 and IEEE 802.11E         155
      *5.11.1 Protocol Overhead*         155
      *5.11.2 Comparison of Features*         155
   5.12 Conclusions         156
5.A Appendix: Analysis of the Frame Error Rate and Backoff Process of EDCA Using
One Dimensional Markov Chain         156
5.A.1 MAC/PHY Layer Overheads         156
5.A.2 Link Model         158
5.A.3 Backoff Procedure         161
5.A.4 Throughput Analysis for EDCA Bursting         165
References         166


**6   Wireless Multimedia Personal Area Networks: An Overview**         **169**

*Minal Mishra, Aniruddha Rangnekar and Krishna*
*M. Sivalingam*

   6.1 Introduction         169
   6.2 Multimedia Information Representation         170
   6.3 Bluetooth® (IEEE 802.15.1)         172
      *6.3.1 The Bluetooth® Protocol Stack*         173
      *6.3.2 Physical Layer Details*         174
      *6.3.3 Description of Bluetooth® Links and Packets*         174
      *6.3.4 Link Manager*         176
      *6.3.5 Secret Discovery and Connection Establishment*         177

    *6.3.6 Bluetooth® Security* 179
    *6.3.7 Application Areas* 180
  6.4 Coexistence with Wireless LANs (IEEE 802.15.2) 180
    *6.4.1 Overview of 802.11 Standard* 181
    *6.4.2 802.11b and Bluetooth® Interference Basics* 181
    *6.4.3 Coexistence Framework* 183
  6.5 High-Rate WPANs (IEEE 802.15.3) 184
    *6.5.1 Physical Layer* 184
    *6.5.2 Network Architecture Basics* 184
    *6.5.3 Piconet Formation and Maintenance* 186
    *6.5.4 Channel Access* 188
    *6.5.5 Power Management* 190
    *6.5.6 Security* 191
    *6.5.7 802.15.3a –Ultra-Wideband* 191
  6.6 Low-rate WPANs (IEEE 802.15.4) 194
    *6.6.1 Applications* 195
    *6.6.2 Network Topologies* 195
    *6.6.3 Overview of 802.15.4 Standard* 196
  6.7 Summary 197
  References 197

**7 QoS Provision in Wireless Multimedia Networks** **199**

*Nikos Passas and Apostolis K. Salkintzis*

  7.1 Introduction 199
  7.2 QoS in WLANs 200
    *7.2.1 QoS During Handover* 200
    *7.2.2 Traffic Scheduling in 802.11e* 206
  7.3 RSVP over Wireless Networks 213
  7.4 QoS in Hybrid 3G/WLAN Networks 217
  7.5 UMTS/WLAN Interworking Architecture 218
    *7.5.1 Reference Points* 220
    *7.5.2 Signaling During UMTS-to-WLAN Handover* 221
  7.6 Interworking QoS Considerations 222
  7.7 Performance Evaluation 224
  7.8 Performance Results 225
    *7.8.1 Contention-Based Scenario* 226
    *7.8.2 Contention-Free Scenario* 228
  7.9 Conclusions 231
  Acknowledgements 232
  References 232

**8 Wireless Multimedia in 3G Networks** **235**

*George Xylomenos and Vasilis Vogkas*

  8.1 Introduction 235
  8.2 Cellular Networks 236
    *8.2.1 First Generation* 236
    *8.2.2 Second Generation* 237
    *8.2.3 Third Generation* 238
  8.3 UMTS Networks 239

      *8.3.1  Services and Service Capabilities*                                      239
      *8.3.2  Core Network*                                                          240
      *8.3.3  Radio Access Network*                                                 241
  8.4  Multimedia Services                                                         242
      *8.4.1  IP Multimedia Subsystem*                                               242
      *8.4.2  Multimedia Broadcast/Multicast Service*                                244
  8.5  IMS Architecture and Implementation                                         245
      *8.5.1  Service Architecture*                                                 245
      *8.5.2  Session Setup and Control*                                             247
      *8.5.3  Interworking Functionality*                                            248
  8.6  MBMS Architecture and Implementation                                        249
      *8.6.1  Service Architecture*                                                 249
      *8.6.2  Service Setup and Control*                                             250
      *8.6.3  Interworking Functionality*                                            251
  8.7  Quality of Service                                                          252
      *8.7.1  Quality of Service Architecture*                                       252
      *8.7.2  Policy-based Quality of Service*                                       254
      *8.7.3  Session Setup and Control*                                             255
  8.8  Summary                                                                     256
  8.9  Glossary of Acronyms                                                        256
  References                                                                        258

**Part Two Wireless Multimedia Applications and Services**

**9    Wireless Application Protocol (WAP)**                                            **263**

*Alessandro Andreadis and Giovanni Giambene*

  9.1  Introduction to WAP Protocol and Architecture                               263
      *9.1.1  WAP-based Multimedia Services: Potentials and Limitations*             266
  9.2  WAP Protocol Stack                                                          267
      *9.2.1  Wireless Application Environment*                                      267
      *9.2.2  Wireless Session Protocol*                                             269
      *9.2.3  Wireless Transaction Protocol*                                         271
      *9.2.4  Wireless Transport Layer Security*                                     273
      *9.2.5  Wireless Datagram Protocol*                                            275
  9.3  WAP languages and Design Tools                                              275
      *9.3.1  WML, WMLScript*                                                        276
      *9.3.2  Complementary Technologies*                                            277
      *9.3.3  Conversion of Existing Wepages to WAP*                                 278
      *9.3.4  Dynamic Content Adaptation for WAP Pages Delivery*                     279
  9.4  WAP Service Design Principles                                               280
  9.5  Performance of WAP over 2G and 2.5G Technologies                            282
      *9.5.1  WAP Traffic Modeling Issues and Performance Evaluation*                282
  9.6  Examples of Experimented and Implemented WAP Services                       286
  References                                                                        290

**10   Multimedia Messaging Service (MMS)**                                             **293**

*Alessandro Andreadis and Giovanni Giambene*

  10.1  Evolution From Short to Multimedia Message Services                         293
  10.2  MMS Architecture and Standard                                              294

      *10.2.1 Detailed Description of MMS Architecture Elements*    295
      *10.2.2 Communications Interfaces*    298
      *10.2.3 MMS Capabilities, limitations and Usability Issues*    301
  10.3  MMS Format    302
      *10.3.1 MMS PDU*    302
  10.4  Transaction Flows    307
  10.5  MMS-based Value-added Services    310
      *10.5.1 Survey of MMS-based Services*    313
  10.6  MMS Development Tools    315
  10.7  MMS Evolution    316
  References    317

**11   Instant Messaging and Presence Service (IMPS)**    **319**

*John Buford and Mahfuzur Rahman*

  11.1  Introduction    319
      *11.1.1 Basic Concepts*    319
      *11.1.2 Brief History*    320
      *11.1.3 Standardization*    321
      *11.1.4 Issues*    323
      *11.1.5 Overview of Chapter*    323
  11.2  Client    323
      *11.2.1 Desktop*    323
      *11.2.2 Mobile*    323
  11.3  Design Considerations    324
      *11.3.1 Basic Functional Elements*    324
      *11.3.2 Basic System Concepts*    325
      *11.3.3 Management Aspects*    326
      *11.3.4 Service Architectures: Client-server and Peer-to-peer*    326
  11.4  Protocols    327
      *11.4.1 OMA Wireless Village*    327
      *11.4.2 IETF SIP/SIMPLE*    331
      *11.4.3 XMPP*    335
      *11.4.4 Functional Comparison*    337
      *11.4.5 Gateways*    339
      *11.4.6 Standard APIs*    339
  11.5  Security and Protocols    339
      *11.5.1 Authentication, Confidentiality/Privacy and Integrity*    339
      *11.5.2 Spam in IM*    340
      *11.5.3 Gateways and End-to-End Services*    340
      *11.5.4 Denial of Service*    341
      *11.5.5 Security Features of Specific Standards*    342
  11.6  Evolution, Direction and Challenges    343
      *11.6.1 Rich Presence*    343
      *11.6.2 Home Appliance Control*    344
      *11.6.3 Context-Aware Instant Messaging*    345
      *11.6.4 Virtual Presence at Websites*    345
      *11.6.5 IMP as Application Middleware*    345
  11.7  Summary    347
  References    347

12  **Instant Messaging Enabled Mobile Payments**                                    **349**

*Stamatis Karnouskos, Tadaaki Arimura, Shigetoshi
Yokoyama and Balázs Csik*

12.1  Introduction                                                                      349
  *12.1.1  Mobile Payments*                                                             349
  *12.1.2  Instant Messaging*                                                           350
  *12.1.3  Instant Messaging Enabled Mobile Payments (IMMP)*                            351
12.2  Instant Messaging Mobile Payment Scenario                                        351
12.3  The Generic MP and IM Platforms of IMMP                                          352
  *12.3.1  The Secure Mobile Payment Service*                                           352
  *12.3.2  Air Series Wireless Instant Messaging*                                       354
12.4  Design of an IM-enabled MP System                                                356
  *12.4.1  Server-based Approach*                                                       358
  *12.4.2  'Cooperating Clients' Approach*                                              359
  *12.4.3  Integrated Module Approach*                                                  359
12.5  Implementation                                                                   360
12.6  Security and Privacy in IMMP                                                     363
12.7  Conclusions                                                                      364
References                                                                             365


13  **Push-to-Talk: A First Step to a Unified Instant
    Communication Future**                                                            **367**

*Johanna Wild, Michael Sasuta and Mark Shaughnessy*

13.1  Short History of PTT                                                             368
13.2  Service Description                                                              370
  *13.2.1  Features*                                                                    370
  *13.2.2  Service Management Functions*                                               374
13.3  Architecture                                                                     375
  *13.3.1  PoC System Diagram and Key Elements*                                         375
  *13.3.2  Interfaces*                                                                  377
13.4  Standardization                                                                  378
  *13.4.1  OMA*                                                                         379
  *13.4.2  3GPP and 3GPP2*                                                              379
  *13.4.3  IETF*                                                                        379
13.5  Service Access                                                                   380
  *13.5.1  Service Control*                                                             380
  *13.5.2  Floor Control*                                                               383
  *13.5.3  Media*                                                                       384
13.6  Performance                                                                      386
  *13.6.1  Voice Delay on GPRS*                                                         387
  *13.6.2  Packet Arrival Jitter*                                                       387
  *13.6.3  Call Setup Delay on GPRS*                                                    388
  *13.6.4  Call Setup Optimizations*                                                    389
  *13.6.5  Talker Arbitration Delay on GPRS*                                            390
  *13.6.6  Capacity Impacts on GPRS Networks*                                           390
  *13.6.7  PTT Capacity Relative to Cellular Voice Calls*                               391
13.7  Architecture Migration                                                           391
13.8  Possible Future, or PTT Evolving to PTX                                          393

**14 Location Based Services** **395**

*Ioannis Priggouris, Stathes Hadjiefthymiades and Giannis Marias*

14.1 Introduction 395
14.2 Requirements 396
14.3 LBS System 397
    *14.3.1 LBS Server* 398
    *14.3.2 Positioning Systems* 405
    *14.3.3 Spatial Data (GIS) Systems* 412
    *14.3.4 Supplementary Systems* 414
    *14.3.5 LBS Clients* 419
14.4 Available LBS Systems 421
Acknowledgement 423
References 423

**Index** **427**

# List of Contributors

**Alessandro Andreadis**
Dipartimento di Ingegneria dell'Informazione, Università degli Studi di Siena, Italy

**Tadaaki Arimura**
NTT Data Corporation, Research and Development Headquarters, Tokyo, Japan

**Pantelis Balaouras**
Department of Informatics and Telecommunications, University of Athens, Greece

**John Buford**
Panasonic Digital Network Lab, Princeton, New Jersey, USA

**Balázs Csik**
NTT Data Corporation, Research and Development Headquarters, Tokyo, Japan

**Anastasios Delopoulos**
Electrical and Computer Engineering Department, Aristotle University of Thessaloniki, Thessaloniki, Greece

**Giovanni Giambene**
Dipartimento di Ingegneria dell'Informazione, Università degli Studi di Siena, Italy

**Antonio F. Gómez-Skarmeta**
Department of Information and Communications Engineering, University of Murcia, Spain

**Stathes Hadjiefthymiades**
Communication Networks Laboratory, Department of Informatics and Telecommunications, University of Athens, Greece

**Stamatis Karnouskos**
Fraunhofer FOKUS, Berlin, Germany

**Giannis Marias**
Communication Networks Laboratory, Department of Informatics and Telecommunications, University of Athens, Greece

**Eduardo Martínez**

University of Murcia, Department of Information and Communications Engineering, Spain

**Minal Mishra**

Department of Computer Science and Electrical Engineering, University of Maryland, USA

**Nikos Passas**

Communication Networks Laboratory, Department of Informatics and Telecommunications, University of Athens, Greece

**Ioannis Priggouris**

Communication Networks Laboratory, Department of Informatics and Telecommunications, University of Athens, Greece

**Mahfuzar Rahman**

Panasonic Digital Network Lab, Princeton, New Jersey, USA

**Aniruddha Rangnekar**

Department of Computer Science and Electrical Engineering, University of Maryland, USA

**Pedro M. Ruiz**

University of Murcia, Department of Information and Communications Engineering, Spain

**Apostolis K. Salkintzis**

Motorola, Athens, Greece

**Juan A. Sánchez**

University of Murcia, Department of Information and Communications Engineering, Spain

**Michael Sasuta**

Motorola, Inc., Illinois, USA

**Sai Shankar N**

Philips Research USA, New York, USA

**Mark Shaughnessy**

Motorola, Inc., Arizona, USA

**Krishna M. Sivalingam**

Department of Computer Science and Electrical Engineering, University of Maryland, USA

**Ioannis Stravrakakis**

University of Athens, Department of Informatics and Telecommunications, Greece

**Vasilis Vogkas**

Mobile Multimedia Laboratory, Department of Informatics, Athens University of Economics and Business, Greece

**Johanna Wild**
Motorola GmbH, Munich, Germany

**George Xylomenos**
Mobile Multimedia Laboratory, Department of Informatics, Athens University of Economics and Business, Greece

**Shigetoshi Yokoyama**
NTT Data Corporation, Research and Development Headquarters, Tokyo, Japan

# 1

# Introduction

Apostolis K. Salkintzis and Nikos Passas

## 1.1 Evolving Wireless Multimedia Networks

The objective of this chapter is to provide a brief and yet comprehensive introduction to the evolving of wireless multimedia networks and the key technological aspects and challenges associated with this evolution. In this context, we aim at defining the appropriate framework for the emerging wireless multimedia technologies and the applications that are described in subsequent chapters of this book.

Undoubtedly, the most widely supported evolving path of wireless networks today is the path towards Internet Protocol-based (IP-based) networks, also known as all-IP networks. The term 'all-IP' emphasizes the fact that IP-based protocols are used for all purposes, including transport, mobility, security, QoS, application-level signaling, multimedia service provisioning, etc. In a typical all-IP network architecture, several wireless and fixed access networks are connected to a common core multimedia network, as illustrated in Figure 1.1. Users are able to use multimedia applications over terminals with (ideally) software-configurable radios, capable of supporting a vast range of radio access technologies, such as Wireless Local Area Networks (WLANs), Wireless Personal Area Networks (WPANs), 3G Cellular such as Universal Mobile Telecommunication System (UMTS), Code Division Multiple Access 2000 (cdma2000), etc. In this environment, seamless mobility across the different access networks is considered to be a key issue. Also, native multimedia support by these networks is very important. For this reason, we devote some of the sections below to providing a brief introduction to the features of these networks in relation to multimedia service provision.

In the all-IP network architecture shown in Figure 1.1, the mobile terminals use the IP-based protocols defined by the Internet Engineering Task Force (IETF) to communicate with the multimedia IP network and perform, for example, session/call control and traffic routing. All services in this architecture are provided on top of IP protocol. As shown in the protocol architecture of Figure 1.2, the mobile networks, such as UMTS, cdma2000, etc., turn into access networks that provide only mobile bearer services. The teleservices in these networks (e.g. cellular voice) are used only to support the legacy 2G and 3G terminals, which do not support IP-based applications (e.g. IP telephony). On the user plane, protocols such as the Real Time Protocol (RTP) and the Real Time Streaming Protocol (RTSP) are employed. These user-plane protocols are addressed extensively in Chapter 3. On the other hand, on the control plane, protocols such as the Session Initiation Protocol (SIP) and Resource Reservation

*Emerging Wireless Multimedia: Services and Technologies*   Edited by A. Salkintzis and N. Passas
© 2005 John Wiley & Sons, Ltd

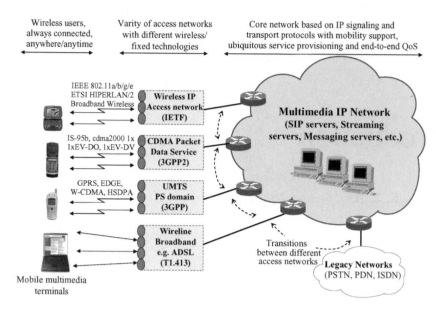

Figure 1.1  Multimedia IP network architecture with various access technologies.

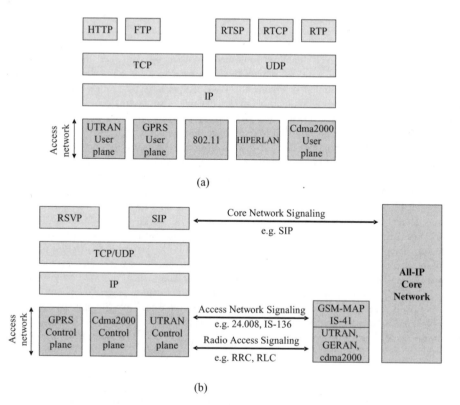

Figure 1.2  Simplified protocol architecture in an all-IP network architecture: (a) user plane, (b) control plane.

Protocol (RSVP) are employed. Chapters 4 and 7 provide a deeper discussion on these control-plane protocols.

For the provision of mobile bearer services, the access networks mainly implement micro-mobility management, radio resource management, and traffic management for provisioning of quality of service. Micro-mobility management in UMTS access networks is based on GPRS Tunneling Protocol (GTP) [1] and uses a hierarchical tunneling scheme for data forwarding. On the other hand, micro-mobility management in cdma2000 access networks is based on IP micro-mobility protocols. Macro-mobility, i.e. mobility across different access networks, is typically based on Mobile-IP, as per RFC 3344 [2].

In the short term, the all-IP network architecture would provide a new communications paradigm based on integrated voice, video and data. You could, for instance, call a user's IP Multimedia Subsystem (IMS) number and be redirected to his web page, where you could have several options, e.g. write an email to him, record a voice message, click on an alternative number to call if he is on vacation, etc. You could also place a SIP call (as discussed in Chapter 4) to a server and update your communication preferences, which could be in the form 'only my manager can call me, all others are redirected to my web page' (or vice versa!). At the same time, you could be on a conference call briefing your colleagues about the outcome of a meeting.

### 1.1.1 Key Aspects of the Evolution

It is instructive at this point to record the key aspects of the evolution towards the wireless multimedia network architecture shown in Figure 1.1. This is because many of these aspects constitute the main focus of this book and are extensively discussed in the following chapters. By briefly referring to these aspects at this point we define an appropriate framework, which entails most of the topics discussed in this book. In short, the most important aspects relevant to the evolution toward the wireless multimedia networks are as follows:

- Wireless networks will evolve to an architecture encompassing an *IP-based multimedia core network* and many wireless access networks (Figure 1.1). As discussed above, the key aspect in this architecture is that signaling with the multimedia core network is based on IP protocols (more correctly, on protocols developed by IETF) and it is independent of the access network (be it UMTS, cdma2000, WLAN, etc.). Therefore, the same IP-based services could be accessed over any access network. An IP-based core network uses IP-based protocols for all purposes, including data transport, networking, mobility, multimedia service provisioning, etc. The first commercial approach towards this IP-based multimedia core network is the co-called IP Multimedia Core Network Subsystem (IMS) standardized by 3GPP and 3GPP2. We further discuss IMS in Chapter 8 along with the Mobile Broadcast/Multicast Service (MBMS).
- The long-term trend is towards *all-IP mobile networks*, where not only the core network but also the radio access network is based solely on IP technology. In this approach, the base stations in a cellular system are IP access routers and mobility/session management is carried out with IP-based protocols (possibly substituting the cellular-specific mobility/session management protocols, such as GTP).
- Enhanced *IP multimedia applications* will be enabled in wireless network by means of application-level signaling protocols standardized by IETF (e.g. SIP, HTTP, etc.). Such protocols are discussed further in Chapters 3 and 4.
- *End-to-end QoS* provisioning will be important for supporting the demanding multimedia applications. In this context, extended interworking between, for example, UMTS QoS and IP QoS schemes is needed or, more generally, interworking between layer-2 QoS schemes and layer-3 QoS (i.e. IP QoS) is required for end-to-end QoS provision. The provision of QoS in multimedia networks is the main topic of Chapter 7.
- *Voice over IP* (VoIP) will be a key technology. As discussed in Chapter 4, several standards organizations are specifying the technology to enable VoIP, e.g. ETSI BRAN TIPHON project, IETF SIP WG, etc.

- The mobile terminals will be based on *software-configurable radios* with capabilities to support many radio access technologies across many frequency bands.
- The ability to move across hybrid access technologies will be an important requirement, which calls for efficient and fast vertical handovers and *seamless mobility*. The IETF working groups SEAMOBY and MOBILE-IP are addressing some of the issues related to seamless mobility. Fast Mobile IP and Micro-mobility schemes are key technologies in this area. We provide more information on seamless mobility in Chapter 7, where we study seamless video continuity across UMTS and WLAN.
- In a highly hybrid access environment, *security* will also play a key role. IEEE 802.11 task group I (TGi) has standardizing new mechanisms for enhanced security in WLANs and also the IETF SEAMOBY group addresses the protocols that deal with (security) context transfer during handovers.
- For extended roaming between different administrative domains and/or different access technologies, *advanced AAA protocols* and *AAA interworking mechanisms* will be implemented.
- *Wireless Personal Area Networks* (WPANs) will play a significant role in the multimedia landscape. WPANs have already start spreading and they will get integrated with the hybrid multimedia network architecture, initially providing services based on the Bluetooth technology (see www. bluetooth.com) and later based on IEEE 802.15.3 high-speed wireless PAN technology, which satisfies the requirement of the digital consumer electronics market (e.g. wireless video communications between a PC and a video camera). WPANs are extensively discussed in Chapter 6.
- *Wireless Local Area Networks* (WLANs) will also contribute considerably to the wireless multimedia provisioning. WLAN technology will evolve further and will support much higher bit rates, in the order of hundreds of Mbps. This is being addressed by the IEEE Wireless Next Generation Standing Committee (see www.ieee802.org/11). WLANs for multimedia services are the main topic of Chapter 5.

As mentioned before, most of the above aspects of the evolution toward the wireless multimedia networks are further discussed in subsequent chapters, mainly Chapters 3–8, which address the emerging multimedia technologies for wireless networks.

## 1.2 Multimedia over Wireless

The evolutionary aspects summarized above call for several technological advances, which are coupled with new technological challenges. These challenges become even tougher when we consider the limitations of wireless environments. One of the most important challenges is the support of multimedia services, such as video broadcasting, video conferencing, combined voice and video applications, etc. The demand for high bandwidth is definitely the key issue for these services, but it is not enough. Other major requirements that should also be considered include seamless mobility, security, context-awareness, flexible charging and unified QoS support, to name but a few.

### 1.2.1 IP over Wireless Networks

Owing to the widespread adoption of IP, most of multimedia services are IP-based. The IP protocol, up to version 4 (IPv4), was designed for fixed networks and 'best effort' applications with low network requirements, such as e-mail and file transfer and, accordingly, it offers an unreliable service that is subject to packet loss, reordering, packet duplication and unbounded delays. This service is completely inappropriate for real-time multimedia services such as video-conference and voice-over-IP, which call for specific delay and loss figures. Additionally, no mobility support is natively provided, making it difficult for pure IP to be used for mobile communications. One of the benefits of version 6 of IP (IPv6) is that it inherently provides some means for QoS and mobility support, but it still needs supporting mechanisms to fulfil the demanding requirements that emerge in the hybrid architecture of Figure 1.1.

The efficient support of IP communications in wireless environments is considered a key issue of emerging wireless multimedia networks and is further addressed in many chapters of this book (see, for example, Chapter 7).

The IP protocol and its main transport layer companions (TCP and UDP) were also designed for fixed networks, with the assumption that the network consists of point-to-point physical links with stable available capacity. However, when a wireless access technology is used in the link layer, it could introduce severe variations on available capacity, and could thus result in low TCP protocol performance. There are two main weaknesses of the IP over wireless links:

(1) *The assumption of reliable communication links.* Assuming highly reliable links (as in fixed networks), the only cause of undelivered IP packets is congestion at some intermediate nodes, which should be treated in higher layers with an appropriate end-to-end congestion control mechanism. UDP, targeted mainly for real-time traffic, does not include any congestion control, as this would introduce unacceptable delays. Instead, it simply provides direct access to IP, leaving applications to deal with the limitations of IP's best effort delivery service. TCP, on the other hand, dynamically tracks the round-trip delay on the end-to-end path and times out when acknowledgments are not received in time, retransmitting unacknowledged data. Additionally, it reduces the sending rate to a minimum and then gradually increases it in order to probe the network's capacity. In WLANs, where errors can occur due to temporary channel quality degradation, both these actions (TCP retransmissions and rate reduction) can lead to increased delays and low utilization of the scarce available bandwidth.

(2) *The lack of traffic prioritization.* Designed as a 'best effort' protocol, IP does not differentiate treatment according to the kind of traffic. For example, delay sensitive real-time traffic, such as VoIP, will be treated in the same way as ftp or e-mail traffic, leading to unreliable service. In fixed networks, this problem can be relaxed with over-provisioning of bandwidth, wherever possible (e.g., by introducing high capacity fiber optics). In WLANs this is not possible because the available bandwidth can be as high as a few tens of Mbps. But even if bandwidth were sufficient, multiple access could still cause unpredictable delays for real-time traffic. For these reasons, the introduction of scheduling mechanisms is required for IP over WLANs, in order to ensure reliable service under all kinds of conditions.

Current approaches for supporting IP QoS over WLANs fall into the following categories [3]:

(1) *Pure end-to-end.* This category focuses on the end-to-end TCP operation and the relevant congestion-avoidance algorithms that must be implemented on end hosts, so as to ensure transport stability. Furthermore, enhancements for fast recovery such as TCP selective acknowledgement (SACK) option and NewReno are also recommended.

(2) *Explicit notification based.* This category considers explicit notification from the network to determine when a loss is due to congestion but, as expected, requires changes in the standard Internet protocols.

(3) *Proxy-based.* Split connection TCP and Snoop are proxy-based approaches, applying the TCP error control schemes only in the last host of a connection. For this reason, they require the AP to act as a proxy for retransmissions.

(4) *Pure link layer.* Pure link layer schemes are based on either retransmissions or coding overhead protection at the link layer (i.e., automatic repeat request (ARQ) and forward error correction (FEC), respectively), so as to make errors invisible at the IP layer. The error control scheme applied is common to every IP flow irrespective of its QoS requirements.

(5) *Adaptive link layer.* Finally, adaptive link layer architectures can adjust local error recovery mechanisms according to the applications requirements (e.g., reliable flows vs. delay-sensitive) and/or channel conditions.

## 1.3 Multimedia Services in WLANs

Wireless Local Area Networks (WLANs) were designed to provide cheap and fast indoor communications. The predominant WLAN standard nowadays, IEEE 802.11 [4], has enjoyed widespread market adoption in the last few years, mainly due to the low price of equipment combined with high bandwidth availability. WLANs can offer high-speed communications in indoor and limited outdoor environments, providing efficient solutions for advanced applications. They can act either as alternative in-building network infrastructures, or complement wide-area mobile networks, as alternative access systems in hot-spots, where a large density of users is expected (e.g., metro stations, malls, airports, etc.). Although the main problems encountered in WLANs for multimedia support are similar to other wireless environments (i.e., security, quality of service and mobility), the particularities of WLANs call for specialized solutions. These particularities include unlicensed operation bands, fast transmission speed, increased interference and fading, combined with node movement. Especially the concern about lack of reliable security mechanisms and interworking solutions with existing and future mobile networks prevent wide adoption of WLANs in commercial applications.

The main characteristics of IEEE 802.11 networks are their simplicity and robustness against failures due to their distributed design approach. Using the Industrial Scientific and Medical (ISM) band at 2.4 GHz, the 802.11b version provides data rates of up to 11 Mbps at the radio interface. Recent advances in the transmission technology provide for transmission speeds up to hundreds of Mbps, facilitating the use of broadband applications. The IEEE 802.11a and 802.11g versions can achieve transmission rates of up to 54 Mbps using the OFDM modulation technique in the unlicensed 5 GHz band and 2.4 GHz band respectively [5]. Today, IEEE 802.11 WLANs can be considered as a wireless version of Ethernet, which supports best-effort service. The mandatory part of the original 802.11 MAC is called Distributed Coordination Function (DCF), and is based on Carrier Sense Multiple Access with Collision Avoidance (CSMA/CA), offering no QoS guarantees. Typically, multimedia services such as Voice over IP, or audio/video conferencing require specified bandwidth, delay and jitter, but can tolerate some losses. However, in DCF mode all mobile stations compete for the resources with the same priorities. There is no differentiation mechanism to guarantee bandwidth, packet delay and jitter for high-priority mobile stations or multimedia flows. Even the optional polling-based Point Coordination Function (PCF), cannot guarantee specific QoS values. The transmission time of a polled mobile station is difficult to control. A polled station is allowed to send a frame of any length between 0 and 2346 bytes, which introduces the variation of transmission time. Furthermore, the physical layer rate of the polled station can change according to the varying channel status, so the transmission time is hard to predict. This makes a barrier for providing guaranteed QoS services for multimedia applications.

The fast increasing interest in wireless networks supporting QoS has led IEEE 802.11 Working Group to define a new supplement called 802.11e to the existing legacy 802.11 Medium Access Control (MAC) sub-layer [6]. The new 802.11e MAC aims at expanding the 802.11 application domain, enabling the efficient support of multimedia applications. The new MAC protocol of the 802.11e is called the Hybrid Coordination Function (HCF), to describe the ability to combine a contention channel access mechanism, referred to as Enhanced Distributed Channel Access (EDCA), and a polling-based channel access mechanism, referred to as HCF Controlled Channel Access (HCCA). EDCA provides differentiated QoS services by introducing classification and prioritization among the different kinds of traffic, while HCCA provides parameterized QoS services to mobile stations based on their traffic specifications and QoS requirements. To perform this operation, the HC has to incorporate a scheduling algorithm that decides on how the available radio resources are allocated to the polled stations. This algorithm, usually referred to as the 'traffic scheduler', is one of the main research areas in 802.11e, as its operation can significantly affect the overall system performance. The traffic scheduler is not part of the 802.11e standard and can thus serve as a product differentiator that should be carefully designed and implemented, as it is directly connected to the QoS provision capabilities of the system. In the open technical literature, only a limited number of 802.11e traffic schedulers have been proposed so far. The latest version of IEEE 802.11e [6] includes an example scheduling algorithm, referred to as the Simple Scheduler, to provide a reference for future, more sophisticated solutions.

From this brief discussion, it is clear that new advances in WLANs provide the required framework for indoor multimedia applications, but there are still a number of open issues to be addressed to guarantee efficiency. A separate chapter in this book (Chapter 5) is dedicated to this area.

## 1.4 Multimedia Services in WPANs

Starting with Bluetooth, Wireless Personal Area Networks (WPANs) became a major part of what we call 'heterogeneous network architectures', mainly due to their ability to offer flexible and efficient ad hoc communication in short ranges, without the need of any fixed infrastructure. This led IEEE 802 group to approve, in March 1999, the establishment of a separate subgroup, namely 802.15, to handle WPAN standardization. Using Bluetooth as a starting point, 802.15 is working on a set of standards to cover different aspects of personal area environments. Today 802.15 consists of four major task groups:

- *802.15.1*   Standardized a Bluetooth-based WPAN, as a first step towards more efficient solutions
- *802.15.2*   Studies coexistence issues of WPANs (802.15) and WLANs (802.11)
- *802.15.3*   Aims at proposing high rate, low power, low cost solutions addressing the needs of portable consumer digital imaging and multimedia applications
- *802.15.4*   Investigates a low data rate solution with multi-month to multi-year battery life and very low complexity, targeted to sensors, interactive toys, smart badges, remote controls and home automation.

One of the main advances for multimedia applications in WPANs is Ultra-Wideband (UWB) communications. The potential strength of the UWB radio technique lies in its use of extremely wide transmission bandwidths, which results in desirable capabilities, including accurate position location and ranging, lack of significant fading, high multiple access capability, covert communications, and possible easier material penetration. The UWB technology itself has been in use in military applications since the 1960s, based on exploiting the wideband property of UWB signals to extract precise timing/ranging information. However, recent Federal Communication Commission (FCC) regulations have paved the way for the development of commercial wireless communication networks based on UWB in the 3.1–10.6 GHz unlicensed band. Because of the restrictions on the transmit power, UWB communications are best suited for short-range communications, namely sensor networks and WPANs. To focus standardization work in this technique, IEEE established subgroup IEEE 802.15.3a, inside 802.15.3, to develop a standard for UWB WPANs. The goals for this new standard are data rates of up to 110 Mbit/s at 10 meters, 200 Mbit/s at 4 meters, and higher data rates at smaller distances. Based on those requirements, different proposals are being submitted to 802.15.3a. An important and open issue of UWB lies in the design of multiple access techniques and radio resource sharing schemes to support multimedia applications with different QoS requirements. One of the decisions that will have to be made is whether to adopt some of the multiple access approaches already being developed for other wireless networks, or to develop entirely new techniques. It remains to be seen whether the existing approaches offer the right capabilities for UWB applications. We have a specific chapter in this book (6), which provides a detailed discussion on multimedia services over WPANs.

## 1.5 Multimedia Services in 3G Networks

Over the past years, there have been major standardization activities undertaken in 3GPP and 3GPP2 for enabling multimedia services over 3G networks (more up-to-date information can be found at www.3gpp.org and www.3gpp2.org). The purpose of this activity was to specify an IP-based multimedia core network, the IMS mentioned before, that could provide a standard IP-based interface to wireless IP

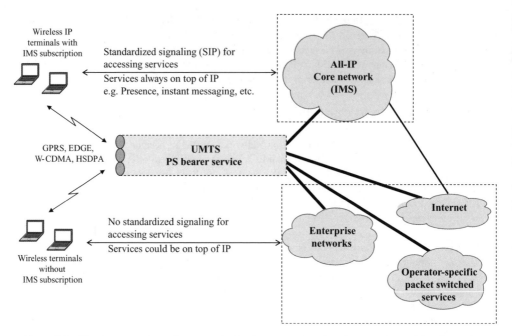

**Figure 1.3** The IMS network provides a standardized IP-based signalling for accessing multimedia services.

terminals for accessing a vast range of multimedia services independently from the access technology (see Figure 1.3). This interface uses the Session Initiation Protocol (SIP) specified by IETF for multimedia session control (see RFC 3261). In addition, SIP is used as an interface between the IMS session control entities and the service platforms, which run the multimedia applications. The initial goal of IMS was to enable the mobile operators to offer to their subscribers multimedia services based on and built upon Internet applications, services and protocols. Note that the IMS architecture of the 3GPP and 3GPP2 is identical, and is based on IETF specifications. Thus, IMS forms a single core network architecture that is globally available and can be accessed through a variety of access technologies, such as mobile data networks, WLANs, fixed broadband (e.g. xDSL), etc. No matter what technology is used to access IMS, the user always employs the same signaling protocols and accesses the same services.

In a way, IMS allows mobile operators to offer popular Internet-alike services, such as instant messaging, Internet telephony, etc. IMS can offer the versatility to develop new applications quickly. In addition, IMS is global (identical across 3GPP and 3GPP2) and it is the first convergence between the mobile world and the IETF world.

The general description of IMS architecture can be found in 3GPP technical specification (TS) 23.228 and the definition of IMS functional elements can be found in 3GPP TS 23.002. All 3GPP specs are available at www.3gpp.org/ftp/specs/. Also more details about IMS can be found in Chapter 8.

### 1.5.1 Multimedia Messaging

Multimedia messaging services are now emerging in 3G cellular networks, providing instant messaging by exploiting the SIP-enabled IMS domain. By combining the support of messaging with other IMS service capabilities, such as Presence (see Chapter 11), new rich and enhanced messaging services for the end users can be created, similar to Yahoo messaging, MSN messaging, AOL messaging and other similar messaging services on the Internet today. The goal of 3G operators is to extend mobile messaging to the IMS, whilst also interoperating with the existing SMS, EMS and MMS wireless messaging solutions as well as SIP-based Internet messaging services. The SIP-based messaging service should support interoperability with the existing 3G messaging services SMS, EMS and MMS, as well as enable development of new messaging services, such as Instant Messaging, Chat, etc.

It should be possible in a standardized way to create message groups (chat rooms) and address messages to a group of recipients as a whole, as well as individual recipients. Additional standardized mechanisms are expected, in order to create and delete message groups, enable and authorize members to join and leave the group and also to issue mass invitations to members of the group.

Apart from Instant Messaging, the support of messaging in the IMS will enable various other messaging services like, Chat, Store and Forward Messaging with rich multimedia components. Chapter 11 elaborates further on Instance Messaging and Presence Service (IMPS).

## 1.6 Multimedia Services for the Enterprise

Figure 1.4 shows an example of how an enterprise could take advantage of an all-IP multimedia core network in order to minimize its communication costs and increase communications efficiency. The typical IP network of the enterprise could be evolved to an IP multimedia network, which would support (i) IP signaling with the end terminals for establishing and controlling multimedia sessions, (ii) provisioning of QoS, (iii) policy-based admission control and (iv) authentication, authorization and possibly accounting. The all-IP network provides an integrated infrastructure for efficiently supporting a vast range of applications with diverse QoS requirements and, in addition, provides robust security mechanisms. The architecture of this all-IP network could be based on the IMS architecture specified by 3GPP/3GPP2 (see specification 3GPP TS 23.228).

In the example shown in Figure 1.4, an employee in the European office could request a voice call to another employee, e.g. in the US office. This request would be routed to the default Proxy Call Service Control Function (P-CSCF) that serves the European office. This P-CSCF relays the request to the

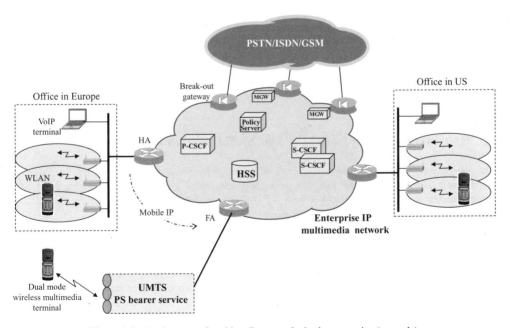

**Figure 1.4**  Deployment of multimedia networks in the enterprise (example).

Serving CSCF (S-CSCF) of the calling employee, i.e. to the CSCF with which this employee has previously registered. This S-CSCF holds subscription information of the calling employee and can verify whether he/she is allowed to place the requested call. In turn, the S-CSCF finds another S-CSCF, with which the called subscriber has registered, and relays the request to this S-CSCF. Note that if the calling employee in the European office were calling a normal PSTN number in US, the call would be routed through the IP network to the break-out gateway that is closest to the called PSTN number. This way, the long-distant charges are saved.

The S-CSCF of the called US employee holds information on the employee's whereabouts and can route the call to the correct location. If the called US employee happens to be roaming in Europe, the call would be routed to the appropriate P-CSCF that currently serves this employee. It is important to note that, although signaling can travel a long path in such a case (e.g. from Europe to the US and then back to Europe), the user-plane path would be the shortest possible.

The support of roaming is another important advantage of the above architecture. For instance, a European employee could take his dual-mode mobile phone to the US office. After powering-on his mobile and registering his current IP address (with his S-CSCF), he would be able to receive calls at his standard number.

Even when the employee is away from his office and cannot directly attach to the enterprise network (e.g. he is driving on the highway), it is still possible to be reached on his standard number. In such a case, the employee uses, for example, the UMTS network to establish a mobile signaling channel to his all-IP enterprise network, which assigns him an IPv6 address. The employee registers this IP address with an S-CSCF (via the appropriate P-CSCF) and thereafter he can receive calls at his standard number. The signaling mobile channel remains activate for as long as the employee uses UMTS to access the enterprise network. In such a scenario, the UMTS network is used only to provide access to the enterprise network and to support the mobile bearers required for IP multimedia services. To establish IP multimedia calls the employee would need to request the appropriate UMTS bearers, each one with the appropriate QoS properties. For instance, to receive an audio-video call, two additional UMTS bearers would be required, one for each media component. The mapping between the application-level QoS and the UMTS QoS, as well as the procedures required to establish the appropriate UMTS bearers are specified in 3GPP Rel-5 specifications, in particularly, in 3GPP TS 24.008 and 3GPP TS 24.229 (available at www.3gpp.org/ftp/specs/).

If the enterprise network supports a macro-mobility protocol, e.g. Mobile-IP, it could be possible to provide session mobility across the enterprise WLAN and the UMTS network. In this case, when the employee moves from the WLAN to the UMTS he uses Mobile-IP to register his new IPv6 address with his home agent. After that, any subsequent terminating traffic would be tunneled from the employee's home agent to the foreign agent that serves the employee over the UMTS network.

## 1.7 Hybrid Multimedia Networks and Seamless Mobility

The integration of existing mobile systems with new wireless access technologies has attracted considerable attention over the last few years and has put a great deal of momentum into the heterogeneous architecture shown in Figure 1.1. Efficient mobility management is considered to be one of the major factors towards seamless provision of multimedia applications across heterogeneous networks. Thus, a large number of solutions have been proposed in an attempt to tackle all the relevant technical issues. Of course, the ultimate goal is to realize a seamless multimedia environment like the one shown in Figure 1.5, where a multimedia session can roam across different radio access technologies seamlessly, i.e. with no user intervention and no noticeable effect on the QoS. There are several studies considering multimedia session continuity with such inter-radio access technology handovers and report interesting results (e.g. [7]).

The high penetration of WLANs, and especially the higher data rates they offer, caused cellular operators to investigate the possibility of integrating them into their systems and support a wider range

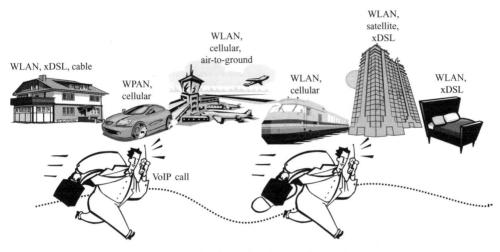

WLAN, xDSL, cable

WLAN,
cellular,
air-to-ground

WLAN,
satellite,
xDSL

WPAN,
cellular

WLAN,
cellular

WLAN,
xDSL

VoIP call

**Figure 1.5**  Multimedia services in a seamless environment.

of services to their users. This integration led to hybrid multimedia systems similar to the one shown in Figure 1.1. The design and implementation of these systems present many technical challenges, such as vertical handover support between, for example, WLAN and UMTS, unified Authentication, Authorization and Accounting (AAA), consistent QoS and security features. Despite the progress in WLAN/ UMTS interworking standardization [8, 9], up to now most attention has been paid to AAA interworking issues and less in mobility and QoS.

Intensive efforts from the research community has gone into trying to identify the unresolved issues and propose specific solutions. The key difference between the proposed solutions is the way the cellular and WLAN networks are coupled. Although several categories have been proposed [10–12], three are the most common types of integration: loose, tight and very tight coupling.

Loose coupling suggests the interconnection of independent networks using Mobile IP [2] mechanisms, it requires almost no modifications in cellular network architecture and is the easiest to deploy. Since minor architectural adjustments are required, these proposals focus mainly on the elaboration of the handover decision process and the introduction of novel mechanisms towards performance improvement. On the other hand, tight and very tight coupling require significant changes regarding the UMTS functionality, since the WLAN appears to the UMTS core network as an alternative access network. Despite the implementation complexity, these two types offer significantly smaller handover latencies. If coupling is done at the level of the core network (i.e., SGSN) tight coupling is considered, while coupling at the access network (i.e., RNC) is identified as very tight.

Owing to the intrinsic differences in mobility management between UMTS and IP-based architectures, a mobility management scheme tailored to heterogeneous environments has to overcome technology-specific particularities and combine their characteristics. An attempt to indicate the required steps towards such a scheme has been made by 3GPP in [9]. In particular, six scenarios have been specified for the evolvement of integration work. Scenario 1 indicates common billing and customer care, scenario 2 provides 3GPP based access control, scenario 3 enables access to 3GPP PS services from WLAN, scenario 4 allows services to continue after inter-system handover, scenario 5 promises seamless functionality in the previous scenario and, finally, scenario 6 provides access to 3GPP CS services from WLAN. These scenarios are extensively discussed in [13] and in Chapter 7.

Based on the above scenarios, a number of proposals aim at offering some degree of seamless integration. Some of them focus on the general framework and the functionalities that these networks should provide for advanced mobility capabilities, using Mobile IP as the basic tool for system integration. This has the advantage of simple implementations, with minimal enhancements on existing components, but at the expense of considerably larger handover execution time. These proposals promise service continuity (scenario 4), but without seamless functionality at all times (scenario 5). This is why they are characterized as loose coupling solutions. The research efforts in this way focus on the design of more complex handover decision algorithms and performance enhancing mechanisms to offer the best quality to the users, while ensuring best network resource usage.

Another group of proposals attempt to integrate UMTS and WLAN in a tighter way, able to add any other IP access network in the future. The focus in this case is on the technical challenges that arise from the extension of current standards in order to interoperate with each other. Fast mobility functions and seamless functionality (scenario 5) are the main targets of these proposals, at the expense of considerable complexity introduced by the enhancement required on existing components. This group includes both the tight and the very tight architectures.

## 1.8 Book Contents

After going through the key aspects of the evolution towards the wireless multimedia technologies and architectures, let us provide a brief outline of this book. The book is organized in two major parts. The first part concentrates on the key enabling technologies that will allow wireless multimedia to become a reality. Following a layered top-down approach, this part starts with the basics of multimedia coding (Chapter 2), focusing on speech and video compressors. The reader becomes familiar with coding techniques, as they are closely related to the requirements that have to be fulfilled by the lower layers. The chapter starts with general information on compression and contains information on the speech and video characteristics, to explain the techniques used for these kinds of traffic. Major compressors are described, such as PCM, MPEG1-4, H.261, etc.

Chapters 3 and 4 describe multimedia transport and control protocols, respectively, and discuss how they are supposed to operate in wireless environments. As most of these protocols were not designed with mobility and wireless interfaces in mind, it is interesting to study their operation and performance in such cases.

Recent advances in wireless local and personal area technologies allow for multimedia applications provision in short-range environments. Chapters 5 and 6 describe these advances for WLANs and WPANs respectively. WLAN discussion moves around 802.11e, as the most promising technique for multimedia support, while the chapter on WPANs contains the major achievements in the context of IEEE 802.15.

Chapter 7 is dedicated to problems and solutions for QoS provision in wireless multimedia networks. It starts with the problem of seamless mobility in WLANs, discussing a solution based on the Inter-Access Point Protocol (IAPP), and continues with traffic scheduling techniques in IEEE 802.11e to guarantee specific QoS values. It then moves to the problem of the supporting IP-based QoS techniques, such as the Resource Reservation Protocol (RSVP), over wireless, and concludes with 3G/WLAN interworking scenarios, describing architectures and common QoS provision schemes.

The last chapter of the first part (Chapter 8) of the book begins with an overview of cellular networks and their evolution, and provides details on the Universal Mobile Telecommunications System (UMTS). It then presents an introductory description of the features and services provided by the two most important parts of UMTS networks, with respect to IP-based multimedia services, namely the IP Multimedia Subsystem (IMS) and the Multimedia Broadcast/Multicast Service (MBMS), followed by more technical information. The chapter concludes with a discussion on the QoS issues for IP-based multimedia services in UMTS, describing the overall QoS concept, the policy-based QoS control scheme and its application to IMS sessions.

The second part of the book is dedicated to major wireless multimedia applications and services. It starts with a description of the protocol that basically established multimedia applications in mobile networks, i.e., the Wireless Application Protocol (WAP) (Chapter 9), and moves to the Mobile

Multimedia Service (MMS) that enabled easy exchange of multimedia content among mobile users (Chapter 10).

Chapter 11 is dedicated to the Instant Messaging and Presence Service (IMPS), which allows a community of users to exchange multimedia messages in real-time and share their online status. This service is a linkage to other services such as file transfer, telephony, and online games. The key technology directions are described, including richer media, security, integration with location-based services and collaboration applications.

Chapter 12 describes one of the most important applications of IMPS, that of mobile payment. The chapter contains the design details of such a system, focusing on the requirements of mobile payment, especially security and privacy, and how these can be addressed by IMPS. It also contains information on a prototype implementation, constructed to prove applicability and effectiveness of such solutions.

Chapter 13 includes the basic information on Push-to-Talk (PTT), an application that is expected to enjoy big success in the years to come, due to its advantages for group communications. PTT allows a user to instantly communicate with one or more participants through an instant voice call. With the heart of the PTT service being IP-based, it is strategically situated to leverage multimedia support for video, multimedia messaging, virtual reality applications (e.g., gaming), etc. Push-to-Talk is a first step to Push-to-'Anything', a suite of 'Push-to' services that enable sending any media, from anywhere to anywhere, in a unified instant communication service set. The chapter provides a view into what PTT is, how it works in the cellular domain, and how this may evolve as an instant communication offering in a multimedia world, fulfilling the vision of a Push-to-Anything future.

Finally, Chapter 14 discusses one of the most rapidly expanding fields of the mobile communications market, that of Location Based Services (LBS). LBS basically include solutions that leverage location information to deliver consumer applications on a mobile device (e.g., navigation, emergency assistance, advertising, etc.). The chapter starts with the requirements for delivering LBS to end users, and moves to the detailed description of an LBS system framework, including positioning, security and billing support. It concludes with available LBS systems, proposed by specific manufacturers or research projects.

# References

[1] 3GPP TS 29.060, GPRS Tunnelling Protocol (GTP) across the Gn and Gp Interface (Release 5), September 2004.

[2] C. Perkins (ed.), IP Mobility Support for IPv4, RFC 3344, August 2002.

[3] P. Mähönen, T. Saarinen, N. Passas, G. Orphanos, L. Muñoz, M. García, A. Marshall, D. Melpignano, T. Inzerilli, F. Lucas and M. Vitiello, Platform-Independent IP Transmission over Wireless Networks: The WINE Approach, *IEEE Personal Communications Mag.*, **8**(6), December 2001.

[4] IEEE 802.11 Wireless LAN Medium Access Control (MAC) and Physical Layer (PHY) specifications, IEEE (1997).

[5] S. Mangold, S. Choi and N. Esseling, An Error Model for Radio Transmissions of Wireless LANs at 5 GHz, *Proc. Aachen Symposium 2001*, Aachen, Germany, pp. 209–214, September 2001.

[6] IEEE Std 802.11e/D13.0, Draft Supplement to Standard for Telecommunications and Information Exchange between Systems – LAN/MAN Specific Requirements. Part 11: Wireless Medium Access Control (MAC) and Physical Layer (PHY) Specifications: Medium Access Control (MAC) Enhancements for Quality of Service (QoS), January 2005.

[7] A. K. Salkintzis, G. Dimitriadis, D. Skyrianoglou, N. Passas and N. Pavlidou, Seamless Continuity of Real-Time Video across UMTS and WLAN Networks: Challenges and Performance Evaluation, *IEEE Wireless Communications*, April 2005.

[8] 3GPP TS 23.234 V6.2.0, 3GPP system to Wireless Local Area Network (WLAN) interworking; System description (Release 6), September 2004.

[9] 3GPP TR 22.934 V6.2.0, Feasibility study on 3GPP system to Wireless Local Area Network (WLAN) interworking (Release 6), September 2003.

[10] A. K. Salkintzis, C. Fors and R. S. Pazhyannur, WLAN-GPRS Integration for Next Generation Mobile Data Networks, *IEEE Wireless Communications*, **9**(5), pp. 112–124, October 2002.

[11] S.-L. Tsao, C.-C. Lin, Design and evaluation of UMTS/WLAN interworking strategies, *Vehicular Technology Conference*, VTC 2002-Fall, 2002.

[12] R. Samarasinghe, V. Friderikos and A.H. Aghvami, Analysis of Intersystem Handover: UMTS FDD & WLAN, *London Communications Symposium*, 8–9 September 2003.

[13] A. K. Salkintzis, Interworking Techniques and Architectures for WLAN/3G Integration Towards 4G Mobile Data Networks, *IEEE Wireless Communications*, **11**(3) pp. 50–61, June 2004.

# Part One

## Multimedia Enabling Technologies

# 2

# Multimedia Coding Techniques for Wireless Networks

Anastasios Delopoulos

## 2.1 Introduction

### 2.1.1 Digital Multimedia and the Need for Compression

All types of information that can be captured by human perception, and can be handled by human made devices are collectively assigned the term multimedia content. This broad definition of multimedia includes text, audio (speech, music), images, video and even other types of signals such as bio-signals, temperature and pressure recordings, etc. Limiting the types of human perception mechanisms to vision and hearing senses yields a somewhat narrower definition of multimedia content that is closer to the scope of multimedia in our everyday or commercial language. If, in addition, we stick to only those information types that can be handled by computer-like devices we come up with the class of digital multimedia content. These include text, digital audio (including digital speech), digital images and video.

Although some of the concepts presented in this chapter are easily extendable to all types of digital multimedia content, we shall focus on speech and video. The reason of adopting this restrictive approach is that these two modalities (i) constitute the vast amount of data transmitted over wireless channels; (ii) they both share the need to be streamed through these channels and (iii) much of the research and standardization activities have been aimed at their efficient transmission.

Speech is generated by the excitation by the vocal track of the air coming from the lungs in a controlled manner. This excitation produces time varying air pressure in the neighborhood of the mouth in the form of propagating acoustic waves that could be captured by human ears. Digital speech is the recording – in the form of a sequence of samples – of this time varying pressure. A microphone followed by an analog-to-digital converter can be used to perform the recording procedure. Conversely, digital speech can converted into acoustic waves by means of a digital-to-analog converter that is used to excite a speaker. More detail about these procedures can be found in Section 2.3.1.

Many design parameters influence the quality of digital speech, that is the accuracy with which the reverse procedure can reproduce the original (analog) speech acoustic waves. The most important of these is the sampling frequency (how often the pressure is measured) and the accuracy of the

*Emerging Wireless Multimedia: Services and Technologies*   Edited by A. Salkintzis and N. Passas
© 2005 John Wiley & Sons, Ltd

representation of each single sample, namely the number of quantization levels used. The latter is closely related to the number of bits used for the discrete representation of the samples.

Typical choices for low quality digital speech include sampling at 8000 samples per second with 8 bits per sample. This sums up to 64 000 bits per second, a bitstream rate that is well above traditional voice communication channels. The situation becomes even more difficult as this rate does not include error correction bits or signaling overhead. The need for compression is apparent.

Unlike speech, video has no structured generation mechanism (analogous to a vocal track). On the contrary, its capturing mechanism is well defined. We may think of video as the time varying recording of light (both luminance and color) on the cells of the retina. In fact, this light corresponds to the idol (image) of scenes as produced on the retina by means of the eye-lens. The human visual perception mechanism is briefly explained in Section 2.6.1. Digital video is an approximation of the eye's perceived information in the form of a three dimensional sample sequence. Two of the dimensions correspond to the location of each sample with respect to the idol coordinate system while the third corresponds to the instant of time at which each sample has been measured. In its simplest form, digital video can be considered as a sequence of still digital images (frames) that, in turn, are fixed-sized two dimensional arrays of samples (light recordings).

The temporal sampling frequency (frame rate), the dimension of frames and the number of quantization levels for each sample determine the quality of digital video and the associated bitstream rate. Typical medium quality choices include 25 frames per second, $576 \times 720$ frame dimensions and 24 bits per sample which translates to 237 Megabytes per second, a bitrate that is far beyond the capacities of available communication channels. Compressing digital video is thus necessary if the signal is to be transmitted.

In view of the previous considerations, it is not surprising that much of the effort of the signal processing community has been devoted to devising efficient compression–decompression algorithms for digital speech and video. A range of standardization bodies is also involved, since the algorithms produced are to be used by diverse manufacturers and perhaps in cross platform environments. This successfully combined effort has already produced a collection of source coders–decoders (another name for compression–decompression algorithms for multimedia content). Most of these codecs are already components of today's multimedia communication products. As expected, though, the research is ongoing and mainly led by the improved processing capabilities of the emerging hardware.

Codec design is guided by a number of, sometimes antagonistic, requirements/specifications:

- Targeted quality. This is determined by the targeted application environment; the quality of video in entertainment applications such as digital television or theaters is certainly more demanding than that in video-conference applications.
- Targeted bitrate. This is mainly determined by the medium used to transmit (or store) the compressed multimedia representations. Transmission of speech over person to person wireless communication channels is usually much more bandwidth parsimonious than its counterpart in wired or broadcast environments.
- Targeted complexity and memory requirements. This is mainly determined by the type of the device (hardware) that hosts the codec. It is also related to the power consumption constraints imposed by these devices.

### 2.1.2 Standardization Activities

Multimedia coding techniques are the subject of standardization activities within multi/inter-national organizations (ITU, ISO, ETSI, 3GPP2) and national authorities (e.g., US Defense Office) and associations (North American TIA, Japanese TTC, etc.)

The International Telecommunication Union (ITU) contributes to the standardization of multimedia information via its Study Group 15 and Study Group 16 including Video Coding Experts Group

(VCEG). A number of standardized speech/audio codecs are included in series G.7xx (Transmission systems and media, digital systems and networks) of ITU-T Recommendations. Video coding standards belong to series H.26x (Audiovisual and multimedia systems).

The International Organization for Standardization (ISO), and particularly its Motion Pictures Experts Group (MPEG), has produced a series of widely accepted video and audio encoding standards like the well known MPEG-1, MPEG-2 and MPEG-4.

The European Telecommunications Standards Institute (ETSI) is an independent, non-profit organization, producing telecommunications standards in the broad sense. A series of speech codecs for GSM communications have been standardized by this organization.

The Third Generation Partnership Project 2 (3GPP2) is a collaborative third generation (3G) telecommunications specifications-setting project comprising North American and Asian interests. Its goal is to develop global specifications for ANSI/TIA/EIA-41 Cellular Radiotelecommunication Intersystem Operations network evolution to 3G and global specifications for the Radio Transmission Technologies (RTTs). The following telecommunication associations participate in 3GPP2:

- ARIB: Association of Radio Industries and Businesses (Japan);
- CCSA: China Communications Standards Association (China);
- TIA: Telecommunications Industry Association (North America);
- TTA: Telecommunications Technology Association (Korea);
- TTC: The Telecommunication Technology Committee (Japan).

### 2.1.3 Structure of the Chapter

The rest of the chapter contains three main parts. First, Section 2.2 offers a unified description of multimedia compression techniques. The fundamental ideas of Entropy Coding, Redundancy Reduction and Controlled Distortion as tools for reducing the size of multimedia representations are introduced in this section.

The second part is devoted to speech coding. Sections 2.3 to 2.5 belong to this part. The nature of speech signals is explored first in order to validate the synthetic models that are then given. The general compression ideas in the field of speech coding are presented, leading to the presentation of the most important speech coding algorithms, including the popular CELP. The last section of this part looks at speech codecs that have been standardized by the ITU, the ETSI and the 3GPP2, linking them to the aforementioned speech coding algorithms.

The third part consists of Sections 2.6 and 2.7 and covers digital video compression aspects. The nature of video signals is explored first and the adaptation of general compression techniques to the case of digital video is then considered. The most important digital video compression standards are presented in the last section.

A bibliography completes the chapter.

## 2.2 Basics of Compression

Compression of speech, audio and video relies on the nature of these signals by modelling their generation mechanism (e.g., vocal track for speech) and/or exploitating human perception limits (e.g., audible spectrum, tone masking, spatiotemporal visual filters).

### 2.2.1 Entropy, Entropy Reduction and Entropy Coding

In their digital form, multimedia modalities can be considered as streams of symbols which are produced by a quantizer. They are all picked from a finite set $S = \{s_0, s_1, \ldots, s_{N-1}\}$ where $N$ is the number of quantization levels used in the discrete representation. Clearly more accurate representations require more quantization levels (a finer quantizer) and thus higher values of $N$. Each symbol $s_i$ may appear in the stream with a certain probability $p_i \in [0, 1]$. In Information Theory language the mechanism that generates the particular stream and the corresponding symbols is called the symbol source and is totally characterized by the set of probabilities $p_i, i = 0, \ldots, N-1$. It turns out [1] that the non-negative quantity,

$$H_s = -\sum_{i=0}^{N-1} p_i \log_2(p_i),\tag{2.1}$$

called entropy, determines the lower bound of the number of bits that are necessary in order to represent the symbols produced by this particular source.

More specifically, let $W = \{w_0, w_1, \ldots, w_{N-1}\}$ be the $N$ different binary (containing 0's and 1's) words used to represent the symbols of $S$ and $|w_i|$ the number of bits of each word (length); the average number of bits used to represent each symbol is

$$L_w = \sum_{i=0}^{N-1} p_i |w_i|.\tag{2.2}$$

It should be clear that $L_w$ multiplied by the rate of symbol production (symbols per second), a parameter that is not essentially controlled by the codec, yields the bitstream rate.

It can be proved that always (even for the cleverest choice of $w_i$'s)

$$L_w \geq H_s.\tag{2.3}$$

In view of Equations (2.1), (2.2) and (2.3), the design of a good code reduces to:

(1) transforming the original sample sequences into an alternative sequence, whose samples can be quantized by symbols in $S'$, with entropy $H_{s'}$ lower than the original $H_s$,
(2) cleverly adopting a binary word representation, $W$, that has average word length, $L_w$, as close to the lower bound $H_{s'}$ as possible.

If the aforementioned transformation is perfectly reversible, the codec is characterized as lossless. In the opposite case, i.e., when the inverse transformation results in an imperfect approximation of the original signal, the codec is lossy. Lossy codecs are popular in multimedia coding scenarios since they can result in significant reduction of entropy and thus allow for extremely low $L_w$'s and, correspondingly, low bitstream rates.

On the other hand, the aforementioned clever selection of binary words for the representation of symbols is a procedure called entropy coding, which does not introduce any information loss. Celebrated representatives of entropy coding techniques are among others, the Huffman entropy coder [2] and the algebraic coder (see e.g., [3]). This chapter does not include any further detail regarding entropy coding. Here our attention focuses on lossy procedures for the alteration of the original speech and video signals into lower entropy representations, since there is the emphasis on (and the differentiation among) source codecs used in wireless multimedia communications.

Methods for transforming the original speech or video signals into sample sequences of lower entropy fall into two complementary categories: controlled distortion and redundancy reduction.

### 2.2.1.1 Controlled Distortion

Some portions of the multimedia signals' content are less important than others in terms of hearing or visual perception. Suppressing these parts may reduce contents entropy without affecting signals perceptual quality. In addition, although some types of distortion are audible or visible, they are considered as unimportant provided they do not alter the signals' semantics; e.g., they do not cause phoneme /a/ to be confused with /e/, etc.

### 2.2.1.2 Redundancy Reduction

Redundancy in source coding is synonymous with temporal or spatial correlation.

Consider for, example, a video sequence produced by a stationary camera capturing a stationary scene. The first captured frame is (almost) identical to all subsequent ones; a redundancy reduction procedure in this simple case would be to encode the first frame and simply inform the decoder that the depicted scene does not change for a certain time interval!

In the case of speech coding, consider a speech segment that is identical to a sinusoid. Provided that the frequency and phase of the sinusoid is identified, it suffices for coding purposes to encode a single sample of the signal and the frequency and phase values. All remaining samples can be recovered using this information.

Although both these examples are too simplistic, they validate our initial argument that links redundancy to correlation. Using pure mathematical arguments the following principle can be established.

Let $y(n) = F[x(n)]$ be a reversible transform of signal $x(n)$ and $s_x(n)$, $s_y(n)$ the quantized versions (symbols) of the original and the transformed signals. If the correlation $E\{y(n)y(n+m)\} < E\{x(n)x(n+m)\}$, then the entropy of the 'source' $s_y(n)$ is less than the entropy of $s_x(n)$.

Application of this principle is extensively used in speech and video coding. Correlation reduction methods include (i) whitening by prediction (see e.g. DPCM and ADPCM methods) used to reduce temporal correlation, (ii) motion estimation and compensation used to reduce spatiotemporal correlation in video coding (see e.g. MPEG coding), (iii) transform coding used to reduce spatial correlation in video encoding (see e.g. H.263 and MPEG algorithms).

### 2.2.2 A General Compression Scheme

Following the previous discussion, compression methods contain the stages of the block diagram of Figure 2.1. Inclusion of the grayed blocks characterizes lossy compression schemes, while in lossless compressors these blocks are not present.

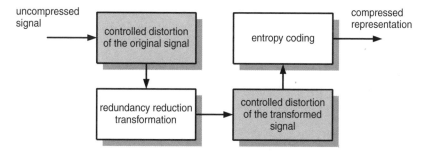

**Figure 2.1**   Basic steps in multimedia source compression.

## 2.3 Understanding Speech Characteristics

### 2.3.1 Speech Generation and Perception

Speech is produced by the cooperation of lungs, glottis (with vocal cords) and articulation tract (mouth and nose cavity). Speech sounds belong to two main categories: voiced and unvoiced.

Voiced sounds (like / a / in bat, / z / in zoo, / l / in let and / n / in net) are produced when the air coming out of the lungs, through the epiglottis, causes a vibration of the vocal cords that, in turn, interrupts the outgoing air stream and produces an almost periodic pressure wave. The pressure impulses are commonly called pitch impulses and the fundamental frequency of the pressure signal is the pitch frequency or simply the pitch. Pitch impulses excite the air in the mouth and/or nose cavity. When these two cavities resonate they radiate the speech sound with a resonance frequency that depends on their particular shape. The principal resonance frequencies of voiced speech sounds are called formants; they may change by reshaping the mouth and nose cavity and determine the type of phoneme produced.

On the other hand, unvoiced speech is characterized by the absence of periodic excitation (pitch impulses). An unvoiced sound wave has either the characteristics of a noise signal (like /s/ in sun), produced by the turbulent flow of the air through the complex of teeth and lips, or the form of a response of the vocal tract to a sudden (impulsive) release of air (like / p / in put)

### 2.3.2 Digital Speech

The human ear acts as a bandpass filtering mechanism. Only frequencies in the range 20 to 20 000 Hz excite the hearing system of the average listener. Although, the human voice may reach or even exceed the upper limits of this range, it has been experimentally recognized that most of the semantics within speech are carried in the frequency range from 200 to 3 200 Hz, while almost the entire speech content comes within 50 to 7000 Hz.

Following these observations, digital speech is produced by A/D converters sampling at either 8000 samples/sec (for telephone quality speech) or 16 000 samples/sec (for high quality speech). Both sampling frequencies satisfy the requirements of Nyquist theorem, i.e., they are more than the double the acquired signal's maximum frequency. Nevertheless, higher sampling frequencies are adequate for capturing high quality music (including voice).

As far as quantization is concerned, speech samples are quantized with either 8 or 16 bits/sample. Three types of quantizers are usually employed:

(1) Linear quantizer. Speech values are normalized in the range of $[-1, 1]$ and uniform quantization to $2^8$ or $2^{16}$ levels is performed next.
(2) A-law quantizer. Original sample values, $x$, are first transformed according to a logarithmic mapping

$$y = \begin{cases} \dfrac{A|x|}{1 + \log A}\,\text{sign}(x) & \text{for } 0 \le |x| \le V/A \\[2mm] \dfrac{V(1 + \log(A|x|/V))}{1 + \log A}\,\text{sign}(x) & \text{for } V/A < |x| \le V \end{cases} \qquad (2.4)$$

and the resulting $y$ values are quantized by a uniform 8-bit quantizer. In the above formula, $V$ is the peak value of $|x|$ and $A$ determines the exact decision levels of the A-law quantizer (typical value for $A = 87.6$).

The A-law quantizers result in more accurate representation of low valued samples.
(3) $\mu$-law quantizer. Similarly to the A-law, original sample values, $x$, are mapped to

$$y = \frac{V \log(1 + \mu|x|/V)}{\log(1 + \mu)}\,\text{sign}(x), \qquad (2.5)$$

and 8-bit uniform quantization is then performed. $V$ is again the peak value of $|x|$ and parameter $\mu$ determines the exact values of the decision levels.

Both the A-law and the $\mu$-law quantizers are prescribed in the G.711 ITU standard ([4]); further details can be found in [5].

### 2.3.3 Speech Modelling and Linear Prediction

Voiced and unvoiced speech are modelled as the output of a linear filter excited by an impulse sequence or a white noise signal. The effect of the vocal tract is approximated by a gain followed by a linear filter, $H(z)$. The exact type of filter involved is significantly different for the two types of speech. Within each type the filter parameters determine the phoneme to be modelled.

In order to imitate the voiced speech generation mechanism, as described in Section 2.3.1, the linear filter $H(z)$ of Figure 2.2 is the cascade of two simpler filters ($H(z) = H_V(z) \equiv H_m(z)H_e(z)$). The first, $H_e(z)$, is a low-pass filter approximating the effect of the epiglottis, which transforms the input impulse sequence into a pulse sequence. The second, $H_m(z)$, approximates the resonance effect of the mouth and nose. The latter is responsible for raising formant frequencies by reshaping the spectrum of the incoming signal.

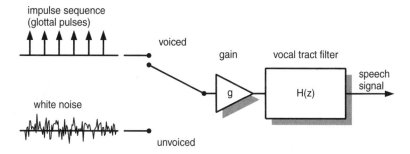

**Figure 2.2**   The fundamental model for voiced and unvoiced speech production.

In the case of unvoiced speech, the random noise sequence is passed through a single filter, $H(z) = H_U(z)$, which approximates the effects of the teeth and lips.

Many compressors employed in wireless speech communication make extensive use of these speech modeling schemes. Based on the spectral characteristics of a particular speech segment, these coders try to estimate (a) the characteristics of input excitation sequence (i.e., the pitch for voiced and noise characteristics for unvoiced speech), (b) the gain $g$ and (c) the parameters of $H_V(z)$ or $H_U(z)$ for voiced and unvoiced speech segments, respectively. Once the estimates of the speech model that best fits the speech segment is in hand, the coders act in one of two ways: (i) they use the model as a linear predictor of the true speech (see e.g., ADPCM methods in Section 2.4.1.3) or (ii) they encode and transmit the model's parameters, rather than speech samples themselves (see Sections 2.4.2 and 2.4.3).

A variety of signal processing algorithms have been adopted by different codecs for appropriately fitting $H_V(z)$ and/or $H_U(z)$ to the speech segment under consideration.

They all fall into the family of Linear Prediction (LP) methods. They attempt to estimate the parameters of an Autoregressive (AR) Model,

$$H(z) = \frac{1}{1 - A(z)} = \frac{1}{1 - \sum_{i=1}^{q} a_i z^{-i}} \tag{2.6}$$

that when fed with a particular input, $x(n)$, produces an output, $y(n)$, as close as possible to a given signal (desired output), $s(n)$. In speech coding applications, the input is either an impulse train with period equal to the pitch (for voiced speech) or a white noise sequence (for unvoiced speech). The desired output, $s(n)$, is the speech segment at hand.

In the time domain, filtering by $H(z)$ is equivalent to the recursion

$$y(n) = \sum_{i=1}^{q} a_i y(n-i) + x(n). \tag{2.7}$$

The optimal selection of $\{a_i\}, (i = 1, \ldots, q)$ AR coefficients is based on the minimization of the mean squared error (MSE) criterion:

$$J = E\{(s(n) - y(n))^2\}, \tag{2.8}$$

which is approximated by its deterministic counterpart

$$J_N = \frac{1}{N} \sum_{n=0}^{N-1} \{(s(n) - y(n))^2\}, \tag{2.9}$$

where $N$ is the number of the available (speech) samples $s(n)$, $(n = 0, \ldots, N-1)$. An exhaustive review of algorithms for estimating $a_i$ can be found in [6].

In the case of unvoiced speech, the assumption of white noise input $x(n)$ is enough for the estimation of the AR parameters. On the other hand, for the case of voiced speech, the pitch period, $p$, should be estimated prior to $a_i$'s estimation. This requires an extra effort and is usually based on the analysis of the autocorrelation sequence $r_{ss}(m) \equiv E\{s(n)s(n+m)\}$ of $s(n)$. Periodicity of $s(n)$ causes periodicity of $r_s(m)$ with the same period $p$ equal to the pitch period. Hence determination of $p$ reduces to finding the first dominant peak of $r_s(m)$ that occurs at lag $m = p > 0$. As an alternative, a rough estimate, $\bar{p}$, of $p$ can be adopted and the exact value of $p$ is chosen in a way that minimizes $J_N$ among different candidate values of $p$ in the neighborhood of $\bar{p}$.

### 2.3.4 General Aspects of Speech Compression

#### 2.3.4.1 Controlled Distortion of Speech

*High frequencies of audio signals.* The average human ear acts as a pass-band filter cutting off frequencies out of the range 20–20 000 Hz. Thus, an almost perfect representation of any audio signal (including speech, music, etc.) is achieved even if we filter out the corresponding frequency components prior to analog-to-digital conversion. This reduces the effective bandwidth of audio signals and allows for sampling rates as low as $2 \times 20000 = 40000$ samples per second, in conformance with the Nyquist Theorem [7].

*High frequencies of speech signals.* Speech signal content is mostly concentrated in an even shorter bandwidth range, namely 50–7000 Hz, which indicates that pre-filtering the remaining frequency components allows for sampling rates as low as 14 000 samples per second. In fact, speech content outside the range 200–3200 Hz hardly affects speech semantics, which means that sampling at 6400 samples per second is sufficient at least for (low quality) human communication. Based on this observation, most speech coding schemes use sampling at the rate of 8000 samples per second.

*Accuracy of sample representation.* It has been experimentally justified that $2^{16}$ quantization levels are more than satisfactory for representing speech samples. Thus, finer sample variation is omitted from speech source encoding schemes and 16 bits per sample are usually used at the quantization stage.

*Tonal masking.* Psychoacoustic experiments have verified that the existence of strong tonal components, corresponding to sharp and high peaks of the audio spectrum, makes the human ear insensitive to frequencies with lower power in the neighborhood of the tone. The tone is characterized as the masker, while the neighboring frequencies are described as masked. After this observation, efficient audio codecs divide the original signal into sub-bands with a narrow bandwidth and fewer bits are used (leading to an increased but still inaudible quantization error) for those bands that are masked by strong maskers.

MPEG audio layer I-II (see e.g., [8]) and layer III (see e.g., [9]) codecs make extensive use of this type of controlled distortion (see Section 2.4.1.4). Also, Analysis by Synthesis speech codecs take into account the masking effect of high energy frequency components (formants) for selecting perceptually optimal parametric models of the encoded speech segments (see Section 2.4.3).

### 2.3.4.2 Redundancy Reduction of Speech

Redundancy of speech signals is due to the almost periodic structure of their voiced parts. Speech codecs attempt to model this periodicity using autoregressive filters whose coefficients are calculated via linear prediction methods, as explained in Section 2.3.3. With these parametric models, redundancy reduction is achieved by two alternative approaches:

(1) decorrelating speech via prediction as described in Section 2.4.1.3 (ADPCM coding); and
(2) encoding model parameters instead of the samples of the corresponding speech segment. This approach is the core idea of Analysis-Synthesis and Analysis-By-Synthesis codecs presented in Sections 2.4.2 and 2.4.3.

## 2.4 Three Types of Speech Compressors

### 2.4.1 Waveform Compression

Waveform compression refers to those codecs that attempt to transform digitized speech sequences (waveforms) into representations that require less bits (lower entropy). The main objective is to be able to reconstruct the original waveform with as little error as possible. The internal structure (pitch, formant) of speech is ignored. Evaluation of the error may take into account the subjective properties of the human ear. Most algorithms of this type are equally applicable to general audio signals (e.g., music).

### 2.4.1.1 Pulse Code Modulation (PCM)

PCM is the simplest speech (actually general audio) codec that basically coincides with the couple of a sampler and a quantizer. Bitrate is determined by controlling only the sampling rate and the number of quantization levels.

The input analog speech signal $x_0(t)$ is first passed through a low pass filter of bandwidth $B$ to prevent aliasing. The output $x(t)$ of this antialising filter is next sampled:

$$x(n) = x(nT)$$

with sampling frequency $f_s = 1/T > 2B$ and quantized by some scalar quantizer $Q[]$:

$$s(n) = Q[x(nT)].$$

Most PCM speech coders use a sampling frequency $f_s$ of 4000 or 8000 samples per second. On the other hand, quantization uses 16 bits or less for each symbol.

### 2.4.1.2 Differential Pulse Code Modulation (DPCM)

Differential Pulse Code Modulation is the oldest compression scheme that attempts to reduce entropy by removing temporal correlation (see Section 2.2.1.2). Similarly to PCM, a sequence of samples $x(n)$ are initially produced after filtering and sampling. A differential signal is next produced as

$$d(n) = x(n) - x(n-1). \tag{2.10}$$

Clearly the original sequence can be reproduced from the initial sample $x(0)$ and the sequence $d(n)$ by recursively using

$$x(n) = x(n-1) + d(n) \quad \text{for} \quad n = 1, \dots \tag{2.11}$$

The idea behind coding sequence $d(n)$ instead of $x(n)$ is that usually $d(n)$ is less correlated and thus according to the observation of Section 2.2.1.2, it assumes lower entropy. Indeed, assuming without loss of generality, that $E\{x(n)\} = 0$, autocorrelation $r_d(m)$ of $d(n)$ can be calculated as follows:

$$
\begin{aligned}
r_d(m) &= E\{d(n)d(n+m)\} \\
&= E\{(x(n) - x(n-1))(x(n+m) - x(n+m-1))\} \\
&= E\{x(n)x(n+m)\} + E\{x(n-1)x(n+m-1)\} \\
&\quad - E\{x(n)x(n+m-1)\} - E\{x(n-1)x(n+m)\} \\
&= 2r_x(m) - r_x(m-1) - r_x(m+1) \\
&\approx 0,
\end{aligned}
\tag{2.12}
$$

where, in the last row of (2.12), we used the assumption that the autocorrelation coefficient $r_x(m)$ is very close to the average of $r_x(m-1)$ and $r_x(m+1)$. In view of Equation (2.12) we may expect that, under certain conditions (not always though), the correlation between successive samples of $d(n)$ is low even in the case where the original sequence $x(n)$ is highly correlated. We thus expect that $d(n)$ has lower entropy than $x(n)$.

In practice, the whole procedure is slightly more complicated because $d(n)$ should be quantized as well. This means that the decoder cannot use Equation (2.11) as it would result in the accumulation of a quantization error. For this reason the couple of expressions (2.10), (2.11) are replaced by:

$$d(n) = x(n) - \hat{x}(n-1), \tag{2.13}$$

where

$$
\begin{aligned}
\hat{x}(n) &= \hat{d}(n) + \hat{x}(n-1) \\
\hat{d}(n) &= \bar{Q}[d(n)].
\end{aligned}
\tag{2.14}
$$

DPCM, as already described, is essentially a one-step ahead prediction procedure, namely $x(n-1)$ is used as a prediction of $x(n)$ and the prediction error is next coded. This procedure can be generalized (and enhanced) if the prediction takes into account more past samples weighted appropriately in order to capture the signal's statistics. In this case, Equations (2.10) and (2.11) are replaced by their generalized counterparts:

$$
\begin{aligned}
d(n) &= x(n) - \mathbf{a}^\mathrm{T}\mathbf{x}(n-1) \\
x(n) &= d(n) + \mathbf{a}^\mathrm{T}\mathbf{x}(n-1)
\end{aligned}
\tag{2.15}
$$

where sample vector $\mathbf{x}(n-1) \triangleq [x(n-1)x(n-2)\cdots x(n-p)]^\mathrm{T}$ contains $p$ past samples and $\mathbf{a} = [a_1 a_2 \dots a_p]^\mathrm{T}$ is a vector containing appropriate weights known also as prediction coefficients.

Again, in practice (2.15) should be modified similarly to (2.14) in order to avoid the accumulation of quantization errors.

### 2.4.1.3 Adaptive Differential Pulse Code Modulation (ADPCM)

In the simplest case, prediction coefficients, $\mathbf{a}$, used in (2.15) are constant quantities characterizing the particular implementation of the ($p$-step) DPCM codec. Better decorrelation of $d(n)$ can be achieved, though, if we adapt these prediction coefficients to the particular correlation properties of $x(n)$. A variety of batch and recursive methods can be employed for this task resulting in the so called Adaptive Differential Pulse Code Modulation (ADPCM).

### 2.4.1.4 Perceptual Audio Coders (MPEG layer III (MP3), etc.)

Both DPCM and ADPCM exploit redundancy reduction to lower entropy and consequently achieve better compression than PCM. Apart from analog filtering (for antialiasing purposes) and quantization, they do not distort the original signal $x(n)$. On the other hand, the family of codecs of this section applies serious controlled distortion to the original sample sequence in order to achieve far lower entropy and consequently much better compression ratios.

Perceptual audio coders, the most celebrated representative being the MPEG-1 layer III audio codec (MP3) (standardized in ISO/IEC 11172-3, [10]), split the original signal into subband signals and use quantizers of different quality depending on the perceptual importance of each subband.

Perceptual coding relies on four fundamental observations validated by extensive psychoacoustic experiments:

(1) Human hearing system cannot capture single tonal audio signals (i.e., signals of narrow frequency content) unless their power exceeds a certain threshold. The same also holds for the distortion of audio signals. The aforementioned audible threshold depends on the particular frequency but is relatively constant among human listeners. Since this threshold refers to single tones in the absence of other audio content, it is called the audible threshold in quiet (ATQ). A plot of ATQ versus frequency is presented in Figure 2.3.

(2) An audio tone of high power, called a masker, causes an increase in the audible threshold for frequencies close to its own frequency. This increase is higher for frequencies close to the masker,

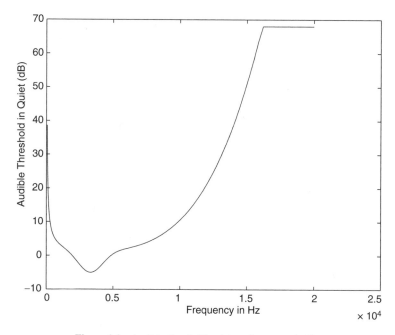

**Figure 2.3** Audible threshold quiet vs. frequency in Hz.

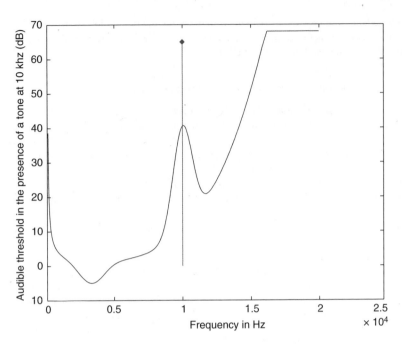

**Figure 2.4**    Audible threshold in the presence of a 10 kHz tone vs. frequency in Hz.

and decays according to a spreading function. A plot of audible threshold in the presence of a masker is presented in Figure 2.4.

(3) The human ear perceives frequency content in an almost logarithmic scale. The Bark scale, rather than linear frequency (Hz) scale, is more representative of the ear's ability to distinguish between two neighboring frequencies. The Bark frequency, $z$, is usually calculated from its linear counterpart $f$ as:

$$z(f) = 13 \arctan(0.00076f) + 3.5 \arctan\left(\left(\frac{f}{7500}\right)^2\right) (\text{bark})$$

Figure 2.5 illustrates the plot of $z$ versus $f$. As a consequence the aforementioned masking spreading function has an almost constant shape when it is expressed in terms of Bark frequency. In terms of the linear frequency (Hz), this leads to a wider spread for maskers with (linear) frequencies residing close to the upper end of the audible spectrum.

(4) By dividing the audible frequency range into bands of one Bark width, we get the so called critical bands. Concentration of high power noise (non-tonal audio components) within one critical band causes an increase in the audible threshold of the neighboring frequencies. Hence, these concentrations of noise resemble the effects of tone maskers and are called Noise Maskers. Their masking effect spreads around their central frequency in a manner similar to their tone counterpart.

Based on these observations, Perceptual Audio Coders: (i) sample and finely quantize the original analog audio signal, (ii) segment it into segments of approximately 1 second duration, (iii) transform each audio segment into an equivalent frequency representation employing a set of complementary

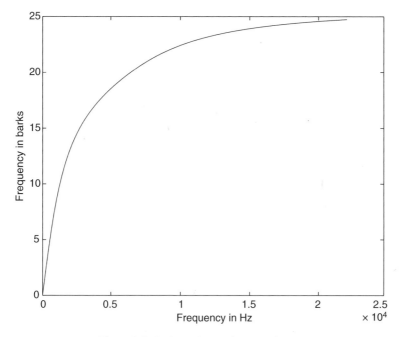

**Figure 2.5**   Bark number vs. frequency in Hz.

frequency selective subband filtters (subband analysis filterbank) followed by a modified version of Discrete Cosine Transform (M-DCT) block, (iv) estimate the overall audible threshold, (v) quantize the frequency coefficients to keep quantization errors just under the corresponding audible threshold. The reverse procedure is performed on the decoder side.

A thorough presentation of the details of Perceptual Audio Coders can be found in [11] or [9] while the exact encoding procedure is defined in ISO standards [MPEG audio layers I, II, III].

### 2.4.2 Open-Loop Vocoders: Analysis – Synthesis Coding

As explained in the previous section, Waveform Codecs share the concept of attempting to approximate the original audio waveform by a copy that is (at least perceptually) close to the original. The achieved compression is a result of the fact that, by design, the copy has less entropy than the original.

Open-Loop Vocoders (see e.g., [12]) of this section and their Closed-Loop descendants, presented in the next section, share a different philosophy initially introduced by H. Dudley in 1939 [13] for encoding analog speech signals. Instead of approximating speech waveforms, they try to dig out models (in fact digital filters) that describe the speech generation mechanism. The parameters of these models are next coded and transmitted. The corresponding encoders are then able to re-synthesize speech by appropriately exciting the prescribed filters.

In particular, Open-Loop Vocoders rely on voiced/unvoiced speech models and use representations of short time speech segments by the corresponding model parameters. Only (quantized versions of) these parameters are encoded and transmitted. Decoders approximate the original speech by forming digital filters on the basis of the received parameter values and exciting them by pseudo-random sequences. This type of compression is highly efficient in terms of compression ratios, and has low encoding and decoding complexity at the cost of low reconstruction quality.

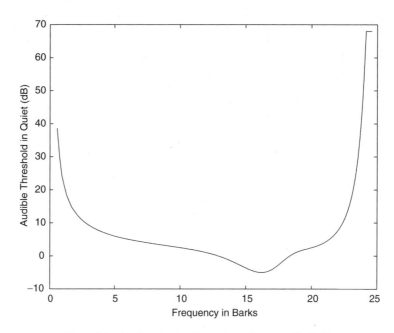

**Figure 2.6**   Audible threshold in quite vs. frequency (in Bark).

### 2.4.3  Closed-Loop Coders: Analysis by Synthesis Coding

This type of speech coder is the preferred choice for most wireless systems. It exploits the same ideas with the Open-Loop Vocoders but improves their reconstruction quality by encoding not only speech model parameters but also information regarding the appropriate excitation sequence that should be used by the decoder. A computationally demanding procedure is employed on the encoder's side in order to select the appropriate excitation sequence. During this procedure the encoder imitates the decoder's synthesis functionality in order to select the optimal excitation sequence from a pool of predefined sequences (known both to the encoder and the decoder). The optimal selection is based on the minimization of audible (perceptually important) reconstruction error.

Figure 2.7 illustrates the basic blocks of an Analysis-by-Synthesis speech encoder. The speech signal $s(n)$ is approximated by a synthetically generated signal $s_e(n)$. The latter is produced by exciting the

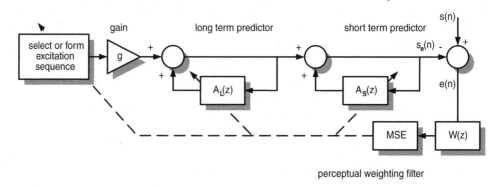

**Figure 2.7**   Basic blocks of an analysis-by-synthesis speech encoder.

cascade of two autoregressive (AR) filters with an appropriately selected excitation sequence. Depending on the type of encoder, this sequence is either selected from a predefined pool of sequences or dynamically generated during the encoding process. The coefficients of the two AR filters are chosen so that they imitate the natural speech generation mechanism. The first is a long term predictor of the form

$$H_L(z) = \frac{1}{1 - A_L(z)} = \frac{1}{1 - az^{-p}} \qquad (2.16)$$

in the frequency domain, or

$$y(n) = a\,y(n - p) + x(n), \qquad (2.17)$$

in the time domain, that approximates the pitch pulse generation. The delay $p$ in Equation (2.16) corresponds to the pitch period. The second filter, a short term predictor of the form,

$$H_S(z) = \frac{1}{1 - A_S(z)} = \frac{1}{1 - \sum_{i=1}^{K} a_i z^{-i}} \qquad (2.18)$$

shapes the spectrum of the synthetic speech according to the formant structure of $s(n)$. Typical values of filter order $K$ are in the range 10 to 16.

The encoding of a speech segment reduces to computing / selecting: (i) the AR coefficients of $A_L(z)$ and $A_S(z)$, (ii) the gain $g$ and (iii) the exact excitation sequence. The selection of the aforementioned optimal parameters is based on minimizing the error sequence $e(n) = s(n) - s_e(n)$. In fact, the Mean Squared Error (MSE) of a weighted version $e_w(n)$ is minimized where $e_w(n)$ is the output of a filter $W(z)$ driven by $e(n)$. This filter, which is also dynamically constructed (as a function of $A_S(z)$) imitates the human hearing mechanism by suppressing those spectral components of $e(n)$ that are close to high energy formants (see Section 2.4.1.4 for the perceptual masking behavior of the ear).

Analysis-by-Synthesis coders are categorized by the exact mechanism that they adopt for generating the excitation sequence. Three major families will be presented in the sequel: (i) the Multi-Pulse Excitation model (MPE) , (ii) the Regular Pulse Excitation model (RPE) and (iii) the Vector or Code Excited Linear Prediction model (CELP) and its variants (ACELP, VSELP).

### 2.4.3.1 Multi-Pulse Excitation Coding (MPE)

This method was originally introduced by Atal and Remde [14]. In its original form MPE used only short term prediction.The excitation sequence is a train of $K$ unequally spaced impulses of the form

$$x(n) = x_0\delta(n - k_0) + x_1\delta(n - k_1) + \cdots + x_{K-1}\delta(n - k_{K-1}), \qquad (2.19)$$

where $\{k_0, k_1, \ldots, k_{K-1}\}$ are the locations of the impulses within the sequence and $x_i$ ($i = 0, \ldots, K - 1$) the corresponding amplitudes. Typically $K$ is 5 or 6 for a sequence of $N = 40$ samples (5 ms at 8000 samples/s). The impulse locations $k_i$ and amplitudes $x_i$ are estimated according to the minimization of the perceptually weighted error, quantized and transmitted to the decoder along with the quantized versions of the short term prediction AR coefficients. Based on this data the decoder is able to reproduce the excitation sequence and pass it through a replica of the short prediction filter in order to generate an approximation of the encoded speech segment synthetically.

In more detail, for each particular speech segment, the encoder performs the following tasks.

*Linear prediction.* The coefficients of $A_S(z)$ of the model in (2.18) are first computed employing Linear Prediction (see end of Section 2.3.3).

*Computation of the weighting filter.* The employed weighting filter is of the form

$$W(z) = \frac{1 - A_S(z)}{1 - A_S(z/\gamma)} = \frac{1 - \sum_{i=1}^{10} a_i z^{-i}}{1 - \sum_{i=1}^{10} \gamma^i a_i z^{-i}}, \tag{2.20}$$

where $\gamma$ is a design parameter (usually $\gamma \approx 0.8$). The transfer function of $W(z)$ of this form has minima in the frequency locations of the formants i.e., the locations where $|H(z)|_{z=e^{j\omega}}$ attains its local maxima. It thus suppresses error frequency components in the neighborhood of strong speech formants; this behavior is compatible with the human hearing perception.

*Iterative estimation of the optimal multipulse excitation.* An all-zero excitation sequence is assumed first and in each iteration a single impulse is added to the sequence so that the weighted MSE is minimized. Assume that $L < K$ impulses have been added so far with locations $k_0, \ldots, k_{L-1}$. The location and amplitude of the $L + 1$ impulse are computed based on the following strategy: If $s^L(n)$ is the output of the short time predictor excited by the already computed $L$-pulse sequence and $k_L$, $x_L$ the unknown location and amplitude of the impulse to be added, then

$$s^{L+1}(n) = s^L(n) + h(n) \star x_L \delta(n - k_L)$$

and the resulting weighted error is

$$e_W^{L+1}(n) = e_W^L(n) - h_\gamma(n) \star x_L \delta(n - k_L) = e_W^L(n) - x_L h_\gamma(n - k_L), \tag{2.21}$$

where $e_W^L(n)$ is the weighted residual obtained using $L$ pulses and $h_\gamma(n)$ is the impulse response of $H(z/\gamma) \equiv W(z)H(z)$. Computation of $x_L$ and $k_L$ is based on the minimization of

$$J(x_L, k_L) = \sum_{n=0}^{N-1} \left( e_W^{L+1}(n) \right)^2. \tag{2.22}$$

Setting $\partial J(x_L, k_L)/\partial x_L = 0$ yields

$$x_L = \frac{r_{eh}(k_L)}{r_{hh}(0)} \tag{2.23}$$

where $r_{eh}(m) \equiv \sum_n e_W^L(n) h_\gamma(n + m)$ and $r_{hh}(m) \equiv \sum_n h_\gamma(n) h_\gamma(n + m)$. By substituting expression (2.23) in (2.21) and the result into (2.22) we obtain

$$J(x_L, k_L)|_{x_L=\text{fixed}} = \sum_n (e_W^L(n))^2 - \frac{r_{eh}^2(k_L)}{r_{hh}(0)}. \tag{2.24}$$

Thus, $k_L$ is chosen so that $r_{eh}^2(k_L)$ in the above expression is maximized. The selected value of the location $k_L$ is next used in (2.23) in order to compute the corresponding amplitude.

Recent extensions of the MPE method incorporate a long term prediction filter as well, activated when the speech segment is identified as voiced. The associated pitch period $p$ in Equation (2.16) is determined by finding the first dominant coefficient of the autocorrelation $r_{ee}(m)$ of the unweighted residual, while the coefficient $a_p$ is computed as

$$a_p = \frac{r_{ee}(p)}{r_{ee}(0)}. \tag{2.25}$$

### 2.4.3.2 Regular Pulse Excitation Coding (RPE)

Regular Pulse Excitation methods are very similar to Multipulse Excitation ones. The basic difference is that the excitation sequence is of the form

$$x(n) = x_0\delta(n - k) + x_1\delta(n - k - p) + \cdots + x_{K-1}\delta(n - k - (K - 1)p), \tag{2.26}$$

i.e., impulses are equally spaced with a period $p$ starting from the location $k$ of the first impulse. Hence, the encoder should optimally select the initial impulse lag $k$, the period $p$ and the amplitudes $x_i$ ($i = 0, \ldots, K - 1$) of all $K$ impulses.

In its original form, proposed by Kroon and Sluyter in [15] the encoder contains only a short term predictor of the form (2.18) and a perceptually weighting filter of the form (2.20). The steps followed by the RPE encoder are summarized next.

*Pitch estimation.* The period $p$ of the involved excitation sequence corresponds to the pitch period in the case of voiced segments. Hence an estimate of $p$ can be obtained by inspecting the local maxima of the autocorrelation function of $s(n)$ as explained in Section 2.3.3.

*Linear prediction.* The coefficients of $A_S(z)$ of the model in (2.18) are computed employing Linear Prediction (see end of Section 2.3.3).

*Impulse lag and amplitude estimation.* This is the core step of RPE. The unknown lag $k$ (i.e., the location of the first impulse) and all amplitudes $x_i$ ($i = 0, \ldots, K - 1$) are jointly estimated. Suppose that the $K \times 1$ vector $\mathbf{x}$ contains all $x_i$'s. Then any excitation sequence $x(n)$ ($n = 0, \ldots, N - 1$) with initial lag $k$ can be written as an $N \times 1$ sparse vector, $\mathbf{x}^k$ with non-zero elements $x_i$ located at $k, k + p, k + 2p, \ldots, k + (K - 1)p$. Equivalently,

$$\mathbf{x}^k = \mathbf{M}^k\mathbf{x}, \tag{2.27}$$

where rows $k + ip$ ($i = 0, \ldots K - 1$) of the $N \times K$ sparse binary matrix $\mathbf{M}$ contain a single 1 at their $i$-th position.

The perceptually weighted error attained by selecting a particular excitation $x(n)$ is

$$\begin{aligned} e(n) &= w(n) \star (s(n) - h(n) \star x(n)) \\ &= w(n) \star s(n) - h_\gamma(n) \star x(n), \end{aligned} \tag{2.28}$$

where $h(n)$ is the impulse response of the short term predictor $H_S(z)$, $h_\gamma(n)$ the impulse response of the cascade $W(z)H(z)$ and $s(n)$ the input speech signal. Equation (2.28) can be rewritten using vector notation as

$$\mathbf{e}^k = \mathbf{s}_w - \mathbf{H}_\gamma\mathbf{M}^k\mathbf{x}, \tag{2.29}$$

where $\mathbf{s}_w$ is an $N \times 1$ vector depending upon $s(n)$ and the previous state of the filters that does not depend on $k$ or $\mathbf{x}$ and $\mathbf{H}$ is an $N \times N$ matrix formed by shifted versions of the impulse response of $H(z/\gamma)$. The influence of $k$ and $x_i$ is incorporated in $\mathbf{M}^k$ and $\mathbf{x}$ respectively (see above for their definition).

For fixed $k$ optimal $\mathbf{x}$ is the one that minimizes

$$\sum_{n=0}^{N-1} e(n)^2 = \left(\mathbf{e}^k\right)^{\mathrm{T}}\left(\mathbf{e}^k\right), \tag{2.30}$$

that is

$$\mathbf{x} = \left[\left(\mathbf{M}^k\right)^{\mathrm{T}}\mathbf{H}_\gamma^{\mathrm{T}}\mathbf{H}_\gamma\mathbf{M}^k\right]^{-1}\left(\mathbf{M}^k\right)^{\mathrm{T}}\mathbf{H}_\gamma^{\mathrm{T}}. \tag{2.31}$$

After finding the optimal $\mathbf{x}$ for all candidate values of $k$ using the above expression the overall optimal combination $(k, x)$ is the one that yields the minimum squared error in Equation (2.30). Although the computational load due to matrix inversion in expression (2.31) seems to be extremely high, the internal structure of the involved matrices allows for fast implementations.

The RPE architecture described above contains only a short term predictor $H_S(z)$. The addition of a long term predictor $H_L(z)$ of the form (2.16) enhances coding performance for high pitch voiced speech segments. Computation of the pitch period $p$ and the coefficient $a$ is carried out by repetitive recalculation of the attained weighted MSE for various choices of $p$.

### 2.4.3.3 Code Excited Linear Prediction Coding (CELP)

CELP is the most distinguished representative of Analysis-by-Synthesis codecs family. It was originally proposed by M. R. Schroeder and B. S. Atal in [16]. This original version of CELP employs both long and short term synthesis filters and its main innovation relies on the structure of the excitation sequences used as input to these filters. A collection of predefined pseudo-Gaussian sequences (vectors) of 40 samples each form the so called Codebook available both to the encoder and the decoder. A codebook of 1024 such sequences is proposed in [16].

Incoming speech is segmented into frames. The encoder performs a sequential search of the codebook in order to find the code vector that produces the minimum error between the synthetically produced speech and the original speech segment. In more detail, each sequence $v_k$ ($k = 0, \ldots, 1023$) is multiplied by a gain $g$ and passed to the cascade of the two synthesis filters (LTP and STP). The output is next modified by a perceptually weighting filter $W(z)$ and compared against an also perceptually weighted version of the input speech segment. Minimization of the resulting MSE allows for estimating the optimal gain for each code vector and, finally, for selecting that code vector with the overall minimum perceptual error.

The parameters of the short term filter ($H_S(z)$) that has the common structure of Equation (2.18) are computed using standard linear prediction optimization once for each frame, while long term filter ($H_L(z)$) parameters, i.e., $p$ and $a$ are recomputed within each sub-frame of 40 samples. In fact, a range $[20, \ldots, 147]$ of integer values of $p$ are examined assuming no excitation. Under this assumption the output of the LTP depends only on past (already available) values of it (see Equation (2.17)). The value of $a$ that minimizes perceptual error is computed for all admissible $p$'s and the final value of $p$ is the one that yields the overall minimum.

The involved perceptual filter $W(z)$ is constructed dynamically as function of $H_L(z)$ in a fashion similar to MPE and LPE.

The encoder transmits: (i) quantized expressions of the LTP and STP coefficients, (ii) the index $k$ of the best fitting codeword, (iii) the quantized version of the optimal gain $g$.

The decoder resynthesizes speech, exciting the reconstructed copies of LTP and STP filters by the code vector $k$.

The descent quality of CELP encoded speech even at low bitrates captured the interest of the scientific community and the standardization bodies as well. Major research goals included; (i) complexity reduction especially for the codebook search part of the algorithm and (ii) improvements on the delay introduced by the encoder. This effort resulted in a series of variants of CELP like VSELP, LD-CELP and ACELP, which are briefly presented in the sequel.

*Vector-Sum Excited Linear Prediction (VSELP)*. This algorithm was proposed by Gerson and Jasiuk in [17] and offers faster codebook search and improved robustness to possible transmission errors.

VSELP assumes three different codebooks; three different excitation sequences are extracted from them, multiplied by their own gains and summed up to form the input to the short term prediction filter. Two of the codebooks are static, each of them containing 128 predifined pseudo-random sequences of length 40. In fact, each of the 128 sequences corresponds to a linear combination of seven basis vectors weighted by $\pm 1$.

On the other hand the third codebook is dynamically updated to contain the state of the autoregressive LTP $H_L(z)$ of Equation (2.16). Essentially, the sequence obtained from this adaptive codebook is equivalent to the output of the LTP filter for a particular choice of the lag $p$ and the coefficient $a$. Optimal selection of $p$ is performed in two stages: an open-loop procedure exploits the autocorrelation of the original speech segment, $s(n)$, to obtain a rough initial estimate of $p$. Then a closed-loop search is performed around this initial lag value to find this combination of $p$ and $a$ that, in the absence of other excitation (from the other two codebooks), produces synthetic speech as close to $s(n)$ as possible.

*Low Delay CELP (LD-CELP)*. This version of CELP is due to J-H. Chen *et al.* [18]. It applies very fine speech signal partitioning into frames of only 2.5 ms consisting of four subframes of 0.625 msec. The algorithm does not assume long term prediction (LTP) and employs a 50th order short term prediction (STP) filter whose coefficients are updated every 2.5 msec. Linear prediction uses a novel autocorrelation estimator that uses only integer arithmetic.

*Algebraic CELP (ACELP)*. ACELP has all the characteristics of the original CELP with the major difference being the simpler structure of its codebook. This contains ternary valued sequences, $c(n)$, ($c(n) \in \{-1, 0, 1\}$), of the form

$$c(n) = \sum_{i=1}^{K} (\alpha_i \delta(n - p_i) + \beta_i \delta(n - q_i)) \tag{2.32}$$

where $\alpha_i, \beta_i = \pm 1$, typically $K = 2, 3, 4$ or 5 (depending on the target bitrate) and the pulse locations $p_i, q_i$ have a small number of admissible values. Table 2.1 includes these values for $K = 5$. This algebraic description of the code vectors allows for compact encoding and also for fast search within the codebook.

**Table 2.1**

| $p_1, q_1 \in$ | {0, | 5, | 10, | 15, | 20, | 25, | 30, | 35} |
|---|---|---|---|---|---|---|---|---|
| $p_2, q_2 \in$ | {1, | 6, | 11, | 16, | 21, | 26, | 31, | 36} |
| $p_3, q_3 \in$ | {2, | 7, | 12, | 17, | 22, | 27, | 32, | 37} |
| $p_4, q_4 \in$ | {3, | 8, | 13, | 18, | 23, | 28, | 33, | 38} |
| $p_5, q_5 \in$ | {4, | 9, | 14, | 19, | 24, | 29, | 34, | 39} |

*Relaxation Code Excited Linear Prediction Coding (RCELP)*. The RCELP algorithm [19] deviates from CELP in that it does not attempt to match the pitch of the original signal, $s(n)$, exactly. Instead, the pitch is estimated once within each frame and linear interpolation is used for approximating the pitch in the intermediate time points. This reduces the number of bits used for encoding pitch values.

## 2.5 Speech Coding Standards

Speech coding standards applicable to wireless communications are briefly presented in this section.

ITU G.722.2 (see [20]) specifies wide-band coding of speech at around 16 kbps using the so called Adaptive Multi-Rate Wideband (AMR-WB) codec. The latter is based on ACELP. The standard describes encoding options targeting bitrates from 6.6 to 23.85 kbps. The entire codec is compatible with the AMR-WB codecs of ETSI-GSM and 3GPP (specification TS 26.190).

ITU G.723.1 (see [21]) uses Multi-Pulse Maximum Likelihood Quantization (MP-MLQ) and the ACELP speech codec. Target bitrates are 6.3 kbps and 5.3 kbps respectively. The coder operates on 30 msec frames of speech sampled at an 8 kHz rate.

ITU G.726 (see [22]) refers to the conversion of linear or $A$-law or $\mu$-law PCM to and from a 40, 32, 24 or 16 kbps bitstream. Some ADPCM coding scheme is used.

ITU G.728 (see [23]) uses LD-CELP to encode speech sampled at 8000 samples/sec with 16 kbps.

ITU G.729 (see [24]) specifies the use of the Conjugate Structure ACELP algorithm for encoding speech at 8 kbps.

ETSI-GSM 06.10 (see [25]) specifies GSM Full Rate (GSM-FR) codec that employs the RPE algorithm for encoding speech sampled at 8000 samples/sec. Target bitrate is 12.2 kbps, i.e., equal to the throughput of GSM Full Rate channels.

ETSI-GSM 06.20 (see [26]) specifies GSM Half Rate (GSM-HR) codec that employs VSELP algorithm for encoding speech sampled at 8000 samples/sec. Target bitrate is 5.6 kbps, i.e., equal to the throughput of GSM Half Rate channels.

ETSI-GSM 06.60 (see [27]) specifies GSM Enhanced Full Rate (GSM-EFR) codec that employs the Conjugate Structure ACELP (CS-ACELP) algorithm for encoding speech sampled at 8000 samples/sec. Target bitrate is 12.2 kbps, i.e., equal to the throughput of GSM Full Rate channels.

ETSI-GSM 06.90 (see [28]) specifies GSM Adaptive Multi-Rate (GSM-AMR) codec that employs the Conjugate Structure ACELP (CS-ACELP) algorithm for encoding speech sampled at 8000 samples/sec. Various target bitrate modes are supported starting from 4.75 kbps up to 12.2 kbps. A newer version of GSM-AMR, GSM WideBand AMR, was adopted by ETSI/GSM for encoding wideband speech sampled at 16 000 samples/sec.

3GPP2 EVRC, adopted by the 3GPP2 consortium (under ARIB: STD-T64-C.S0014-0, TIA: IS-127 and TTA: TTAE.3G-C.S0014), specifies the so called Enhanced Variable Rate Codec (EVRC) that is based on RCELP speech coding algorithm. It supports three modes of operation, targeting bitrates of 1.2, 4.8 and 9.6 kbps.

3GPP2 SMV, adopted by 3GPP2 (under TIA: TIA-893-1), specifies the Selectable Mode Vocoder (SMV) for Wideband Spread Spectrum Communication Systems. SMV is CELP based and supports four modes of operation targeting bitrates of 1.2, 2.4, 4.8 and 9.6 kbps.

## 2.6 Understanding Video Characteristics

### 2.6.1 Video Perception

Color information of a point light source is represented by a $3 \times 1$ vector $\mathbf{c}$. This representation is possible due to the human visual perception mechanism. In particular, color sense is a combination of the stimulation of three different types of cones (light sensitive cells on the retina). Each cone type has a different frequency response when it is excited by the visible light (with wavelength $\lambda \in [\lambda_{\min}, \lambda_{\max}]$ where $\lambda_{\min} \approx 360$ nm and $\lambda_{\max} \approx 830$ nm). For a light source with spectrum $f(\lambda)$ the produced stimulus reaching the vision center of the brain is equivalent to the vector

$$\mathbf{c} = \begin{bmatrix} c_1 \\ c_2 \\ c_3 \end{bmatrix}, \quad \text{where } c_i = \int_{\lambda_{\min}}^{\lambda_{\max}} s_i(\lambda) f(\lambda) \mathrm{d}\lambda, \quad i = 1, 2, 3. \tag{2.33}$$

Functions $s_i(\lambda)$ attain their maxima in the neighborhoods of Red (R), Green (G) and Blue (B) as illustrated in Figure 2.8.

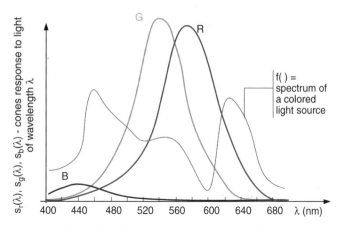

**Figure 2.8**  Tri-stimulus response to colored light.

*2.6.2  Discrete Representation of Video – Digital Video*

Digital Video is essentially a sequence of still images of fixed size, i.e.,

$$\mathbf{x}(n_c, n_r, n_t), n_c = 0, \ldots, N_c - 1, n_r = 0, \ldots, N_r - 1, n_t = 0, 1, \ldots, \tag{2.34}$$

where $N_c$, $N_r$ are the numbers of columns and rows of each single image in the sequence and $n_t$ determines the order of the particular image with respect to the very first one. In fact, if $T_s$ is the time interval between capturing or displaying two successive images of the above sequence, $T_s n_t$ is the time elapsed between the acquisition/presentation of the first image and the $n_t$-th one.

The feeling of smooth motion requires presentation of successive images at rates higher than 10 to 15 per second. An almost perfect sense of smooth motion is attained using 50 to 60 changes per second. The latter correspond to $T_s = 1/50$ or $1/60$ sec. Considering, for example, the European standard PAL for the representation of digital video $N_r = 576$, $N_c = 720$ and $T = 1/50$ sec. Simple calculations indicate that an overwhelming amount of approximately $20 \times 10^6$ samples should be captured/displayed per second. This raises the main issue of digital video handling: extreme volumes of data. The following sections are devoted to how these volumes can be represented in compact ways, particularly for video transmission purposes.

**2.6.2.1  Color Representation**

In the previous paragraph we introduced the representation $\mathbf{x}(n_c, n_r, n_t)$ associating it with the somehow vague notion of video sample. Indeed digital video is a 3-D sequence of samples, i.e., measurements and, more precisely, measurements of color. Each of these samples is essentially a vector usually of length 3, corresponding to a transformation

$$\mathbf{x} = \mathbf{Tc} \tag{2.35}$$

of the color vector of Equation (2.33).

*RGB representation*. In the simplest case

$$\mathbf{x} = \begin{bmatrix} r \\ g \\ b \end{bmatrix}, \tag{2.36}$$

where $r$, $g$, $b$ are normalized versions of $c_1$, $c_2$ and $c_3$ respectively, as defined in (2.33). In fact, since digital video is captured by video cameras rather than the human eye, the exact shape of $s_i(\lambda)$ in (2.33) depends on the particular frequency response of the acquisition sensors (e.g. CCD cells). Still, though, they are frequency selective and concentrated around the frequencies of Red, Green and Blue light. RGB representation is popular in the computer world but not that useful in video encoding/transmission applications.

*YCrCb representation.* The preferred color representation domain for video codecs is YCrCb. Historically this choice was due to compatibility constraints originating from moving from black and white television to color television; in this transition luminance (Y) is represented as a discrete component and color information (Cr and Cb) is transmitted through an additional channel providing backward compatibility. Digital video encoding and transmission stemming from their analog antecedents favor representations that decouple luminance from color. YCrCb is related to RGB through the transformation

$$
\begin{bmatrix} Y \\ Cr \\ Cb \end{bmatrix} = \begin{bmatrix} 0.299 & 0.587 & 0.114 \\ 0.500 & -0.4187 & -0.0813 \\ -0.1687 & -0.3313 & 0.500 \end{bmatrix} \begin{bmatrix} r \\ g \\ b \end{bmatrix},
\tag{2.37}
$$

where $Y$ represents the luminance level and $Cr$, $Cb$ carry the information of color.

Other types of transformation result in alternative representations like the YUV and YIQ, which also contain a separate luminance component.

### 2.6.3 Basic Video Compression Ideas

#### 2.6.3.1 Controlled Distortion of Video

*Frame size adaptation.* Depending on the target application, video frame size varies from as low as $96 \times 128$ samples per frame for low quality multimedia presentations up to $1080 \times 1920$ samples for high definition television. In fact, video frames for digital cinema reach even larger sizes. The second column of Table 2.2 gives standard frame sizes of most popular video formats.

**Table 2.2**   Characteristics of common standardized video formats

| Format | Size | Framerate (fps) | Interlaced | Color representation |
|--------|------|-----------------|------------|----------------------|
| CIF | $288 \times 352$ | | NO | 4:2:0 |
| QCIF | $144 \times 176$ | | NO | 4:2:0 |
| SQCIF | $96 \times 128$ | | NO | 4:2:0 |
| SIF-625 | $288 \times 352$ | 25 | NO | 4:2:0 |
| SIF-525 | $240 \times 352$ | 30 | NO | 4:2:0 |
| PAL | $576 \times 720$ | 25 | YES | 4:2:2 |
| NTSC | $486 \times 720$ | 29.97 | YES | 4:2:2 |
| HDTV | $720 \times 1280$ | 59.94 | NO | 4:2:0 |
| HDTV | $1080 \times 1920$ | 29.97 | YES | 4:2:0 |

Most cameras capture video in either PAL (in Europe) or NTSC (in the US) and subsampling to smaller frame sizes is performed prior to video compression.

The typical procedure for frame size reduction contains the following steps:

(1) abortion of odd lines,
(2) application of horizontal decimation i.e., low pass filtering and 2 : 1 subsampling.

*Frame Rate Adaptation.* The human vision system (eye retina cells, nerves and brain vision center) acts as a low pass filter regarding the temporal changes of the captured visual content. A side effect of this incapability is that by presenting to our vision system sequences with still images every 50–60 times per second is enough to generate the sense of smooth scene change. This fundamental observation is behind the idea of approximating moving images by sequences of still frames. Well before the appearance of digital video technology the same idea was (and is still) used in traditional cinema.

Thus, using frame rates in the range 50–60 fps yields a satisfactory visual quality. In certain bitrate critical applications, such as video conference, frame rates as low as 10–15 fps are used, leading to obvious degradation of the quality.

In fact, psycho-visual experiments led to halving the 50–60 fps rates using an approach that cheats the human vision system. The so called interlaced frames are split into even and odd fields that contain the even and odd numbered rows of samples of the original frame. By successively altering the content of only the even or the odd fields 50–60 times per second, a satisfactory smoothness results, although this corresponds to an actual frame rate of only 25–30 fps.

The third column of Table 2.2 lists the standardized frame rates of popular video formats. Missing framerates are not subject to standardization. In addition, the fourth column of the table shows whether the corresponding video frames are interlaced.

*Color subsampling of video.* Apart from backwards compatibility constraints that forced the use of YCrCb color representation for video codecs, an additional advantage of this representation has been identified. Psychovisual experiments showed that the human vision system is more sensitive to high spatial frequencies of luminance than in the same range of spatial frequencies of color components Cr and Cb. This allowed for subsampling of Cr and Cb (i.e., using less chrominance samples per frame) without serious visible deterioration of the visual content. Three main types of color subsampling have been standardized: (i) 4:4:4 where no color subsampling is performed, (ii) 4:2:2 where for every four samples of Y only two samples of Cr and two samples of Cb are encoded, (iii) 4:2:0 where for every four samples of Y only one sample of Cr and one sample of Cb is encoded. The last column of Table 2.2 refers to the color subsampling scheme used in the included video formats.

*Accuracy of color representation.* Usually both luminance $(Y)$ and color $(Cr$ and $Cb)$ samples are quantized to $2^8$ levels and thus 8 bits are used for their representation.

### 2.6.3.2 Redundancy Reduction of Video

*Motion estimation and compensation.* Motion estimation aims at reducing temporal correlation between successive frames of a video sequence. It is a technique analogous to prediction used in DPCM and ADPCM. Motion estimation is applied to selected frames of the video sequence in the following way.

(1) *Macroblock grouping.* Pixels of each frame are grouped into macroblocks usually consisting of four $8 \times 8$ luminance $(Y)$ blocks and from a single $8 \times 8$ block for each chrominance component $(Cr$ and $Cb$ for the YCrCb coloor representation). In fact this grouping is compatible to with the 4:2:0 color subsampling scheme. If 4:4:4 or 4:2:2 is used, grouping is modified accordingly.

(2) *Motion estimation.* For motion estimation for the macroblocks of a frame corresponding to current time index $n$ a past or future frame corresponding to time $m$ is used as a reference. For each macroblock, say $B_n$, of the current frame a search procedure is employed to find some $16 \times 16$ region, say $M_m$, of the reference frame whose luminance best matches the $16 \times 16$ luminance samples of $B_n$. Matching is evaluated on the basis of some distance measure such as the sum of the squared differences or the sum of the absolute differences between the corresponding luminance samples.

The outcome of motion estimation is a motion vector for every macrobloock i.e., a $2 \times 1$ vector, **v** equal to the relative displacement between $B_n$ and $M_m$.

(3) *Calculation of Motion Compensated Prediction Error (Residual).* In the sequel, instead of coding the pixel values of each macroblock, the difference macroblock

$$E_n \triangleq B_n - M_m$$

is computed and coded. The corresponding motion vector is also encoded Figure 2.9 illustrates this procedure.

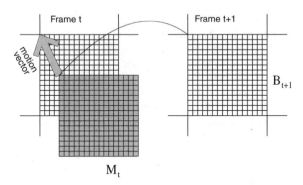

**Figure 2.9** Motion estimation of a $16 \times 16$ macroblock, $B_{t+1}$ of the $t+1$-frame using the $t$-frame as a reference. In this example, the resulting motion vector is $\mathbf{v} = (-4, 8)$.

Every few frames, motion estimation is interrupted and particular frames are encoded rather than their motion compensated residuals. This prevents the accumulation of errors and offers access points for restarting decoding. In general, video frames are categorized into three types: I, P and B.

- Type I or Intra frames are those that are independently encoded. No motion estimation is performed for the blocks of this type of frames.
- Type P or Predicted frames are those that are motion compensated using as reference the most recent of the past Intra or Predicted frames. Time index $n > m$ in this case for P frames.
- Type B or Bidirectionnaly Interpolated frames are those that are motion compensated with reference to past and/or future I and P frames. Motion estimation results in this case in two different motion vectors: one for each of the past and future reference frames pointing to best matching regions $M_{m^-}$ and $M_{m^+}$, respectively. The motion error macroblock (which is passed to the next coding stages) is computed as

$$E_n \triangleq B_n - \frac{1}{2}(M_{m^-} + M_{m^+}).$$

Usually video sequence is segmented into consecutive Groups of Pictures (GOP), starting with an I frame followed by types P and B frames located in predefined positions. The GOP of Figure 2.10 has the structure IBBPBBPBBPBB. During decoding frames are reproduced in a different order, since decoding of a B frame requires subsequent I or P reference frames. For the previous example the following order should be used: **I**BBPBBPBBPBB**I**BB, where the bold face B frames belong to the previous GOP and the bold face I frame belongs to the next GOP.

*Transform coding – the Discrete Cosine Transform (DCT).* While motion estimation techniques are used to remove temporal correlation, DCT is used to remove spatial correlation. DCT is applied on $8 \times 8$ blocks of luminance and chrominance. The original sample values are transformed in the case of I frames and the prediction error blocks for P and B frames. If $X$ represents any of these blocks the

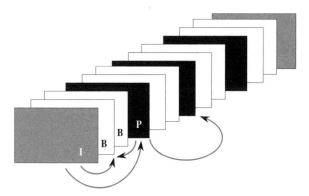

**Figure 2.10**   Ordering and prediction reference of I, P and B frames within a GOP of 12 frames.

resulting transformed block, $Y$, is also $8 \times 8$ and is obtained as

$$Y = FX\mathbb{F}^{\mathrm{T}} \tag{2.38}$$

where the real valued $8 \times 8$ DCT transformation matrix $F$ is of the form

$$F_{kl} = \begin{cases} \dfrac{1}{2}\cos\left(\dfrac{\pi}{8}k\left(l+\dfrac{1}{2}\right)\right), & k = 1,\ldots 7 \\[3mm] \dfrac{1}{2\sqrt{2}}\cos\left(\dfrac{\pi}{8}k\left(l+\dfrac{1}{2}\right)\right), & k = 0. \end{cases} \tag{2.39}$$

Decorrelation properties of DCT have been proved by extensive experimentation on natural image data. Beyond decorrelation DCT exposes excellent energy compaction properties. In practice this means that the most informative DCT coefficients within $Y$ are positioned close to the upper-left portion of the transformed block. This behavior is demonstrated in Figure 2.11 where a natural image

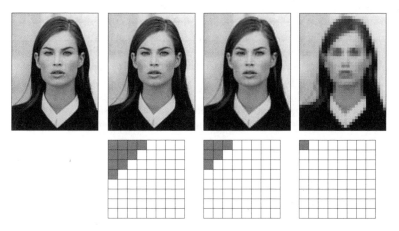

**Figure 2.11**   Result of applying block DCT. Discard least significant coefficients and the inverse DCT. The original image is the left-most one. Dark positions on the $8 \times 8$ grids in the lower part of the figure indicate the DCT coefficients that were retained before inverse DCT.

(top-left) is transformed using block DCT, least significant coefficients were discarded (set to 0) and an approximation of the original image was produced by inverse DCT.

Apart from its excellent decorrelation and energy compaction properties, DCT is preferred in coding applications because a number of fast DCT implementations (some of them in hardware) are available.

## 2.7  Video Compression Standards

### 2.7.1  H.261

The ITU video encoding international standard H.261 [29] was developed for use in video-conferencing applications over ISDN channels allowing bitrates of $p \times 64$ Kbps, $p = 1 \cdots 30$. In order to bridge the gap between the European pal and the North American NTSC video formats, H.261 adopts the Common Interchange Format (CIF) and for lower bitrates the QCIF (see Table 2.2). It is interesting to notice that even using QCIF / 4:2:0 with a framerate of 10 frames per second requires a bitrate of approximately 3 Mbps, which means that a compression ratio of $48 : 1$ is required in order to transmit it over a 64 kbps ISDN channel.

H.261 defines a hierarchical data structure. Each frame consists of 12 GOB (group-of-blocks). Each GOB contains 33 MacroBlocks (MB) that are further split into $8 \times 8$ Blocks (B). Encoding parameters are assumed unchanged within each macroblock. Each MB consists of four Luminance (Y) blocks and two chrominance blocks in accordance with the 4:2:0 color subsampling scheme.

The standard adopts a hybrid encoding algorithm using the Discrete Cosine Transform (DCT) and Motion Compensation between successive frames.

Two modes of operation are supported as follows.

*Interframe coding.* In this mode, already encoded (decoded) frames are buffered in the memory of the encoder (decoder). Motion estimation is applied on Macroblocks of the current frame using the previous frame as a reference (Type P frames). The motion compensation residual is computed by subtracting the best matching region of the previous frame from the current Macroblock. The six $8 \times 8$ blocks of the residual are next DCT transformed. DCT coefficients are quantized and quantization symbols are entropy encoded. Non-zero motion vectors are also encoded. The obtained bitstream containing (i) the encoded quantized DCT coefficients, (ii) the encoded non-zero motion vectors and (iii) the parameters of the employed quantizer, is passed to the output buffer that guarantees constant outgoing bitrate. Monitoring the level of this same buffer, a control mechanism determines the quantization quality for the next macroblocks in order to avoid overflow or underflow.

*Intraframe coding.* In order (i) to avoid accumulation of errors, (ii) to allow for (re)starting of the decoding procedure at arbitrary time instances and (iii) for improving image quality in the case of abrupt changes of video content (where motion compensated prediction fails to offer good estimates) the encoder supports block DCT encoding of selected frames (instead of motion compensation residuals). These Type I frames may appear in arbitrary time instances; it is a matter for the particular implementation to decide when and under which conditions an Intra frame will be inserted.

Either Intra blocks or motion compensation residuals are DCT transformed and quantized.

The H.261 decoder follows the inverse procedure in a straightforward manner.

### 2.7.2  H.263

The H.263 ITU standard [30] is a descendant of H.261, offering better encoding quality especially for low bitrate applications, which are its main target. In comparison to H.261 it incorporates more accurate motion estimation procedures, resulting in motion vectors of half-pixel accuracy. In addition, motion estimation can switch between $16 \times 16$ and $8 \times 8$ block matching; this offers better performance especially in high detail image areas. H.263 supports bi-directional motion estimation (B frames) and the use of arithmetic coding of the DCT coefficients.

### 2.7.3 MPEG-1

The MPEG-1 ISO standard [10], produced by the Motion Pictures Expert Group is the first in a series of video (and audio) standards produced by this group of ISO. In fact, the standard itself describes the necessary structure and the semantics of the encoded stream in order to be decodable by an MPEG-1 compliant decoder. The exact operation of the encoder and the employed algorithms (e.g. motion estimation search method) are purposely left as open design issues to be decided by developers in a competitive manner.

The MPEG-1 targets video and audio encoding at bitrates in the range 1.5 Mbits/sec. Approximately 1.25 Mbits/sec are assigned for encoding SIF-625 or SIF-525 non-interlaced video and 250 Kbits/sec for stereo audio encoding. MPEG-1 was originally designed for storing/playing back video to/from single speed CD-ROMs.

The standard assumes a CCIR 601 input image sequence i.e., images with 576 lines with 720 luminance samples and from 360 samples for Cr and Cb. The incoming frame rate is up to 50 fps.

Input frames are lowpass filtered and decimated to 288 lines of 360 (180) luminance (chrominance) samples.

Video codec part of the standard relies on the use of:

- decimation for downsizing the original frames to SIF; interpolation at the decoder's side;
- motion estimation and compensation as described in Section 2.6.3.2;
- block DCT on $8 \times 8$ blocks of luminance and chrominance;
- quantization of the DCT coefficients using a dead zone quantizer. Appropriate amplification of the DCT coefficients prior to quantization results in finer resolution for the most significant of them and suppression of the weak high-frequency ones;
- Run Length Encoding (RLE) using zig-zag scanning of the DCT coefficients (see Figure 2.12). In particular, if $s(0)$ is the symbol assigned to the dead zone (to DCT coefficients around zero) and $s(i)$ any other quantization symbol, RLE represents symbol strings of the form

$$\underbrace{s(0) \quad \cdots \quad s(0)}_{n} s(i), \quad n \geq 0,$$

with new shortcut symbols $A_{ni}$ indicating that $n$ zeros ($s(0)$) are followed by the symbol $s(i)$;
- entropy coding of the run symbols $A_{ni}$ using the Huffman Variable Length Encoder.

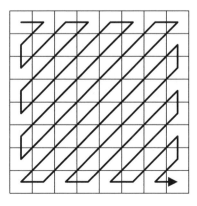

**Figure 2.12**   Reordering of the $8 \times 8$ quantized DCT coefficients into a $64 \times 1$ linear array uses the zig-zag convention. Coefficients corresponding to DC and low spatial frequencies are scanned first, while highest frequencies are left for the very end of the array. Normally this results in long zero-runs after the first few elements of the latter.

MPEG-1 encoders support bitrate control mechanisms. The produced bitstream is passed to control a FIFO buffer that empties with a rate equal to the target bitrate. When the level of the buffer exceeds a predefined threshold, the encoder forces its quantizer to reduce quantization quality (e.g., increase the width of the dead zone) leading of course to deterioration of encoding quality as well. Conversely, when the control buffer level becomes lower than a certain bound, the encoder forces finer quantization of the DCT coefficients. This mechanism guarantees an actual average bitrate that is very close to the target bitrate. This type of encoding is known as the Constant Bit Rate (CBR). When the control mechanism is absent, or activated only in very extreme situations, the encoding quality remains almost constant but the bitrate strongly changes, leading to the so-called Variable Bitrate (VBR) encoding.

### 2.7.4 MPEG-2

The MPEG-2 ISO standard [31] has been designed for high bitrate applications typically starting from 2 and reaching up to 60 Mbps. It can handle a multitude of input formats, including CCIR-601, HDTV, 4K, etc. Unlike MPEG-1, MPEG-2 allows for interlaced video, which is very common in broadcast video applications, and exploits the redundancy between odd and even fields. Target uses of the standard include digital television, high definition television, DVD and digital cinema.

The core encoding strategies of MPEG-2 are very close to those of MPEG-1; perhaps its most important improvement relies on the so-called scalable coding approach. Information scaling refers to the subdivision of the encoded stream into separate sub-streams that carry different levels of information detail. One of them transports the absolutely necessary information that is available to all users (decoders) while the others contain complementary data that may upgrade the quality of the received video. Four types of scalability are supported:

(1) Data partitioning where, for example, the basic stream contains information only for low frequency DCT coefficients while high frequency DCT coefficients can be retrieved only through the complementary streams.
(2) SNR scalability, where the basic stream contains information regarding the most significant bits of the DCT coefficients (equivalent to coarse quantization), while the other streams carry least significant bits of information.
(3) Spatial scalability, where the basic stream encodes a low resolution version of the video sequence and complementary streams may be used for improving image analysis.
(4) Temporal scalability, where complementary streams encode time decimated versions of the video sequence, while their combination may increase temporal resolution.

### 2.7.5 MPEG-4

Unlike MPEG-1/2, which introduced particular compression schemes for audio-visual data, the MPEG-4 ISO standard [32] concentrates on the combined management of versatile multimedia sources. Different codecs are supported within the standard for optimal compression of each of these sources. MPEG-4 adopts the notion of an audiovisual scene that is composed of multiple Audiovisual Objects (AVO) that evolve both in space and time. These objects may be:

- moving images (natural video) or space/time segments of them;
- synthetic (computer generated) video;
- still images or segments of them;
- synthetic 2-D or 3-D objects;
- digital sound;
- graphics or
- text.

MPEG-4 encoders use existing encoding schemes (such as MPEG-1 or JPEG) for encoding the various types of audiovisual objects. Their most important task is to handle AVO hierarchy (e.g., the object newscaster comprises the lower level objects: moving image of the newscaster and voice of the newscaster). Beyond that, MPEG-4 encodes the time alignment of the encoded AVOs.

A major innovation of MPEG-4 is that it assigns the synthesis procedure of the final form of the video to the end viewer. The viewer (actually the decoder parametrized by the viewer) receives the encoded information of the separate AVOs and is responsible for the final synthesis, possibly in accordance with an instructions stream distributed by the encoder. Instructions are expressed using an MPEG-4 specific language called Binary Format for Scenes – BIFS, which is very close to VRML. A brief presentation of MPEG-4 can be found in [33], while a detailed description of BIFS is presented in [34].

The resulting advantages of MPEG-4 approach are summarized below.

(1) It may offer better compression rates of natural video by adapting compression quality or other encoding parameters (like the motion estimation algorithm) to the visual or semantic importance of particular portions of it. For example, background objects can be encoded with less bits than the important objects of the foreground.

(2) It provides a genuine handling of different multimedia modalities. For example, text or graphics need not be encoded as pixels rasters superimposed on video pixels; each of them can be encoded separately using their native codecs, postponing superposition till the synthesis procedure at decoding stage.

(3) It offers advanced levels of interaction since it assigns to the end user the task of (re-) synthesizing the transmitted audiovisual objects into an integrated scene. In fact, instead of being dummy decoders MPEG-4 players can be interactive multimedia applications. For example, consider a scenario of a sports match where, together with the live video, MPEG-4 encoders streams the game statistics in textual form, gives information regarding the participating players, etc.

### 2.7.6 H.264

H.264 is the first video (and audio) compression standard [35] to be produced by the combined standardization efforts of ITU's Video Coding Experts Group (VCEG) and ISO's Motion Pictures Experts Group (MPEG). The standard, released in 2003, defines five different profiles and, overall, 15 levels distributed with these profiles. Each profile determines *a* subset of the syntax used by H.264 to represent encoded data. This allows for adapting the complexity of the corresponding codecs to the actual needs of particular applications. For the same reason, different levels within each profile limit the options for various parameter values (like the size of particular look-up tables). Profiles and levels are ordered according to the quality requirements of targeted applications. Indicatively, Level 1 (within Profile 1) is appropriate for encoding video with up to 64 kbps. At the other end, Level 5.1 within profile 5 is considered for video encoding at bitrates up to 240 000 kbps.

H.264 achieves much better video compression – two or three times lower bitrates for the same quality of decoded video – than all previous standards, at the cost of increased coding complexity. It uses all the tools described in Section 2.6.3 for controlled distortion, subsampling, redundancy reduction (via motion estimation) transform coding and entropy coding also used in MPEG-1/2 and H.261/3 with some major innovations. These innovative characteristics of H.264 are summarized below.

*Intra prediction.* Motion estimation as presented within Section 2.6.3.2 was described as a means for reducing temporal correlation in the sense that macroblocks of a B or P frame, with time index $n$, are predicted from blocks of equal size (perhaps in different locations) of previous or/and future reference frames, of time index $m \neq n$. H.264 recognizes that a macroblock may be similar to another macroblock within the same frame. Hence, motion estimation and compensation is extended to intra frame processing (the reference coincides with the current frame, $m = n$) searching for self similarities. Of course, this search is limited to portions of the same frame that will be available to the decoder prior to the currently encoded macroblock. On top of that, computation of the residual,

$$E_n \triangleq B_n - \hat{M}_n,$$

is based on a decoded version of the reference region. Using this approach, H.264 achieves not only temporal decorrelation (with the conventional inter motion estimation) but also spatial decorrelation.

In addition, macroblocks are allowed to be non-square shaped and non-equally sized (apart from $16 \times 16$, sizes of $16 \times 8$, $8 \times 16$, $8 \times 8$ are allowed). In Profile 1, Macroblocks are further split into blocks of $4 \times 4$ luminance samples. and $2 \times 2$ chrominance samples; Larger $8 \times 8$ blocks are allowed in higher profiles. This extra degree of freedom offers better motion estimation results, especially for regions of high detail where matching fails for large sized macroblocks.

*Integer arithmetic transform.* H.264 introduces a deviation of the Discrete Cosine Transform with integer valued transformation matrix $\mathbf{F}$ (see Equation (2.38)). In fact, entries of $\mathbf{F}$ assume slightly different values depending on whether the transform is applied on intra blocks or residual (motion compensated) blocks, luminance or chrominance blocks. Entries of $\mathbf{F}$ are chosen in a way that both the direct and inverse transform can be implemented using only bit-shifts and additions (multiplication free).

*Improved lossless coding.* Instead of the conventional Run Length Encoding (RLE) followed by Huffman or Arithmetic entropy coding, H.264 introduces two other techniques for encoding the transformed residual sample values, the motion vectors, etc., namely,

(1) Exponential Golomb Code is used for encoding single parameter values while Context-based Adaptive Variable Length Coding (CAVLC) is introduced as an improvement of the conventional RLE.
(2) Context-based Adaptive Binary Arithmetic Coding (CABAC) is introduced in place of Huffman or conventional Arithmetic coding.

# References

[1] C. Shannon, Communication in the presence of noise, *Proceedings of the IRE*, **37**, pp. 10–21, 1949.
[2] D. Huffman, A method for the construction of minimum-redundancy codes, *Proceedings of the IRE*, **40**, pp. 1098–1101, September, 1952.
[3] P. G. Howard and J. S. Vitter, Analysis of arithmetic coding for data compression, *Information Processing and Management*, **28**(6), 749–764, 1992.
[4] ITU-T, Recommendation G.711 – pulse code modulation (PCM) of voice frequencies, Gneva, Switzerland, 1988.
[5] B. Sklar, *Digital Communications: Fundamentals and Applications*, Englewood Cliffs, NJ, Prentice-Hall, 1988.
[6] S. Haykin, *Adaptive Filter Theory.* Upper Saddle River, NJ, Prentice-Hall, 3rd edn, 1996.
[7] H. Nyquist, Certain topics in telegraph transmission theory, *Trans. AIEE*, **47**, 617–644, April 1928.
[8] C. Lanciani and R. Schafer, Psychoacoustically-based processing of MPEG-I Layer 1-2 encoded signals, 1997.
[9] D. Pan, A tutorial on MPEG/audio compression, *IEEE MultiMedia*, **2**(2), 60–74, 1995.
[10] ISO/IEC, MPEG-1 coding of moving pictures and associated audio for digital storage media at up to about 1.5 mbit/s, *ISO/IEC 11172*, 1993.
[11] A. Spanias, Speech coding: A tutorial review, *Proceedings of the IEEE*, **82**, 1541–1582, October 1994.
[12] R. M. B. Gold and P. E. Blankenship, New applications of channel vocoders, *IEEE Trans. ASSP*, **29**, 13–23, February 1981.
[13] H. Dudley, Remarking speech, *J. Acoust. Soc. Am.* **11**(2), 169–177, 1939.
[14] B. Atal and J. Remde, A new model for LPC excitation for producing natural sound speech at low bit rates, *Proc. ICASSP-82*, **1**, pp. 614–617, May 1982.
[15] E. D. P. Kroon and R. Sluyter, Regular-pulse excitation: A novel approach to effective and efficient multi-pulse coding of speech, *IEEE Trans. ASSP*, **34**, 1054–1063, October 1986.
[16] M. Schroeder and B. Atal, Code-excited linear prediction (CELP): High-quality speech at very low bit rates, *Proc. ICASSP*, pp. 937–940, March 1985.

[17] I. Gerson and M. Jasiuk, Vector sum excited linear prediction (VSELP) speech coding at 8 kbit/s, *Proc. ICASSP-90, New Mexico*, April 1990.

[18] J-H., Chen, R. V. Cox, Y.-C. Lin, N. S. Jayant and M. Melchner, A low delay CELP order for the CCITT 16 kbps speech coding standard, *IEEE J. Selected Areas in Communications*, **10**, 830–849, June 1992.

[19] W. B. Kleijn, Kroon, P. and D. Nahumi, The RCELP speech-coding algorithm, *European Trans. on Telecommunications*, **5**, 573–582, September/October 1994.

[20] ITU-T, Recommendation G.722.2 – wideband coding of speech at around 16 kbit/s using adaptive multi-rate wideband (AMR-WB), Geneva, Switzerland, July 2003.

[21] ITU-T, Recommendation G.723.1 – dual rate speech coder for multimedia communications, Geneva, Switzerland, March 1996.

[22] ITU-T, Recommendation G.726 – 40, 32, 24, 16 kbit/s adaptive differential pulse code modulation (ADPCM), Geneva, Switzerland, December 1990.

[23] ITU-T, Recommendation G.728 – coding of speech at 16 kbit/s using low-delay code excited linear prediction, Geneva, Switzerland, September 1992.

[24] ITU-T, Recommendation G.729 – coding of speech at 8 kbit/s using conjugate-structure algebraic-code-excited linear-prediction (CS-ACELP), Geneva, Switzerland, March 1996.

[25] ETSI EN 300 961 V8.1.1, GSM 6.10 – digital cellular telecommunications system (phase 2+); full rate speech; transcoding, Sophia Antipolis Cedex, France, November 2000.

[26] ETSI EN 300 969 V8.0.1, GSM 6.20 – digital cellular telecommunications system (phase 2+); half rate speech; half rate speech transcoding, Sophia Antipolis Cedex, France, November 2000.

[27] ETSI EN 300 726 V8.0.1, GSM 6.60 – digital cellular telecommunications system (phase 2+); enhanced full rate (EFR) speech transcoding, Sophia Antipolis Cedex, France, November 2000.

[28] ETSI EN 300 704 V7.2.1, GSM 6.90 – digital cellular telecommunications system (phase 2+); adaptive multi-rate (AMR) speech, transcoding, Sophia Antipolis Cedex, France, April 2000.

[29] ITU-T, Recommendations H.261 – video codec for audiovisual services at p × 64 kbit/s, Geneva, Switzerland, 1993.

[30] ITU-T, Recommendation H.263 – video coding for low bit rate communication, Geneva, Switzerland, February 1998.

[31] ISO/IEC, MPEG-2 generic coding of moving pictures and associated audio information, *ISO/IEC 13818*, 1996.

[32] ISO-IEC, Overview of the MPEG-4 standard, *ISO/IEC JTC1/SC29/WG11 N2323*, July 1998.

[33] Rob Koenen, MPEG-4 multimedia for our time, *IEEE Spectrum*, **36**, 26–33, February 1999.

[34] Julien Signes, Binary Format for Scene (BIFS): Combining MPEG-4 media to build rich multimedia services, *SPIE Proceedings*, 1998.

[35] ITU-T, Recommendation H.264 – advanced video coding (avc) for generic audiovisual services, Geneva, Switzerland, May 2003.

# 3

# Multimedia Transport Protocols for Wireless Networks

Pantelis Balaouras and Ioannis Stavrakakis

## 3.1 Introduction

Audio and video communication over the wired Internet is already popular and has an increasing degree of penetration to the Internet users. The rapid development of broadband wireless networks, such as wireless Local Area Networks (WLANs), third generation (3G) and fourth generation (4G) cellular networks, will probably result in an even more intensive penetration of multimedia-based services. The provision of multimedia-based services over the wired Internet presents significant technical challenges, since these services require Quality of Service (QoS) guarantees from the underlying wired network. In the case of wireless networks these challenges are even greater due to the characteristics of a wireless environment, such as:

- highly dynamic channel characteristics;
- high burst error rates and resulting packet losses;
- limited bandwidth;
- delays due to handoffs in case of user mobility.

In Section 3.2, we present a classification of the media types (discrete and continuous media), and the multimedia-based services (non real-time and real-time). The requirements of the multimedia services for preserving the intra-media and inter-media synchronizations – which are quite challenging over packet-switched networks – are also discussed. A short discussion on Constant Bit Rate (CBR) and Variable Bit Rate (VBR) encoding and the transmission of VBR content over CBR channels follows. In Section 3.3, we classify the real-time multimedia based services into three categories (one-way streaming, on demand delivery and conversational communication) and present their specific QoS requirements. In Section 3.4, we discuss issues regarding the adaptation at the encoding level, that is, non-adaptive, adaptive and, scalable/layered encoding. In Section 3.5, we discuss QoS issues for real-time services and, specifically, how the continuous media flows can (i) adjust their rate to match the available bandwidth; (ii) cope with the induced network delay and delay variation (jitter); and (iii) recover lost packets. In Section 3.6, we discuss why TCP is not suitable for real-time continuous media

*Emerging Wireless Multimedia: Services and Technologies*   Edited by A. Salkintzis and N. Passas
© 2005 John Wiley & Sons, Ltd

services, not only for wireless but also for wired networking environments and describe the widely adopted RTP/UDP/IP protocol stack. The Real-time Transport Protocol (RTP) and its control protocol RTCP are presented in detail in Section 3.7, whereas in Section 3.8 the multimedia (RTP payload) types that could be transferred by RTP are presented. In Section 3.9, the supported media types in 3G wireless networks and RTP implementation issues for 3G wireless networks are summarized.

## 3.2 Networked Multimedia-based Services

### 3.2.1 Time Relations in Multimedia

In multimedia systems different types of media, such as audio, video, voice, text, still images and graphics, are involved yielding in multimedia compositions. The involved media may be continuous or discrete depending on their relation to time.

Continuous media (CM), such as audio and video, require continuous play-out and explicit timing information for correct presentation; for this reason they are also referred to as time-dependent media. A multimedia composition includes intra-media and inter-media timing information produced during the media encoding process (see Figure 3.1). The series of media units/samples for each media is called a stream. The time relations or dependencies between successive media units/samples of a media stream (e.g., the sampling frequency and sequence) determine the intra-media time constraints. If, for instance, the sequence of the media units/samples or the presentation frequency is changed when the media stream is played-out, the media and its perceived meaning is altered. The time synchronization across the involved media, e.g., video and audio streams, is referred to as inter-media synchronization. If the synchronization is changed, the multimedia is negatively affected, e.g., lip synchronization fails. Therefore, the intra-media and inter-media synchronization are essential for the proper reconstruction of a multimedia composition at the player.

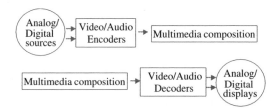

**Figure 3.1**   Producing and reconstructing a multimedia composition.

Discrete media are composed of time-independent media units/samples. Although discrete media such as images, text files or graphics may be animated according to a specific time sequence to be sensible to the users, time is not part of the semantics of discrete media.

### 3.2.2 Non-Real-time and Real-time Multimedia Services

Networked multimedia services in packet-switched networks, like the Internet, transmit a multimedia stream from a source end, where the stream is generated or stored, to the receiver end, where the stream is reconstructed; streams are transmitted as a series of packets. Multimedia services may be non-real-time or real-time.

In the non-real-time (non-RT) case, the multimedia file (composition) is played (reconstructed) after it is downloaded (transferred and stored as a whole) at the receiver side (see Figure 3.2). In the non-RT case, there are no strict time constraints for delivering a packet and packets are retransmitted, if lost. Therefore, the non-RT services do not require any QoS guarantee from the underlying network

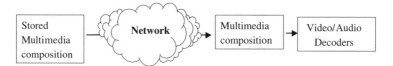

**Figure 3.2** Non-real-time transfer.

regarding the delay, jitter and packet losses but only reliable packet delivery and congestion control. The main requirement is for the receiver to have enough storage space for storing the multimedia file. It is easily concluded that non-RT services can be deployed over a wireless environment, especially if the size of the multimedia files is relatively small. That is why the multimedia message service (MMS), a typical example of a non-real-time service that allows the mobile users to exchange multimedia files of small size, is the first multimedia-based service deployed for mobile users.

A real-time (RT) multimedia transmission system is shown in Figure 3.3. At the source side, the encoder is responsible for the real-time encoding of the analog or digital source information. The encoded data are placed in a buffer before they are encapsulated in the packets and transmitted over the network. Since the encoder generates data continuously and at a possibly variable rate, buffer over-flow may occur if the buffer is not sized properly.

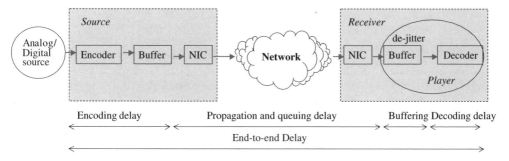

**Figure 3.3** Real-time transmission.

Real Time (RT) multimedia streams are played out at the receiver in order and while being trans-ferred to it. Thus, in order for the playout to be done properly, timing relations between consecutive packets of a single media as well as between different inter-related media (e.g. video and audio) must be preserved. Since the network typically induces diverse inter and intra delays to transmitted streams, it is necessary that these delays and their variation (jitter) be corrected at the receiver. This is accomplished – at the expense of some additional playout delay – by placing the received packet to a de-jitter buffer before forwarding them to the decoder.[a] If the decoder finds the buffer empty, due to a delayed packet arriving after its playout time, the output will be distorted. Finally, it should be noted that packet losses affect the quality at the reception as well, since packet retransmission is typically not possible in (delay-constrained) RT multimedia services. Packet losses may occur within the network or due to the end system buffer overflows. The interested reader in networked multimedia systems may refer to [1].

---

[a] The decoder reads the data from the de-jitter buffer, decodes and renders the data in real time.

### 3.2.3 CBR vs. VBR Encoding for Video

As mentioned in [2], the video encoding schemes can be classified into constant bit rate (CBR) and variable bit rate (VBR).

CBR encoding can be applied for transmission over constant rate links (e.g., ISDN) under strict end-to-end delay constraints. Conversational multimedia services, such as videoconference, typically employs a CBR encoding scheme, such as H.263. The low delay constraint of such applications requires the encoder to generate a video stream that is transmitted over a constant bit rate channel and can be decoded and displayed at the receiver virtually without any pre-decoder or post-decoder buffering. In order to maintain a picture rate as constant as possible, CBR encoding schemes try to limit the number of bits, representing each picture in a video sequence, regardless of the complexity of the picture. Therefore, the final quality of the compressed video stream depends mainly on the complexity of the content, typically determined by the overall amount of motion and also by the level of detail in each particular picture. CBR coding for video works fine, as long as the complexity of the scene is more or less constant, as is the case for head-and-shoulder scenes with little motion.

VBR encoding schemes can be used if the low-delay or constant transmission rate constraint of the application is relaxed. VBR allows for video bitrate variation (i.e., the number of bytes decoded per defined period can vary over different periods of the stream). VBR video in general can provide for a more consistent visual quality by not eliminating (as was done in the CBR case) the inherent variable video rate. The variation of bit rate can also be controlled to adhere the channel throughput limitations and pre-decoder and post-decoder buffering constraints. One way streaming services, such as live video streaming, typically employ VBR encoding scheme (e.g., MPEG-4).

### 3.2.4 Transmission of VBR Content over Constant Rate Channels

Real-time transmission of a variable rate encoded video stream would require a transport channel capable of accommodating this variable bit rate at each time instant. However, many typically used Internet access channels are characterized by a certain bottleneck link rate, which cannot be exceeded (e.g., analog modem speeds, ISDN and so on). A UMTS WCDMA bearer with strict QoS guarantees is

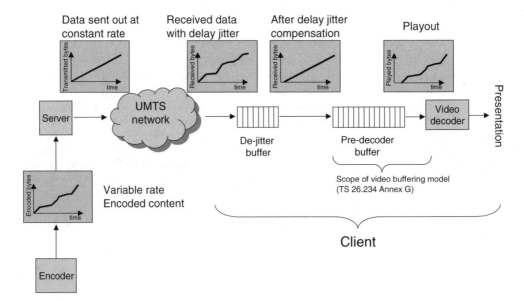

**Figure 3.4**  Transport of VBR streams over UMTS.

another example for such a bottleneck link. Therefore, rate-smoothing techniques are required to enable streaming variable rate encoded content at a constant transmission rate [3].

The transmission of variable rate encoded video content over UMTS is illustrated in Figure 3.4 (taken from [2]). The encoder generates variable rate encoded video streams. The transmission rate of the streaming server is adjusted to the available bandwidth on the UMTS streaming bearer, (in the example this is a constant rate), which corresponds to the negotiated guaranteed bitrate. Delivery over UMTS introduces delay jitter, which needs to be compensated for at the streaming client in the de-jitter buffer. In addition to delay jitter compensation, the streaming client buffer is required to compensate the variation of the incoming rate (i.e. pre-decoder buffer). The interested reader may refer to [2] for a more detailed description.

Clearly, the provision of multimedia-based services over wired Internet is quite challenging, since these services require Quality of Service (QoS) guarantees. Specifically, upper bounds on the delay, jitter and packet loss rate have to be guaranteed for a multimedia application. In the case of wireless networks, these challenges are greater since wireless links have highly dynamic channel character-istics, higher error rates and limited bandwidth, compared with those of the wired links. Delays due to handoffs in case of user mobility are also possible.

## 3.3  Classification of Real-time Services

The real-time multimedia-based services could be classified as follows:

- one-way streaming;
- media on demand delivery;
- conversational audio or/and video communication.

### 3.3.1  One-way Streaming

One-way streaming may concern live events or stored content. In this case, the media stream(s) is delivered to one or more destinations (usually it is addressed to a large audience). In the case of multiple destinations, the packets of the media streams are replicated either by the network layer or the appli-cation layer. Under the network layer-based replication, the media stream packets are replicated by intermediate network devices, routers or switches, according to the multicast or point-to-multipoint capabilities of the underlying IP or ATM network (see Figure 3.5). In the application layer-based

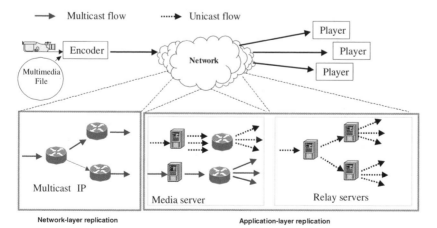

**Figure 3.5**  One-way streaming of live events.

replication, the media encoder module transmits the stream to a root media server, which delivers the stream to the receivers either directly or through a content delivery network (CDN) consisting of inter-mediate media servers, referred to as relay servers. The communication between the root media server, the relay servers and the receivers is point-to-point (pseudo-multicasting), therefore there is no need for the underlying network to support multicast or point-multipoint communications (see Figure 3.5). In the one-way streaming, a limited set of VCR-like interactivity options/commands is available to the user: pause/resume, start/stop. Rewind and record options are possible if these characteristics are sup-ported by the streaming server or the player. Clearly, fast forward, although highly desirable, is impos-sible. Regarding the requirements of the one-way streaming, the end user may tolerate loss ratios of 1% and 2% for audio and video, respectively, and a start-up delay up to 10 seconds. The relatively long start-up delay may allow for the retransmission of lost packets, when the loss ratio thresholds are exceeded.

When the audience is relatively large, it is possible that heterogeneous users with conflicting band-width and decoding capabilities are participating. In order to avoid restricting the encoding and trans-mitting rates to the least-common denominator, the encoder may produce the content at multiple rates or in a complementary way; in this case, the root media server should be equipped with a mechanism for determining the proper sending rate for each user.

One-way streaming of stored content is similar to the streaming of live events with the difference being that the media concerns a prerecorded and stored multimedia file. Theoretically, this gives the option of fast forwarding to an authorized receiver.

Apart from the high error rate, the major issue, when a wireless network is involved, is the limited bandwidth compared with that of a wired network. When the number of receivers in the same wireless network is relatively high, the aggregate bandwidth of the multimedia streams may exceed the link bandwidth. In this case the low rate content production according to specific standards (e.g., 3gpp and 3gpp2 formats) is adopted, enabling the content delivery over wireless/mobile networks and play-out by mobile terminals. The delays due to handoffs are not considered to be a major problem since the users typically tolerate a start-up delay that can absorb the delay due to handoffs.

### 3.3.2 Media on Demand (MoD) Delivery

This service allows the users to select the media file they wish to play from a media server. The media stream is transmitted from the media server directly to the receiver in a point to point manner. VCR-like functionality enables interactivity between the user and the media server. The user can control the media playout by invoking pause, rewind, fast forward or stop commands. A start-up delay of up to 10 seconds and a VCR-like command induced delay of up to 2 seconds may be tolerated by the user. A start-up delay also allows for the retransmission of lost packets.

When a specific content is popular, it may be requested by a number of heterogeneous users. To accomplish such a content delivery to a diverse set of users two approaches may be followed (see Figure 3.6). The first – applicable when content is created for the first time – is to adopt a video encoding

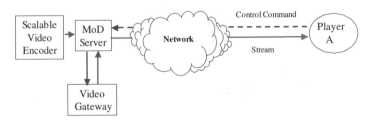

**Figure 3.6**  MoD services.

scheme that enables the video encoding of a file in multiple rates or scalable encoding. The Media on Demand (MoD) server is usually equipped with a mechanism that determines the proper sending rate for each receiver. The second alternative – applicable when content is already available – is to use a video gateway that transforms, on the fly, the original multimedia stream to another stream with the proper rate. The original multimedia file may be encoded at a quality level or rate according to a specific encoding scheme (e.g., MPEG-2), whereas the new stream may be in another format (e.g., MPEG-4) of lower quality level or rate.

The major issue in the MoD services, when a wireless network is involved, is again the limited bandwidth compared with that of a wired network. The same solution as for the one-way streaming service can be followed (see Figure 3.7). The content in this case may be created according to specific low rate standards (e.g., 3gpp and 3gpp2 formats) enabling the content delivery over wireless/mobile networks and content playout by mobile terminals.

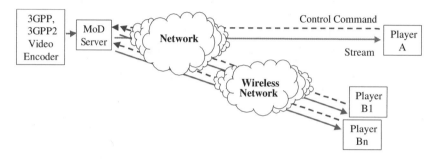

**Figure 3.7**  MoD over wireless.

### 3.3.3 Conversational Communication

Examples of conversational applications are the IP telephony and videoconferencing (videophone). Clearly, at least two participants are involved in an audio/video communication. Each participant transmits an audio and optionally a video stream to the remote participant, and receives an audio and optionally a video stream by the remote participant (see Figure 3.8). The terminal of each participant is equipped with an encoder and decoder capable of encoding and decoding the video and audio formats involved in the session. A session initiation is required in order for the participants to exchange information regarding the network addresses, port, encoding/decoding algorithms and terminal capabilities. Regarding the QoS requirements, high end-to-end delays are noticeable and negatively affect the quality of the communication. A delay of up to 150 msec allows for a good quality conversational communication. Delays higher than 300–400 msec are not acceptable. When both video and audio are transmitted, lip synchronization is also required. The requirement for low end-to-end delay prohibits the retransmission of lost packets and restricts the de-jitter buffer size. The end-user may tolerate loss ratios of 1% and 2% for audio and video, respectively. Clearly, the conversational services are very demanding in terms of QoS.

**Figure 3.8**  Videoconference session.

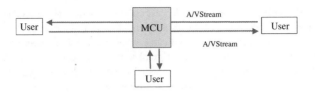

**Figure 3.9**   Centralized videoconference session.

A special case of conversational communication is the multipoint conference where more than two participants are involved in interactive video/audio conferencing. A multipoint conference can be established as centralized, decentralized or hybrid. In a centralized conference, a central entity, the Multipoint Control Unit (MCU), is required (see Figure 3.9). Each participant transmits their video and audio stream to the MCU in a point-to-point manner. The MCU processes the media streams and distributes the mixed media or the streams of the participant who has the floor back to the participants in a point-to-point manner. In a decentralized conference, each participant multicasts their audio and video streams to all receivers, eliminating the need to engage a central point. A hybrid conference combines elements of both a centralized and a decentralized conference. Clearly, the centralized approach is preferable for the wireless environment since under this approach only one audio–video pair of flows has to be transmitted to each participant.

As already mentioned, the conversational services are very demanding in terms of QoS. Therefore, these services, and especially the videophone, are the most difficult services to deploy over a wireless environment.

## 3.4 Adaptation at the Video Encoding Level

A video encoder may or may not adjust on the fly the rate of the produced video stream, or it may produce a number of complementary streams. In the latter case, the more streams the receiver gets the better the perceptual quality. The encoding schemes may be classified into the following categories with respect to their adaptation capabilities: non-adaptive encoding, adaptive encoding and scalable or layered encoding.

### 3.4.1 Non-adaptive Encoding

Under non-adaptive encoding the encoder produces a non-elastic stream with a preset average encoding rate; no encoding parameters may be modified during the transmission. The stream can be CBR or VBR with a fixed average bit rate. Issues regarding CBR and VBR encoding were discussed in Section 3.2.3.

### 3.4.2 Adaptive Encoding

Under adaptive encoding, the encoder can adjust certain parameters (such as the quantization level and bit rate) to adapt to the bandwidth fluctuations. Adaptive encoding is proper for conversational video communication and the delivery of stored video files using a video gateway. This technique may be used in conjunction with source-based rate adaptation schemes for elastic flows (see Section 3.5.1), which utilize a mechanism to feed network status information back to the source. No negotiation is required in this case between the terminal and the encoder. In the case of the involvement of a wireless network, the feedback information may be used in conjunction with a wireless channel model to predict the future wireless channel conditions. Source-based rate adaptation algorithms are discussed in Section 3.5.1. In the case of VBR encoding, smoothing techniques may be used in order to transmit the stream in CBR mode, as briefly discussed in Section 3.2.4.

This technique is unsuitable for one-way streaming to a large audience because of the conflicting bandwidth and decoding requirements of heterogeneous receivers. If applied, it would result in a least-common denominator scenario; that is, the smallest pipe in the network mesh would dictate the quality and fidelity of the overall live multimedia 'broadcast'. For one-way streaming, the scalable/layered encoding, discussed in the sequel, is more suitable.

### 3.4.3 Scalable/Layered Encoding

Scalable/layered encoding techniques produce a base layer and one or more enhancement layers, each transmitted as a separate complementary stream. The base layer provides for the basic video quality, whereas the enhancement layers improve the video quality. The type of layers accessed by the terminals depends on their decoding capabilities (hardware, processing power, power, viewing displays) and the network conditions. This technique is suitable for one-way live streaming and MoD.

In the context of MoD, the encoder may produce a scalable video file so that the MoD server may adjust the transmission rate of the pre-encoded stream according to feedback from the streaming client. The server may also change other traffic characteristics of the application, such as the packet size, according to the characteristics of the network.

## 3.5 Quality of Service Issues for Real-time Multimedia Services

As already mentioned, real-time multimedia services are sensitive to (i) the end-to-end delay, (ii) jitter and (iii) packet loss rate. If the underlying network provides best-effort services, as does the Internet, it cannot guarantee an upper bound on the above critical QoS parameters.

### 3.5.1 Bandwidth Availability

The current Internet provides services on a best-effort basis and cannot reserve bandwidth for a specific flow. The bandwidth is shared by a large number of flows that serve different types of applications. The available bandwidth fluctuates dynamically as flows are initiated and terminated. If the aggregate load of the transmitted flows is higher than the available bandwidth, then congestion occurs and the flows experience delays and packet losses. In the case of continuous media (CM) flows, the packet losses may negatively affect the quality of the transmitted media, e.g., the video quality. If the aggregate load is lower than the available bandwidth, then the flows' rate could be increased, enhancing the quality of the transmitted media. Thus, it is important that the aggregate rate of the CM flows matches the available bandwidth. To enable this rate adaptation, algorithms are employed to ensure that the flows obtain a fair share of the available bandwidth and the aggregate load is close to the overall capacity under the condition that the induced packet losses are kept low. As described below, the rate adaptation control may be applied at the source or the receiver.

In the source-based rate control, the source can explicitly adapt the current transmission rate to a new rate. The new rate is determined based on feedback information sent by the receiver that indicates the presence or absence of congestion. Feedback information may be non-binary (continuous), e.g., the packet loss rate. There are rate adaptation control algorithms that exploit the packet loss rate for shaping the new rate when congestion is indicated. The rate control algorithms at the source can be distinguished as probe-based and model-based (or equation-based). The probe-based algorithms are very simple. They probe for more bandwidth (increase the transmission rate) when the loss rate is less than a threshold and they release bandwidth (decrease the transmission rate) when the loss rate is higher than the threshold. The model-based algorithms are applied in environments where CM flows co-exist with TCP flows. They aim to achieve a long-term throughput for a CM flow similar to the throughput of a TCP flow, so that CM flows are fair and friendly towards TCP flows.

In the receiver-based rate control, the source transmits the media in separate and complementary CM flows, e.g., scalable or layered coded video streams. An example is the Receiver-driven Layered Multicast [23], where the source transmits different layers, i.e., base, enhanced 1, enhanced 2 layers, over different multicast groups. Each receiver estimates its access capability in terms of rate and joins an appropriate number of multicast groups. Each receiver can dynamically join/drop layers by joining/ dropping multicast groups, adjusting, this way, the aggregate reception rate and quality.

### 3.5.2 Delay and Jitter

Packet delay jitter introduced by the network is at least partially eliminated by adding a buffer at the decoder, referred to as the de-jitter buffer. Packets are not played out immediately upon arrival but they are placed in the de-jitter/playout buffer instead. This corresponds to adding a time-offset to the playout time of each packet. The time-offset depends on the buffer size. If the packet delay is less than the offset, then it is buffered until its playout time. However, if the packet delay is higher than the offset, then the packet will arrive later than its playout time, causing a problem to the operation of the decoder. The playout buffer is typically between 5 and 15 seconds for the video streaming applications. Apart from the jitter compensation, the buffering enables the retransmission of lost packets.

Clearly, there is a trade-off between playout delay and the probability of finding the buffer empty (starvation of the decoding process); the higher the former, the lower the latter. Video streaming service can tolerate delays of up to 5–15 seconds, or even more. Real-time conversational video communication cannot tolerate delays longer than 150 msec. Since the delay jitter is typically variable, fixed playout delay approaches would be less effective than adaptive ones. Adaptive playout algorithms estimate the delay jitter and adapt the playout delay accordingly.

### 3.5.3 Recovering Losses

In the case of streamed media, the packet sequence numbers can be used for loss detection. Losses may be recovered by invoking the following application/transport layer approaches:

- Forward Error Correction (FEC);
- retransmission;
- interleaving.

Note that the FEC and retransmission mechanisms discussed in this section refer to the application and transport layers and not to lower layers, such as the wireless MAC layer. This subsection summarizes RFC 2354 [5].

In the wired Internet mostly single packet losses occur. The frequency of bursts of losses of two or more packets is less by an order of magnitude than that of single packet losses. Longer burst losses (of the order of tens of packets) occur infrequently. Therefore, the primary focus of a repair scheme must be the correction of single packet losses, since this is the most frequent occurrence. It is desirable that losses of a relatively small number of consecutive packets may also be repaired, since such losses represent a small but noticeable fraction. The correction of large bursts of losses is of considerably less importance. This is also valid for wireless networks where interference and path loss usually cause single packet losses.

### 3.5.3.1 Forward Error Correction (FEC)

The goal of FEC is to add specialized redundancy data that can be used by the receiver to recover from errors, e.g., overcoming packet losses in a packet network, without any further involvement of the source. The redundant data can be independent of the content of the stream or use knowledge of the stream to improve the repair process, classifying the FEC mechanisms as media-independent or media-specific, respectively.

## Media-Independent FEC

A number of media-independent FEC schemes have been proposed for use with streamed media. These techniques add redundant data, transmitted in separate packets.

The redundant FEC data is typically calculated using the mathematics of finite fields. The simplest finite field is GF(2) where addition is just the eXclusive-OR operation. Basic FEC schemes transmit $k$ data packets with $n-k$ parity (redundant) packets allowing the reconstruction of the original data from any $k$ of the total $n$ transmitted packets. Budge *et al.* [10] proposed applying the XOR operation across different combinations of the media data with the redundant data transmitted separately as parity packets. These vary the pattern of packets over which the parity is calculated, and hence have different bandwidth, latency and loss repair characteristics.

*Parity-based FEC* based techniques have the significant advantage that they are media independent and provide for the exact repair of the lost packets. In addition, the processing requirements are relatively light, especially when compared with some media-specific FEC (redundancy) schemes, which use very low bandwidth but high complexity encoding schemes. The disadvantage of parity based FEC is that the coding has higher latency than the media-specific schemes, discussed in the following section.

A number of FEC schemes exist that are based on higher-order finite fields, for example Reed–Solomon (RS) codes, which are more sophisticated and computationally demanding. These are usually structured so that they have good burst loss protection, although this typically comes at the expense of increased latency. Dependent on the observed loss patterns, such codes may give improved performance, compared with parity-based FEC.

Error bursts (e.g., in wireless) may produce more than $n-k$ successive lost packets. A possible solution is to combine FEC with interleaving to spread out the lost packets. A potential problem is that long error bursts need a large interleaving depth, which results in large delays.

## Media-Specific FEC

The basic idea behind a media-specific FEC is to use knowledge of the specific media compression scheme employed to allow for a more efficient repair of a stream. To repair a stream that has been subject to packet loss, it is necessary to add redundancy to that stream: some information is added that is not required in the absence of packet loss, but that can be used to recover from that loss. Depending on the nature of the media stream, different ways to add the redundancy may be considered.

- If a packet includes one or more media units[b] (audio or video samples), it is logical to use the media unit as the redundancy unit and send duplicate units. By recoding the redundant copy of a media unit, significant bandwidth savings may be achieved, at the expense of additional computational complexity and approximation in the repair. This approach has been advocated for streaming audio [4, 14] showing acceptable performance.
- If media units are split into multiple packets, it is sensible to include redundancy directly within the output of a codec. For example the proposed RTP payload for H.263+ [9] includes multiple copies of key portions of the stream, separated to enhance robustness against packet losses. The advantages of this second approach is efficiency, since the codec designer knows exactly which portions of the stream are most important to protect, and low complexity, since each unit is coded only once.

An alternative approach is to apply media-independent FEC techniques to the most significant bits of a codec output, rather than applying them over the entire packet. Several codec descriptions include special bit field that make this feasible. This approach has low computational cost and can be tailored to represent an arbitrary fraction of the transmitted data.

---

[b]A unit is defined to be a timed interval of media data, typically derived from the workings of the media coder. A packet comprises one or more units, encapsulated for transmission over the network. For example, many audio coders operate at 20 msec units, which are typically combined to produce 40 msec or 80 msec packets for transmission.

The use of media-specific FEC has the advantage of low-latency, with only a single-packet delay being added. This makes it suitable for conversational applications, where large end-to-end delays cannot be tolerated. In a one-way streaming environment it is possible to delay sending the redundant data, achieving improved performance in the presence of bursty losses [15], at the expense of additional latency.

### 3.5.3.2 Retransmission

Packet loss can be repaired by retransmitting lost packets. This requires a feedback mechanism in order for the receiver to inform the source as to which packets were lost and have to be retransmitted. The advantage of this approach is that the bandwidth is effectively used since no redundant packets but only lost packets are retransmitted. The disadvantages are the introduced latency that is at least equal to the round-trip-time and the need for a feedback mechanism that is not feasible in broadcast, multicast or point-to-point transmissions without a feedback channel.

There are variations in the retransmission-based schemes. For the video streaming with time-sensitive data, two kinds of retransmissions can be applied:

(1) delay-constrained retransmission – only packets that can arrive in time for being playout are retransmitted;
(2) priority-based retransmission – important packets are retransmitted before less important ones.

The above schemes make sense for non-interactive applications with relaxed delay bounds but they are inefficient for conversational applications due to the delay introduced by the retransmissions.

In addition to the latency, retransmissions may introduce potentially large bandwidth overhead. Not only are units of data sent multiple times, but additional control traffic must flow to request the retransmissions. It has been shown that, in a large multicast session, most packets are lost by at least one receiver [12]. In this case, the overhead associated with the retransmission of the lost packets may be higher than that introduced by forward error correction schemes. This leads to a natural synergy between the two mechanisms: (a) a forward error correction scheme can be used to repair all single packet losses, and (b) a retransmission based repair as an additional loss recovery mechanism for those receivers that experience bursty losses, and are willing to accept the additional latency introduced by the retransmissions. Such mechanisms have been used in a number of reliable multicast schemes [13, 16].

In order to reduce the retransmissions overhead, the retransmitted units may be piggy-backed onto ongoing transmissions, using a payload format such as that described in [17]. This also allows for the retransmission to be re-coded in a different format, to reduce the bandwidth overhead further. As an alternative, FEC information may be sent in response to retransmission requests [16], allowing for a single retransmission to potentially repair several losses.

### 3.5.3.3 Interleaving

When the unit size is smaller than the packet size and the end-to-end delay is not of concern, interleaving [18] may be employed to reduce the effects of loss. Units are re-sequenced before transmission, so that originally adjacent units are separated by a guaranteed distance in the transmitted stream; their original order is restored at the receiver. Interleaving disperses the effect of packet losses. If, for example, units are 5 ms in length and packets 20 ms (i.e., four units per packet), then the first packet could contain units 1, 5, 9, 13; the second packet could contain units 2, 6, 10, 14 and so on. It can be seen that the loss of a single packet from an interleaved stream results in multiple small gaps in the reconstructed stream, as opposed to the single large gap that would occur in a non-interleaved stream. In many cases, it is easier to reconstruct a stream with such loss patterns, although this is clearly media and codec dependent. Note that the size of the gaps is dependent on the degree of interleaving used, and can be made arbitrarily small at the expense of additional latency.

The obvious disadvantage of interleaving is that it increases latency. This limits the use of this technique for interactive applications, although it performs well for non-interactive use. The major advantage of interleaving is that it does not increase the bandwidth requirements of a stream.

#### 3.5.3.4 Criteria for Selecting a Repair Scheme

In the case of one-way streaming, such as for a radio or television broadcast, latency is considerably less important than reception quality. In this case, any of the previously mentioned approaches may be used (interleaving, retransmission-based repair, FEC). If an approximate repair is acceptable, interleaving is clearly the preferred solution, since it does not increase the bandwidth of a stream. Media independent FEC is typically the next best option, since a single FEC packet has the potential of repairing multiple lost packets.

In a conversational session, the delay introduced by interleaving or retransmission-based schemes is typically not acceptable. A low-latency FEC scheme is the only suitable approach to repairing. The choice between media independent and media specific forward error correction is a less clear-cut: media-specific FEC can be more efficient but requires modifying the output of the codec. This is specified when defining the packet format for the codec. If an existing codec is to be used, a media independent forward error correction scheme is usually easier to implement, it can perform well while at the same time allowing for interactivity. A media stream protected this way may be augmented with a retransmission-based repair scheme with minimal overhead, providing improved quality for those receivers willing to tolerate an additional delay. While the addition of FEC data to a media stream is an effective means for protecting a stream against packet losses, application designers should also take into account that the addition of large amounts of repair data when a loss is detected will increase network congestion, and hence packet loss, leading to a worsening of the problem which the use of error correction coding was intended to solve.

## 3.6 Protocols for Multimedia-based Communication Over the Wireless Internet

### 3.6.1 Why TCP is not Suitable for Real-time CM Services

The Transmission Control Protocol (TCP) [7] is the most-commonly employed transport layer protocol in today's Internet and is largely responsible for its robustness and stability. TCP has been designed – and successfully used – for unicast reliable data transfer. However, TCP is not suitable for real-time CM services because of its congestion control and retransmission mechanisms, and the lack of multicast support.

The congestion control mechanism of TCP is based on the Additive Increase Multiplicative Decrease (AI/MD) algorithm [6, 8], which is a linear algorithm. In the linear control algorithms the flows adjust their rate to a value that is determined by a portion of the current rate plus a fixed amount, as shown below, depending on whether a binary feedback $f$ indicates congestion ($f > 0$) or not ($f = 0$)

$$New\ rate = a_1 + b_1(Current\ rate) \quad if\ f = 0 \quad OR \quad a_D + b_D(Current\ rate) \quad if\ f > 0$$

Typical values for the control parameters $a_I$, $b_I$, $a_D$, $b_D$ of a linear congestion control algorithm are: $a_I > 0$, $b_I = 1$, $a_D = 0$, $b_D < 1$ (AI/MD).

One of TCP's primary goals for fast recovery from congestion is achieved by halving (multiplicative decrease factor $b_D$ equal to 0.5) the transmission rate upon congestion. The rate halving affects the TCP flow's smoothness and throughput, making TCP – and generally all AI/MD schemes with a small multiplicative decrease factor – unsuitable for the continuous media (CM) streaming applications.

Furthermore, TCP's retransmission mechanism, which guarantees data delivery via retransmissions of the lost packets, introduces a typically high and unacceptable end-to-end delay and delay variation (jitter) for CM applications. The requirement for the timely delivery prohibits the recovery of the lost packets through a retransmission scheme, as the recovery mechanism typically fails to meet the decoding deadlines. Note that CM flows do not require 100% reliability, since they are loss tolerant.

When more than two participants are involved in a multimedia-based communication, multiple TCP connections are required since TCP does not support multicast. Thus, TCP is not suitable for one-way streaming or interactive multi-participant communication. Moreover, TCP flows have very low throughput over wireless links due to the high and variable packet losses. Thus, deployment of multimedia-based services over wireless links based on TCP is not viable.

### 3.6.2 RTP/UDP/IP

The User Datagram Protocol (UDP), which lacks a retransmission mechanism and supports multicast, is mainly used to support the CM streaming applications in conjunction with the Real Time Protocol (RTP) and Real Time Control Protocol (RTCP). Applications that use UDP as a transport protocol should also implement an end-to-end congestion control scheme, like TCP, to maintain the stability of the Internet. Otherwise, all supported applications would suffer and eventually the network would collapse [11].

Unlike TCP, UDP does not enforce any congestion control. Therefore the congestion control (i.e., rate adaptation) is implemented at the application level, based on the profile of the implemented service. RTP/UDP/IP streams are divided into inelastic and elastic streams.

RTP streams are often inelastic, that is, they are generated at a fixed or controlled rate. For instance, audio streams are generated at a fixed rate, whereas H.263 video streams may be generated at rates multiple of 64 kbps, subject to a maximum rate specified at the session initiation. This inelasticity reduces the risk of congestion when a relatively low number of RTP streams are transmitted over a link, because inelastic RTP streams will not increase their rates to consume all the available bandwidth as can TCP streams. However, inelastic RTP streams cannot arbitrarily reduce their load on the network to overcome congestion when it occurs.

RTP streams may be elastic for some service profiles, in which case specific methods, such as data rate adaptation based on RTCP feedback are required (see Section 3.7.5.4).

Since RTP may be used for a wide variety of applications in many different contexts, there is no single congestion control mechanism that will efficiently work for all. Therefore, the appropriate congestion control is defined separately in each RTP profile. For some profiles, it may be sufficient to include an applicability statement restricting the use of that profile to environments where congestion is avoided by engineering other mechanisms, e.g., resource reservation or Explicit Congestion Notification (ECN) [49].

## 3.7 Real-time Transport Protocol (RTP)

The Real-time Transport Protocol (RTP) and its control protocol RTCP were specified by the Audio/ Video Transport (AVT) working group within the Internet Engineering Task Force (IETF) as a framework for supporting real-time traffic. It has been also adopted by the International Telecommunication Union (ITU-T) as the transport protocol for multimedia in recommendations H.323 and H.225.0. RTP and RTCP are specified in RFC1889 and its revised specification RFC 3550 [26].

RTP provides end-to-end delivery services that are suitable for applications transmitting real-time data, such as video, audio or simulation data, over unicast or multicast network services. It should be noted that RTP does not guarantee any QoS requirement regarding the delivery of data packets nor does it assume that the underlying network is reliable and delivers packets on time and in sequence. RTP packets may be delivered delayed, be out of order or get lost. RTCP packets may also be lost. For any QoS guarantees, RTP/RTCP relies on lower-layer services like Integrated [50] or Differentiated Services [51] for this.

RTCP provides for delivery monitoring, whereas RTP provides for the following delivery services.

- Payload type identification. RTP packets indicate the media encoding type of the transmitted media. This enables a receiver to distinguish flows of different encoded media and engage the proper decoder.
- Sequence numbering. The sequence numbers included in RTP packets allow the receiver to reconstruct the sender's packet sequence, but sequence numbers might also be used to determine the proper location of a packet, for example in video decoding. It also allows for detecting lost packets.
- Time stamping. RTP packets indicate the sampling time instant at which the first byte of the media of the RTP packet was produced (encoded). Time stamps are required in order to reconstruct directly the intra-media timing. They can also enable inter-media synchronization when a reference clock that represents the absolute time and is shared by all the involved media in a session is available.

Applications typically run RTP/RTCP over UDP to make use of its multiplexing and checksum services; both protocols contribute to the transport protocol functionality. However, RTP may be used with other suitable underlying network or transport protocols. RTP supports data transfer to multiple destinations using multicast distribution, provided that multicast is supported by the underlying network.

While RTP is primarily designed to satisfy the needs of multi-participant multimedia conferences, its applicability is not limited to that particular application. Storage of continuous data, interactive distributed simulation and control and measurement applications may also find RTP applicable.

As already mentioned, RTP is a protocol framework and, thus, not complete. Each particular RTP/RTCP-based application is accompanied by one or more documents:

- a profile specification document, which defines a set of payload type codes and their mapping to payload formats (e.g., media encodings) – a profile for audio and video data may be found in the companion RFC3551 [27];
- payload format specification documents, which define how a particular payload, such as an audio or video encoding, is to be carried in RTP.

RTCP monitors and provides feedback about the quality of service and information about the session participants. The latter aspect of RTCP may be sufficient for 'loosely controlled' sessions, i.e., where there is no explicit membership control and set-up. RTCP is not necessarily intended to support all of an application's control communication requirements. The latter may be fully or partially provided by a separate session control protocol like SIP [52] or H.323 [53]. RTCP allows for monitoring of the data delivery in a manner that is scalable to large multicast networks and provides for minimal control and identification functionality.

### 3.7.1 Multimedia Session with RTP or how RTP is Used

When a user starts using a real-time multimedia service, a multimedia session is established between the user's end-system and the end-systems of the other participants or server. The application that generates the content and sends the RTP packets and/or consumes the content of the received RTP packets is referred to as the end system in RFC3550. A multimedia session consists of a set of concurrent RTP sessions among a common group of participants/end-systems. According to RFC3530, an RTP session is defined as 'an association among a set of participants communicating with RTP'. The end-system that generates the content and transmits the RTP packets, is called the source. A source sends the RTP packets to the destination transport address. A transport address is the combination of a network address, e.g., an IP address, and the port that identifies a transport-level endpoint, e.g., a UDP port. Packets are transmitted from a source transport address to a destination transport address. An end system can act as one or more sources in a particular RTP session, but typically contains only one source.

The source of a stream of RTP packets is identified by a 32-bit numeric Synchronization SouRCe (SSRC) identifier. The SSRC identifier is carried in the RTP header and is independent of the network address. All packets from a synchronization source are part of the same timing and sequence number space, so a receiver groups the received packets based on the synchronization source for playback. Examples of synchronization sources include the sender of a stream of packets derived from a signal source such as a microphone or a camera. A synchronization source may change its data format, e.g., audio encoding, over time. The SSRC identifier is a randomly chosen value meant to be globally unique within a particular RTP session. A participant does not use the same SSRC identifier for all the RTP sessions in a multimedia session; the binding of the SSRC identifiers is provided through RTCP. If a participant generates multiple streams in one RTP session, for example from separate video cameras, each stream is identified as a different SSRC.

As already mentioned, an end-system may be involved in multiple RTP sessions, e.g., send or receive two RTP flows at the same time. For example, a videoconference may contain an audio RTP session and a video RTP session. In a multimedia session, each media is typically carried in a separate RTP session with its own RTCP packets unless the encoding itself multiplexes multiple media into a single data stream.

The end-system distinguishes the multiple RTP sessions in which it participates from the different transport addresses it sends/receives the RTP streams. All participants in an RTP session may share a common destination transport address pair, as in the case of IP multicast, or the pairs may be different for each participant, as in the case of individual unicast network addresses and port pairs. In the unicast case, a participant may receive from all other participants in the session using the same pair of ports, or may use a distinct pair of ports for each.

A source packetizes the media content in RTP packets and sends them to the destination. Each RTP packet is a data packet consisting of the RTP header and the payload data referred to as the RTP payload. The latter are data transported by RTP in a packet, for example audio samples or compressed video data. Typically one packet of the underlying protocol, e.g., UDP, contains a single RTP packet, but several RTP packets may be contained if permitted by the encapsulation method. Some underlying protocols may require an encapsulation of the RTP packet to be defined.

Apart from the RTP packets, the source and receiver(s) exchange RTCP packets. Each RTCP packet is a control packet that consists of a fixed header part similar to that of the RTP data packets, followed by structured elements that vary depending upon the RTCP packet type. Typically, multiple RTCP packets are sent together as a compound RTCP packet in a single packet of the underlying protocol.

In addition to the participating end-systems, intermediate systems may participate in an RTP session. RFC3550 defines two types of intermediate systems, the translators and the mixers. A translator is an intermediate system that forwards RTP packets with their synchronization source identifier unaltered. Examples of translators include devices that convert encodings without mixing, e.g. video gateways, replicators from multicast to unicast, and application-level filters in firewalls. Mixer is an intermediate system that receives RTP packets from one or more source, possibly changes the data format, e.g., video or audio transcoding, combines the packets in some manner and then forwards a new RTP packet. Since multiple input sources are not generally synchronized, a mixer performs timing adjustments among the streams and generates its own timing for the combined stream. Thus, all data packets originating from a mixer will be identified as having the mixer as their synchronization source. The mixer inserts the original list of the SSRC identifiers of the sources that contributed to the generation of a particular packet into the RTP header of the new packet. This list is called the Contributing source (CSRC) list. An example application is audio conferencing where a mixer indicates all the talkers whose speech was combined to produce the outgoing packet, allowing the receiver to indicate the current talker, even though all the audio packets contain the same SSRC identifier (that of the mixer).

### 3.7.2 RTP Fixed Head Fields

An RTP packet consists of the RTP header and the payload (see Table 3.1).

The first twelve octets (the first three rows in Table 3.1) are present in every RTP packet, while the list of CSRC identifiers is present only when inserted by a mixer. The fields have the following meanings.

**Table 3.1** RTP packet format

| V | P | X | CC | M | PT | Sequence number |
|---|---|---|----|----|----|---|
| Timestamp | | | | | | |
| Synchronization source (SSRC) identifier | | | | | | |
| Contributing source (CSRC) identifier #1 | | | | | | |
| . . . | | | | | | |
| Contributing source (CSRC) identifier #n | | | | | | |
| Payload | | | | | | |

- The version (V) number (2 bits) identifies the version of RTP. The current version defined by RFC3550 is two (2).
- The padding (P) flag (1 bit) indicates whether the packet contains one or more additional padding octets at the ends that are not part of the payload (P = 1) or not (P = 0). Padding maybe used for encryption algorithms with fixed block sizes or for carrying several RTP packets in a lower-layer protocol data unit.
- The extension (X) flag (1 bit) indicates whether the fixed header is followed by exactly one content/ profile-dependent header extension.
- The CSRC count (CC) (4 bits) identifies the number of CSRC identifiers that follow the fixed header.
- The marker (M) flag (1 bit); its interpretation is profile-dependent. It is intended to allow significant events such as frame boundaries to be marked in the packet stream.
- The payload type (PT) (7 bits) identifies the format of the RTP payload types discussed in Section 3.7.6. A set of default mappings for audio and video is specified in the RFC 3551 [27].
- The sequence number (16 bits) is incremented by one for each RTP data packet sent, and is used by the receiver to detect packet losses and to restore packet sequence. The initial value of the sequence number is randomly selected in order to be unpredictable and to make attacks on encryption, if applied, more difficult.
- The timestamp (32 bits) reflects the sampling time instant of the first octet in the RTP data packet. This time instant is derived from a clock that increments monotonically and linearly in time to allow the synchronization and jitter calculations required for ensuring the correct playout at the receiver side.
- The SSRC field (32 bits) identifies the synchronization source. This identifier is chosen randomly, so that it is unique in the RTP session.
- The CSRC list – 0 to 15 items (32 bits each) – identifies the contributing sources for the payload contained in this packet. The number of identifiers is given by the CC field. If there are more than 15 contributing sources, only 15 can be identified. CSRC identifiers are inserted by mixers using the SSRC identifiers of contributing sources.

### 3.7.3 RTCP Packet Format

Five different RTCP packet types are defined in RFC3550 to exchange a variety of control information and serve different control functions, as follows.

(1) Sender Reports (SR) are generated by the active sender and sent to the receivers. RTCP SR include information required for the proper playout of the media at the receiver side.
(2) Receiver Reports (RR) are generated by the receivers and sent to the senders. RTCP RR include information about the quality of the packet delivery, such as loss ratio, jitter, delay, etc.
(3) Source DEScription (SDES) packets include information about the participants, such as canonical and real names, phone numbers, e-mails, etc.

(4) GoodBYE (BYE) packets are used to indicate the end of a participation.

(5) Application-specific functions (APP) packets are defined for experimental use.

Each RTCP packet begins with a fixed part similar to that of RTP data packets, followed by structured elements of variable length according to the packet type.

### 3.7.3.1 Sender Report RTCP Packet

The sender report packet consists of three sections (see Table 3.2), possibly followed by a fourth profile-specific extension section if defined.

**Table 3.2**   RTCP SR packet format

| V = 2 | P | RC | PT = SR = 200 | Length |
|-------|---|----|---------------|--------|
| SSRC of sender | | | | |
| NTP timestamp, most significant word | | | | |
| NTP timestamp, least significant word | | | | |
| RTP timestamp | | | | |
| sender's packet count | | | | |
| sender's octet count | | | | |
| SSRC_1 (SSRC of first source) | | | | |
| fraction lost \| cumulative number of packets lost | | | | |
| extended highest sequence number received | | | | |
| interarrival jitter | | | | |
| last SR (LSR) | | | | |
| delay since last SR (DLSR) | | | | |
| SSRC_n (SSRC of last source) | | | | |
| . . .. | | | | |
| profile-specific extensions | | | | |

In the first section, the header is eight octets long and contains the following fields.

- The version (V) number (2 bits) and the padding (P) flag (1 bit) are the same as in the RTP packets.
- The reception report count (RC) (5 bits) identifies the number of reception report blocks contained in the packet.
- The packet type (PT) (8 bits) contains the constant 200 to identify that the type is a SR.
- The length (16 bits) of the RTCP packet in 32-bit words minus one, including the header and any padding.
- The SSRC field (32 bits) contains the synchronization source identifier for the originator of the SR packet.

The second section contains the following fields.

- The Network Time Protocol (NTP) timestamp (64 bits) indicates the 'wallclock' time (absolute date and time) when the SR is sent so that it may be used in combination with timestamps returned in RTCP RRs by other receivers to measure round-trip time (RTT) to those receivers.
- The RTP timestamp (32 bits) corresponds to the same time as the NTP timestamp (above), but in the same units and with the same random offset as the RTP timestamps in data packets. This

correspondence may be used for intra and inter-media synchronization for sources whose NTP timestamps are synchronized, and may be used by media-independent receivers to estimate the nominal RTP clock frequency.

- The sender's packet count (32 bits) contains the total number of RTP data packets transmitted by the sender since starting transmission until the time the SR packet was generated.
- The sender's octet count (32 bits) contains the total number of payload octets (i.e., not including header or padding) transmitted in RTP data packets by the sender since transmission's start up until the time the current SR packet was generated.

The third section contains none or more reception report blocks depending on the number of other sources heard by this sender since the last report. Each reception report block contains statistics on the reception of RTP packets from a single synchronization source. These statistics are:

- The SSRC_n (source identifier) (32 bits) contains the SSRC identifier of the source to which the information in this reception report block pertains.
- The fraction lost (8 bits) contains the fraction of RTP data packets from source SSRC_n lost since the previous SR or RR packet was sent.
- The cumulative number of packets lost (24 bits) contains the total number of RTP data packets from source SSRC_n that have been lost since the starting of transmission.
- The extended highest sequence number received (32 bits) contain the highest sequence number received in an RTP data packet from source SSRC_n (the lower 16 bits), and the count of sequence number cycles of resets (most significant 16 bits).
- The interarrival jitter (32 bits) contains an estimate of the statistical variance of the RTP data packet interarrival time, measured in timestamp units.
- The last SR timestamp (LSR) (32 bits) and delay since last SR (DLSR) (32 bits) contain the time instant at which the last SR sent by the source SSRC_n was received and the time passed since the last SR was received, respectively. They are used for the estimation of RTT and jitter. Details on their usage are given in Section 3.7.5.

### 3.7.3.2 Receiver Report RTCP Packet

The format of the receiver report (RR) packet is the same as that of the SR packet (see Table 3.3) except that the packet type field contains the constant 201 and the five words of sender information are omitted (these are the NTP and RTP timestamps and sender's packet and octet counts). The remaining fields have the same meaning as for the SR packet.

**Table 3.3**  RTCP RR packet format

| V = 2 | P | RC | PT = 201 | Length |
|-------|---|----|----------|--------|
| SSRC of sender | | | | |
| SSRC_1 (SSRC of first source) | | | | |
| fraction lost \| cumulative number of packets lost | | | | |
| extended highest sequence number received | | | | |
| interarrival jitter | | | | |
| last SR (LSR) | | | | |
| delay since last SR (DLSR) | | | | |
| SSRC_n (SSRC of last source) | | | | |
| . . .. | | | | |
| profile-specific extensions | | | | |

**Table 3.4** RTCP SDES packet format

| V=2 | P | SC | PT=202 | Length |
|---|---|---|---|---|
| SSRC or CSRC _1 of sender | | | | |
| type &#124; length &#124; text | | | | |
| extended highest sequence number received | | | | |
| text continued | | | | |
| . . .. | | | | |
| SSRC or CSRC _n of sender | | | | |
| type &#124; length &#124; text | | | | |
| extended highest sequence number received | | | | |
| text continued | | | | |

### 3.7.3.3 Source Description RTCP Packet

The SDES packet is a three-level structure composed of a header and zero or more chunks (see Table 3.4). Each chunk is composed of items describing the source identified in that chunk. End systems send one SDES packet containing their own source identifier (the same as the SSRC in the fixed RTP header) whereas a mixer sends one SDES packet containing a chunk for each contributing source from which it is receiving SDES information, or multiple complete SDES packets in the format above if there are more than 31 such sources. The items are as follows.

- The version (V) number (2 bits) and the padding (P) flag (1 bit) are the same as in the RTP packets.
- The source count (SC) (5 bits) identifies the number of SSRC/CSRC chunks contained in the packet.
- The packet type (PT) (8 bits) contains the constant 202 to identify that the type is a SDES.

Each chunk consists of an SSRC/CSRC identifier followed by a list of one or more items, which carry information about the SSRC/CSRC. Each item consists of a type field (8-bit), describing the length of the text (8-bit) and the text itself.

The SDES item types defined in RFC3550 are the following:

- Canonical End-Point Identifier (CNAME), which should follow the format 'user@host' or 'host' if a user name is not available as on single-user systems. The CNAME should be fixed for a participant to provide a binding across multiple media tools used by one participant in a set of related RTP sessions. Also, the CNAME should be suitable for either a program or a person to locate the source to facilitate third-party monitoring.
- User Name (NAME) contains the real name used to describe the source, e.g., 'John Doe, Network Engineer' and it may follow any form. For applications, such as conferencing, this form of name may be the most desirable for display in participant lists, and therefore might be sent most frequently of those items other than CNAME.
- Electronic Mail Address (EMAIL) contains the email address formatted according to RFC 2822 [9].
- Phone Number (PHONE) contains the phone number formatted with the plus sign replacing the international access code.
- Geographic User Location (LOC) contains detail about the user location.
- Application or Tool Name (TOOL) contains the name and version of the application generating the stream.

- Notice/Status (NOTE) contains messages describing the current state of the source, e.g., 'on the phone, can't talk'.
- Private Extensions (PRIV) is used to define experimental or application-specific SDES extensions.

From the above items only the CNAME item type is mandatory in a SDES packet.

### 3.7.3.4 Goodbye RTCP Packet

The BYE packet (see Table 3.5) indicates that one or more sources are no longer active. If a BYE packet is received by a mixer, the mixer forwards the BYE packet with the SSRC/CSRC identifier(s) unchanged. If a mixer shuts down, it sends a BYE packet listing all contributing sources it handles, as well as its own SSRC identifier.

**Table 3.5**   RTCP BYE packet format

| V=2 | P | SC | PT=203 | Length |
|-----|---|----|--------|--------|
| SSRC or CSRC of sender | | | | |
| length   |   reason for leaving | | | |
| extended highest sequence number received | | | | |
| . . . . | | | | |
| last chunk | | | | |

The items are described individually in subsequent sections.

- The version (V) number (2 bits) and the padding (P) flag (1 bit) are the same as in the RTP packets.
- The source count (SC) (5 bits) identifies the number of SSRC/CSRC chunks contained in the packet.
- The packet type (PT) (8 bits) contains the constant 203 to identify that the type is a GOODBYE.

### 3.7.3.5 Application-defined RTCP Packet

The APP packet (see Table 3.6) is intended for experimental use as new applications and new features are developed, without requiring packet type value registration.
  The fields have the following meanings.

- The version (V) number (2 bits) and the padding (P) flag (1 bit) are the same as in the RTP packets.
- The packet type (PT) (8 bits) contains the constant 203 to identify that the type is a APP.
- The subtype (5 bits) may be used as a subtype to allow a set of APP packets to be defined under one unique name, or for any application-dependent data.
- The name (4 octets) contains a name chosen by the person defining the set of APP packets to be unique with respect to other APP packets this application might receive.

**Table 3.6**   RTCP APP packet format

| V=2 | P | subtype | PT=204 | Length |
|-----|---|---------|--------|--------|
| SSRC or CSRC of sender | | | | |
| name (ASCII) | | | | |
| application-dependent data | | | | |

- The application-dependent data (variable length) is not mandatory and may contain application-dependent data.

### 3.7.4 How Intra-media and Inter-media Synchronization is Achieved

The timestamp (32 bits) contained in the RTP header field timestamp reflects the sampling time instant at which the first octet in the RTP data packet was produced. The sampling time instant is derived from a clock that increments monotonically and linearly in time to allow synchronization and jitter calculations required for the proper playout of the stream at the receiver(s) side(s). The resolution of the clock is defined in order to be sufficient for the desired synchronization accuracy and for measuring packet arrival jitter. For instance, one tick per video frame is typically not sufficient. The clock frequency is content-dependent and is specified statically in the profile or payload format specification that defines the format of the content.

Consecutive RTP packets will have equal timestamps only if they are (logically) generated at once, e.g., belong to the same video frame. Consecutive RTP packets may contain timestamps that are not monotonic, if the data are not transmitted in the order they were sampled, as in the case of MPEG interpolated video frames. However, the sequence numbers of the packets as transmitted will still be monotonic.

RTP timestamps from different media streams may advance at different rates and usually have independent, random offsets. Therefore, although these timestamps are sufficient to reconstruct the timing of a single stream (intra-media timing), directly comparing RTP timestamps from different media is not effective for inter-media synchronization. Instead, for each media the RTP timestamp is related to the sampling time instant by pairing it with a timestamp from a reference clock (wallclock) that represents the absolute time when the data corresponding to the RTP timestamp were sampled. All media to be synchronized share the same reference clock. The timestamp pairs are not transmitted in every data packet, but at a lower rate in RTCP SR packets.

The sampling time instant is chosen as the point of reference for the RTP timestamp because it is known to the transmitting source endpoint and has a common definition for all media, independent of encoding delays or other processing. The purpose is to allow synchronized presentation of all the involved media sampled at the same time.

Applications transmitting stored data rather than data sampled in real time typically use a virtual presentation timeline, derived from wallclock time during the encoding process, to determine when the next frame or some other unit of each involved media in the stored data should be presented. In this case, the RTP timestamp would reflect the presentation time for each unit. That is, the RTP timestamp for each unit would be related to the wallclock time at which the unit becomes current on the virtual presentation timeline. Actual presentation occurs some time later as determined by the receiver.

### 3.7.5 Monitoring RTT, Jitter and Packet Loss Rate

#### 3.7.5.1 Estimation of Round Trip Time

The Round Trip Time (RTT) is estimated in the source by exploiting the RTCP RR. Recall that each RTCP Sender Report contains in the field NTP timestamp (64 bits) the absolute sending time of the SR from source SSRC$i$. This sending time (middle 32-bits of the NTP timestamp) is replicated in the field LSR (timestamp from the last SR reception) of the RTCP RR block for source SSRC$i$. In the field DLSR of the same block the time passed since the last reception of the SR from SSRC$i$ is added. Thus, any source that receives the RTCP RR from the receiver at time instant $t$ can estimate the RTT between SSRC$i$ and the receiver sends the report as follows:

$$RTT = t - LSR - DSLR$$

### 3.7.5.2 Estimation of Inter-arrival Jitter

The receivers monitor continuously the variance of delay of the receiving RTP packets. The one-way delay and the jitter of the incoming RTP packets are estimated as follows. Let $S_i$ and $R_i$ denote the RTP timestamp and the time of arrival in RTP timestamp units for packet $i$, respectively. The one-way delay (in RTP timestamp units) from the source to the receiver for packet $i$ and the delay variation for packets $i$ and $j$ are given by $D(i) = (R_i - S_i)$ and $DV(i,j) = D(j) - D(i) = (R_j - S_j) - (R_i - S_i)$, respectively.

The interarrival jitter is defined as the smoothed absolute value of DV and is calculated continuously as each data packet $i$ is received from source SSRC_n, using this difference DV for that packet and the previous packet $i - 1$ in order of arrival (not necessarily in sequence), according to the formula:

$$\text{Jitter}(i) = \text{Jitter}(i - 1) + |DV(i - 1, i)| - \text{Jitter}(i - 1))/16$$

This value is contained in the RTCP RR header field interarrival jitter.

### 3.7.5.3 Packet Loss Rate

The packet loss rate is calculated by the receiver as the number of packets lost divided by the number of packets expected. The latter is defined to be the extended last sequence number received less the initial sequence number received.

### 3.7.5.4 RTCP Based Information as Input to the Encoding–Transmission–Reception–Decoding Process

The packet loss rate, the delay, the jitter and the round trip time could be used as feedback to several algorithms that aim to the adjustment of the overall encoding–transmission–reception–decoding process. For instance, the RTCP could be served as a feedback mechanism for rate adaptation control algorithms, in the case of source-based rate control algorithms, where the loss rate is taken into account for deciding the reaction (rate increase or decrease) and the shaping of the new decreased rate [21, 22]. Also, receiver-based rate control algorithms could use the loss rate for deciding a layer join/drop action [23].

### 3.7.6 RTP and Loss Repairing Techniques

In the case of streamed media over RTP/UDP/IP, loss detection is provided by the sequence numbers in RTP packets. Some RFCs that deal with repairing issues are as follows.

- *Retransmission.* There is no standard protocol framework for requesting retransmission of streaming media. An experimental RTP profile extension for SRM-style retransmission requests has described in [5].
- *Media-independent FEC.* An RTP payload format for generic FEC, suitable for both parity based and Reed–Solomon encoded FEC is defined in [19].
- *Media-specific FEC.* The RTP payload format for this form of redundancy has been defined [17]. If media units span multiple packets, for instance video, it is sensible to include redundancy directly within the output of a codec. For example the proposed RTP payload for H.263+ [9] includes multiple copies of key portions of the stream, separated to avoid the problems of packet loss.
- *Interleaving.* A potential RTP payload format for interleaved data is a simple extension of the redundant audio payload [17]. That payload requires that the redundant copy of a unit is sent after the primary. If this restriction is removed, it is possible to transmit an arbitrary interleaving of units with this payload format.

## 3.8 RTP Payload Types

As already mentioned, RTP is a protocol framework that allows the support of new encodings and features. Each particular RTP/RTCP-based application is accompanied by one or more documents:

- a profile specification document, which defines a set of payload type codes and their mapping to payload formats (e.g., media encodings) – a profile for audio and video data may be found in the companion RFC3551 [27];
- payload format specification documents, which define how a particular payload, such as an audio or video encoding, is to be carried in RTP.

### 3.8.1 RTP Profiles for Audio and Video Conferences (RFC3551)

RFC3551 lists a set of audio and video encodings used within audio and video conferences with minimal, or no session control. Each audio and video encoding comprises:

- a particular media data compression or representation called payload type, plus
- a payload format for encapsulation within RTP.

RFC3551 reserves payload type numbers in the ranges 1–95 and 96–127 for static and dynamic assignment, respectively. The set of static payload type (PT) assignments is provided in Tables 3.7 and 3.8 (see column *PT*).

Payload type 13 indicates the Comfort Noise (CN) payload format specified in RFC 3389.

Some of the payload formats of the payload types are specified in RFC3551, while others are specified in separate RFCs. RFC3551 also assigns to each encoding a short name (see column short encoding name) which may be used by higher-level control protocols, such as the Session Description Protocol (SDP), RFC 2327 [25], to identify encodings selected for a particular RTP session.

Mechanisms for defining dynamic payload type bindings have been specified in the Session Description Protocol (SDP) and in other protocols, such as ITU-T Recommendation H.323/H.245. These mechanisms associate the registered name of the encoding/payload format, along with any additional required parameters, such as the RTP timestamp clock rate and number of channels, with a payload type number. This association is effective only for the duration of the RTP session in which the dynamic payload type binding is made. This association applies only to the RTP session for which it is made, thus the numbers can be reused for different encodings in different sessions so the number space limitation is avoided.

#### 3.8.1.1 Audio

##### RTP Clock Rate

The RTP clock rate used for generating the RTP timestamp is independent of the number of channels and the encoding; it usually equals the number of sampling periods per second. For $N$-channel encodings, each sampling period (say, 1/8000 of a second) generates $N$ samples. If multiple audio channels are used, channels are numbered left-to-right, starting at one. In RTP audio packets, information from lower-numbered channels precedes that from higher-numbered channels.

Samples for all channels belonging to a single sampling instant must be within the same packet. The interleaving of samples from different channels depends on the encoding. The sampling frequency is drawn from the set: 8000, 11 025, 16 000, 22 050, 24 000, 32 000, 44 100 and 48 000 Hz. However, most audio encodings are defined for a more restricted set of sampling frequencies.

For packetized audio, the default packetization interval has a duration of 20 ms or one frame, whichever is longer, unless otherwise noted in Table 3.7 (column Default 'ms/packet'). The packetization interval determines the minimum end-to-end delay; longer packets introduce less header overhead

**Table 3.7** Payload types (PT) and properties for audio encodings (n/a: not applicable)

| PT | Short encoding name | Sample or frame | Bits/sample | Sampling (clock) rate (Hz) | ms/ frame | Default ms/ packet | Channels |
|---|---|---|---|---|---|---|---|
| 0 | PCMU | Sample | 8 | var. | | 20 | 1 |
| 1 | Reserved | | | | | | |
| 2 | Reserved | | | | | | |
| 3 | GSM | Frame | n/a | 8000 | 20 | 20 | 1 |
| 4 | G723 | Frame | n/a | 8000 | 30 | 30 | 1 |
| 5 | DVI4 | Sample | 4 | 8000 | | 20 | 1 |
| 6 | DVI4 | Sample | 4 | 16 000 | | 20 | 1 |
| 7 | LPC | Frame | n/a | 8000 | 20 | 20 | 1 |
| 8 | PCMA | Sample | 8 | 8000 | | 20 | 1 |
| 9 | G722 | Sample | 8 | 16 000 | | 20 | 1 |
| 10 | L16 | Sample | 16 | 44 100 | | 20 | 2 |
| 11 | L16 | Sample | 16 | 44 100 | | 20 | 1 |
| 12 | QCELP | Frame | n/a | 8000 | 20 | 20 | 1 |
| 13 | CN | | | 8000 | | | 1 |
| 14 | MPA | Frame | n/a | 90 000 | var. | | |
| 15 | G728 | Frame | n/a | 8000 | 2.5 | 20 | 1 |
| 16 | DVI4 | Sample | 4 | 11 025 | | 20 | 1 |
| 17 | DVI4 | Sample | 4 | 22 050 | | 20 | 1 |
| 18 | G729 | Frame | na | 8000 | 10 | 20 | 1 |
| 19 | Reserved | | | | | | |
| 20 | Unassigned | | | | | | |
| 21 | Unassigned | | | | | | |
| 22 | Unassigned | | | | | | |
| 23 | Unassigned | | | | | | |
| dyn | G726-40 | Sample | 5 | 8000 | | 20 | 1 |
| dyn | G726-32 | Sample | 4 | 8000 | | 20 | 1 |
| dyn | G726-24 | Sample | 3 | 8000 | | 20 | 1 |
| dyn | G726-16 | Sample | 2 | 8000 | | 20 | 1 |
| dyn | G729D | Frame | n/a | 8000 | 10 | 20 | 1 |
| dyn | G729E | Frame | n/a | 8000 | 10 | 20 | 1 |
| dyn | GSM-EFR | Frame | n/a | 8000 | 20 | 20 | 1 |
| dyn | L8 | Sample | 8 | Variable | | 20 | Variable |
| dyn | RED | | | | | | |
| dyn | VDVI | Sample | Variable | Variable | | 20 | 1 |

**Table 3.8** Payload types (PT) for video and combined encodings

| PT | Short encoding name | Clock rate (Hz) | PT | Short encoding name | Clock rate (Hz) |
|---|---|---|---|---|---|
| 24 | Unassigned | | 32 | MPV | 90 000 |
| 25 | CelB | 90 000 | 33 | MP2T | 90 000 |
| 26 | JPEG | 90 000 | 34 | H263 | 90 000 |
| 27 | Unassigned | | 35–71 | Unassigned | |
| 28 | nv | 90 000 | 72–76 | Reserved | |
| 29 | Unassigned | | 77–95 | Unsigned | |
| 30 | Unassigned | | 96–127 | Dynamic | |
| 31 | H261 | 90 000 | Dyn | h263-1998 | 90 000 |

but higher delay and make packet loss more noticeable. For non-interactive applications such as lectures or for links with severe bandwidth constraints, a higher packetization delay may be used. A receiver should accept packets representing between 0 and 200 ms of audio data. This restriction allows reasonable buffer sizing for the receiver.

### Sample and Frame-based Encodings

In sample-based encodings, each audio sample is represented by a fixed number of bits. An RTP audio packet may contain any number of audio samples, subject to the constraint that the number of bits per sample times the number of samples per packet yields an integral octet count.

The duration of an audio packet is determined by the number of samples in the packet. For sample-based encodings producing one or more octets per sample; samples from different channels sampled at the same sampling instant are packed in consecutive octets. For example, for a two-channel encoding, the octet sequence is (left channel, first sample), (right channel, first sample), (left channel, second sample), (right channel, second sample) . . .. The packing of sample-based encodings producing less than one octet per sample is encoding-specific.

The RTP timestamp reflects the instant at which the first sample in the packet was sampled, that is, the oldest information in the packet.

Frame-based encodings encode a fixed-length block of audio into another block of compressed data, typically also of fixed length. For frame-based encodings, the sender may choose to combine several such frames into a single RTP packet. The receiver can tell the number of frames contained in an RTP packet, provided that all the frames have the same length, by dividing the RTP payload length by the audio frame size that is defined as part of the encoding.

For frame-based codecs, the channel order is defined for the whole block. That is, for two-channel audio, right and left samples are coded independently, with the encoded frame for the left channel preceding that for the right channel.

All frame-oriented audio codecs are able to encode and decode several consecutive frames within a single packet. Since the frame size for the frame-oriented codecs is given, there is no need to use a separate designation for the same encoding, but with different number of frames per packet.

RTP packets contain a number of frames which are inserted according to their age, so that the oldest frame (to be played first) is inserted immediately after the RTP packet header. The RTP timestamp reflects the instant at which the first sample in the first frame was sampled, that is, the oldest information in the packet.

### Silence Suppression

Since the ability to suppress silence is one of the primary motivations for using packets to transmit voice, the RTP header carries both a sequence number and a timestamp to allow a receiver to distinguish between lost packets and periods of time when no data are transmitted. Discontinuous transmission (silence suppression) is used with any audio payload format. In the sequel, the audio encodings are listed:

- DVI4: DVI4 uses an adaptive delta pulse code modulation (ADPCM) encoding scheme that was specified by the Interactive Multimedia Association (IMA) as the 'IMA ADPCM wave type'. However, the encoding defined in RFC3551 here as DVI4 differs in three respects from the IMA specification.
- G722: G722 is specified in ITU-T Recommendation G.722, '7 kHz audio-coding within 64 kbit/s'. The G.722 encoder produces a stream of octets, each of which shall be octet-aligned in an RTP packet.
- G723: G723 is specified in ITU Recommendation G.723.1, 'Dual-rate speech coder for multimedia communications transmitting at 5.3 and 6.3 kbit/s'. The G.723.1 5.3/6.3 kbit/s codec was defined by the ITU-T as a mandatory codec for ITU-T H.324 GSTN videophone terminal applications.
- G726-40, G726-32, G726-24 and G726-16: ITU-T Recommendation G.726 describes, among others, the algorithm recommended for conversion of a single 64 kbit/s A-law or mu-law PCM channel encoded at 8000 samples/sec to and from a 40, 32, 24, or 16 kbit/s channel.

- G729: G729 is specified in ITU-T Recommendation G.729, 'Coding of speech at 8 kbit/s using conjugate structure-algebraic code excited linear prediction (CS-ACELP)'.
- GSM: GSM (Group Speciale Mobile) denotes the European GSM 06.10 standard for full-rate speech transcoding, ETS 300 961, which is based on RPE/LTP (residual pulse excitation/long term prediction) coding at a rate of 13 kbit/s.
- GSM-EFR: GSM-EFR denotes GSM 06.60 enhanced full rate speech transcoding, specified in ETS 300 726.
- L8: L8 denotes linear audio data samples, using 8 bits of precision with an offset of 128, that is, the most negative signal is encoded as zero.
- L16: L16 denotes uncompressed audio data samples, using 16-bit signed representation with 65 535 equally divided steps between minimum and maximum signal level, ranging from $-32\,768$ to $32\,767$.
- LPC: LPC designates an experimental linear predictive encoding.
- MPA: MPA denotes MPEG-1 or MPEG-2 audio encapsulated as elementary streams. The encoding is defined in ISO standards ISO/IEC 11172-3 and 13818-3. The encapsulation is specified in RFC 2250.
- PCMA and PCMU: PCMA and PCMU are specified in ITU-T Recommendation G.711. Audio data is encoded as eight bits per sample, after logarithmic scaling. PCMU denotes mu-law scaling, PCMA A-law scaling.
- QCELP: The Electronic Industries Association (EIA) and Telecommunications Industry Association (TIA) standard IS-733, 'TR45: High Rate Speech Service Option for Wideband Spread Spectrum Communications Systems', defines the QCELP audio compression algorithm for use in wireless CDMA applications.
- RED: The redundant audio payload format 'RED' is specified by RFC 2198. It defines a means by which multiple redundant copies of an audio packet may be transmitted in a single RTP stream.
- VDVI: VDVI is a variable-rate version of DVI4, yielding speech bit rates between 10 and 25 kbit/s.

### 3.8.1.2 Video

This section describes the video encodings that are defined in RFC3551 and give their abbreviated names used for identification. These video encodings and their payload types are listed in Table 3.8. All of these video encodings use an RTP timestamp frequency of 90 000 Hz, the same as the MPEG presentation time stamp frequency. This frequency yields exact integer timestamp increments for the typical 24 (HDTV), 25 (PAL), and 29.97 (NTSC) and 30 (HDTV) Hz frame rates and 50, 59.94 and 60 Hz field rates. While 90 Hz is the recommended rate for future video encodings used within this profile, other rates may be used as well. However, it is not sufficient to use the video frame rate (typically between 15 and 30 Hz) because that does not provide adequate resolution for typical synchronization requirements when calculating the RTP timestamp corresponding to the NTP timestamp in an RTCP SR packet. The timestamp resolution must also be sufficient for the jitter estimate contained in the receiver reports.

For most of these video encodings, the RTP timestamp encodes the sampling instant of the video image contained in the RTP data packet. If a video image occupies more than one packet, the timestamp is the same on all of those packets. Packets from different video images are distinguished by their different timestamps.

Most of these video encodings also specify that the marker bit of the RTP header is set to one in the last packet of a video frame and otherwise set to zero. Thus, it is not necessary to wait for a following packet with a different timestamp to detect that a new frame should be displayed. In the sequel, the video encodings are listed:

- CelB: The CELL-B encoding is a proprietary encoding proposed by Sun Microsystems. The byte stream format is described in RFC 2029.
- JPEG: The encoding is specified in ISO Standards 10918-1 and 10918-2. The RTP payload format is as specified in RFC 2435.

**Table 3.9**  RFC for RTP profiles and payload format

**Protocols and payload formats**

| | |
|---|---|
| RFC 1889 | RTP: A transport protocol for real-time applications (obsoleted by RFC 3550) |
| RFC 1890 | RTP profile for audio and video conferences with minimal control (obsoleted by RFC 3551) |
| RFC 2035 | RTP payload format for JPEG-compressed video (obsoleted by RFC 2435) |
| RFC 2032 | RTP payload format for H.261 video streams |
| RFC 2038 | RTP payload format for MPEG1/MPEG2 video obsoleted by RFC 2250 |
| RFC 2029 | RTP payload format of Sun's CellB video encoding |
| RFC 2190 | RTP payload format for H.263 video streams |
| RFC 2198 | RTP payload for redundant audio data |
| RFC 2250 | RTP payload format for MPEG1/MPEG2 video |
| RFC 2343 | RTP payload format for bundled MPEG |
| RFC 2429 | RTP payload format for the 1998 version of ITU-T Rec. H.263 Video (H.263+) |
| RFC 2431 | RTP payload format for BT.656 video encoding |
| RFC 2435 | RTP payload format for JPEG-compressed video |
| RFC 2733 | An RTP payload format for generic forward error correction |
| RFC 2736 | Guidelines for writers of RTP payload format specifications |
| RFC 2793 | RTP payload for text conversation |
| RFC 2833 | RTP payload for DTMF digits, telephony tones and telephony signals |
| RFC 2862 | RTP payload format for real-time pointers |
| RFC 3016 | RTP payload format for MPEG-4 audio/visual streams |
| RFC 3047 | RTP payload format for ITU-T Recommendation G.722.1 |
| RFC 3119 | A more loss-tolerant RTP payload format for MP3 audio |
| RFC 3158 | RTP testing strategies |
| RFC 3189 | RTP payload format for DV format video |
| RFC 3190 | RTP payload format for 12-bit DAT, 20- and 24-bit linear sampled audio |
| RFC 3267 | RTP payload format and file storage format for the Adaptive Multi-Rate (AMR) and Adaptive Multi-Rate Wideband (AMR-WB) audio codecs |
| RFC 3389 | RTP payload for comfort noise |
| RFC 3497 | RTP payload format for Society of Motion Picture and Television Engineers (SMPTE) 292M video |
| RFC 3550 | RTP: A transport protocol for real-time applications |
| RFC 3551 | RTP profile for audio and video conferences with minimal control |
| RFC 3555 | MIME type registration of RTP payload formats |
| RFC 3557 | RTP payload format for European Telecommunications Standards Institute (ETSI) European Standard ES 201 108 distributed speech recognition encoding |
| RFC 3558 | RTP payload format for Enhanced Variable Rate Codecs (EVRC) and Selectable Mode Vocoders (SMV) |
| RFC 3640 | RTP payload format for transport of MPEG-4 elementary streams |
| RFC 3711 | The secure real-time transport protocol |
| RFC 3545 | Enhanced compressed RTP (CRTP) for links with high delay, packet loss and reordering |
| RFC 3611 | RTP Control Protocol Extended Reports (RTCP XR) |

**Repairing losses**

| | |
|---|---|
| RFC 2354 | Options for repair of streaming media |

**Others**

| | |
|---|---|
| RFC 3009 | Registration of parity FEC MIME types |
| RFC 3556 | Session Description Protocol (SDP) bandwidth modifiers for RTP Control Protocol (RTCP) bandwidth |
| RFC 2959 | Real-time transport protocol management information base |
| RFC 2508 | Compressing IP/UDP/RTP headers for low-speed serial links |
| RFC 2762 | Sampling of the group membership in RTP |

- H261: The encoding is specified in ITU-T Recommendation H.261, 'Video codec for audiovisual services at p × 64 Kbit/s'. The packetization and RTP-specific properties are described in RFC 2032.
- H263: The encoding is specified in the 1996 version of ITU-T Recommendation H.263, 'Video coding for low bit rate communication'. The packetization and RTP-specific properties are described in RFC 2190.
- H263-1998: The encoding is specified in the 1998 version of ITU-T Recommendation H.263, 'Video coding for low bit rate communication'. The packetization and RTP-specific properties are described in RFC 2429.
- MPV: MPV designates the use of MPEG-1 and MPEG-2 video encoding elementary streams as specified in ISO Standards ISO/IEC 11172 and 13818-2, respectively. The RTP payload format is as specified in RFC 2250. The MIME registration for MPV in RFC 3555 specifies a parameter that may be used with MIME or SDP to restrict the selection of the type of MPEG video.
- MP2T: MP2T designates the use of MPEG-2 transport streams, for either audio or video. The RTP payload format is described in RFC 2250.
- nv: The encoding is implemented in the program 'nv', version 4, developed at Xerox PARC.

Table 3.9 summarizes the RFCs defined for RTP profiles and payload format.

## 3.9 RTP in 3G

This section summarizes the supported media types in 3G and RTP implementation issues for 3G, as reported in 3GPP TR 26.937 [2], TR 26.234 [33] and TR 22.233 [34].

Figure 3.10 shows the basic entities involved in a 3G Packet-Switch Streaming Service (PSS). Clients initiate the service and connect to the selected Content Server. Content Servers, apart from prerecorded content, can generate live content, e.g. video, from a concert or TV (see Table 3.10: Potential services over PSS.). User Profile and terminal capability data can be stored in a network server and will be accessed at the initial set up. User Profile will provide the PSS service with the user's preferences. Terminal capabilities will be used by the PSS service to decide whether or not

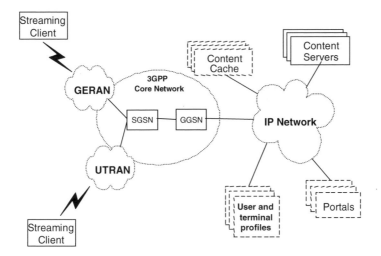

**Figure 3.10**  Network elements involved in a 3G packet switched streaming service.

**Table 3.10**  Potential services over PSS

| **Infotainment** |
| --- |
| Video on demand, including TV |
| Audio on demand, including news, music, etc. |
| Multimedia travel guide |
| Karaoke – song words change colour to indicate when to sing |
| Multimedia information services: sports, news, stock quotes, traffic |
| Weather cams – gives information on other part of country or the world |
| **Edutainment** |
| Distance learning – video stream of teacher or learning material together with teacher's voice or audio track. |
| How to ? service – manufacturers show how to program the VCR at home |
| **Corporate** |
| Field engineering information – junior engineer gets access to online manuals to show how to repair, say, the central heating system |
| Surveillance of business premises or private property (real-time and non-real-time) |
| **M-commerce** |
| Multimedia cinema ticketing application |
| On line shopping – product presentations could be streamed to the user and then the user could buy on line. |

the client is capable of receiving the streamed content. Portals are servers allowing for a convenient access to streamed media content. For instance, a portal might offer content browse and search facilities. In the simplest case, it is simply a Web/WAP-page with a list of links to streaming content. The content itself is usually stored in content servers, which can be located elsewhere in the network.

### 3.9.1 Supported Media Types in 3GPP

In the 3GPP's Packet-Switched streaming Service (PSS), the communication between the client and the streaming servers, including session control and transport of media data, is IP-based. Thus, the RTP/UDP/IP and HTTP/TCP/IP protocol stacks have been adopted for the transport of continuous media and discrete media, respectively. The supported continuous media types are restricted to the following set:

- AMR narrow-band speech codec RTP payload format according to RFC3267 [28],
- AMR-WB (WideBand) speech codec RTP payload format according to RF3267 [28],
- MPEG-4 AAC audio codec RTP payload format according to RFC 3016 [29],
- MPEG-4 video codec RTP payload format according to RFC 3016 [29],
- H.263 video codec RTP payload format according to RFC 2429 [30].

The usage scenarios of the above continuous data are:

(1)  voice only streaming (AMR at 12.2 kbps),
(2)  high-quality voice/low quality music only streaming (AMR-WB at 23.85 kbps),
(3)  music only streaming (AAC at 52 kbps),
(4)  voice and video streaming (AMR at 7.95 kbps + video at 44 kbps),
(5)  voice and video streaming (AMR at 4.75 kbps + video at 30 kbps).

During streaming, the packets are encapsulated using RTP/UDP/IP protocols. The total header overhead consists of: IP header: 20 bytes for IPv4 (IPv6 would add a 20 bytes overhead); UDP header: 8 bytes; RTP header: 12 bytes.

The supported discrete media types (which use the HTTP/TCP/IP stack) for scene description, text, bitmap graphics and still images, are as follows:

- Still images: ISO/IEC JPEG [35] together with JFIF [36] decoders are supported. The support for ISO/IEC JPEG only apply to the following modes: baseline DCT, non-differential, Huffman coding, and progressive DCT, non-differential, Huffman coding.
- Bitmap graphics: GIF87a [40], GIF89a [41], PNG [42].
- Synthetic audio: The Scalable Polyphony MIDI (SP-MIDI) content format defined in Scalable Polyphony MIDI Specification [45] and the device requirements defined in Scalable Polyphony MIDI Device 5-to-24 Note Profile for 3GPP [46] are supported. SP-MIDI content is delivered in the structure specified in Standard MIDI Files 1.0 [47], either in format 0 or format 1.
- Vector graphics: The SVG Tiny profile [43, 44] shall be supported. In addition SVG Basic profile [43, 44] may be supported.
- Text: The text decoder is intended to enable formatted text in a SMIL presentation. The UTF-8 [38] and UCS-2 [37] character coding formats are supported. A PSS client shall support:
  - text formatted according to XHTML Mobile Profile [32, 48];
  - rendering a SMIL presentation where text is referenced with the SMIL 2.0 'text' element together with the SMIL 2.0 'src' attribute.
- Scene description: The 3GPP PSS uses a subset of SMIL 2.0 [39] as format of the scene description. PSS clients and servers with support for scene descriptions support the 3GPP PSS SMIL Language Profile (defined in 3GPP TS 26.234 specification [33]). This profile is a subset of the SMIL 2.0 Language Profile, but a superset of the SMIL 2.0 Basic Language Profile. It should be noted that not that all streaming sessions are required to use SMIL. For some types of sessions, e.g. consisting of one single continuous media or two media synchronized by using RTP timestamps, SMIL may not be needed.
- Presentation description: SDP is used as the format of the presentation description for both PSS clients and servers. PSS servers shall provide and clients interpret the SDP syntax according to the SDP specification [25] and appendix C of [24]. The SDP delivered to the PSS client shall declare the media types to be used in the session using a codec specific MIME media type for each media.

### 3.9.2 RTP Implementation Issues for 3G

#### 3.9.2.1 Transport and Transmission

Media streams can be packetized using different strategies. For example, video encoded data could be encapsulated using:

- one slice of a target size per RTP packet;
- one Group of Blocks (GOB), that is, a row of macroblocks per RTP packet;
- one frame per RTP packet.

Speech data could be encapsulated using an arbitrary (but reasonable) number of speech frames per RTP packet, and using bit- or byte alignment, along with options such as interleaving. The transmission of RTP packets take place in two different ways:

(1) VBRP (Variable Bit Rate Packet) transmission – the transmission time of a packet depends solely on the timestamp of the video frame to which the packet belongs, therefore, the video rate variation is directly reflected to the channel;
(2) CBRP (Constant Bit Rate Packet) transmission – the delay between sending consecutive packets is continuously adjusted to maintain a near constant rate.

### ·3.9.2.2 Maximum and Minimum RTP Packet Size

The RFC 3550 (RTP) [26] does not impose a maximum size for RTP packets. However, when RTP packets are sent over the radio link of a 3GPP PSS, limiting the maximum size of RTP packets can be advantageous.

Two types of bearers can be envisaged for streaming using either acknowledged mode (AM) or unacknowledged mode (UM) Radio Link Control (RLC). The AM uses retransmissions over the radio link, whereas the UM does not. In UM mode, large RTP packets are more susceptible to losses over the radio link compared with small RTP packets, since the loss of a segment may result in the loss of the entire packet. On the other hand in AM mode, large RTP packets will result in a larger delay jitter compared with small packets, as it is more likely that more segments have to be retransmitted.

Fragmentation is one more reason for limiting packet sizes. It is well known that fragmentation causes:

- increased bandwidth requirement, due to additional header(s) overhead;
- increased delay, because of operations of segmentation and re-assembly.

Implementers should consider avoiding/preventing fragmentation at any link of the transmission path from the streaming server to the streaming client.

For the above reasons it is recommended that the maximum size of RTP packets is limited, taking into account the wireless link. This will decrease the RTP packet loss rate particularly for RLC in UM. For RLC in AM the delay jitter will be reduced, permitting the client to use a smaller receiving buffer. It should also be noted that too small RTP packets could result in too much overhead if IP/UDP/RTP header compression is not applied or unnecessary load at the streaming server. While there are no theoretical limits for the usage of small packet sizes, implementers must be aware of the implications of using too small RTP packets. The use of such packets would result in three drawbacks.

(1) The RTP/UDP/IP packet header overhead becomes too large compared with the media data.
(2) The bandwidth requirement for the bearer allocation increases, for a given media bit rate.
(3) The packet rate increases considerably, producing challenging situations for server, network and mobile client.

As an example, Figure 3.11 shows a chart with the bandwidth partitions between RTP payload media data and RTP/UDP/IP headers for different RTP payload sizes. The example assumes IPv4. The space

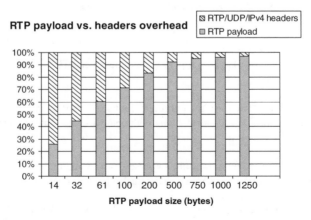

**Figure 3.11**  Bandwidth of RTP payload and RTP/UDP/IP header for different packet sizes.

occupied by RTP payload headers is considered to be included in the RTP payload. The smallest RTP payload sizes (14, 32 and 61 bytes) are examples related to minimum payload sizes for AMR at 4.75 kbps, 12.20 kbps and for AMR-WB at 23.85 kbps (1 speech frame per packet). As Figure 3.11 shows, too small packet sizes ($\leq$100 bytes) yield an RTP/UDP/IPv4 header overhead from 29 to 74%. When using large packets ($\geq$750 bytes) the header overhead is 3 to 5%.

When transporting video using RTP, large RTP packets may be avoided by splitting a video frame into more than one RTP packet. Then, to be able to decode packets following a lost packet in the same video frame, it is recommended that synchronization information is inserted at the start of such an RTP packet. For H.263, this implies the use of GOBs with non-empty GOB headers and, in the case of MPEG-4 video, the use of video packets (resynchronization markers). If the optional Slice Structured mode (Annex K) of H.263 is in use, GOBs are replaced by slices.

# References

[1] S. V. Raghavan and S. K. Tripathi. *Networked Multimedia Systems: Concepts, Architecture and Design*, Prentice Hall, 1998.

[2] 3GPP TR26.937. Technical Specification Group Services and System Aspects; Transparent end-to-end PSS; RTP usage model (Rel.6, 03-2004).

[3] V. Varsa and M. Karczewicz, Long Window Rate Control for Video Streaming, *Proceedings of 11th International Packet Video Workshop*, Kyungju, South Korea.

[4] J. -C. Bolot and A. Vega-Garcia, The case for FEC based error control for packet audio in the Internet, ACM Multimedia Systems.

[5] IETF RFC 2354. Options for Repair of Streaming Media, C. Perkins and O. Hodson, June 1998.

[6] V. Jacobson, Congestion avoidance control. In *Proceedings of the SIGCOMM '88 Conference on Communications Architectures and Protocols*, 1988.

[7] IETF RFC 2001. TCP Slow Start, Congestion Avoidance, Fast Retransmit, and Fast Recovery Algorithms.

[8] D. M. Chiu and R. Jain, Analysis of the increase and decrease algorithms for congestion avoidance in computer networks, *Computer Networks and ISDN Systems*, **17**, 1989, 1–14.

[9] C. Bormann, L. Cline, G. Deisher, T. Gardos, C. Maciocco, D. Newell, J. Ott, S. Wenger and C. Zhu, RTP payload format for the 1998 version of ITU-T reccomendation H.263 video (H.263+).

[10] D. Budge, R. McKenzie, W. Mills, W. Diss and P. Long, Media-independent error correction using RTP.

[11] S. Floyd and K. Fall, Promoting the use of end-to-end congestion control in the internet, *IEEE/ACM Transactions on Networking*, August 1999.

[12] M. Handley, An examination of Mbone performance, USC/ISI Research Report: ISI/RR-97-450, April 1997.

[13] M. Handley and J. Crowcroft, Network text editor (NTE): A scalable shared text editor for the Mbone. In *Proceedings ACM SIGCOMM'97*, Cannes, France, September 1997.

[14] V. Hardman, M. A. Sasse, M. Handley, and A. Watson, Reliable audio for use over the Internet. In *Proceedings of INET'95*, 1995.

[15] I. Kouvelas, O. Hodson, V. Hardman and J. Crowcroft. Redundancy control in real-time Internet audio conferencing. In *Proceedings of AVSPN'97*, Aberdeen, Scotland, September 1997.

[16] J. Nonnenmacher, E. Biersack and D. Towsley. Parity-based loss recovery for reliable multicast transmission. In *Proceedings ACM SIGCOMM'97*, Cannes, France, September 1997.

[17] IETF RFC 2198. RTP Payload for Redundant Audio Data, C. Perkins, I. Kouvelas, O. Hodson, V. Hardman, M. Handley, J-C. Bolot, A. Vega-Garcia, and Fosse-Parisis, S. September 1997.

[18] J. L. Ramsey, Realization of optimum interleavers. *IEEE Transactions on Information Theory*, **IT-16**, 338–345.

[19] J. Rosenberg and H. Schulzrinne, An A/V profile extension for generic forward error correction in RTP.

[20] M. Yajnik, J. Kurose and D. Towsley, Packet loss correlation in the Mbone multicast network. In *Proceedings IEEE Global Internet Conference*, November 1996.

[21] I. Busse, B. Defner and H. Schulzrinne, Dynamic QoS Control of Multimedia Application based on RTP, May.

[22] J. Bolot and T. Turletti, Experience with rate control mechanisms for packet video in the Internet, *ACM SIGCOMM Computer Communication Review*, **28**(1), 4–15.

[23] S. McCanne, V. Jacobson and M. Vetterli, Receiver-driven Layered Multicast, *Proc. of ACM SIGCOOM*, Stanford, CA, August 1996.

[24]  IETF RFC 2326: Real Time Streaming Protocol (RTSP), H. Schulzrinne, A. Rao, and R. Lanphier, April 1998.

[25]  IETF RFC 2327: SDP: Session Description Protocol, M. Handley and V. Jacobson, April 1998.

[26]  IETF RFC 3550: RTP: A Transport Protocol for Real-Time Applications, H. Schulzrinne *et al.*, July 2003.

[27]  IETF RFC 3551: RTP Profile for Audio and Video Conferences with Minimal Control, H. Schulzrinne and S. Casner, July 2003.

[28]  IETF RFC 3267: Real-Time Transport Protocol (RTP) Payload Format and File Storage Format for the Adaptive Multi-Rate (AMR) Adaptive Multi-Rate Wideband (AMR-WB) Audio Codecs, J. Sjoberg *et al.*, June 2002.

[29]  IETF RFC 3016: RTP Payload Format for MPEG-4 Audio/Visual Streams, Y. Kikuchi *et al.*, November 2000.

[30]  IETF RFC 2429: RTP Payload Format for the 1998 Version of ITU-T Rec. H.263 Video (H.263+), C. Bormann *et al.*, October 1998.

[31]  IETF RFC 2046: Multipurpose Internet Mail Extensions (MIME) Part Two: Media Types, N. Freed and N. Borenstein, November 1996.

[32]  IETF RFC 3236: The 'application/xhtml+xml' Media Type, M. Baker and P. Stark, January 2002.

[33]  3GPP TR26.234: Technical Specification Group Services and System Aspects; Transparent end-to-end PSS; Protocols and codecs (Rel.6.1.0, 09-2004).

[34]  3GPP TR22.233: Technical Specification Group Services and System Aspects; Transparent end-to-end PSS; Stage 1 (Rel.6.3, 09-2003).

[35]  ITU-T Recommendation T.81 (1992) | ISO/IEC 10918-1:1993: Information Technology – Digital Compression and Coding of Continuous-tone Still Images – Requirements and Guidelines.

[36]  C-Cube Microsystems: JPEG File Interchange Format, Version 1.02, September 1, 1992.

[37]  ISO/IEC 10646-1:2000: Information Technology – Universal Multiple-Octet Coded Character Set (UCS) – Part 1: Architecture and Basic Multilingual Plane.

[38]  The Unicode Consortium: The Unicode Standard, Version 3.0 Reading, MA, Addison-Wesley Developers Press, 2000.

[39]  W3C Recommendation: Synchronized Multimedia Integration Language (SMIL 2.0), http://www.w3.org/TR/2001/REC-smil20-20010807/, August 2001.

[40]  CompuServe Incorporated: GIF Graphics Interchange Format: A Standard Defining a Mechanism for the Storage and Transmission of Raster-based Graphics Information, Columbus, OH, USA, 1987.

[41]  CompuServe Incorporated: Graphics Interchange Format: Version 89a, Columbus, OH, USA, 1990.

[42]  IETF RFC 2083: PNG (Portable Networks Graphics) Specification Version 1.0, T. Boutell, *et al.*, March 1997.

[43]  W3C Recommendation: Scalable Vector Graphics (SVG) 1.1 Specification, http://www.w3.org/TR/2003/REC-SVG11-20030114/, January 2003.

[44]  W3C Recommendation: Mobile SVG Profiles: SVG Tiny and SVG Basic, http://www.w3.org/TR/2003/REC-SVGMobile-20030114/, January 2003.

[45]  Scalable Polyphony MIDI Specification Version 1.0, RP-34, MIDI Manufacturers' Association, Los Angeles, CA, February 2002.

[46]  Scalable Polyphony MIDI Device 5-to-24 Note Profile for 3GPP Version 1.0, RP-35, MIDI Manufacturers Association, Los Angeles, CA, February 2002.

[47]  Standard MIDI Files 1.0, RP-001. In *The Complete MIDI 1.0 Detailed Specification, Document Version 96.1*, The MIDI Manufacturers Association, Los Angeles, CA, USA, February 1996.

[48]  WAP Forum Specification: XHTML Mobile Profile, http://www1.wapforum.org/tech/terms.asp?doc=WAP-277-XHTMLMP-20011029-a.pdf, October 2001.

[49]  IETF RFC 3168: The Addition of Explicit Congestion Notification (ECN) to IP, K. Ramakrishnan and S. Floyd. September 2001.

[50]  IETF RFC 2210: The Use of RSVP with IETF Integrated Services, J. Wroclawski, September 1997.

[51]  IETF RFC 2475: An Architecture for Differentiated Services, S. Blake, December 1998.

[52]  IETF RFC 2543 - SIP: Session Initiation Protocol, M. Handley *et al.*, March 1999.

[53]  ITU-T Rec. H.323: Visual Telephone Systems and Terminal Equipment for Local Area Networks which Provide a Non-Guaranteed Quality of Service, 1996.

# 4

# Multimedia Control Protocols for Wireless Networks

Pedro M. Ruiz, Eduardo Martínez, Juan A. Sánchez
and Antonio F. Gómez-Skarmeta

## 4.1 Introduction

The previous chapter was devoted to the analysis of transport protocols for multimedia content over wireless networks. That is, it mainly focused on how the multimedia content is delivered from multimedia sources to multimedia consumers. However, before the data can be transmitted through the network, a multimedia session among the different parties has to be established. This often requires the ability of control protocols to convey the information about the session that is required by the participants. For instance, a multimedia terminal needs to know which payload types are supported by the other participants, the IP address of the other end (or the group address in case of multicast sessions), the port numbers to be used, etc. The protocols employed to initiate and manage multimedia sessions are often called multimedia control protocols, and these are the focus of this chapter.

The functions performed by multimedia control protocols usually go beyond establishing the session. They include, among others:

- session establishment and call setup,
- renegotiation of session parameters,
- the definition of session parameters to be used by participating terminals,
- control delivery of on-demand multimedia data,
- admission control of session establishments,
- multimedia gateway control, for transcoding and interworking across different standards.

The multimedia control protocols that are being considered in wireless networks are mostly the same as those that the Internet Engineering Task Force (IETF) has standardized for fixed IP networks. The main reason for this is the great support that 'All-IP' wireless networks are receiving from within the research community. Since the Release 5 of UMTS multimedia, services are going to be offered by the IP Multimedia Subsystem (IMS), which is largely based on IETF multimedia control protocols. However,

*Emerging Wireless Multimedia: Services and Technologies*   Edited by A. Salkintzis and N. Passas
© 2005 John Wiley & Sons, Ltd

in many cases these protocols require adaptations and extensions, which we shall address later in this chapter.

The remainder of the chapter is organized as follows: in Section 4.2, we introduce the different multimedia control protocols that have been used in datagram-based networks. We also analyze why only a subset of these have been considered for wireless networks. Sections 4.3 to 4.5 describe the main control protocols considered in wireless networks. In particular, Section 4.3 explains the details of the Session Description Protocol (SDP), which is widely used to represent the parameters that define a multimedia session. Section 4.4 describes the Real-Time Streaming Protocol (RTSP), which is an application-level protocol for controling the delivery of multimedia data. In addition, in Section 4.5 we discuss the basic operation of the Session Initiation Protocol (SIP). This protocol has also been proposed by the IETF, but it is now the 'de facto' standard for session establishment in many existing and future wireless networks. In Section 4.6 we describe the advanced SIP functionalities that have recently been incorporated into the basic specification to support additional services that are relevant to wireless networks, such as roaming of sessions, multiconferencing, etc. In Section 4.7 we discuss the particular uses of all these protocols within the latest UMTS specifications. In particular we focus on the description of features and adaptations that have been introduced into these protocols to incorporate them into the specification. Finally, Section 4.8 gives some ideas for future research.

## 4.2 A Premier on the Control Plane of Existing Multimedia Standards

With the advent of new networking technologies that can provide higher network capacities, during the 90's many research groups started to investigate the provision of multimedia services over packet oriented networks. At that time the Audio/Video Transport (AVT) working group of the IETF was defining the standards (e.g. RTP, RTCP, etc.) for such services.

The International Telecommunications Union (ITU) was also interested in developing a standard for videoconferencing on packet switched networks. By that time most of the efforts in the ITU-T were focused on circuit switched videoconferencing standards such as H.320 [1], which was approved in 1990. The new ITU standard for packet switched networks grew out of the H.320 standard. Its first version was approved in 1996 and it was named H.323 [2]. Two subsequent versions adding improvements were also approved in 1998 and 1999, respectively. Currently there is also a fourth version but most of the implementations are based on H.323v3.

Since the mid-90's both IETF and ITU videoconferencing protocols have been developed in parallel although they have some common components. For instance, the data plane is in both cases based on RTP/RTCP [4] (see previous chapter) over UDP. As a consequence, all the payload formats defined in H.323 are common to both approaches. However, the control plane is completely different in the two approaches, and the only way in which applications from both worlds can interoperate is by using signaling gateways.

In this section we introduce the basic protocols in the architecture proposed by each standardization body, and then analyze why IETF protocols are being adopted for wireless networks rather than the ITU-T ones. Given that the data transport protocols in both cases are similar to those presented in the previous chapter, we focus our discussion on the control plane.

### 4.2.1 ITU Protocols for Videoconferencing on Packet-switched Networks

As mentioned above, H.323 is the technical recommendation from ITU-T for real-time videoconferencing on packet-switched networks without guarantees of quality of service. However, H.323 rather than being a technical specification, is like an umbrella recommendation which defines how to use different protocols to establish a session, transmit multimedia data, etc. In particular, H.323 defines which protocols must be used for each of the following functions.

- Establishment of point-to-point conferences. When a multipoint control unit (MCU) is available, H.323 also defines how to use it for multiparty conferences.
- Interworking with other ITU conferencing systems like H.320 (ISDN), H.321 (ATM), H.324 (PSTN), etc.
- Negotiation of terminal capabilities. For instance, if one terminal has only audio capabilities, both terminals can agree to use only audio. The sessions are represented using ASN.1 grammar.
- Security and encryption providing authentication, integrity, privacy and non-repudiation.
- Audio and video codecs. H.323 defines a minimum set of codecs that each terminal must have. This guarantees that at least a communication can be established. However, the terminals can agree to use any other codec supported by both of them.
- Call admission and accounting support. Defines how the network can enforce admission control based on the number of ongoing calls, bandwidth limitations, etc. In addition, it also defines how to perform accounting for billing purposes.

In addition, H.323 defines different entities (called endpoints) depending on the functions that they perform. Their functions and names are shown in Table 4.1.

Figure 4.1 shows the protocols involved in the H.323 recommendation, including both the control and the data planes. As we see in the figure, H.323 defines the minimum codecs that need to be supported

**Table 4.1** H.323 entities and their functionalities

| H.323 entity | Functions performed |
|---|---|
| Terminal | User equipment that captures multimedia data, and originates and terminates data and signaling flows. |
| Gateway | Optional component required for the interworking across different network types (e.g. H.323-H.320) translating both data and control flows as required. |
| Gatekeeper | It is also an optional component that is used for admission and access control, bandwidth management, routing of calls, etc. When present, every endpoint in its zone must register with it, and they must send all control flows through it. |
| MCU | It is used to enable multiconferences among three or more endpoints. |

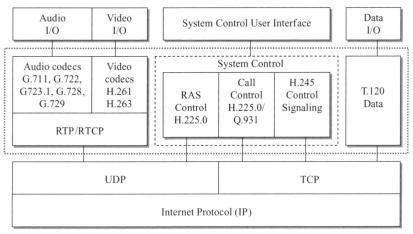

·············· Covered by H.323 Recommendation

**Figure 4.1** H.323 protocol stack including multimedia control protocols.

both for audio and video communications. However it does not include any specification regarding the audio and video capture devices. According to H.323 recommendation, audio and video flows must be delivered over the RTP/RTCP protocol as described in the previous chapter. In addition, the recommendation defines a data conferencing module based on the T.120 ITU-T standard [3]. Unlike many IETF data conferencing protocols, T.120 uses TCP at the transport layer, rather than reliable multicast. So, we can see that there are no big differences in the data plane between IETF and ITU-T standards.

Regarding the control plane, H.323 is largely different from the multimedia control protocols defined by IETF. In H.323 the control functions are performed by three different protocols. The encryption and security features are provided by the H.235 protocol [5], which we have not included in the figure for the sake of simplicity. The other two protocols in charge of controlling H.323 multimedia sessions are H.225.0 [6], which takes care of the call signalling and the admission control, and H.245 [7] which is responsible for the negotiation of capabilities such as payload types, codecs, bit rates and so forth.

The H.225.0 protocol has two different components, commonly named H.225.0 Registration Admission Status (RAS), and H.225.0 call signaling (a subset of the standard ISDN call control protocol Q.931). The H.225.0 RAS component uses UDP at the transport layer, whereas the H.225.0 call signaling is performed reliably using TCP as the underlying protocol.

H.225.0 call signaling provides the basic messages to set up and tear down multimedia connections. Unlike IETF session set up protocols, it can be used only to set up point-to-point connections. When multiparty sessions are required, each terminal establishes a point-to-point connection to an MCU, and the MCU replicates the messages from each sender to the rest of terminals. The protocol uses four basic messages as follows.

(1) Setup. A setup message is sent to initiate a connection to another terminal.
(2) Alerting. This message is sent by the callee to indicate that it is notifying the user.
(3) Connect. It is also sent by the callee to indicate that the user accepted the call.
(4) Release. This message can be sent by any of the parties to tear down the connection.

After the Connect message is received by the terminal originating the call, both terminals use the H.245 protocol to interchange session capabilities (described using ASN.1) to agree on the set of parameters to be used for the session. The main functions provided by H.245 are as follows.

- Capability exchange. Each terminal describe their receive and sends capabilities in ASN.1 and sends them in a *termCapSet* message. These messages are acknowledged by the other end. The description of capabilities includes, among others, the audio and video codecs supported, and data rates.
- Opening and closing of logical channels. A logical channel is basically a pair (IP address, port) identifying a flow between both terminals. Data channels, by relaying on TCP, are naturally bi-directional. Media channels (e.g. audio) are unidirectional. H.245 defines a message call *openReq* to request the creation of such a channel. The *endSession* message is also used to close the logical channels of the session.
- Flow control. In the event of any problem, the other end can receive notifications.
- Changes in channel parameters. There are messages than can be used by terminals to notify other events such as change in the codec being used.

Finally, H.225.0 RAS defines all the messages that are needed to communicate terminals and gatekepers. Its main functionalities include the following.

- Discovery of gatekeepers. The *Gatekeeper Request (GRQ)* message is multicast by a terminal to the well-known multicast address of all the gatekeepers (224.0.1.41) whenever it needs to find a gatekeeper. Gatekeepers answer with a *Gatekeeper Confirm (GCF)* message, which includes the transport-layer address (i.e. UDP port) of its RAS channel.

- Registration of endpoints. These are used by terminals to join the zone administered by a gate-keeper. The terminals inform the gatekeeper about their IP and alias addresses. H.225.0 RAS provides messages for requesting registration (RRQ), confirming registration (RCF), rejecting a regis-tration (RRJ), requesting being un-registered (URQ), confirming unregistrations (UCF) and rejecting unregistrations (URJ).
- Admission control. Terminals send *Admission Request (ARQ)* messages to the gatekeeper to initiate calls. The gatekeeper can answer with an *Admission Confirm (ACF)* message to accept the call, or an *Admission Reject (ARJ)* message to reject the call. These messages may include bandwidth requests associated with them. In addition, if the bandwidth requirements change during a session, this can be notified by specific H.225.0 RAS messages.
- Endpoint location and status information. These messages are interchanged between gatekeepers. They are used to gather information about how to signal a call to an endpoint in the zone of the other gatekeeper as well as to check whether an endpoint is currently online (i.e. registered to any gatekeeper) or off-line.

As we have seen, the main multimedia control functionalities covered by H.323 are (i) the negotiation of capabilities, (ii) a description of capabilities in ASN.1, (iii) the call setup and tear down and (iv) call admission control. We shall see the functionalities provided by IETF control protocols in the next section.

### 4.2.2 IETF Multimedia Internetworking Protocols

The multimedia architecture proposed by the IETF also consists of a set of protocols that, combined together, form the overall multimedia protocol stack. In addition, they can also be easily divided into a control plane and a data plane.

As mentioned before, the data plane consists basically of the same RTP/RTCP over UDP approach that the ITU-T borrowed from the IETF for the H.323 recommendation. However, there is a difference on the transport of data applications. As we can see in Figure 4.2, in the proposal from the IETF these data applications use reliable multicast protocols as an underlying transport. This is because most of these protocols were designed to be used in IP multicast networks in the early stages of the MBone [8]. Thus, rather than using TCP as a transport protocol (which cannot work with IP multicast), the research community decided to investigate protocols to provide reliable delivery over unreliable UDP-based multicast communications.

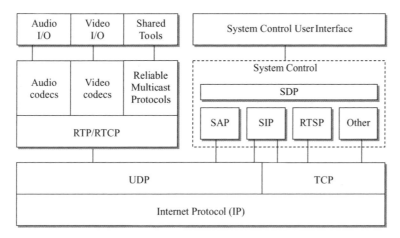

**Figure 4.2**   IETF multimedia protocol stack.

Regarding the control plane, we can see that the protocols proposed by the IETF are completely different from those recommended in H.323. However, the functions that they perform are largely the same.

Similarly to H.323, the IETF defined a protocol that describes the parameters to be used in multimedia sessions. This protocol is called Session Description Protocol (SDP) [9] and it is the equivalent to the ASN.1 descriptions used in H.323. However, rather than relying on such a complicated binary format, SDP employs a very easy-to-understand text-based format that makes the whole protocol very extensible, human readable and easy to parse in a variety of programming languages. SDP descriptions are designed to carry enough information so that any terminal receiving such a description can participate in the session. Another important advantage of its textual and simple format is that it can easily be carried as MIME-encoded data. Thus, any other Internet applications that are able to deal with MIME [10] information (e.g. email, HTTP) can be used as a potential session establishment application. This clearly adds a lot of flexibility to the whole stack in contrast to the extremely coarse H.323 stack. Needless to say that the SDP protocol is the core of all the session establishment protocols. As can been seen from the figure, all the control protocols carry SDP descriptions in their packets.

As explained, almost any Internet application that can transfer SDP descriptions is a candidate for the session establishment approach. In fact, practices such as publishing SDP descriptions in web pages or sending them by email are perfectly valid. However, the IETF also defined session establishment protocols that provide some additional functionality such as security or advanced call setup features. The Session Announcement Protocol (SAP) [11] is such a protocol and is specifically designed to advertise information about existing sessions. This protocol was initially designed as the underlying mechanism of a distributed directory of sessions similar to a TV program guide. Thus, it is specifically designed for multiparty sessions and it uses IP multicast messages to periodically advertise existing sessions. To start a session, all the interested parties just process the SDP description associated with that session, which must be stored in the local session directory. Because of its requirements of wide-area multicast deployment it is used only in experimental multicast networks nowadays.

However, the IETF realized that this approach was not sufficiently suitable to accommodate very common scenarios such as the case where one user wants to establish a session with another user, or wants to invite another user to an already ongoing session. To support these requirements a new session setup protocol called Session Initiation Protocol (SIP) [12] was proposed. SIP also uses a very simple and extensible text-based packet format. In addition, the protocol supports call control functions (e.g. renegotiation of parameters) similar to those offered by H.245 as well as location and registration functions similar to those that H.225.0 RAS offers. The SIP specification has suffered a great deal of modifications over the last few years. Most of these are adaptations to enable it to operate in many different environments, such as VoIP and future wireless networks. These are described in detail in the following sections.

In addition to these control protocols, the IETF has also standardized a protocol to control the delivery of multimedia data. This protocol is called the Real Time Streaming Protocol (RTSP) [13] and there is no such protocol in the H.323 recommendation. Following the same philosophy of simplicity and extensibility of SDP, and SIP, the RTSP protocol is based on text-formatted messages that are reliably delivered from clients (receivers) to servers (multimedia sources) and vice versa. The reliability of these messages is achieved by using TCP as the transport layer. The RTSP protocol is specifically designed for streaming services in which there can be a large playout buffer at the receiver when receiving data from the streaming server. RTSP messages are used by the client to request a multimedia content from the server, ask the server to send more data, pause the transmission, etc. An example of this kind of streaming services is video on-demand. The detailed operation of the RTSP protocol is explained in Section 4.4.

In the next subsection we compare both approaches and give some insight on the key properties that made SIP the winning candidate for future IP-based wireless networks.

### 4.2.3 Control Protocols for Wireless Networks

Over the last few years there has been a tough competition between SIP and H.323 for the voice over IP (VoIP) market. In addition, the widespread support that packet-switched cellular networks have received within the research community, expanded this debate to the arena of mobile networks. When the 3rd Generation Partnership project (www.3gpp.org), 3GPP, moved towards an 'All-IP' UMTS network architecture, a lot of discussion was needed after an agreement on the single multimedia control standard to be considered.

The 3GPP needed to support some additional services that, at that point, were not supported by any of the candidates. Thus, the extensibility of the protocols was one of the key factors affecting the final decision. Table 4.2 compares the alternatives according to some of the key factors to demonstrate why IETF multimedia protocols are the ones that were finally selected for wireless networks.

**Table 4.2**  Comparison of SIP and H.323 multimedia control

| Function | H.323 | SIP | Evaluation comments |
| --- | --- | --- | --- |
| Session description | Binary encoding | Textual | SDP is easier to decode and requires less CPU. ASN.1 consumes a little lessband-width, but that is not a big advantage considering multimedia flows. |
| Complexity | High | Moderate | ASN.1 and the other protocols are hard to process and program. |
| Extensibility | Extensible | More extensible | ASN.1 is almost vendor specific and it is hard to accommodate new options and extensions. On the other hand, SIP can be easily extended with new features. |
| Architecture | Monolithic | Modular | SIP modularity allows for an easy addition of components and a simple interworking with existing services (e.g. billing) that are already in use by the operator. |
| Interdomain call routing | Static | Hierarchically based on DNS | SIP, by relying on existing DNS domain names is able to route calls across domains by simply resolving the names of the SIP server of the callee domain name. |
| Debugging | Difficult | Simple | The textual and human-readable format of SIP messages makes it easy for developers to understand. In the case of H.323, special tools are required. |
| Size of protocol's stack | Bigger | Lower | The SIP stack is smaller and allows for a reduction in the memory required by the devices. |
| Web services | Requires changes | Directly supported | The ability for SDP messages and SIP payloads to be transmitted as MIME encoded text allows for a natural integration with web-based services. |
| Billing and accounting | Performed by the Gatekeeper | SIP Authorization header | SIP can easily be integrated with existing AAA mechanisms used by the operator (i.e. Radius or Diameter). |
| Personal mobility | Not naturally supported | Inherently supported | SIP is able to deliver a call to the terminal that the user is using at that particular time. It also supports roaming of sessions. H.323 can redirect calls, but this needs to be configured through user-to-user signaling. |

As we can see from the table, the main reason why H.323 is considered complex is because of the binary format that is used to describe sessions. ASN.1 is hard to decode, compared with the simplicity of an SDP decoder, which can be written in a few lines of code in any scripting language. However, one of the most important factors was the excellent extensibility of the SIP protocol. First of all, the default processing of SIP headers by which unknown headers are simply ignored facilitates a simple backward compatibility as well as an easy way to include operator-specific features. Secondly, it is very easy to create new SIP headers and payloads because of the simplicity offered by its text encoding.

In the case of cellular networks in which terminals have limited capabilities, it is also very important that the SIP protocol stack has a smaller size. This allows for a reduction of the memory required by the terminal to handle the SIP protocol. In addition, the lower CPU required to decode SIP messages compared with H.323 also makes SIP attractive from the same point-of-view.

Thus, given that wireless networks are expected to employ IETF protocols to control multimedia sessions, the rest of the chapter will focus on giving a detailed and comprehensive description of SDP, RTSP and SIP. Special attention will be paid to functionalities related to wireless networks and an example will be given on how they will be used to provide multimedia services in the latest releases of the UMTS specification.

## 4.3 Protocol for Describing Multimedia Sessions: SDP

In the context of the SDP protocol, a session is defined in [9] as 'a set of media streams that exist for some duration of time'. This duration might or might not be continuous. The goal of SDP is to convey enough information for a terminal receiving the session description to join the session. In the case of multicast sessions, at the same time, the reception of the SDP message serves to discover the existence of the session itself.

We have seen above that the multimedia control protocols defined by the IETF use, in some form or another, session descriptions according to the SDP protocol. To be more specific, SDP messages can be carried into SAP advertisements, SIP messages, RTSP packets, and any other application understanding MIME extensions (using the MIME-type application/sdp) such as email or HTTP. In this subsection we take a deeper look at SDP specifications and give some examples of session descriptions.

### 4.3.1 The Syntax of SDP Messages

The information conveyed by SDP messages can be categorized into media information (e.g. encoding, transport protocol, etc.), timing information regarding start and end times as well as repetitions and, finally, some additional information about the session, such as who created it, what the session is about, related URLs, etc. The format of SDP messages largely follows this categorization.

As mentioned above, an SDP session is encoded using plain text (ISO 10646 character set with UTF-8 encoding). This allows for some internationalization regarding special characters. However, field names and attributes can only use the US-ASCII subset of UTF-8.

An SDP session description consists of several lines of text separated by a CRLF character. However, it is recommended that parsers also accept an LF as a valid delimiter. The general format of each of the lines is of the form:

$$< \textbf{type} >=< \textbf{value} >$$

where $<$type$>$ is always a single-character, case-sensitive field name, and $<$value$>$ can be either a number of field values separated by white spaces or a free format string. Please note that no whitespaces are allowed on either side of the '$=$' sign.

SDP fields can be classified into session-level fields or media-level fields. The former are those fields whose values are relevant to the whole session and all media streams. The latter refer to values that only apply to a particular media stream. Accordingly, the session description message consists of one session-level section followed by zero or more media level sections. There is no specific delimiter among sections, but the names of the fields themselves. This is because in order to simplify SDP parsers, the order in which SDP lines appear is strict. Thus, the first media-level field ('m=') in the SDP message indicates the starting of the media-level section.

The general format of an SDP message is given in Figure 4.3. Fields marked with * are optional, wheras the others are mandatory. We explain below the use of the fields which are needed by most applications. We refer the reader to [9] for full deails on the protocol.

**Figure 4.3**   General format of an SDP message.

As we see in the figure, the session description starts with version of SDP. For the time being, '**v** = **0**' is the only existing version. Next field is the originator of the session. It consists of a username, e.g. pedrom or '=' in the case in which the operating system of the machine generating the advertisement doesn't have the concept of user-ids. The < session-id > is a numerical identifier so that the tuple (<username>,<session-id>,<net-type>,<addr-type>,<addr>) is unique. It is recommended that one use an NTP timestamp at the session creation time, although it is not mandatory. An additional field called <version> is included to assess which description of the same session is the most recent. It is sufficient to increment the counter every time the session description is modified, although it is also recommended that an NTP timestamp at the modification time is used. The <net-type> refers to the type of network. Currently the value 'IN' is used to mean the Internet. The <addr-type> field identifies the type of address where the network has different types of addresses. Currently defined values for IP networks are 'IP4' for IPv4 and 'IP6' for the case of IPv6. Finally addr represents the address of the host from which the user announced the session. Whenever it is available, the fully qualified domain name should be included. In addition, each session must have one and only one name which is defined using the '**s** =' field followed by a string corresponding to the name.

Optionally, a session description can also include additional information after the '**i** =' field. This field can be present both at the media level and at the session level. In any case having more than one session level information field or more than one per media is not allowed. The session level field is usually used as a kind of abstract about the session, whereas at the media level it is used to label different media flows. The optional fields 'u =', 'e =' and 'p =' are just followed by strings that convey information about a URL with additional information, the e-mail address and the phone number of the owner of the session, respectively.

Regarding the connection information (field 'c =') there can be either individual connection fields for each media or a single connection field at the session level. Another possible option is having a general session-level connection field that is valid for all media but the ones having their own connection information field. In both cases, a connection field is followed by a <net-type> and an <addr-type> attributes with the same format as was explained for the 'o=' field. In addition, a <connection-addr> attribute is required, which may correspond to either a multicast address (either IPv4 or IPv6) for the session or an unicast address. In the latter case, an 'a=' attribute will be used to indicate whether that unicast address (or fully-qualified domain name) corresponds to the source or the data sink. For an IP multicast address the TTL must be appended using a slash separator. For example, c = IN IP4 224.2.3.4/48.

Another important element that is mandatory for any SDP description is the timing of the session. In its simplest form, it consists of a 't =' field followed by the start and end times. These times are codified as the decimal representation of NTP time values in seconds [14]. To convert these values to UNIX time, subtract 2208988800. There may be as many 't =' fields as starting and ending times of a session. However, when repetitions are periodic, it is recommended that one use the optional 'r =' field to specify them. In this case, the start-time of the 't =' field corresponds to the start of the first repetition, whereas the end-time of the same field must be the end-time of the last repetition. Each 'r =' field defines the periodicity of the session <repeat-interval>, the duration of each repetition <active-duration> and several <offset> values that define the start time of the different repetitions before the next <repeat-interval>. For example, if we want to advertise a session which takes place every Monday from 8:00 am to 10:00 am and every Wednesday from 9:00 am to 11:00 am every week for 2 months, it will be coded as:

$$t = 3034429876 \quad 3038468288$$
$$r = 7d \quad 2h \quad 0 \quad 25h$$

Where 3034429876 is the start time of the first repetition, 3038468288 is the end time of the last repetition after the 2 months, and the 'r =' indicates that these sessions are to be repeated every 7 days, lasting for 2 hours being the first of the repetition at multiples of 7 days of the start-time plus 0 hours, and the other repetition at multiples of 7 days of the start-time plus 25 h.

Encryption keys are used to provide multimedia applications with the required keys to participate in the session. For instance, when encrypted RTP data is expected for that session. Another interesting feature for extending SDP are attributes. Attributes, specified with the 'a =' field, can be of two types: property and value. Property attributes are of the type 'a =<flag>' where flag is a string. They are used to specify properties of the session. On the other hand, value attributes are used like property attributes in which the property can take different values. An example of a property attribute is 'a = recvonly', indicating that users are not allowed to transmit data to the session. An example of a value attribute is 'a = type:meeting', which specifies the type of session. User defined attributes start with 'X-'.

Finally, the most important part of SDP messages is the media descriptions. As mentioned, media descriptions are codified using the 'm =' field. A session description can have many media descriptions, although generally there is one for each medium used in the session such as audio or video. The first attribute after the '=' sign is the media type. Defined media types are audio, video, application, data and control. Data refers to raw data transfer, whereas application refers to application data such us whiteboards, shared text editors, etc. The second sub-field is the transport-layer port to which the media is to be sent. In the case of RTP, the associated RTCP port is usually automatically obtained as the next port to the one in this sub-field. In the cases in which the RTCP port does not follow that rule, it must be specified according to RFC-3605 [15]. The port value is used in combination with the transport type, which is given in the third sub-field. Possible transports are 'RTP/AVP' (for IETF's RTP data) and 'udp' for data sent out directly over UDP. Finally, the fourth sub-field is the media format to be used for audio and video. This media format is an integer that represents the codecs to be used according to RTP-AV profiles described in the previous chapter. For instance, 'm = audio 51012 RTP/ AVP 0' corresponds to a u-law PCM coded audio sampled at 8 KHz being sent using RTP to port 51012.

When additional information needs to be provided to identify the coding parameters fully, we use the 'a = rtpmap' attribute, with the following format:

**a=rtpmap: < payload-type >< encoding > / <clock >[/encoding parameters]**

Encoding represents the type of encoding, clock is the sampling rate, and encoding parameters is usually employed to convey information about the number of audio channels. Encoding parameters have not been defined for video. An example for 16-bit linearly encoded stereo audio stream sampled at 16 khz we use 'a = rtpmap:98 L16/16000/2'.

In the next subsection we give an example of an unicast and a multicast IPv4 session description. For IPv6 sessions, one only has to change IP4 to IP6 and the IPv4 addresses to the standard IPv6 address notation. The detailed ABNF syntax for IPv6 in SDP is defined in [16]. Another interesting document for readers needing all the details of SDP operation is RFC-3388 [17], which describes extensions to SDP that allow for the grouping of several media lines for lip synchronization and for receiving several media streams of the same flow on different ports and host interfaces.

### 4.3.2 SDP Examples

In Figure 4.4 we show an example of a description for an IP multicast session. As we can see, the order in which fields appear has to strictly follow the SDP standard. An interesting aspect of the example is that it illustrates the difference between session-level fields and media-level fields. For instance, the first 'c =' field informs about the multicast address that the applications must use. However, within the media description for the whiteboard, we override that information. It could have got the same effect by not having a session-level connection information and replicating this both in the audio and video media descriptions. Note that the TTL must be included after the IP multicast address in the 'c =' field.

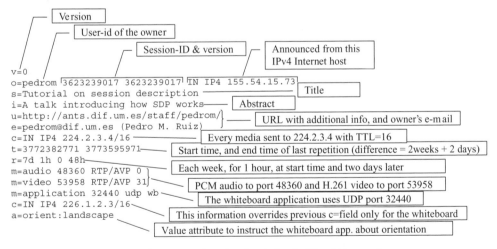

**Figure 4.4** Annotated session description for IP multicast session.

As we can see in Figure 4.5, for unicast sessions the TTL is not used. The main difference compared with the previous example is in the connection information part ('c =' field). As you can see, the IP address now corresponds to the IP unicast address to which the terminal receiving the SDP message has to send multimedia data. Note that media-level connection information is not usually needed unless the originator of the SDP message uses different terminals to receive different media.

```
v=0
o=pedrom 3623239017 3623239017 IN IP4 155.54.15.73
s=One to one session
i=This session is intended to anyone willing to contactme
c=IN IP4 155.54.15.73
t=3772382771 3773595971
r=7d 1h 0 48h
m=audio 48360 RTP/AVP 0
m=video 53958 RTP/AVP 31
m=application 32440 udp wb
a=orient:landscape
```

Unicast address without TTL.
It happens to be the same host from which the
session was advertised. But it might be different.

**Figure 4.5**   Example of description for a unicast session.

## 4.4  Control Protocols for Media Streaming

One-way streaming and media on demand delivery real-time services (Section 3.3) are characterized by the provision of some form of VCR-like control to select media contents and to move forward and backward within the content. This functionality can be implemented with a high degree of independence with respect to the actual transport of the continuous data from the server to the client. The main justification for the separation of control and transport duties is extensibility: a single control protocol, designed with extensibility in mind, acts as a framework prepared to work with current and future media formats and transport protocols. In addition, this control protocol may provide value-added services that improve the mere control (start/stop) of continuous data transport, such as the description of media contents or the adaptation of those contents to client preferences or player capabilities. The protocol developed by the IETF to control the delivery of real-time data is the Real-Time Streaming Protocol (RTSP), currently defined in RFC 2326 [13], and revised in a submitted Internet Draft. Both documents can be found in [18].

RTSP is an out-of-band protocol, focused on the control of one or several time-synchronized streams (audio and video tracks of a movie, for instance), although it is prepared to interleave media data with control information. RTSP messages can use both TCP and UDP at the transport layer, whereas the transmission of media streams controlled by RTSP may use several protocols, such as TCP, UDP or RTP (Section 3.7). RTSP is complemented by a protocol to describe the characteristics of the streams that make up the media streaming session. Usually, SDP (Section 4.3) is the choice, but RTSP is general enough to work with other media description syntaxes.

RSTP messages are intentionally similar in syntax and operation to HTTP/1.1 messages [19]. The successful experience of HTTP as an extensible framework to request and transfer discrete media data (images, text, files) had a strong influence on this decision. However, there are some important differences in RTSP:

- RTSP defines new methods and headers;
- RTSP servers maintain the state of media sessions across several client connections (when using TCP) or messages (when using UDP) while HTTP is stateless;
- RTSP uses UTF-8 rather than ISO 8859-1;
- the URI contained in a RTSP request message, which identifies the media object, is absolute, while HTTP request messages carry only the object path and put the host name in the Host header;
- RTSP includes some methods that are bi-directional, so both servers and clients can send requests.

### 4.4.1  RSTP Operation

Before giving descriptions of RTSP messages in detail, it is interesting to take a look at the overall operation of a RTSP session between a streaming client and a server (see Figure 4.6). The common

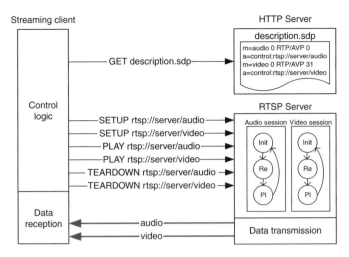

**Figure 4.6**  Overall RTSP operation.

scenario begins with the streaming client retrieving a description of the media streaming session. This description specifies the characteristics of the streams that make up the session. The streaming client may retrieve the description directly from the media server or use other means such as HTTP or email. In this description, each stream is identified by an RTSP URL that acts as a handle to control the stream. Note that each RTSP URL may point to a different streaming server. With respect to transport parameters, like network destination address and port, RTSP specification describes two modes of operation: unicast and multicast. Unicast mode corresponds to media on demand delivery: media data is transmitted directly from the media server to the source of the RTSP request using the port number chosen by the client. Multicast mode corresponds to one-way streaming with network layer-based replication (Section 3.3). In this mode, the selection of the multicast address and port can be made at the server or at the client. If it is made at the server, the scenario corresponds to a TV-like live transmission, with clients tuning channels using the media session description that, in this case, will include the multicast address and port required for each stream. If the server is to participate in an existing multicast conference, the multicast address and port could be chosen by the client. These sessions are usually established by other protocols, although SIP is the most common. Note that one-way streaming with application layer-based replication can be implemented though point-to-point unicast mode connections between a root media server, relays and receivers.

Once the client has a description of media streams, it issues setup orders to media servers. Upon reception of these requests, the server allocates resources for the streams and creates RTSP sessions. A server responds to a setup request with a message that includes an RTSP session identifier. The session identifier will be included in subsequent requests from the client until the session is terminated. For simplicity, session identifiers are not represented in Figure 4.6. Servers use session identifiers to demultiplex commands that apply to different sessions. With a setup message, a session (usually implemented as a state machine) enters in a ready state that lets the client issue play commands to trigger data transmission from the server. Usually, a play request message specifies the time range of the media data that will be transmitted. This means that play orders employ a seeking mechanism. To freeze data transmission, clients send pause messages. The transmission restarts with new play messages. Finally, a client terminates a session with a teardown command, letting the server release resources allocated to the session.

As with HTTP, RTSP requests and responses can cross one or more proxy servers in its end-to-end way from a client to a final server and vice versa.

```
Request = Request-Line
          *(general-header|request-header|entity-header)
          CRLF
          [message-body]
Request-Line = Method SP Request-URI SP RTSP-Version CRLF
Response = Status-Line
          *(general-header|response-header|entity-header)
          CRLF
          [message-body]
Status-Line = RTSP-Version SP Status-Code SP Reason-Phrase CRLF
```

**Figure 4.7**   Format of RTSP messages.

### 4.4.2 RTSP Messages

RTSP messages are UTF-8 text-based, and very similar to HTTP messages. Lines are ended with CRLF, but the specification also accepts CR and LF as line terminators. Request and response messages must comply with the syntax shown in Figure 4.7. The two kinds of messages have similar structures: a first line contains the fundamental information about the request or the response, and subsequent lines complete this information with pieces of data, called headers, whose end is marked with an empty line. In some cases, request or responses carry a message-body after the empty line, and some headers refer to the characteristics of the content included in the message-body (MIME type, length, etc.). The set made up by the message-body and the headers that provide information about it is called an entity.

Headers can be classified into four different categories: general headers, request headers, response headers and entity headers. General headers can be used in requests and responses, and they specify some basic characteristics about the communication act: date, sequence number, or HTTP-like connection control (keep-alive, close). Obviously, request headers are specific to request messages, whereas response headers are specific to response messages. Entity headers provide information about the entity-body or, if no body is present, about the resource identified by the request. RTSP employs some HTTP headers, and defines new ones. To avoid tedious descriptions of each RTSP header, we will give an explanation of the most important headers in examples of message interchanges, and we refer the reader to the protocol specification.

#### 4.4.2.1 Request Messages

The first line of a request message is called the request line. It consists of three substrings: method, request-uri and rtsp-version. The method identifies the type of operation that the client is requesting the server to perform over the resource identified with the request-uri (an absolute URI). Some methods can be applied over a general resource, as the whole server. In that case an asterisk '*' is used as the request-uri. The last component of the request line is the string that identifies the version of the protocol that the client application is using. It must be 'RTSP/1.0' if the software follows the current RTSP specification.

#### 4.4.2.2 Response Messages

A response message gives information about the result of an operation that a client previously requested. Remember that RTSP could use UDP as the transport protocol, meaning that there is no guarantee that responses will arrive in order. To relate requests and responses, RTSP provides a field called Cseq in the general-header. This field must be present in all requests and responses and it specifies the sequence number of request–response pairs. The header will be included in request and response related messages, with the same number, and is monotonically incremented by one for each new request.

The first line in a response message is called the status line. It contains three substrings: rtsp-version, status-code and reason-phrase. The first one identifies the version of the protocol that the server is using. If there is a mismatch with the version used by the client this field may help to figure out which characteristics are or are not supported by both ends. If RTSP follows HTTP progression, new protocol versions will be backwards compatible. Returning to the status line of response messages, the key information is the status-code. It is a 3 digit code that specifies the result of the related request. It is used by applications to figure out if that request was successful. The reason-phrase is intended for human users. RTSP employs most of HTTP/1.1 status codes and adds additional ones. Status codes are classified into the following five groups according to the value of the first digit.

(1) 1xx: Informational. Request received, continuing process.
(2) 2xx: Success. The action was successfully received, understood, and accepted.
(3) 3xx: Redirection. Further action must be taken in order to complete the request.
(4) 4xx: Client Error. Request with bad syntax or cannot be fulfilled.
(5) 5xx: Server Error. The server failed to fulfil an apparently valid request.

### 4.4.3 RTSP Methods

Table 4.3 shows the complete list of RTSP methods with a summary of information on their use. The column headed 'Direction' indicates the application that sends request messages in each method (C = Client, S = Server). The third column clarifies if a * request-uri applies to the method. The fourth column states whether the entity-body is used in RTSP request and response messages of the method. The column headed 'Object' indicates whether the method operates on the whole presentation (P) or on the stream (S) specified with the request-URI. The last two columns show if it is required, optional or recommended to implement the method in the server and the client respectively. As an example, the case of OPTION shows that it is required to be implemented in the C→S direction, but optional in the S→C direction.

**Table 4.3**  Summary of RTSP methods

| Method | Direction | * URI | Entity-body | Object | Server required | Client required |
|---|---|---|---|---|---|---|
| OPTIONS | C ↔ S | Yes | No | P,S | response = req / request = opt | request = req / response = opt |
| DESCRIBE | C → S | No | No | P,S | recommended | recommended |
| SETUP | C → S | No | No | S | required | required |
| PLAY | C → S | No | No | P,S | required | required |
| PAUSE | C → S | No | No | P,S | recommended | recommended |
| TEARDOWN | C → S | Yes | No | P,S | required | required |
| PING | C ↔ S | Yes | No | P,S | recommended | optional |
| REDIRECT | C ← S | No | No | P,S | optional | optional |
| GET_PARAMETER | C ↔ S | No | Yes | P,S | optional | optional |
| SET_PARAMETER | C ↔ S | No | Yes | P,S | optional | optional |

### 4.4.3.1 OPTIONS: Capability Handling

The RTSP specification defines methods that are recommended or optional in server and client implementations. Moreover, RTSP is extensible through the addition of new headers or methods. RTSP provides the OPTIONS method to allow protocol entities (clients, servers and proxies) to discover the capabilities of each other. This kind of request can be issued at any time and in both directions.

Depending on the URI, it applies to a certain media resource, the whole server (when the URL contains only the host address) or the next hop only (using '*'). The receiver must include a Public header in the response, with the list of RTSP methods that can be applied over the resource with the URI. A simple use of the OPTIONS method is shown below.

```
C->S:   OPTIONS * RTSP/1.0
        CSeq: 1
        User-Agent: BasicRTSPClient/1.0
        Require: play.basic
        Supported: play.basic, play.scale
S->C:   RTSP/1.0 200 OK
        CSeq: 1
        Public: DESCRIBE, SETUP, TEARDOWN, PLAY, PAUSE
        Server: BasicRTSPServer/1.1
        Supported: play.basic, play.scale
```

To handle functionality additions, called features, the last draft specification describes a procedure to register feature-tags with the Internet Assigned Numbers Authority (IANA). These features can range from a simple header to a block of headers and methods. Some feature-tags are defined in the latest version of the draft: 'play.basic' identifies the minimal implementation for playback operations, whereas 'play.scale' and 'play.speed' refer to the functionality for changing the playback scale or speed data delivery. 'con.persistent' is used to indicate the support of persistent TCP connections, and 'setup.playing' establishes that SETUP and TEARDOWN methods can be used in the Play state. To determine which features are implemented, the RTSP headers below can be included in OPTIONS requests and responses.

- Supported. This header is included in requests and responses. Its value is the list of feature-tags of requestor and receiver.
- Proxy-Supported. This is similar to Supported, but provides the functionality implemented by the proxy-chain between client and server.
- Require. This header is included in requests. Its value is the list of feature-tags that the receiver is required to implement. When some feature is not supported, responses must use 551 status (option not supported) and the Unsupported header must include the feature-tags not implemented.
- Proxy-Require. This is similar to Require, except that it applies only to proxies.
- Unsupported. This header includes a list of features that are required but not supported by the end point or the proxy chain.

### 4.4.3.2 DESCRIBE: Media Descriptions

Although a streaming client can obtain a description of a media object by HTTP, email or similar procedures outside RTSP, the DESCRIBE method can be used to request the description of a resource directly to the streaming server that contains it. Using a well-known HTTP technique, the client can include an Accept header in the request, with the list of MIME types of the description formats that it understands. An example of the DESCRIBE method is given below.

```
C->S:   DESCRIBE rstp://rstp.um.es/movie.mp4 RTSP/1.0
        CSeq: 1
        User-Agent: BasicRTSPClient/1.0
        Accept: application/sdp, application/rtsl, application/mheg
S->C:   RTSP/1.0 200 OK
        CSeq: 1
```

```
Date: Thu, 20 Feb 2003 11:00:34 GMT
Server: BasicRTSPServer/1.1
Content-Type: application/sdp
Content-Length: 475
Content-Base: rtsp://rtsp.um.es/movie.mp4/

v=0
o=UMServer 3254727685 1025882408000 IN IP4 155.54.210.13
s=Film example
e=adm@um.es
c=IN IP4 0.0.0.0
t=0 0
a=control:*
m=video 0 RTP/AVP 96
a=rtpmap:96 MP4V-ES/90000
a=control:trackID=1
a=fmtp:96 profile-level-id=1;config=000001[...]
m=audio 0 RTP/AVP 97
a=rtpmap:97 mpeg4-generic/11025/2
a=control:trackID=2
a=fmtp:97 profile-level-id=1;mode=AAC-hbr;sizelength=13;
indexlength=3;indexdeltalength=3;config=15000[...]
```

SDP (Section 4.3) has some features when used with RTSP that need some explanation. We can start with the connection information specified in the 'c' field, which contains the destination address for the media. In unicast and some multicast sessions where the client specifies the destination address, the 'c' field contains the null IP address '0.0.0.0', whereas the destination port of each media is 0 ('m = video 0' and 'm = audio 0'). The client specifies the transport information using the SETUP method described in the next section. The 't =' field may also contain 0 as start and stop times, meaning that the session is always available.

An important feature given in the SDP description is the mechanism to control independent media streams. The entire session and each media section may have a control URL specified with the 'a = control' field. The attribute may contain either relative or absolute URLs. When using relative URLs, the base URL is determined following the order Content-Base header, Content-Location header and request-uri. If 'a = control' contains only an asterisk '*', the URL is treated as a relative empty URL, hence it represents the entire base URL. When there is a session-level 'a = control' attribute, the client can use aggregate control, that is, it can control all the session streams as a whole with the session-level URL.

It is very common to use dynamic payload types in SDP descriptions for streams. The field 'rtpmap' specifies the format of the stream, by means of an encoding name. In the previous example, 'MP4V-ES' refers to video MPEG4, according to RFC 3016 [20], while 'mpeg4-generic' refers to audio AAC, according to RFC 3640 [21]. In this context, the field 'fmtp' is used to carry initialization data to properly start the decoding engine.

### 4.4.3.3 SETUP: Transport Mechanism

Once the RTSP client knows the format of the different streams that make up a streaming presentation, a decision has to be made to start the transmission of the whole presentation or some of the streams. The decision depends on the availability of adequate components to extract the payload (when using RTP) and decode it. Even if the needed components are not present in the system, the SDP description gives a chance to streaming clients to query software repositories automatically about components able to treat some format and, if the search is successful, they can download and install them. A simple software repository can be implemented with an HTTP server.

The SETUP method is used to interchange the transport parameters of each stream. Transport is the header that contains the parameters in a SETUP request and the corresponding response. There are several decisions that the streaming entities must take about the transport of a stream. First of all, the SDP description may specify several transport mechanisms in an 'm =' field, so that the client can make a choice. Secondly, the client must specify whether to operate in unicast or multicast mode. Finally, when using unicast (media on demand), the server implicitly knows the client IP address (using sockets to examine the origin of RTSP requests), but the destination and origin ports of RTP, UDP or TCP (depending on the transport mechanism selected) still need to be determined. When using multicast (one-way streaming), if the client invites the server to a conference, the Transport header of the SETUP request includes the destination, ports and time to live. On the other hand, if the server is transmitting data to a multicast address (live presentation), the server provides the destination, ports and time to live in the Transport header of the response, so that the client can tune the transmission.

A SETUP response also includes the identifier of an RTSP session. It is carried in the Session header. This header and the identifier are required to perform subsequent operations on the stream before the teardown. When no aggregation control is used, each stream is prepared with a SETUP request; hence there is one session per stream. When aggregation control is used (with the precondition of the existence of a session-level control URL in the media description), the first SETUP response provides a session identifier that is included in subsequent SETUP requests for the rest of the streams. The server interprets that it must bundle the stream into the existing session; hence there is a unique session for the whole presentation. In the next example we continue the previous example from Section 4.4.3.2, showing the use of SETUP to prepare the aggregate control of video and audio streams.

```
C->S:   SETUP rtsp://rtsp.um.es/movie.mp4/trackID=1
        RTSP/1.0
        CSeq: 2
        Transport: RTP/AVP;unicast;client_port=50312-50313
        User-Agent: BasicRTSPClient/1.0
S->C:   RTSP/1.0 200 OK
        Server: BasicRTSPServer/1.1
        CSeq: 2
        Session: 21784074157144
        Date: Thu, 20 Feb 2003 11:00:34 GMT
        Transport: RTP/AVP;unicast;client_port=50312-50313;
                server_port=6972-6973;ssrc=00007726
C->S:   SETUP rtsp://rtsp.um.es/movie.mp4/trackID=2
        RTSP/1.0
        CSeq: 3
        Session: 21784074157144
        Transport: RTP/AVP;unicast;client_port=50314-50315
        User-Agent: BasicRTSPClient/1.0
S->C:   RTSP/1.0 200 OK
        Server: BasicRTSPServer/1.1
        CSeq: 3
        Session: 21784074157144
        Date: Thu, 20 Feb 2003 11:00:34 GMT
        Transport: RTP/AVP;unicast;client_port=50314-50315;
                server_port=6972-6973;ssrc=00002891
```

### 4.4.3.4 PLAY and PAUSE: Transmission Control

To actually start media transmission, the client must issue a PLAY request. If aggregate control is being used, the request-URI in the request line must be the session-level URL; PLAY can use any media-level URL if there is no aggregation. By default, the transmission starts immediately from the beginning of the media, but the client can specify one or more ranges of media fragments and the time in UTC at which the transmission of media data within the specified ranges should start. The specification uses the Range header, which is in charge of carrying that timing information of PLAY requests and responses. The Range header in responses indicates which ranges are actually being played. There are three different ways to specify a range:

- SMPTE Relative Timestamps. Time is expressed relative to the start of the media. Its format is: hours:minutes:seconds:frames.subframes. For example, 'smpte = 10:12:33:20-' indicates the rage that starts at the 20th frame that occurs 10 hours, 12 minutes and 33 seconds after the beginning of the media, till the end.
- Normal Play Time (NPT). Time is indicated with a stream absolute position, relative to the beginning of the media, using a decimal fraction. The left part expresses seconds or hours, minutes and seconds, while the right part measures fractions of a second. For example, 'npt = 125.25-130' represents the range that starts at 125 and a quarter seconds and ends at 130 seconds.
- Absolute time. Used for live presentations, time is expressed with the ISO 8601 format using UTC. For example, a range that starts at November 25, 2004, at 14h00 and 20 and a quarter seconds UTC is represented as 'clock = 20041125T140020.25Z-'.

A PLAY response shall contain the RTP-Info header, whose value is fundamental to the adequate inter-media synchronization at the receiver when RTP is used. As explained in Section 3.7.4, because streams are transported separately across the network, they may have different network delays. This forces the receiver to synchronize the play out times using the RTP timestamps. To avoid distortions at the rendering of media, the receiver uses a de-jitter buffer. The size of that buffer affects the interactivity of PLAY streaming operations, because it must be filled before starting the decoding process. It is accepted that this buffering delay for streaming purposes can be in the order of several seconds. The synchronization mechanism needs a way to relate RTP timestamps of different streams to a common presentation clock. This can be done if the receiver knows the RTP timestamp and sequence number of the first RTP packet of each stream that is transmitted as a consequence of a PLAY operation. The server can provide that information in the response of a PLAY operation; this is the purpose of the RTP-Info header. When streaming from an already compressed media resource, the server uses the same RTP packetization routines as for live captures. Upon receiving the first PLAY command of a session, the server generates random initial values for RTP timestamps and sequence numbers. It then includes those values in the RTP-Info header of the response. The NPT time of the data contained in packet $n$ after a PLAY request may be inferred using the formula:

$$npt_n = npt_{Range} + (timestamp_n - timestamp_{RTP\text{-}Info})/Freq_{RTPClock},$$

where $ntp_n$ is the NTP time of packet $n$, $ntp_{Range}$ is the NTP start time of the Range header in the last PLAY response, $timestamp_n$ is the RTP timestamp in packet $n$ (starting counting with the first RTP packet after the last PLAY response), $timestamp_{RTP\text{-}Info}$ is the RTP timestamp for the first packet of the stream, specified in the RTP-Info header of the last PLAY response and $Freq_{RTPClock}$ is the frequency in hertz of the RTP clock specified for the stream payload type (Section 3.8). To perform synchronization, a stream is selected as the time reference that other streams must follow. Usually, audio is the choice, and video is reproduced making adjustments (for example, dropping frames) to maintain its NTP rendering time as similar as possible to the audio NTP rendering time.

The PAUSE request causes the transmission of streams to be halted temporarily, keeping server resources and leading to a session transition from the Play state to the Ready state. Similarly to PLAY requests, PAUSE commands must include the session-level URL when using aggregate control or any media-level URL if there is no aggregation. PAUSE requests may contain a Range header to indicate the pause point, expressed as the beginning value of an open range. A subsequent PLAY request without Range header resumes from the pause point and plays until the end of the media. The RTSP specification states that the RTP layer at the receiver side should not be affected by jumps in the play time, so that traditional RTP client programs (like vic or rat) can be used to receive and render streams transmitted from a RTSP server. To achieve this objective, both RTP sequence numbers and timestamps must be continuous and monotonic across jumps, and the RTP marker bit of the first packet after a new PLAY request must be set if the stream contains continuous audio, in order to perform playout delay adaptation [22].

The following example shows the use of PLAY with aggregate control (RTP-Info contains RTP data of the streams aggregated to the session with previous SETUP requests).

```
C->S:   PLAY rtsp://rtsp.um.es/movie.mp4
        RTSP/1.0
        CSeq: 4
        Session: 21784074157144
        Range: npt=0-
        User-Agent: BasicRTSPClient/1.0
S->C:   RTSP/1.0 200 OK
        Server: BasicRTSPServer/1.1
        CSeq: 4
        Session: 21784074157144
        Date: Thu, 20 Feb 2003 11:00:34 GMT
        RTP-Info: url= rtsp://rtsp.um.es/movie.mp4/trackID=1;
                       seq=2120;rtptime=10536,
                       url= rtsp://rtsp.um.es/movie.mp4/trackID=2;
                       seq=1553;rtptime=14326
```

### 4.4.3.5 TEARDOWN: Session End

The TEARDOWN method stops the delivery of streams identified with the request-URI, causing the server to free resources associated with them. If the request specifies a session-level URL with aggregate control, the whole session is removed. On the other hand, individual streams can be removed from a session by simply specifying the stream control URI in the TEARDOWN request. Note that a dropped stream can be re-bundled later on with a SETUP command. This allows a streaming client to drastically perform adaptations according to bandwidth availability.

### 4.4.3.6 GET_PARAMETER and SET_PARAMETER: Parameter Handling

GET_PARAMETER and SET_PARAMETER methods add another RTSP extension mechanism to feature identifiers and the OPTION method. The standard does not define a concrete set of parameters, but specifies the way to use RTSP to transfer parameter information between clients and servers (both can send requests). Parameters being queried or set are held in the body of requests and responses, implying the use of the Content-Type HTML header to determine the format being used to transfer parameter names and values.

### 4.4.3.7 PING: Checking Availability

This method is used to check the availability of a session or a proxy. A PING request can be issued either by clients or servers, and they must include the Session header. If the request-uri is '*', the command checks a single-hop liveness. The command has no side effects on the session state.

### 4.4.3.8 REDIRECT: Server-driven Location Change

A streaming server may send a REDIRECT message to a streaming client in order to provide a new URL that the client must use instead of the current session URL. The new URL is the value of the Location header. If the client wants to continue to receive media from the media identified with the request-uri, it must issue a TEARDOWN for the current session and begin a new session with the host designated by the new URL.

## 4.5 Session Setup: The Session Initiation Protocol (SIP)

SIP is an application level control protocol designed to help create and manage multimedia sessions for one or more participants. A multimedia session could be a telephone call over the Internet, distributed conferences, a peer-to-peer instant messaging conversation, or, in general, every kind of data communication between two or more peers. Here we are talking about the last version of the protocol that is specified in RFC 3261 [12].

We have already seen in the previous chapter some protocols used to transport multimedia contents such as audio or video over IP networks. SIP works in conjunction with those protocols, making it possible for the end points (terminals used by the end user) to find out other's end points and services, initiate sessions, interchange device capabilities and user preferences, subscribe to events, etc.

SIP provides at an application level the nowadays well-known peer-to-peer model, which means that it might be direct communication between every pair of SIP clients. Every peer can initiate the communication with an other peer by establishing a new session. However, SIP is actually a text based client/server protocol that is very similar to HTTP. It differs from HTTP in that every end point works as client and as server.

The SIP protocol is a transport protocol that runs equally over IPv4 and IPv6. It also has the possibility of working with UDP, TCP and even TLS when security becomes a big concern.

### 4.5.1 Components

#### 4.5.1.1 User Agents

These components are logical entities that can act as both a user agent client and user agent server.

- User Agent Client (UAC). A user agent client is a logical entity that creates a new request, and then sends it to the destination, creating and storing at the same time: what in SIP is called a transaction. The role of UAC lasts only for the duration of that transaction. In other words, if a piece of software initiates a request, it acts as a UAC for the duration of that transaction. If it receives a request later, it assumes the role of a user agent server for the processing of that transaction.
- User Agent Server (UAS). A user agent server is a logical entity that generates a response to a SIP request. The response accepts, rejects or redirects the request. Similarly to the UAC, this role is only assumed for the duration of that transaction. After answering a request, it can become an UAC as soon as it needs to send a request.

### 4.5.1.2 SIP Proxy

A SIP Proxy is an intermediary entity that acts as both a server and a client for the purpose of making requests on behalf of other clients. A proxy server primarily plays the role of routing, which means its job is to ensure that a request is sent to another entity that is 'closer' to the targeted user. Proxies are also useful for enforcing policies (for example, making sure a user is allowed to make a call). A proxy interprets and, if necessary, rewrites specific parts of a request message before forwarding it.

There are two kinds of proxies: Stateful and Stateless.

(1) Stateful proxy. A logical entity that maintains the client and server transaction state machines during the processing of a request; it is also known as a transaction stateful proxy.
(2) Stateless proxy. A logical entity that does not maintain the client or server transaction state machines when it processes requests. A stateless proxy forwards every request it receives downstream and every response it receives upstream.

### 4.5.1.3 Redirect Server

A redirect server is a user agent server that generates 3xx responses to the requests it receives, directing the client to contact an alternate set of URIs. There is an important difference between redirect servers and SIP proxies. Whereas the first one will remain in the middle of the message flow during the session establishment and further session operations, redirect servers do not. That subtle difference allows redirect servers to have lower load levels than proxies. Redirect servers will be discussed later in the section about SIP and mobility.

### 4.5.1.4 Registrar Server

A registrar is a server that accepts REGISTER requests and places the information it receives in those requests into the location service for the domain it handles. In Section 4.5.2.2 we will see some other request that the SIP protocol can use. At present it is enough to know that REGISTER is a message containing information to bind a unique id (a SIP URI) to an IP address.

### 4.5.1.5 Location Server

A location service is used by a SIP redirect or proxy server to obtain information about a callee's possible location(s). It contains a list of bindings of address-of-record keys to zero or more contact addresses. The bindings can be created and removed in many ways; this specification defines a REGISTER method that updates the bindings. A location Server does not need to be a physical server. Instead, it may consist only of a table inside the register or proxy server.

### 4.5.2 SIP Messages

### 4.5.2.1 Message Format

A SIP message is a set of lines ending in CRLF (Carriage Return, Line Feed). These lines are coded using HTTP/1.1 (RFC 2068) syntax and the ISO 10646 character set with UTF-8 (RFC 2279) codification. There are two types of messages, requests sent by clients and responses sent by servers and both share the same format as shown in Figure 4.8.

- First line. This line in a request message will have a different format, as if it were a response message. It is used to differentiate the two kinds of messages.
- Headers. This part of the message is a group of Headers similar to those used in email messages. In fact, some of them have the same meaning. Although there are several headers defined in SIP

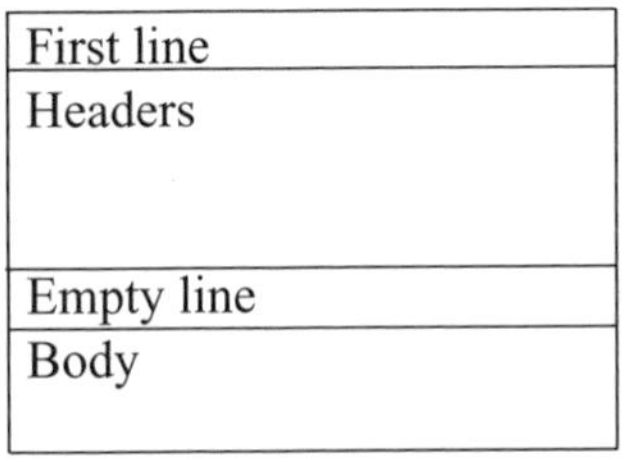

| First line |
|---|
| Headers |
| |
| Empty line |
| Body |
| |

**Figure 4.8**   SIP message format.

RFC, only a few are mandatory: To, From, CSeq, Call-ID, Max-Forwards and Via. These headers are mandatory because they are the minimum set of information needed to transport a message from the source (From) to the destination (To).

- Empty line. The CRLF has a similar function as in HTTP, and it is used by a parser to determine where the end of the headers set is. This is specially useful when the transport is TCP.
- Body. This part of the message is a chunk of data in a format specified by a group of headers as Content-Type, Content-Length, etc. Bodies might be binary or text encoded. The most common body is the one used during the Session Establishment, i.e. the SDP message describing the session.

### 4.5.2.2 Requests and Responses

The first line of a SIP request starts with the name of the method. A method is the function that the caller is requiring from the callee. A list of the available methods can be found in Table 4.4. After the method we can find the destination of the message and, finally, SIP version (SIP/2.0) used by the end point sending the message.

From the wireless networking point of view, the most important methods are REGISTER and REFER, even though the others are necessary to establish and control a session (e.g. there is no session without the INVITE method). These two methods will be very useful as we will see in Section 4.6.

**Table 4.4**   SIP request methods

| Method | Description |
|---|---|
| INVITE | Session setup |
| ACK | Acknowledgment of final response to INVITE |
| BYE | Session termination |
| CANCEL | Pending session cancellation |
| REGISTER | Registration of a user's URL |
| OPTIONS | Query of options and capabilities |
| INFO | Midcall signaling transport |
| PRACK | Provisional response acknowledgment |
| COMET | Preconditions met notification |
| REFER | Transfer user to a URL |
| SUBSCRIBE | Request notification of an event |
| UNSUSCRIBE | Cancel notification of an event |
| NOTIFY | Transport of subscribed event notification |
| MESSAGE | Transport of an instant message body |

Let us see now an example of a complete INVITE message.

```
INVITE sip:pedrom@um.es SIP/2.0
Via: SIP/2.0/UDP istar.dif.um.es;branch=z9hG4bK776asdhds
Max-Forwards: 70
To: Pedro M. Ruiz <sip:pedrom@um.es>
From: Juan A. Sanchez <sip:jlaguna@um.es>;tag=1928301774
Call-ID: a84b4c76e66710@istar.dif.um.es
CSeq: 314159 INVITE
Contact: <sip:jlaguna@istar.dif.um.es>
Content-Type: application/sdp
Content-Length:. ..

v=0
o=origin 5532355 2323532334 IN IP4 128.3.4.5
c=IN IP4 istar.dif.um.es
m=audio 3456 RTP/AVP 0 3 5
```

As we can see, a SIP message contains enough information to allow the components of the SIP architecture to transport it from the source (jlaguna) to the destination (pedrom). In this example, the session that jlaguna is trying to initiate with pedrom uses only audio.

Every SIP message needs some kind of answer or confirmation. These messages are known as responses. The first line of a response starts with a SIP version followed by a response code and a reason phrase. Response codes are classified into six groups as is shown in Table 4.5. Every group has several messages, and all of them share the same meaning as the group but explain the problem more precisely.

**Table 4.5**   SIP response codes

| Class | Description |
| --- | --- |
| 1xx | Provisional or Informational: request is progressing but is not yet complete |
| 2xx | Success: request has been completed successfully |
| 3xx | Redirection: request should be tried at another location |
| 4xx | Client error: request was not completed due to error in request, it can be retried when corrected |
| 5xx | Server error: request was not completed due to error in recipient, it can be retried at another location |
| 6xx | Global failure: request has failed and should not be retried |

### 4.5.3 Addresses

All messages in SIP have a destination and an originator. The way in which SIP specifies these addresses is by the use of elements called SIP and SIPS URIS. A SIP or SIPS URI identifies a communications resource. Like all URIs, SIP and SIPS URIs may be placed in web pages, email messages, or printed literature. They contain sufficient information to initiate and maintain a multimedia session with the resource.

The only difference between the two schemes is that the second one (SIPS) is used when a client wants to establish a secure session. In this case, TLS must be used as a transport system in order to ensure a totally private delivery.

A SIP URI has a form similar to this one:

**sip[s]:user:password@host:port;uri-parameters?headers**

**sips:jlaguna@istar.dif.um.es**

As it can be seen, the first part identifies the scheme (in this case SIP). Following the scheme appears an optional part known as user-info, which includes user, password and the 'at' symbol. User-info is optional, allowing us to send a message to a host without specifying the user. This is the case, for example, when we want to send a message asking for a service and we do not care which user will receive and process our request. The last part is the host providing the SIP resource. It contains either numeric IPv4 or IPv6 address or, when possible, a fully-qualified domain name.

The 'port', 'uri-parameters' and 'headers' are optional fields. They are useful in some situations where it is desirable to concentrate all the information required to form the message into one single resource such as a hyperlink in a web page.

### 4.5.4 Address Resolution

The process through which a SIP client is able to determine where to send a message is called address resolution. RFC 3263 [23] presents the way this process should be accomplished. We have already studied how the address of a peer is described in terms of SIP or SIPS URIs and it is very similar to that used for the well-known email system. So, the process to resolve a SIP URI into the IP address, port, transport and protocol should be very similar. In fact, as RFC 2363 presents, it also uses DNS procedures to find the necessary information.

In SIP the address resolution is used not only to find the destination of a message. It is also used to allow a server to send a response to a backup client if the primary client has failed. Normally, a user agent has configured a SIP proxy to send all the messages. This implies that the user agent does not have to resolve any SIP URI; it is the proxy that will have the responsibility for that. In order to ensure the correct return of the response messages using the same proxies as for the request, it is possible to include a mark in the message identifying every hop the message has made. This list of hops is used to reconstruct the path backwards,which is possible due to the Via header included in every request message.

### 4.5.5 Session Setup

The first of the SIP functionalities related to session management is the session set up function is depicted in Fig. 4.9. SIP uses a three-way handshake protocol to accomplish this task. The rest of the functions do not use three-way handshakes but, as the initiation process may require some time as well as human responses, the use of a third message is introduced as a reliability mechanism. It is

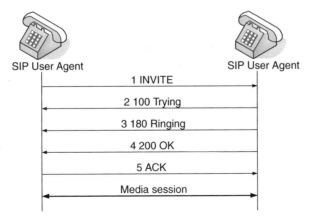

**Figure 4.9** Session set up.

also possible, due the nature of the protocol, that a single attempt of establishment results in different responses from different SIP user agents. In general, only one of these responses should be answered.

When a user agent wants to establish a SIP session, it sends an INVITE request. This message contains all the information needed to describe the type of session in which the user wants to take part. In addition, it also contains the media and ports to be used, codecs supported, and so forth. Normally, this information is described using SDP. Owing to the intrinsic terminals' differences, it is possible that some of the preferences or characteristics of the terminals differ. In order to solve these problems it is recommended that one follows the offer-response model specified in RFC 3264 [24]. This document defines a mechanism by which two entities can make use of the Session Description Protocol (SDP) to reach a common agreement on the settings to use for a particular multimedia session.

The callee receiving the INVITE request answers immediately with a provisional response (100 Trying). This will allow the callee to know that, at least, there is someone at the other side of the line. The second provisional response (180 Ringing) will let the caller know that the user agent contacted is running and trying to alert the callee. Once the user has answered the call, a 200 OK response is sent from the callee user agent to inform the caller that the call is accepted. When the caller receives such a final response the three-way handshake ends by sending an ACK request. This last message does not need response. In fact, it is the only request that does not need it.

Finally, media transmission can start. Both terminals could send and receive multimedia flows to the correct address and ports and use a set of codecs that the correspondent may understand. Although this media interchange is usually made by using UDP, it is important to bear in mind that messages involved in session establishment might have been transported over TCP or even over TLS.

An established SIP session will remain in that state until one of the peers starts the termination or the modification of it. There is no necessity to send any message periodically to maintain the session. The only requirement for the peers is to save the parameters used to establish it: To, From and Call-Id headers.

### 4.5.6 Session Termination and Cancellation

It is very important to differentiate between terminating and cancelling a session. Terminating a session means that the user wants to stop sending and receiving media to and from the peer or peers taking part in the session. Cancellating means the desire to not complete the current session establishment process.

When a SIP user agent wants to terminate a session, it creates a BYE request. This request must include the same headers used to create the session, i.e. the session state variables. We have already said that a user agent must maintain some information related to the session (To, From, Call-ID). This is because SIP uses the concept of Dialog. A dialog represents a peer-to-peer SIP relationship between two user agents that persists for some time. The dialog facilitates sequencing of messages between the user agents and proper routing of requests between both of them. The dialog represents a context that enables the interpretation of SIP messages. A BYE request will be answered with a final response, but even in the absent of response, the user agent must stop sending and receiving multimedia flows.

A cancellation is usually applied when, for example, the user has made a mistake when they introduced the destination of the call. As a result, there is an unfinished wait for response. In that moment, the cancellation might be useful. The user agent sends a CANCEL request before the call setup is completed. CANCEL requests are answered by a user agent with a 200 OK final response. However, one important difference between BYE and CANCEL is that BYE is an end-to-end request, whereas CANCEL is a hop by hop request. That means that a proxy receiving a CANCEL responds with a 200 OK and then starts sending CANCELs to the set of destinations that it included in the initial INVITE.

## 4.6 Advanced SIP Features for Wireless Networks

SIP was not designed for supporting wireless networks. However, thanks to its extensible nature and its features, it has become the 'de facto' standard for session establishment in wireless environments. The most important fact is that even the Third Generation Partnership Project (3GPP) has included the use of SIP for call setup in its Internet Protocol Multimedia Subsystem (IMS).

### 4.6.1 Support of User Mobility

SIP supports user mobility by proxying and redirecting requests to the user's current location. Users can register their current location in the Proxy of the foreign subnet. Furthermore, a user can register more than one location at the same time: its PC's address, its mobile phone address (e.g. phone number) and, for instance, its email. Such features are clearly the most important ones from the point of view of wireless networks and mobility-related features.

One of the characteristics of wireless networks is that the devices can change from one subnet to another. There is nowadays a solution to allow devices to continue working even in these conditions. This solution is called Mobile IP and it allows terminals to roam across networks without disrupting ongoing sessions.

As SIP is normally used to establish multimedia sessions, and this kind of session is usually based on the use of RTP over UDP, using Mobile IP might sometimes degrade the QoS. This degradation is due to the increase in latency. This is not acceptable for delay sensitive traffic such as audio or video flows. There is also an overhead of typically 20 bytes (IP header) added during the encapsulation needed to forward data-packets.

In any case, the use of Mobile IP is necessary to keep the usability of other TCP-based applications such as telnet, ftp, web, etc. When we have Mobile IP and applications based on SIP we must articulate a mechanism by which the application could be able to inform the other peer of the session established before the IP change. In this way, the mobile host could change the address where it is sending its media flows.

### 4.6.2 Personal Mobility

Personal mobility is the ability of end users to originate and receive calls and access subscribed telecommunication services on any terminal in any location, and the ability of the network to identify end users as they move. Personal mobility is based on the use of a unique personal identity called the SIP URI.

When a user agent starts up, the first action it makes is to register the user in the subnet proxy/register server. This register might consist only in binding the user SIP URI to its current IP address, but it might also make several bindings at the same time. For instance, Pedro arrives at the office and connects his laptop and mobile phone. His laptop obtains an IP address through DHCP.

When Pedro starts a softphone that has configured his preferences, the user agent registers not only his current IP address but also the mobile phone number and his e-mail.

Then, when someone wants to contact Pedro, the INVITE reaching the proxy will be first redirected to his softphone, but if this is not up and running the proxy could send the INVITE to his mobile phone. When Pedro leaves the office, his user agent running in the softphone will deregister its address but not the one bound to the mobile phone. So, Pedro will continue being reachable while he is moving to another office where he could repeat the process to have his softphone running with a new IP address.

At this point we need to remember the differences between proxies and redirect servers. Basically, the difference is that, whereas proxies remain as one of the nodes taking part in the message interchange, redirect servers only answer petitions with a redirect message informing of the user current and correct address. Therefore, if we estimate a high number of session establishments, it is better to have a redirect server for minimizing the work load. Sometimes this is called a SIP precall mobility because the user has already moved before the call is made.

### 4.6.3 Session Modification

As will be seen later, session modification is the key to allow SIP Midcall mobility. Yet, first of all, what does session modification mean and what is it used for? Session modification is the ability that a user agent has to change session parameters in the middle of a previously established session. Examples of parameters that could be changed are contact address, addresses and ports to receive media, codecs, etc.

Concerning usability, this in-session modification will be used in several scenarios. For example, as can be read in [25], the adaptivity of the session parameters plays a key role in improving the quality of service in wireless environments. In this paper the authors focus on detecting variations in network conditions based on packet losses and, when this occurs, they correct the problem by applying a new configuration of codecs and/or codec parameters. These modifications are a clear example of session modification.

When a user agent wants to make the kind of change described above, they only need what in SIP is called a re-INVITE. A re-INVITE consists of starting a new session setup within the peers already involved in a session. It is necessary to use the same session state parameters that were used in the first establishment, i.e. the same Call-ID, To and From headers. When a user agent receives a new INVITE request while it is in the middle of a session, it first examines the mentioned headers in order to know whether this INVITE comes from the same partner that it is currently connected to or not. In this is the case, it stops sending media and starts again, using the new parameters.

### 4.6.4 Session Mobility

As a result of combining Session Modification with Mobile IP it is possible to support network changes during a session that is already started. A way to do this is described in [26]. Session Mobility ensures that sessions are not disrupted when the user's terminal changes the point of attachment to the network or whenever a session is transferred from one terminal to another, or even from one user to another.

Figure 4.10 depicts an example where user <pedrom@dif.um.es> has established a session with user <jlaguna@dif.um.es> and the concept of Session Mobility will be necessary. This establishment has been made by trying to contact the user with a previous INVITE. The redirect server answers pedrom's user agent with a 302 redirect response informing of the real address of jlaguna. With this information, pedrom's user agent sends a new invite and then the session starts interchanging audio and video flows over RTP. During the session jlaguna starts downloading a large file. Once the downloading is in progress, he moves physically from his desktop to another room two floors higher, where the sales department is located. This movement originates a change in his IP because now he is out of the wireless coverage of his own department. The sales department access point gives him a new IP address (Mobile IP is working) transparently, so the download of a large file continues almost without any interruption. But what about the videoconference that he was having with Pedro?

It is not efficient to let Mobile IP do the tunnelling with RTP flows. Therefore, the best option is to do a Session Modification informing Pedro's user agent that there has been a change in session parameters. More concretely, we have to send a re-INVITE with our new IP address so the multimedia flows can be sent directly without any kind of encapsulation.

In the next section we show how SIP and the other control protocols are used in the framework of future IP-based UMTS networks.

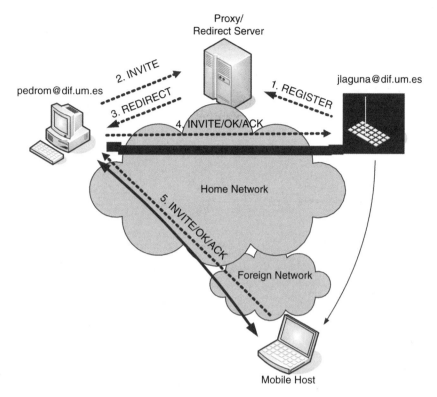

**Figure 4.10**   Example of session mobility.

## 4.7  Multimedia Control Plane in UMTS: IMS

One of the most relevant examples of the advantages of IETF's multimedia control protocol suite (and SIP in particular) is the fact that they have been selected as the basic building block for the control of the IP Multimedia Subsystem (IMS [27, 28]) of UMTS. During this section we shall explain the general architecture of the IMS, and we will give examples of the use of these protocols within UMTS.

### 4.7.1 IMS Architecture

The initial releases of UMTS had two different subnetworks: the packet switched plane, and the circuit switched plane. The former was similar to the architecture provided by the GPRS service, whereas the latter was used for the traditional circuit-switched phone calls. However, since the Release 5 of UMTS, the system has evolved into an 'All-IP' network in which all the services use a single converged packet-based network. The main reasons for this are the lower infrastructure cost and the lower maintenance cost, as well as the possibility of offering a wider range of services such as the combination of voice and data services. In such a kind of packet based network, the natural step in providing the traditional call services was the use of existing VoIP protocols. In fact, the 3GPP has chosen SIP as the basic call control protocol.

The 3GPP created a new core network subsystem to deal with all the issues related to the provision of multimedia services in UMTS. It is often referred to as IMS and stands for IP Multimedia Subsystem. As Figure 4.11 shows, the IMS consists of the following elements:

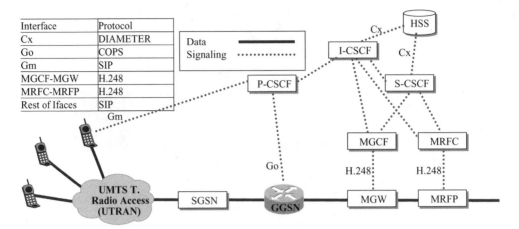

**Figure 4.11** Overview of IMS architecture and protocols.

- Call Session Control Functions (CSCF);
- Home Subscriber Server (HSS);
- Media Gateways (MGW);
- Media Gateway Control Function (MGCF);
- Multimedia Resource Function (MRF).

As we see in the figure, the signalling protocols being used are all defined by the IETF, except H.248 which is standardized both by the IETF (under the name MEGACO) and ITU-T (under the name H.248). DIAMETER [29] is used to provide authentication, authorization and accounting (AAA). COPS [30] is used to establish and configure policies at the SGSN regarding the traffic profiles that are allowed to individual users based on their contracts (which are stored in the HSS). So, as we can see here, in addition to defining the multimedia session establishment and control, IMS also defines how the session control is integrated with AAA and QoS functions by using the COPS and DIAMETER protocols. These protocols are not directly related to the multimedia control plane and are outside the scope of this chapter. Interested readers can find additional information in [29, 30, 31].

The CSCF is functionally split into different entities communicating among them using SIP. They are signaling proxies (i.e. SIP proxies) in charge of the establishment, modification and tear-down of multimedia sessions, with guaranteed QoS and AAA. As we just mentioned, a CSCF can play three different roles within the IMS, although several of these functional entities may be implemented in a single hardware equipment.

(1) Proxy CSCF (P-CSCF). It acts on behalf of the terminal equipment. By analogy with IETF SIP architecture, this will be equivalent to a SIP proxy.
(2) Interrogating CSCF (I-CSCF). It is basically a SIP proxy which is used as the entrance point at the IMS domain for signalling messages coming from other operators' CSCFs. They are used to hide the real internal topology of the local IMS domain to other operators.
(3) Serving CSCF(S-CSCF). They are the ones in charge of performing the user registration and the session control. Usually, when an I-CSCF receives a request it forwards it to one of the internal S-CSCF which is the one really attending that request.

The HSS is a database containing the user profiles. It is mainly used to perform user authentication, authorization and accounting (AAA). It is similar to the HLR database in GSM networks. The records in this database clearly define which services and associated parameters a user can request based on his profile. When I-CSCF or S-CSCF processes a multimedia session establishment or modification, they use the DIAMETER protocol to check if the user is allowed to use the requested settings for that session (e.g. bandwidth, QoS guarantees, etc.).

When sessions are established among IMS terminals, these are the only elements required. However, for backward compatibility, the 3GPP defined additional entities to provide interworking between IMS devices and other networks such as the existing fixed telephone infrastructures and GSM networks. The entities in charge of providing such interworking functions are commonly known as 'media gateways'. These functions are usually split between a control part and a processor part. For instance, the media gateway is divided into the MGCF (Media Gateway Control Function), which receives SIP commands from the I-CSCF or S-CSCF and sends control commands to the MGW (Media Gateway) using H.248/ MEGACO [32]. So, the MGCF is a control plane entity whereas the MGW is in the data plane. The media resource function is also split into the MRFC (Media Resource Function Controller) and the MRFP (Media Resource Function Processor). The former is in charge of controlling the processing of the media streams, whereas the latter does the real process of the stream. To sum up, these four entities are in charge of providing interworking functions with circuit-switched and other existing networks not supporting IMS, and they can be implemented either as separate equipment or into the same hardware device.

In the next subsection we give some examples of operation to help the reader understand the analogies between the IMS control plane and the control protocols explained in previous sections.

### 4.7.2 Session Establishment in IMS

Before an UMTS UE can establish a SIP session it needs to get access to the network. For an UMTS terminal to be able to send IP packets, it has first to establish a data connection or 'pipe' to the GGSN after attaching to a SGSN. This process is commonly known as the establishment of a PDP Context. In response to that PDP Context request, the UE receives the address of the P-CSCF that it has to use. Once a terminal has established such a context, an IP address is configured and it is able to send IP packets towards the GGSN that will act as its default gateway.

In the next examples, we assume that the terminal already has a PDP Context for SIP signalling activated, and it already knows its P-CSCF from the final response in the PDP activation. We illustrate below how a terminal registers into the IMS, as well as the process of establishing a session.

#### 4.7.2.1 Registration into the IMS

As we show in Section 4.5, before a SIP user agent can be contacted, its SIP proxy server must know about its availability and how to contact it. Something similar happens in IMS. Although the terminal is already registered into the network (and has an active PDP context), it has to register into the IMS before it can receive or perform SIP calls. This process is commonly known as registration.

Unlike the standard SIP registration in IP networks, in UMTS the registration is also used to perform AAA functions on the subscriber and to send the subscriber profile to its S-CSCF so that it can verify that the user's requests fulfil his contract. In Figure 4.12 we give an example of the message flow when a UE located in a visited network registers with the IMS of its operator. Of course, the P-CSCF in the figure refers to the proxy CSCF in the visited network whereas the I-CSCF and S-CSCF refer to the CSCF entities in the network of its operator. In the case of registration within a UE with its own operator the process is similar, but the P-CSCF in that case will also be in the local network.

As the figure depicts, once the UE has established its PDP Context for SIP signaling, it sends a SIP Register message to its P-CSCF, similar to that which the IETF specified for IP networks. The P-CSCF uses the domain name in SIP URI of the UE to determine through DNS the I-CSCF of its operator. Once that DNS resolution is finished, the P-CSCF forwards the SIP Register to the I-CSCF. Upon receiving the message, the I-CSCF sends a Cx Query message to the HSS to verify that the user is allowed to register in the visited network. The HSS answers with a Cx Response message. If the user

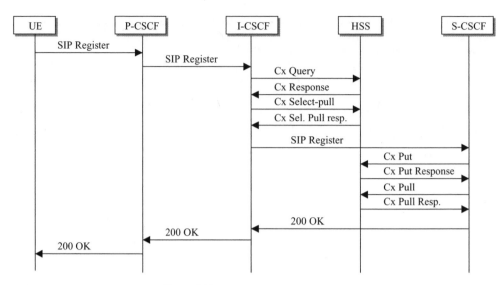

**Figure 4.12**   Registration process in IMS.

is allowed to register, then the I-CSCF sends a Cx Select-pull message to the HSS to ask for the required S-CSCF capabilities. The HSS sends the capabilities to the I-CSCF. These capabilities allow the I-CSCF to select a proper S-CSCF for this user as well as a contact point for the visited network. Then, the I-CSCF sends the SIP Register to the selected S-CSCF, including the address of the P-CSCF and the contact name in the visited network. Before accepting the registration, the S-CSCF contacts the HSS with a Cx Put message to inform it that it is the selected S-CSCF for this user. After receiving the acknowledgment from the HSS then the S-CSCF sends a Cx Pull message to retrieve the profile information about this user, so that in the future it can verify which services the user can request according to its profile. Once the profile is received in the Cx Pull Response message from the HSS, the S-CSCF sends the SIP (200 OK) answer back to the I-CSCF which, in turn, sends it to the P-CSCF which ultimately delivers it to the UE.

As you can tell, this is exactly like the SIP registration process with some additional authentication messages in between. Below, we illustrate the process of establishing a session from the UE.

### 4.7.2.2 Originating a Call Setup from a Visited Network

To make the example simple, we illustrate the establishment of a voice call. However, the general process is also valid for a multimedia session including video and possibly other media. Figure 4.13 illustrates a session establishment between a mobile device in a visited network and a destination which is in the home network of the caller. If the destination is in another network, the only difference would be that the S-CSCF in the home network would contact the I-CSCF of that network rather than directly contacting the UE. If the destination is a traditional phone number, then the S-CSCF in the home network would have contacted the MGCF instead. So, in the example, we assume that the P-CSCF is in the visited network, and both the I-CSCF and the S-CSCF are in the home network.

As Figure 4.13 shows, we again have a basic SIP session setup, to which some additional messages have been added for the particular case of a UMTS network. In particular we refer to the Resource Reservation messages. One additional difference is that a P-CSCF is allowed to remove media descriptions from the SDP description of the INVITE message. Thus, the operator can control which particular media formats and data rates it is going to permit. The same applies to the S-CSCF regarding the SDP response which is sent out by the callee.

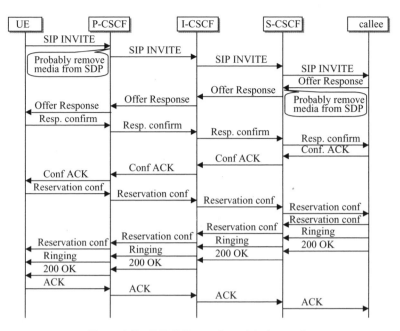

**Figure 4.13**  IMS Call setup from visited network.

The process is very simple, the caller sends the SIP invite message including the SDP description of the media to support as well as the SIP URI of the callee in the 'To:' field. The P-CSCF remembers from the registration process (described above) the I-CSCF in the home network of the caller to which the invite message needs to be propagated. In addition, before propagating the message, it can remove from the SDP description those media that the visited network is not willing to support.

Once the I-CSCF in the home network receives the SIP INVITE, it sends it to the corresponding S-CSCF that was assigned to that caller during the registration process. The S-CSCF is responsible for validating that the caller is allowed to perform that operation and checks that the media included in the SDP are authorized in his profile.

The S-CSCF also processes the 'To:' header to determine the destination of the call. In this case, the destination is in the same home network, thus the INVITE message can be sent directly to the callee. If the destination is in another IMS domain, the S-CSCF would have used DNS to find the next I-CSCF of the other operator to forward the SIP INVITE.

After receiving the SIP INVITE, the called party responds with a provisional response that includes an SDP, indicating which of the proposed media it accepts. Again, the S-CSCF upon receiving that message can remove undesired media descriptions which the home network is not willing to support. This (probably modified) provisional response is sent back to the caller through the I-CSCF and the P-CSCF.

Upon receipt of this provisional response, the caller decides the final media to be used and sends a Response Confirmation to the destination indicating the final set of media to use. This message will be propagated again through the same CSCF entities in between. When the caller acknowledges this message, the intermediate CSCF will start reserving the required resources by issuing COPS messages to establish the reservation in the relevant GGSNs. After reservation is completed, a Reservation Confirmation message is sent in both directions. When the caller receives that confirmation, it knows that the resources in the network are reserved so that it can finally accept the call (from the SIP point of view).

The rest of messages are just the standard SIP session establishment explained before without any difference from the IETF specification. The caller just sends a provisional response (Ringing), followed by a SIP 200 OK message indicating that the session is accepted. When the caller receives this message, it acknowledges the session acceptance (ACK) and, at that point in time, it can start sending media.

The establishment of a session towards a traditional phone is similar except for the fact that the S-CSCF will forward the SIP INVITE to a media gateway control (MGCF) rather than to the destination mobile phone. The MGCF takes care of translating from SIP signaling to SS7 or whatever signaling mechanisms are required.

IMS specification is still work in progress. For readers interested in the details of the latest changes we recommend [33]. In the next subsection, we describe how streaming services are provided within UMTS networks.

### 4.7.3 Streaming Services in UMTS

The 3rd Generation Partnership Project (3GPP) offers technical specifications that detail the use of RTSP for the implementation of an end-to-end packet-switched streaming service (PSS). One of these, 3GPP TR 26.234 [34], contains careful descriptions of extensions of RTSP and SDP to help in the adaptation of continuous media transmission in wireless environments. The object of this specification is to define control mechanisms to achieve the maximum quality of experience, minimizing the impact of interrupts due to transmission problems in wireless links. The strategy followed by the specification is to estimate network resources and adapt server transmission rates to available link rates, preventing underflow and overflow problems in the receiver buffer. The extensions of RTSP and SDP define link and buffer client feedback procedures in order to help the server to decide at which rate corrections are needed.

#### 4.7.3.1 SDP for Streaming Sessions in 3GPP

When an SDP document is generated, the specification indicates the use of a set of SDP fields (some of these are SDP standard fields, while others are defined in the TR 26.234 document) to help in the implementation of the rate control mechanism. A SDP field that should be used in SDP specifications is the bandwidth field 'b=' at the media level. The field can specify the bandwidth for RTP traffic ('b = AS:' in kbps), RTCP Sender Reports ('b = RS:' in bps) or RTCP Receiver Reports ('b = RR:' in bps) [35]. Another important media level field is 'a = 3GPP-Adaptation-Support:freq', which indicates that the server supports bit-rate adaptation for a media stream. The value of freq specifies to an adaptation aware client that it will include an adaptation report in every freq RTCP RR compound packet. The adaptation report must comply with the Extended RTP Profile for RTCP-based Feedback (RTP/AVPF) [36], and includes the current playout delay and free space of the client buffer. The following SDP example shows the use of these media level fields.

```
m=audio 0 RTP/AVP 97
b=AS:13
b=RR:350
b=RS:300
a=rtpmap:97 AMR/8000
a=fmtp:97 octet-align=1
a=control: streamID=0
a=3GPP-Adaptation-Support:2
```

#### 4.7.3.2 RTSP 3GPP Headers

The TR 26.234 specification defines the 3GPP-Link-Char header that should be included in a SETUP or PLAY request by the client to give the initial values for the link characteristics. The client can update the value of this header with SET_PARAMETER requests that affect a playing session. The header value contains three parameters with information about the client wireless link:

(1) GBW: the link's guaranteed bit-rate in kbps;
(2) MBW: the link's maximum bit-rate in kbps;
(3) MTD: the link's maximum transfer delay in milliseconds.

The next example shows the 3GPP-Link-Char header that a client uses to tell the server that all streams aggregated by the indicated session-level URL will have the wireless link QoS settings specified by the GBW, MBW and MTD parameters:

```
3GPP-Link-Char: url=''rtsp://rtsp.um.es/movie.3gp'';
                GBW=32;MBW=128;MTD=2000
```

Another adaptation header defined in the TR 26.234 specification is 3GPP-Adaptation. It is used by a RTSP client in a SETUP or SET_PARAMETER request to inform the server about the size of the reception and de-jittering buffer, and the desired amount of playback time in milliseconds to guarantee interrupt-free playback (target time). The next example shows how a client tells the server the buffer size and target time of a concrete stream.

```
3GPP-Adaptation: url=''rtsp://rtsp.um.es/movie.3gp/streamID=0'';
                 size=14500;target-time=5000
```

### 4.7.3.3 3GPP Quality of Experience (QoE)

The TR 26.234 specification provides a combined extension of RTSP and SDP to allow the optional gathering and report of client QoE metrics. The extension allows the negotiation of which metrics will be measured and the period of the reports sent from the client to the server. One of the proposed metrics is Corruption_Duration, which indicates the time period from the NPT time (Section 4.4.3.4) of the last good frame before the corruption (lost of part or the entire frame) to the NPT time of the first subsequent good frame. Rebuffering_Duration is the duration of a stall period due to any involuntary event at the client side. Initial_Buffering_Duration is the time from the reception of the first RTP packet until playing starts. Successive_Loss indicates the number of RTP packets lost in succession per media channel. Framerate_Deviation is the difference between the pre-defined frame rate and the actual playback frame rate. The last metric is Jitter_Duration: time duration of the playback jitter. Taking into account the fact that new measures could be defined in the future, any unknown metrics will be ignored. QoE metrics negotiation starts with the SDP description that the client receives in a DESCRIBE response or retrieves using any mechanism external to RTSP. The SDP description can include 'a = 3GPP-QoE-Metrics' fields at session or media level, stating the metrics and the rate of client reports suggested by the server. The following example shows the use of this field (some lines are formatted):

```
v=0
...
a=control:*
a=3GPP-QoE-Metrics:{Initial_Buffering_Duration,
                    Rebuffering_Duration};rate=End
...
m=video 0 RTP/AVP 96
a=control:trackID=3
a=3GPP-QoE-Metrics:{Decoded_Bytes};rate=15;range:npt=0-40
```

```
...
m=audio 0 RTP/AVP 98
a=control:trackID=5
a=3GPP-QoE-Metrics:{Corruption_Duration};rate=20
```

The session level QoE field indicates that the initial buffering and the rebuffering duration should be monitored and reported once at the end of the session. The video specific metrics (decoded bytes) will be reported every 15 seconds until 40 seconds of NPT time. Finally, audio specific metrics (corruption duration) will be reported every 20 seconds during all the session.

A QoE aware client that receives a SDP description with QoE metrics fields may continue the negotiation with a SETUP request that includes a 3GPP-QoE-Metrics. This header allows the client to propose QoE metrics modifications. The value of the header can contain session and media metrics separated with the session-level and media-level URLs:

```
3GPP-QoE-Metrics: url=''rtsp://rtsp.um.es/movie.3gp/trackID=3'';
                  metrics={Decoded_Bytes};rate=10;range:npt=0-40,
                  url=''rtsp://rtps.um.es/movie.3gp'';
                  metrics={Rebuffering_Duration};rate=End
```

The server can accept the modifications of the client, echoing them in the SETUP response, or the server can deny modifications, continuing the re-negotiation until the PLAY request. QoE metrics reports can be disabled by the server using SET_PARAMETER requests with the 3GPP-QoE-Metrics header containing 'Off'. To send QoE metrics feedback, the client will issue SET_PARAMETER requests with the 3GPP-QoE-Feedback header:

```
3GPP-QoE-Feedback: url=''rtsp://rtsp.um.es/movie.3gp/trackID=5'';
                   Corruption_Duration={ 200 1300}
```

## 4.8 Research Challenges and Opportunities

As the user might have noticed, the work in control protocols for multimedia communications is far from complete. In particular, during the last few years with the advent of wireless and mobile networks, a lot of new modifications and enhancements are being engineered to fulfill the wide range of new requirements that these networks are bringing up.

In addition, future 4G wireless networks consisting of IP core networks to which different wireless access technologies will be interconnected are posing even stronger requirements, given the heterogeneous nature of those networks. For instance, it is expected that the terminal capabilities might be completely different among devices. New concepts like session roaming, session transfers, service discovery and many others require new control-plane functions, which in the majority of the cases are not fully available.

As a terminal roams across different access technologies in 4G networks its network connectivity might vary strongly. This kind of heterogeneous and variable scenarios is posing additional requirements on multimedia internetworking technologies. For instance, applications should be able to adapt their operating settings to the changes in the underlying network. All these changing events, which are not currently notified by existing control protocols, will need to be considered by (in many cases) extending existing control protocols or even designing new ones.

Moreover, location-aware and user-aware services are expected to be delivered in those networks. This means that signaling and control protocols will require extensions in order to be able to convey contextual information to multimedia applications. The paradigm shifts from the concept of establishing a session to the concept of establishing a session automatically configuring the session parameters according to the user's preferences, location, contextual information, network capabilities, etc.

These new requirements are opening up a number of research opportunities and areas which include among others:

- context-aware and personalized applications and services;
- adaptive applications and services that can self-configure;
- middleware architectures for context-aware applications;
- enhanced highly-descriptive capability negotiation mechanisms;
- semantic technologies to abstract and model contextual information.

In conclusion, we have explained how existing multimedia control protocols work, and why IETF proposals have been considered as the 'de facto' standard for existing and future wireless and mobile networks. We have given a detailed description of the main protocols (SDP, RTSP and SIP). Finally, we have described the multimedia control plane of UMTS (IMS), giving examples that allow the reader to understand the basic principles and operations.

For those readers interested in obtaining detailed specifications, a great deal of relevant literature is cited below.

## Acknowledgment

The work of Pedro M. Ruiz was partially funded by the Spanish Ministry of Science and Technology by means of the Ramón and Cajal work programme.

## References

[1] ITU-T Rec. H.320, Narrow-band Visual Telephone Systems and Terminal Equipment, 1990.

[2] ITU-T Rec. H.323, Visual Telephone Systems and Terminal Equipment for Local Area Networks which Provide a Non-Guaranteed Quality of Service, November, 1996.

[3] ITU-T Rec. T.120, Data Protocols for Multimedia Conferencing, July 1996.

[4] H. Schulzrinne, S. Casner, R. Frederick and V. Jackobson, IETF Request For Comments, RFC-3550: RTP: A Transport Protocol for Real-Time Applications, July 2003.

[5] ITU-T Rec. H.235, Security and encryption of H-Series (H.323 and other H.245-based) multimedia terminals, February, 1998.

[6] ITU-T Rec. H.225.0, Call Signaling Protocols and Media Stream Packetization for Packet-based Multimedia Communication Systems, February, 1998.

[7] ITU-T Rec. H.245, Control Protocol for Multimedia Communication, September, 1998.

[8] M. R. Macedonia and D. P. Brutzman, MBone provides audio and video across the Internet, *IEEE Computer*, **27**(4), 30–36, April 1994.

[9] M. Handley and V. Jacobson, IETF Request For Comments, RFC-2327: SDP: Session Description Protocol, April, 1998.

[10] N. Freed and N. Borenstein, IETF Request For Comments, RFC-2045: Multipurpose Internet Mail Extensions (MIME) Part One: Format of Internet Message Bodies, November, 1996.

[11] M. Handley, C. Perkins, E. Whelan, IETF Request For Comments, RFC-2974: Session Announcement Protocol, October, 2000.

[12] J. Rosenberg, H. Schulzrinne, G. Camarillo, A. Hohnston, J. Peterson, R. Sparks, M. Handley and E. Schooler, IETF Request For Comments RFC-3261, SIP Session Initiation Protocol, June, 2002.

[13] H. Schulzrinne, A. Rao, R. Kanphier, M. Westerlund and A. Narasimhan, IETF Request for Comments RFC-2326: Real Time Streaming Protocol (RTSP), April, 1998.

[14] D. Mills, IETF Request for Comments RFC-1305, Network Time Protocol (version 3) specification and implementation, March 1992.

[15] C. Huitema, Request For Comments RFC-3605, Real Time Control Protocol (RTCP) attribute in Session Description Protocol (SDP), October, 2003.

[16] S. Olson, G. Camarillo and A. B. Roach, IETF Request For Comments RFC-3266: Spport for IPv6 in Session Description Protocol (SDP), June 2002.

[17] G. Camarillo, G. Eriksson, J. Holler, H. Schulzrine, IETF Request For Comments RFC-3388: Grouping of Media Lines in the Session Description Protocol (SDP), December, 2002.

[18] IETF Multiparty MUltimedia SessIon Control (MMUSIC) Working Group. http://www. ietf.org/html.charters/mmusic-charter.html.

[19] IETF RFC 2616, Hypertext Transfer Protocol, R. Fielding, J. Gettys, J. Mogul, H. Frystyk, L. Masinter, P. Leach and T. Berners-Lee, June 1999.

[20] IETF RFC 3016, RTP Payload Format for MPEG-4 Audio/Visual Streams, Y. Kikuchi, T. Nomura, S. Fukunaga, Y. Matsui and H. Kimata. November 2000.

[21] IETF RFC 3640, RTP Payload Format for Transport of MPEG-4 Elementary Streams, J. van der Meer, D. Mackie, V. Swaminathan, D. Singer and P. Gentric, November 2003.

[22] Sue B. Moon, Jim Kurose and Don Towsley, Packet audio playout delay adjustment: performance bounds and algorithms. In *Multimedia Systems*, **6**, 1998, 17–28, Springer-Verlag.

[23] J. Rosenberg and H. Schulzrinne, IETF RFC 3263, Session Initiation Protocol (SIP): Locating SIP Servers, June 2002.

[24] J. Rosenberg and H. Schulzrinne, IETF RFC 3264, An Offer/Answer Model with the Session Description Protocol (SDP), June 2002.

[25] Pedro M. Ruiz, Antonio F. Gómez-Skarmeta, Pedro Martínez, Juan A. Sánchez and Emilio García, Effective multimedia and multi-party communications on multicast MANET extensions to IP access networks, In *Proc. 16th IEEE International Conference on Information Networking ICOIN-2003*, Jeju Island, Korea, pp 870–879, February 2003.

[26] E. Wedlund and H. Schulzrinne, Mobility support using SIP, In *Proc. of 2nd ACM International Workshop on Wireless Mobile Multimedia*, Seattle, WA, August 1999.

[27] 3GPP TS 23.002, Network Architecture, v6.2.0, September 2003.

[28] 3GPP TS 23.228, IP Multimedia Subsystem (IMS) v6.5.0, March 2004.

[29] P. Calhoun, L. Loughney, M. E. Guttman, G. Zorn and V. Jacobsen, Diameter Base Protocol, IETF-RFC 3588, September 2003.

[30] D. Durham, J. Boyle, R. Cohen, S. Herzog, R. Rajon and A. Sastry, The COPS (Common Open Policy Service) Protocol, IETF-RFC 2748, January 2000.

[31] J. Loughney, Diameter Command Codes for Thrid Generation Partnership Project (3GPP) Release 5, IETF-RFC 3589, September 2003.

[32] ITU-T, Technical Recommendation H.248.1, Media Gateway Control Protocol, May 2002.

[33] 3GPP TS 24.228, IP Multimedia Subsystem (IMS) Stage 3.

[34] 3GPP TR26.234, Technical Specification Group Services and Aspects; Transparent end-to-end PSS; Protocols and codecs (Rel 6.1.0, 09-2004).

[35] IETF RFC 3556, Session Description Protocol (SDP) Bandwidth Modifiers for RTP Control Protocol (RTCP) Bandwidth, S. Casner, July 2003.

[36] IETF Internet Draft draft-ietf-avt-rtcp-feedback-11.txt, Extended RTP Profile for RTCP-based Feedback (RTP/AVPF), Joerg Ott, Stephan Wenger, Noriyuki Sato, Carsten Burmeister, José Rey, expires February 2005.

# 5

# Multimedia Wireless Local Area Networks

Sai Shankar N

## 5.1 Introduction

Wireless networking has made a significant impact with the invention of wide area cellular networks based on different standards, e.g. the Global System for Mobile communications (GSM), Advanced Mobile Phone System (AMPS), etc. They have been defined with the main purpose of supporting voice, though some also offer data communication services at very low speed (10 kbit/s). With the invention of the WLAN, data communication services in a residential and office environments have undergone significant changes. WLAN products based on the different flavors of 802.11 are available from a range of vendors. Depending on the transmission scheme, products may offer bandwidths ranging from about 1 Mbit/s up to 54 Mbit/s. There is a significant interest in transmission of multimedia over wireless networks. These range from the low rate video transmissions for the mobile phones to the high rate Audio/Video (AV) streaming from an Digital Video Disk (DVD) player to a flat panel television inside the home. Typically, supporting the AV applications over the networks requires Quality of Service (QoS) supports such as bounded packet delivery latency and guaranteed throughput. While the QoS support in any network can be a challenging task, supporting QoS in wireless networks is even more challenging due to the limited bandwidth compared with the wired counterpart and error-prone wireless channel conditions. Thanks to the emerging broadband WLAN technologies, it is becoming possible to support the QoS in the indoor environment. In this chapter, we introduce and review two distinct broadband WLAN standards, namely, IEEE 802.11e and ETSI BRAN HiperLAN/2, especially, in the context of QoS support for the AV applications.

### 5.1.1 ETSI's HiperLAN

HiperLAN/1 is a standard for a WLAN defined by the ETSI [1]. HiperLAN supports the ad hoc topology along with the multihop routing capability to forward packets from a source to a destination that cannot communicate directly [1]. The HiperLAN/1 MAC protocol explicitly supports a quality of service (QoS) for packet delivery that is provided via two mechanisms: the user priority and the frame lifetime.

*Emerging Wireless Multimedia: Services and Technologies*   Edited by A. Salkintzis and N. Passas
© 2005 John Wiley & Sons, Ltd

HiperLAN/2 is a European 5 GHz WLAN standard developed within ETSI BRAN. The standardization effort started in Spring 1997, and addressed specifications on both the physical (PHY) layer, the data link control (DLC) layer and different convergence layers as interfaces to various higher layers including the Ethernet, IEEE 1394, and Asynchronous Transfer Mode (ATM). HiperLAN/2 was designed to give wireless access to the Internet and future multimedia at speeds of up to 54 Mbit/s for residential and corporate users.

The Mobile Terminals (MT) communicate with the Access Points (AP) as well as directly with each other to transfer information. An MT communicates with only one AP to which it is associated. The APs ensure that the radio network is automatically configured by using dynamic frequency selection, thus removing the need for manual frequency planning.

HiperLAN/2 has a very high transmission rate of 54 Mbit/s. It uses Orthogonal Frequency Digital Multiplexing (OFDM) to transmit the analog signals. OFDM is very efficient in time-dispersive environments, where the transmitted radio signals are reflected from many points, leading to different propagation times before they eventually reach the receiver. Above the physical layer, the Medium Access Control (MAC) layer implements a dynamic time-division duplex (TDD) for most efficient utilization of radio resources.

The following are the essential features of HiperLAN/2.

- Connection-oriented. In a HiperLAN/2 network, data transmission is connection oriented. In order to accomplish this, the MT and the AP must establish a connection prior to the transmission using signalling functions of the HiperLAN/2 control plane. Connections are Time Division Multiplexed (TDM) over the air interface. There are two types of connections: point-to-point and point-to-multipoint. Point-to-point connections are bidirectional whereas point-to-multipoint is unidirectional and is in the direction towards the MT. In addition, there is also a dedicated broadcast channel through which traffic reaches all terminals transmitted from AP.
- Quality-of-Service (QoS) support. The connection-oriented nature of HiperLAN/2 makes it straightforward to implement support for QoS. Each connection can be assigned a specific QoS, for instance in terms of bandwidth, delay, jitter, bit error rate, etc. It is also possible to use a more simplistic approach, where each connection can be assigned a priority level relative to other connections. This QoS support in combination with the high transmission rate facilitates the simultaneous transmission of many different types of data streams.
- Dynamic Frequency Selection. In a HiperLAN/2 network, there is no need for manual frequency planning as in cellular networks like GSM. The APs in the HiperLAN/2, have a built-in support for automatically selecting an appropriate radio channel for transmission within each AP's coverage area. An AP scans all the channels to determine if there are neighboring APs and chooses an appropriate channel that minimizes interference.
- Security support. The HiperLAN/2 network has support for both authentication and encryption. With authentication, both the AP and the MT can authenticate each other to ensure authorized access to the network. Authentication relies on a supporting function, such as a directory service, that is not in the scope of HiperLAN/2. The user traffic is encrypted on established connections to prevent eaves-dropping.
- Mobility support. The MT tries to associate with the AP that has the best radio signal. When the MT moves, it may detect that there is an alternative AP with better radio transmission performance than the associated AP. The MT then initiates a hand over to this AP. All established connections from this MT will be moved to this new AP. During handover, some packet loss may occur. If an MT moves out of radio coverage for a certain time, the MT may loose its association to the HiperLAN/2 network resulting in the release of all connections.
- Last Mile Access. The HiperLAN/2 protocol stack has a flexible architecture for easy adaptation and integration with a variety of fixed networks. A HiperLAN/2 network can, for instance, be used as the last hop wireless segment of a switched Ethernet, but it may also be used in other configurations, e.g. as an access network to third generation cellular networks.

- Power save. In HiperLAN/2, the MT may request the AP for entering into sleep mode. The MT sends a request to the AP about its intention to enter a low power state for a specific period. At the expiration of the negotiated sleep period, the MT searches for the presence of any wake up indication from the AP. In the absence of the wake up indication the MT reverts back to its low power state for the next sleep period. An AP will defer any pending data to the MT until the corresponding sleep period expires. Different sleep periods are supported to allow for either short latency requirement or low power requirement.

### 5.1.2 IEEE 802.11

In recent years, IEEE 802.11 WLAN [5] has emerged as a prevailing technology for the indoor broadband wireless access for the mobile/portable devices. Today, IEEE 802.11 can be considered as a wireless version of Ethernet by virtue of supporting a best-effort service (not guaranteeing any service level to users/applications). IEEE 802.11b is an extension to the original 802.11 to support up to 11 Mbps at 2.4 GHz, is the most popular WLAN technology in the market. The other extensions, called IEEE 802.11a and IEEE 802.11g, support up to 54 Mbps at 5 GHz and 2.4 Ghz respectively. IEEE 802.11 today is known as the wireless Ethernet, and is becoming very popular to replace and/or complement the popular Ethernet in many environments including corporate, public, and home. The mandatory part of the original 802.11-99 MAC is called the Distributed Coordination Function (DCF), which is based on Carrier Sense Multiple Access with Collision Avoidance (CSMA/CA).

However, as with Ethernet, the current 802.11 is not suitable to support QoS. Since early 2000, the IEEE 802.11 Working Group (WG) has been working on another extension to enhance the MAC to support QoS; the extension is to be called IEEE 802.11e. The overview of 802.11e in this paper is based on the draft specification [7]. The new standard is scheduled to be finalized by the end of 2004. The new MAC protocol of the upcoming 802.11e is called the Hybrid Coordination Function (HCF). The HCF is called 'hybrid' as it combines a contention channel access mechanism, referred to as Enhanced Distributed Channel Access (EDCA), and a polling-based channel access mechanism, referred to as HCF Controlled Channel Access (HCCA), each of which operates simultaneously and continuously within the Basic Service Set (BSS)[a]. This is different from the legacy 802.11-1999 standard [5], which specifies two coordination functions, one mandatory, the Distributed Coordination Function (DCF) and one optional, the Point Coordination Function (PCF). These two operate disjointedly during alternating subsets of the beacon interval. All of today's products in the market only implement the mandatory DCF, which is based on Carrier Sense Multiple Access with Collision Avoidance (CSMA/CA).

The two access mechanisms of HCF provide two distinct levels of QoS, namely, prioritized QoS and parameterized QoS. EDCA is an enhanced version of the legacy DCF MAC and it is used to provide the prioritized QoS service. With EDCA a single MAC can have multiple queues that work independently, in parallel, for different priorities. Frames with different priorities are transmitted using different CSMA/CA contention parameters. With the EDCA, a station cannot transmit a frame that extends beyond the EDCA Transmission Opportunity (TXOP) limit. A TXOP is defined as a period of time during which the STA can send multiple frames.

HCCA is used to provide a parameterized QoS service. With HCCA, there is a negotiation of QoS requirements between a Station (STA) and the Hybrid Coordinator (HC). Once a stream for an STA is established, the HC allocates TXOPs via polling to the STA, in order to guarantee its QoS requirements. The HC enjoys free access to the medium during both the Contention Free Period (CFP) and the Contention Period (CP)[b], in order to (1) send polls to allocate TXOPs and (2) send downlink parameterized traffic. HCCA guarantees that the QoS requirements are met once a stream has been

---

[a] A BSS is composed of an Access Point (AP) and multiple stations (STA) associated with the AP.
[b] During the CP, the HC uses the highest EDCA priority and its access to the medium is guaranteed once it becomes idle.

admitted into the network, while EDCA only provides a QoS priority differentiation via a random distributed access mechanism.

## 5.2 Overview of Physical Layers of HiperLAN/2 and IEEE 802.11a

The transmission format on the physical layer consists of a preamble part and a data part. The channel spacing is 20 MHz, which allows high bit rates per channel. The physical layer for both the IEEE 802.11a and HiperLAN/2 is based on Orthogonal Frequency Division Multiplexing (OFDM). OFDM uses 52 subcarriers per channel, where 48 subcarriers carry actual data and 4 subcarriers are pilots that facilitate phase tracking for coherent demodulation. The duration of the guard interval is equal to 800 ns, which is sufficient to enable good performance on channels with delay spread of up to 250 ns. An optional shorter guard interval of 400 ns may be used in small indoor environments. OFDM is used to combat frequency selective fading and to randomize the burst errors caused by a wide band fading channel. The PHY layer modes with different coding and modulation schemes are shown in Table 5.1. The MAC selects any of the available rates for transmitting its data based on the channel condition. This algorithm is called link adaptation and it is not specified by the standard as to how it should be performed, thus enabling product differentiation between different vendors.

**Table 5.1**   Different modulation schemes of IEEE 802.11a and HiperLAN/2 physical layer

| Mode scheme | Modulation | Coding rate, $R$ | Bit rate (Mb/s) | Coded bits/ subcarrier | Coded bits/ OFDM Symbol | Data bits/ OFDM Symbol |
|---|---|---|---|---|---|---|
| 1 | BPSK | 1/2 | 6 | 1 | 48 | 24 |
| 2 | BPSK | 3/4 | 9 | 1 | 48 | 36 |
| 3 | QPSK | 1/2 | 12 | 2 | 96 | 48 |
| 4 | QPSK | 3/4 | 18 | 2 | 96 | 72 |
| 5 | 16 QAM (H/2 only) | 9/16 | 27 | 4 | 192 | 108 |
| 5 | 16 QAM (IEEE only) | 1/2 | 24 | 4 | 192 | 96 |
| 6 | 16 QAM | 3/4 | 36 | 4 | 192 | 144 |
| 7 | 64 QAM | 3/4 | 54 | 6 | 288 | 216 |
| 8 | 64 QAM (IEEE only) | 2/3 | 48 | 6 | 288 | 192 |

   Data for transmission is supplied to the PHY layer in the form of an input Protocol Data Unit (PDU) train or Physical Layer Convergence Procedure (PLCP) Protocol Data Unit (PPDU) frame. This is then passed to a scrambler that prevents long runs of 1s and 0s in the input data. Although both 802.11a and HiperLAN/2 scramble the data with a length 127 pseudorandom sequence, the initialization of the scrambler is different. The scrambled data is then passed to a convolutional encoder. The encoder consists of a 1/2 rate mother code and subsequent puncturing. The puncturing schemes facilitate the use of code rates 1/2, 3/4, 9/16 (HiperLAN/2 only), and 2/3 (802.11a only). In the case of 16-Quadrature Amplitude Modulation (QAM), HiperLAN/2 uses rate 9/16 instead of rate 1/2 in order to ensure an integer number of OFDM symbols per PDU train. The rate 2/3 is used only for the case of 64-QAM in 802.11a. Note that there is no equivalent mode for HiperLAN/2. HiperLAN/2 also uses additional puncturing in order to keep an integer number of OFDM symbols with 54-byte PDUs. The coded data is interleaved in order to prevent error bursts from being input to the convolutional decoding process in the receiver. The interleaved data is subsequently mapped to data symbols according to either a Binary Phase Shift Keying (BPSK), Quadrature PSK (QPSK), 16-QAM, or 64-QAM constellation. OFDM modulation is implemented by means of an inverse Fast Fourier Transform (FFT). 48 data symbols and four pilots are transmitted in parallel in the form of one OFDM symbol. Numerical values for the OFDM parameters are given in Table 5.2. In order to prevent Inter Symbol Interference (ISI) and Inter Carrier

**Table 5.2**  Overhead calculation for HiperLAN transmission

| Channel | OFDM symbols |
|---|---|
| BCH + Preamble$_{BCH}$ | $5 + 4 = 9$ |
| FCH$_{min}$ | 6 |
| ACH | 3 |
| RCH + Preamble$_{RCH}$ | $3 + 4 = 7$ |
| Uplink overhead | $4 + \left\lceil \dfrac{9}{BpS} \right\rceil$ |

Interference (ICI) due to delay spread, a guard interval is implemented by means of a cyclic extension. Thus, each OFDM symbol is preceded by a periodic extension of the symbol itself. The total OFDM symbol duration is $T_{total} = T_g + T$, where $T_g$ represents the guard interval and $T$ the useful OFDM symbol duration. When the guard interval is longer than the excess delay of the radio channel, ISI is eliminated. The OFDM receiver basically performs the reverse operations of the transmitter. However, the receiver is also required to perform Automatic Gain Control (AGC), time and frequency synchronization, and channel estimation. Training sequences are provided in the preamble for the specific purpose of supporting these functions. Two OFDM symbols are provided in the preamble in order to support the channel estimation process. A prior knowledge of the transmitted preamble signal facilitates the generation of a vector defining the channel estimate, commonly referred to as the Channel State Information (CSI). The channel estimation preamble is formed such that the two symbols effectively provide a single guard interval of length 1.6 ms. This format makes it particularly robust to ISI. By averaging over two OFDM symbols, the distorting effects of noise on the channel estimation process can also be reduced. HiperLAN/2 and 802.11a use different training sequences in the preamble. The training symbols used for channel estimation are the same, but the sequences provided for time and frequency synchronization are different. Decoding of the convolutional code is typically implemented by means of a Viterbi decoder. The physical layer modes (PHY modes) are specified in Table 5.1.

## 5.3  Overview of HiperLAN/1

The PHY layer of HiperLAN/1 uses 200 MHz at 5.15–5.35 GHz. This band is divided into five channels with channel spacing of 40 MHz in the European Union and six channels of 33 MHz spacing in the USA. The transmission power can go up to 1 W. The modulation scheme is single carrier Gaussian Minimum Shift Keying (GMSK) that can support up to 23 Mbps. Decision Feedback Equalizer (DFE) is employed at the receiver because of the high data rate and it consumes more power [3].

### 5.3.1  MAC Protocol of HiperLAN/1

The HiperLAN/1 channel access mechanism is based on channel sensing and a contention resolution scheme called Elimination Yield – Non-preemptive Priority Multiple Access (EY-NPMA). In this scheme, channel status is sensed by each node that has a data frame to transmit. If the channel is sensed idle for at least 1700 bit-periods, then the channel is considered free, and the node is allowed to start transmission of the data frame immediately. Each data frame transmission must be explicitly acknowledged by an acknowledgement (ACK) transmission from the destination node. If the channel is sensed busy, a channel access with synchronization has to precede before frame transmission. Synchronization is performed to the end of the current transmission interval according to the EY-NPMA scheme.

The channel access cycle consists of three phases: the prioritization phase, the contention phase and the transmission phase. Figure 5.1 shows the channel access using EY-NPMA. The aim of the

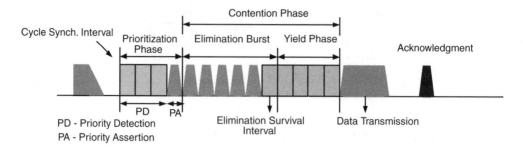

**Figure 5.1**   EY-NPMA MAC Protocol of HiperLAN/1.

prioritization phase is to allow only nodes with the highest channel access priority frame, among the contending ones, to participate in the next phase. In HiperLAN/1 a priority level $h$ is assigned to each frame. Priority level 0 represents the highest priority. The prioritization phase consists of at most $H$ prioritization slots, each 256 bit-periods long. Each node that has a frame with priority level $h$ senses the channel for the first $h$ prioritization slots. If the channel is idle during this interval, then the node transmits a burst in the $(h + 1)$th slot and it is admitted to the contention phase, otherwise it stops contending and waits for the next channel access cycle.

The contention phase starts immediately after the transmission of the prioritization burst, and it further consists of two phases: the elimination phase and the yield phase. The elimination phase consists of at most $n$ elimination slots, each 256 bit-periods long, followed by a 256 bit-periods long elimination survival verification slot. Starting from the first elimination slot, each node transmits a burst for a number $B$ $(0 \leq B \leq n)$ of subsequent elimination slots, according to the truncated geometric probability distribution function:

$$\Pr(B = b) = \begin{cases} (1 - q)q^b & : \quad 0 \leq b < n \\ q^n & : \quad i = q. \end{cases}$$

After the end of the burst transmission, each node senses the channel for the duration of the elimination survival verification slot. If the channel is sensed idle, the node is admitted to the yield phase, otherwise it drops itself from contention and waits for the next channel access cycle. The yield phase starts immediately after the end of the elimination survival verification interval and consists of at most $m$ yield slots, each 64 bit-periods long. Each node listens to the channel for a number $D$ $(0 \leq D \leq m)$ of yield slots before beginning transmission. $D$ is an rv with truncated geometric distribution as follows:

$$\Pr(D = d) = \begin{cases} (1 - p)p^d & : \quad 0 \leq d < m \\ p^m & : \quad d = m. \end{cases}$$

If the channel is sensed idle during the yield listening interval, the node is allowed to begin the transmission phase, otherwise the node loses contention and waits for the next channel access cycle.

The operation parameter settings, according to HiperLAN, are reported in Table 5.1. Elimination and yield phases are complementary to each other. Elimination phase drastically reduces the number of nodes taking part to the channel access cycle, and this result is remarkably achieved almost independently of number of nodes [2].

The yield phase reduces the number of nodes allowed to transmit possibly to one. In EY-NPMA at least one node will always be allowed to transmit. Real time traffic transmission is supported in HiperLAN by dynamically varying the CAM priority depending upon the user priority and the

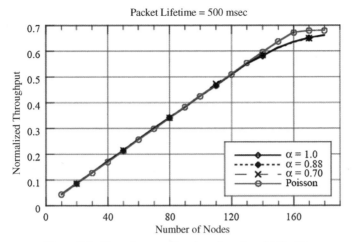

**Figure 5.2**   Throughput as a function of number of MTs [2].

packet residual lifetime as reported in Table 5.2. The user priority is an attribute that is assigned to each packet according to the type of traffic carried and it determines the maximum CAM priority value the packet may eventually reach. The residual packet lifetime is the time interval within which the transmission of the packet must occur before the packet has to be discarded. Figure 5.2 shows the aggregate throughput achieved by HiperLAN/2 vs. the number of data nodes. The throughput increases as the number of sources increase and stabilizes beyond a certain point.

## 5.4 Overview of HiperLAN/2

Figure 5.3 shows the protocol reference model for the HiperLAN/2 radio. The protocol stack is divided into a control plane part and a user plane. The user plane includes functions for transmission of traffic over established connections, and the control plane includes functions for the control of connection establishment, release and supervision. The HiperLAN/2 protocol has three basic layers: the Physical layer (PHY), the Data Link Control layer (DLC) and the Convergence layer (CL). The hiperLAN consists of an Access Point (AP) and the Mobile Terminals (MTs) that are associated with the AP. The

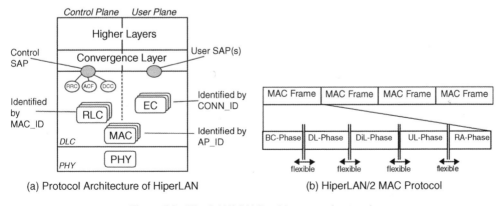

(a) Protocol Architecture of HiperLAN                        (b) HiperLAN/2 MAC Protocol

**Figure 5.3**   HiperLAN2 MAC architecture and protocol.

AP acts like a central controller and coordinates the data and control information transmission over the wireless channel from all the MTs.

### 5.4.1 Data Link Layer

The Data Link Control (DLC) layer includes user plane functions, such as both medium access and transmission, as well as control plane functions such as connection handling. Thus, the DLC layer consists of a set of sub layers:

- Medium Access Control (MAC) protocol;
- Error Control (EC) protocol;
- Radio Link Control (RLC) protocol with the associated signalling entities, DLC Connection Control (DCC), the Radio Resource Control (RRC) and the Association Control Function (ACF).

### 5.4.2 MAC Protocol

The control is centralized to the AP (also called the central controller (CC)), which inform the MTs of which point of time in the MAC frame they are allowed to transmit their data, which adapts according to the request for resources from each of the MTs. The air interface is based on time-division duplex (TDD) and dynamic time-division multiple access (TDMA). The basic MAC frame has a fixed duration of 2 ms and comprises transport channels for broadcast control, frame control, access control, downlink (DL) and uplink (UL) data transmission and random access (see Figure 5.3(b)). All data from both AP and the MTs is transmitted in dedicated time slots, except for the random access channel where contention for the same time slot is allowed. The duration of broadcast control is fixed, whereas the duration of other fields is dynamically adapted to the current traffic situation. The MAC frame and the transport channels form the interface between DLC and the physical layer.

Each MAC frame consists of following five phases.

(1) Broadcast (BC) phase. The BC phase carries the BCCH (broadcast control channel) and the FCCH (frame control channel). The BCCH contains general announcements and some status bits announcing the appearance of more detailed broadcast information in the downlink phase (DL). The announcement includes information about transmission power levels, starting point and length of the FCCH and the Random CHannel (RCH), wake-up indicator, and identifiers for identifying both the HiperLAN/2 network and the AP. The FCCH, transmitted by AP, carries the information about the structure of the ongoing frame, and contains an exact description of how resources have been allocated within the current MAC frame in the Down Link (DL), the Up Link (UL)-phase and for the RCH.

(2) Downlink (DL) phase. The DL phase carries user specific control information and user data, transmitted from AP/CC to MTs. Additionally, the DL phase may contain further broadcast information that does not fit in the fixed BCCH field.

(3) Uplink (UL) phase. The UL phase carries control and user data from the MTs to the AP/CC. The MTs have to request capacity for one of the following frames in order to get resources granted by the AP/CC.

(4) Direct Link (DiL) phase. The DiL phase carries user data traffic between MTs without the direct involvement of the AP/CC. However, for control traffic, the AP/CC is indirectly involved by receiving Resource Requests from MTs for these connections and transmitting Resource Grants in the FCCH.

(5) Random Access (RA) phase. The RA phase carries a number of RCH (random access channels). MTs to which no capacity has been allocated in the UL phase use this phase for the transmission of control information. Non-associated MTs use RCHs for the first contact with an AP/CC. This phase is also used by MTs performing handover to have their connections switched over to a new AP/CC.

Additionally it is used to convey some RLC signalling message. When the request for more transmission resources increase from the MTs, the AP will allocate more resources for the RCH. The RCH is composed entirely of contention slots that all the MTs associated to the AP compete for. Collisions may occur and the results from RCH access are reported back to the MTs in ACH.

- *Logical channels.* The transport channels (SCH, LCH and RCH) are used as an underlying resource for the logical channels. The slow broadcast channel (SBCH) is transmitted by the AP to all MTs and conveys broadcast control information concerning the whole radio cell. The information is transmitted only when necessary, which is determined by the AP. Following information may be sent in the SBCH:
- broadcast RLC messages;
- conveys an assigned MAC-ID to a none-associated MT;
- handover acknowledgements;
- convergence Layer (higher layer) broadcast information;
- seed for encryption.

All terminals have access to the SBCH. SBCH shall be sent once per MAC frame per antenna element. The Dedicated Control Channel (DCCH) is bidirectional and conveys RLC sublayer signals between an MT and the AP. Within the DCCH, the RLC carries messages defined for the DLC connection control and association control functions. The DCCH forms a logical connection and is established implicitly during association of a terminal without any explicit signaling by using predefined parameters. The DCCH is realized as a DLC connection. Each associated terminal has one DCCH per MAC-ID. This means that when an MT has been allocated its MAC-ID it will use this connection for control signaling.

The User Data Channel (UDCH) is also bidirectional and conveys user data (DLC PDU for convergence layer data) between the AP and an MT. The DLC guarantees in sequence delivery of SDUs to the convergence layer. A DLC user connection for the UDCH is set up using signaling over the DCCH. Parameters related to the connection are negotiated during association and connection setup. In the uplink, the MT requests transmission slots for the connection related to UDCH, and then the resource grant is announced in a following FCH. In downlink, the AP can allocate resources for UDCH without the terminal request. ARQ is by default applied to ensure reliable transmission over the UDCH. There may be connections that are not using the ARQ, e.g. connections for multicast traffic.

The Link Control Channel (LCCH) is also bidirectional and conveys information between the error control (EC) functions in the AP and the MT for a certain UDCH. The AP determines the needed transmission slots for the LCCH in the uplink and the resource grant is announced in an upcoming FCH.

The Association Control Channel (ASCH) is transmitted by the MTs only and conveys new association request and re-association request messages. These messages can only be sent during handover and by a disassociated MT.

A DLC connection is used for either unicast, multicast or broadcast. A connection is uniquely defined by the combination of the MAC identifier and the DLC connection identifier. This combination is also referred to as a DLC User Connection (DUC). For the purpose of transmission of unicast traffic, each MT is allocated a MAC identifier (local significance, per AP) and one or more DLC connection identifiers depending on the number of DUCs. In case of multicast, HiperLAN/2 defines two different modes of operation: N*unicast and MAC multicast. With N*unicast, the multicast is treated in the same way as unicast transmission in which case Automatic Repeat reQuest (ARQ) applies. Using MAC multicast, a separate MAC-ID (local significance, per AP) is allocated for each multicast group. ARQ can't be used in this case, i.e. each U-PDU is only transmitted once. All multicast traffic for that group is mapped to the same single DLC connection. HiperLAN/2 allows for up to 32 multicast groups to be mapped to separate MAC identifiers. In the case where the associated MTs like to join more than 32 multicast groups, one of the MAC identifiers will work as an overflow MAC identifier, meaning that two or more multicast groups may be mapped to that identifier. Broadcast is also supported. As in the case with multicast, the ARQ doesn't apply, but as the transmission of broadcast is many times more critical for the overall system performance, a scheme with repetition of the broadcast U-PDUs have been defined. This means that the same U-PDU is retransmitted a number of times (configurable) within the same MAC-frame, to increase the probability of a successful transmission. It is worth noticing that reception of broadcast will not change the sleep state of an MT.

### 5.4.3 Error Control Protocol

Selective repeat (SR) ARQ is the Error Control (EC) mechanism that is used to increase the reliability over the radio link. EC in this context means detection of bit errors, and the resulting retransmission of UPDU(s) if such errors occur. EC also ensures that the U-PDU's are delivered in-sequence to the convergence layer. The method for controlling this is by giving each transmitted U-PDU a sequence number per connection. The ARQ ACK/NACK messages are signalled in the LCCH. An errored U-PDU can be retransmitted a number of times. To support QoS for delay critical applications such as voice in an efficient manner, a U-PDU discard mechanism is defined. If the data become obsolete (e.g. beyond the playback point), the sender entity in the EC protocol can initiate a discard of a U-PDU and all U-PDUs that have lower sequence numbers and that haven't been acknowledged. The result is that the transmission in DLC allows for holes (missing data) while retaining the DLC connection active. It is up to higher layers, if need be, to recover from missing data.

### 5.4.4 Association Control Function (ACF)

#### 5.4.4.1 Association

This starts with the MT listening to the BCH from different APs and selects the AP with the best radio link quality. Part of the information provided in the BCH works as a beacon signal in this stage. The MT then continues by listening to the broadcast of a globally unique network operator id in the SBCH so as to avoid association with a network that is not able or allowed to offer services to the user of the MT. If the MT decides to continue the association, the MT will request and be given a MAC-ID from the AP. This is followed by an exchange of link capabilities using the ASCH, starting with the MT providing information about the following: (i) supported PHY modes, (ii) supported Convergence layers and (iii) supported authentication and encryption procedures and algorithms. The AP will respond with a subset of supported PHY modes, a selected Convergence layer, and a selected authentication and encryption procedure. If encryption has been negotiated, the MT will start the key exchange to negotiate the secret session key for all unicast traffic between the MT and the AP. In this way, the following authentication procedure is protected by encryption. HiperLAN/2 supports both the use of the DES and the 3-DES algorithms for strong encryption. Broadcast and multicast traffic can also be protected by encryption through the use of common keys (all MTs associated with the same AP use the same key). Common keys are distributed encrypted through the use of the unicast encryption key. All encryption keys must be periodically refreshed to avoid flaws in the security. There are two alternatives for authentication: one is to use a pre-shared key and the other is to use a public key. When using a public key, HiperLAN/2 supports a Public Key Infrastructure by means of generating a digital signature. The authentication algorithms supported are MD5, HMAC and RSA. Also bidirectional authentication is supported for authentication of both the AP and the MT. HiperLAN/2 supports a variety of identifiers for identification of the user and/or the MT, e.g. Network Access Identifier (NAI), IEEE address, and X.509 certificate. After association, the MT can request for a dedicated control channel (i.e. the DCCH) that it uses to setup radio bearers (within the HiperLAN/2 community, a radio bearer is referred to as a DLC user connection). The MT can request multiple DLC user connections where each connection has the unique support for QoS.

#### 5.4.4.2 Disassociation

An MT may disassociate explicitly or implicitly. When disassociating explicitly, the MT will notify the AP that it no longer wants to communicate via the HiperLAN/2 network. Implicitly means that the MT has been unreachable for the AP for a certain time period. In both cases, the AP will release all resources allocated for that MT.

### 5.4.5 Signaling and Radio Resource Management

The Radio Link Control (RLC) protocol gives a transport service for the signalling entities such as ACF, Radio Resource Control function (RRC), and the DLC user Connection Control function (DCC). These four entities comprise the DLC control plane for the exchange of signalling messages between the AP and the MT.

#### 5.4.5.1 DLC User Connection Control (DCC)

In the HiperLAN network, the MT, as well as the AP, request DLC user connection by transmitting signaling messages over the DCCH. The DCCH controls the resources for one specific MAC entity. No traffic in the user plane can be transmitted until there is at least one DLC user connection between the AP and the MT. The signaling is quite simple with a request followed by an acknowledgment if a connection can be established. For each request, the connection characteristics are given. If the AP determines that the connection characteristics can be satisfied, then it acknowledges the acceptance of the connection to the MT. The established connection is identified with a DLC connection identifier, allocated by the AP. A connection is subsequently released using a procedure similar to the establishment. HiperLAN/2 also supports modification of the connection characteristics for an established connection.

#### 5.4.5.2 Radio Resource Control (RRC)

Handover, Dynamic Frequency Selection, synchronization with MT periodically and power save are the key functions of Radio Resource Control. These are briefly explained below.

The handover starts when the MT determines that the current radio signal quality with the AP is not good and is not the result of short term fading. If the MT determines that the signal quality is bad and is not the result of fading, it requests a handover. There are two types of handover: reassociation and handover via the support of signaling across the fixed network. Reassociation basically means starting over again with an association as described above, which may take some time, especially in relation to ongoing traffic. The alternative means that the new AP to which the MT has requested a handover, will retrieve association and connection information from the old AP by transfer of information across the fixed network. The MT provides the new AP with a fixed network address (e.g. an IP address) to enable communication between the old and new AP. This alternative results in a fast handover, minimizing loss of user plane traffic during the handover phase.

RRC supports DFS by letting the AP have the possibility to instruct the associated MTs to perform measurements on radio signals received from neighboring APs. Owing to changes in the environment and network topology, RRC also includes signaling for informing associated MTs that the AP will change frequency.

The AP supervises inactive MTs which don't transmit any traffic in the uplink by sending an alive message to the MT for the MT to respond to. This process is called MT Alive. As an alternative, the AP may set a timer for how long an MT may be inactive. If there is no response from the alive messages or, alternatively, if the timer expires, the MT will be disassociated.

Power save is responsible for entering or leaving low consumption modes and for controlling the power of the transmitter. This function is MT initiated. After a negotiation on the sleeping time ($N$ number of frames where N in [2, 216]) the MT goes to sleep. After $N$ frames there are four possible scenarios.

(1) The AP wakes-up the MT (cause, e.g., data pending in AP).
(2) The MT wakes-up (cause, e.g., data pending in MT).

(3) The AP tells the MT to continue to sleep (again for $N$ frames).

(4) The MT misses the wake-up messages from the AP. It will then execute the MT Alive sequence.

### 5.4.6 Convergence Layer

The convergence layer (CL) has two main functions: adapting service request from higher layers to the service offered by the DLC and converting the higher layer packets (SDUs) with variable or possibly fixed size into a fixed size that is used within the DLC. The padding, segmentation and reassembly function of the fixed size DLC SDUs is one key issue that makes it possible to standardize and implement a DLC and PHY that is independent of the fixed network to which the HiperLAN/2 network is connected. The generic architecture of the CL makes HiperLAN/2 suitable as a radio access network for a diversity of fixed networks, e.g. Ethernet, IP, ATM, UMTS, etc. There are currently two different types of CLs defined: cell-based and packet-based as depicted in Figure 5.4. The former is intended for interconnection to ATM networks, whereas the latter can be used in a variety of configurations depending on the fixed network type and how the interworking is specified.

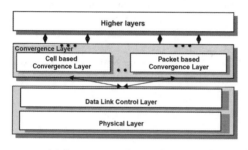

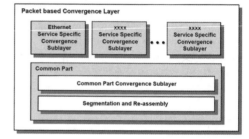

(a) Convergence Layer of HiperLAN                              (b) Packet based Convergence layer

**Figure 5.4**   Convergence layer architecture for HiperLAN2.

The structure of the packet-based CL with a common and service-specific part allows for easy adaption to different configurations and fixed networks. From the beginning though, the HiperLAN/2 standard specifies the common part and a service specific part for interworking with a fixed ethernet network. The packet-based CL is depicted in Figure 5.4(b).

(1) Common part. The main function of the common part of the Convergence layer is to segment packets received from the SSCS, and to reassemble segmented packets received from the DLC layer before they are handed over to the SSCS. Included in this sublayer is also to add/remove padding octets as needed to make a Common Part PDU being an integral number of DLC SDUs.

(2) Ethernet SSCS. The Ethernet SSCS makes the HiperLAN/2 network look like wireless segments of a switched Ethernet. Its main functionality is the preservation of Ethernet frames. Both, IEEE 802.31 frames and tagged IEEE802.3ac2 frames are supported. The Ethernet SSCS offers two Quality of Service schemes: The best effort scheme is mandatory supported and treats all traffic equally. The IEEE 802.1p based priority scheme is optional and separates traffic in to different priority queues as described in IEEE 802.1p. As a benefit, the DLC can treat the different priority queues in an optimized way for specific traffic types.

### 5.4.7 Throughput Performance of HiperLAN/2

The HiperLAN/2 MAC protocol provides the flexibility to accommodate a large variety of MTs with different QoS requirements. The actual data rate supported by the MAC protocol can be defined by an

AP for each MT connection individually over time by defining the size of a PDU train and the PHY modes. The throughput HiperLAN/2 is calculated by first summing up the length of the channels and the overhead for their transmission as listed in Table 5.2. The length of the OFDM symbol is 4 µs. With a super frame size of 2 milliseconds (ms), we can transmit 500 OFDM symbols in a super frame. With a total number of 500 OFDM symbols per MAC frame, the total number of user PDUs (NPDU) per MAC frame is given by:

$$N_{PPDU} = \left(471 - \left\lceil \frac{9}{BpS} \right\rceil\right) \frac{BpS}{54}.$$

The throughput is given by:

$$S = N_{PPDU} \frac{x}{\left\lceil \frac{x}{48} \right\rceil} \frac{8}{t_{frame}}. \tag{5.1}$$

## 5.5 IEEE 802.11 MAC

The IEEE 802.11 legacy MAC [5] is based on the logical functions, called the coordination functions, that determine when a station operating within a Basic Service Set (BSS) is permitted to transmit and may be able to receive frames via the wireless medium. Two coordination functions are defined, namely, the mandatory DCF based on CSMA/CA and the optional Point Coordination Function (PCF) based on a poll-and-response mechanism. Most of today's 802.11 devices operate in the DCF mode only.

### 5.5.1 Distributed Coordination Function

The fundamental access method of the IEEE 802.11 MAC is a DCF known as carrier sense multiple access with collision avoidance (CSMA/CA). The DCF shall be implemented in all STAs, for use within both Independent Basic Service Set (IBSS) and infrastructure network configurations. The 802.11 MAC works with a single first-in-first-out (FIFO) transmission queue in each station. The CSMA/CA constitutes a distributed MAC based on a local assessment of the channel status, i.e., whether the channel is busy (i.e., a station is transmitting a frame) or idle (i.e., no transmission). The CSMA/CA of DCF works as follows.

- When a frame (or an MSDU) arrives at the head of the transmission queue, if the channel is busy, the MAC waits until the medium becomes idle, then defers for an extra time interval, called the DCF Inter Frame Space (DIFS). If the channel stays idle during the DIFS deference, the MAC then starts the backoff process by selecting a random backoff counter (or BC). How to select the value of BC is explained in the next paragraph. For each slot time[c] interval, during which the medium stays idle, the random BC is decremented. When the BC reaches zero, the frame is transmitted.
- On the other hand, when a frame arrives at the head of the queue, if the MAC is in either the DIFS deference or the random backoff process, the processes described above are again applied. That is, the frame is transmitted only when the random backoff has finished successfully.
- When a frame arrives at an empty queue with backoff value being zero and the medium has been idle longer than the DIFS time interval, the frame is transmitted immediately.

---

[c] The slot time for an 802.11 PHY shall be the sum of receiver-to-transmitter turnaround time and the energy detect time. The propagation delay is considered as part of the energy detect time.

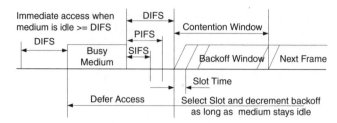

**Figure 5.5**   IEEE 802.11 DCF channel access.

Each station maintains a contention window (*CW*), which is used to select the random backoff counter. The BC is determined as a random integer drawn from a uniform distribution over the interval $[0, CW]$. How to determine the *CW* value is further detailed below. If the channel becomes busy during a backoff process, the backoff is suspended. When the channel becomes idle again, and stays idle for an extra DIFS time interval, the backoff process resumes with the latest BC value. The timing of DCF channel access is illustrated in Figure 5.5. For each successful reception of a frame, the receiving station immediately acknowledges the frame reception by sending an acknowledgment (ACK) frame. The ACK frame is transmitted after a short IFS (SIFS), which is shorter than the DIFS. Other stations resume the backoff process after the DIFS idle time. Thanks to the SIFS interval between the data and ACK frames, the ACK frame transmission is protected from other stations' contention. If an ACK frame is not received after the data transmission, the frame is retransmitted after another random backoff. The *CW* size is initially assigned $CW_{min}$, and increases when a transmission fails, i.e., the transmitted data frame has not been acknowledged. After any unsuccessful transmission attempt, another backoff is performed using a new *CW* value updated by

$$CW \Leftarrow 2 \cdot (CW + 1) - 1$$

with the an upper bound of $CW_{max}$. This reduces the collision probability where there are multiple stations attempting to access the channel. After each successful transmission, the *CW* value is reset to $CW_{min}$, and the transmission-completing station performs another DIFS deference and a random backoff, even if there is no other pending frame in the queue. This is often referred to as 'post' backoff, as this backoff is done after, not before, a transmission. This post backoff ensures there is at least one backoff interval between two consecutive MAC Service Data Unit (MSDU) transmissions. The *CW* value is also reset when the retransmission limit is reached. All of the MAC parameters including *SIFS, DIFS, Slot Time*, $CW_{min}$, and $CW_{max}$ are dependent on the underlying physical layer (PHY). Table 5.3 shows these values for the 802.11a PHY [6]. Irrespective of the PHY, DIFS is determined by *SIFS* + 2*SlotTime*, and another important IFS, called PCF IFS (PIFS), is determined by *SIFS* + *SlotTime*.

**Table 5.3**   MAC parameters for 802.11a PHY

| Parameters | aSIFStime (µs) | aDIFStime (µs) | aSLOTtime (µs) | $aCW_{min}$ | $aCW_{max}$ |
|---|---|---|---|---|---|
| 802.11a PHY | 16 | 34 | 9 | 15 | 1023 |
| 802.11b PHY | 10 | 50 | 20 | 31 | 1023 |

## 5.5.2 *Point Coordination Function*

The IEEE 802.11 MAC has an optional access method called a PCF, which is only usable on infrastructure network configurations. This access method uses a Point Coordinator (PC), which shall operate at the AP of the Basic Services Set (BSS), to determine which STA currently has the right to transmit. The operation is essentially that of polling, with the PC performing the role of the polling master. The operation of the PCF may require additional coordination to permit efficient operation in cases where multiple point-coordinated BSSs are operating on the same channel at the same physical space. The PCF uses a virtual carrier-sense (CS) mechanism aided by an access priority mechanism. The PCF shall distribute information within Beacon management frames to gain control of the medium by setting the Network Allocation Vector (NAV) in STAs. In addition, all frame transmissions under the PCF may use an Inter Frame Space (IFS) that is smaller than the IFS for frames transmitted via the DCF. The use of a smaller IFS implies that point-coordinated traffic shall have priority access to the medium over STAs in overlapping BSSs operating under the DCF access method. The access priority provided by a PCF may be utilized to create a Contention Free (CF) access method. The PC controls the frame transmissions of the STAs so as to eliminate contention for a limited period of time.

All STAs inherently obey the medium access rules of the PCF, because these rules are based on the DCF, and they set their NAV at the beginning of each Contention Free Period (CFP). A STA that is able to respond to CF Polls sent by the PC located in the AP is referred to as being CF-Pollable, and may request to be polled by an active PC. CF-Pollable STAs and the PC do not use RTS/CTS in the CFP. When polled by the PC, a CF-Pollable STA may transmit only one MPDU, which can be to any destination (not just to the PC), and may piggyback the acknowledgment of a frame received from the PC using particular data frame subtypes for this transmission. If the data frame is not in turn acknowledged, the CF-Pollable STA shall not retransmit the frame unless it is polled again by the PC, or it decides to retransmit during the CP. If the addressed recipient of a CF transmission is not CF Pollable, that STA acknowledges the transmission using the DCF acknowledgment rules, and the PC retains control of the medium. A PC may use CF frame transfer solely for delivery of frames to STAs, and never to poll non-CF-Pollable STAs.

The PCF controls frame transfers during a CFP. The CFP shall alternate with a CP, when the DCF controls frame transfers, as shown in Figure 5.6. Each CFP shall begin with a Beacon frame that contains a DTIM element. The CFPs shall occur at a defined repetition rate, which shall be synchronized with the beacon interval. The PC generates CFPs at the CF repetition rate (CFPRate), which is defined as the number of DTIM intervals. The PC shall determine the CFPRate to use from the CFPRate parameter in the CF Parameter Set. This value, in units of DTIM intervals, shall be communicated to other STAs in the BSS in the CFPPeriod field of the CF Parameter Set element of Beacon frames. The CF Parameter Set element shall only be present in Beacon and Probe Response frames transmitted by STAs containing an active PC. From Figure 5.6, it may be possible that the start of the CFP may be postponed if a STA operating in CP starts transmission just before the beacon period, thus delaying the start of CFP until the completion of the frame transmission/collision time plus the PIFS time. In this case the CFP is shortened.

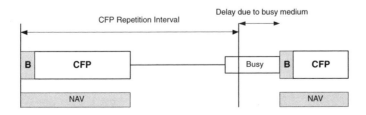

**Figure 5.6** Super Frame Structure of IEEE 802.11 MAC (B represents Beacon and CFP represents the contention free priod).

## 5.6 Overview of IEEE 802.11 Standardization

As already explained IEEE 802.11 is an industry standard set of specifications for WLANs developed by the Institute of Electrical and Electronics Engineers (IEEE). IEEE 802.11 defines the physical layer and media access control (MAC) sub-layer for wireless communications. The first standard for 802.11 came out in 1997, in this the MAC layer was defined with three different PHY layers based on Infrared, Direct Sequence Spread Spectrum (DSSS) and Frequency Hopping (FH). Those PHY layers supported only 1 and 2 Mbps. The following extensions were then developed to enhance the performance of IEEE 802.11 WLANs.

**802.11a.** IEEE 802.11a operates at a data transmission rate as high as 54 megabits per second (Mbps) and uses a radio frequency of 5.8 GHz. Instead of DSSS, 802.11a uses orthogonal frequency-division multiplexing (OFDM). OFDM allows data to be transmitted by sub-frequencies in parallel. This modulation mode provides better resistance to interference and improved data transmission.

**802.11b.** IEEE 802.11b, an enhancement to IEEE 802.11, provides standardization of the physical layer to support higher bit rates. IEEE 802.11b uses 2.45 GHz, the same frequency as IEEE 802.11, and supports two additional speeds: 5.5 Mbps and 11 Mbps. It uses the DSSS modulation scheme to provide higher data transmission rates. The bit rate of 11 Mbps is achievable in ideal conditions. In less-than-ideal conditions, the slower speeds of 5.5 Mbps, 2 Mbps and 1 Mbps are used.

**802.11c.** 802.11c provides required information to ensure proper bridge operations. This is very important for implementation of APs as they have to bridge between wired and wireless LANs.

**802.11d.** When 802.11 was launched in the late 1990s, only a handful of regulatory domains (e.g., USA, Europe and Japan) had rules in place for the operation of 802.11 wireless LANs. In order to support a widespread adoption of 802.11, the 802.11d task group has an ongoing charter to define PHY requirements that satisfy regulatory within additional countries.

**802.11e.** The current 802.11 is just the wireless version of Ethernet and there was a strong push from the industry to develop a MAC that would deliver QoS. Without strong quality of service (QoS), the existing version of the 802.11 standard doesn't optimize the transmission of voice and video. There's currently no effective mechanism to prioritize traffic within 802.11. As a result, the 802.11e task group is currently refining the 802.11 MAC (Medium Access Layer) to improve QoS for better support of audio and video (such as Moving Pictures Expert Group (MPEG-2)) applications. This will be the main focus of this chapter.

**802.11f.** The existing 802.11 standard doesn't specify the communications between access points in order to support users roaming from one access point to another. In order to make this communication possible, 802.11 defined 802.11 f that defines the rules for communication between different APs. This becomes very important to optimize the performance of TCP/UDP communications when mobility happens. In the absence of 802.11 f, you should utilize the same vendor for access points to ensure inter-operability for roaming users.

**802.11g.** The charter of the 802.11 g task group is to develop a higher speed extension (up to 54 Mbps) to the 802.11 b PHY, while operating in the 2.4 GHz band. 802.11 g will implement all mandatory elements of the IEEE 802.11 b PHY standard. This also uses OFDM to increase its channel rate to 54 Mbps. In the case of the existence of 802.11b stations, the 802.11g stations use RTS/CTS exchange to prevent 802.11 b stations from accessing the medium.

**802.11h.** 802.11 h addresses the requirements of the European regulatory bodies. It provides Dynamic Channel Selection (DCS) and Transmit Power Control (TPC) for devices operating in the 5 GHz band (802.11a). In Europe, there's a strong potential for 802.11a interfering with satellite communications, which have 'primary use' designations. Most countries authorize WLANs for 'secondary use' only. Through the use of DCS and TPC, 802.11 h will avoid interference with the primary user.

**802.11i.** 802.11 i is actively defining enhancements to the MAC Layer to counter the issues related to Wired Equivalent Privacy (WEP). The existing 802.11 standard specifies the use of relatively weak, static encryption keys without any form of key distribution management. This makes it possible for hackers to access and decipher WEP-encrypted data on your WLAN. 802.11 i will incorporate 802.1 x and stronger encryption techniques, such as AES (Advanced Encryption Standard).

**802.11j.** The purpose of Task Group J is to enhance the 802.11 standard and amendments, to add channel selection for 4.9 GHz and 5 GHz in Japan, to conform to the Japanese rules on operational mode, operational rate, radiated power, spurious emissions and channel sense.

**802.11k.** The IEEE 802.11 standard for wireless LANs enables inter-operability between different vendors' access points and switches, but it does not let WLAN systems assess a client's radio frequency resources. Consequently, this limits administrators' ability to manage their networks efficiently. As a proposed standard for radio resource measurement, 802.11 k aims to provide key client feedback to WLAN access points and switches. The proposed standard defines a series of measurement requests and reports that detail Layer 1 and Layer 2 client statistics. In most cases, access points or WLAN switches ask clients to report data, but in some cases clients might request data from access points.

**802.11m.** The purpose of this task group is maintenance. It will look for any editorial changes in the other 802.11 standards and will also answer any specific questions raised by implementors.

**802.11n.** This task group was formed recently and the purpose of this task group is to design a MAC that will provide a base throughput of 100 Mbps at the MAC layer. This will have IEEE 802.11e as the base MAC and the Multiple Input Multiple Output (MIMO) as its physical layer. The call for proposals have started recently and the standardization is expected to be complete by 2005.

## 5.7 IEEE 802.11e HCF

The new MAC protocol of the upcoming 802.11e is called the Hybrid Coordination Function (HCF). This HCF has two subfunctions called Enhanced Distributed Channel Access (EDCA) and HCF Coordination Channel Access (HCCA). Both these access mechanisms are explained below.

### 5.7.1 EDCA

EDCA is an enhanced version of the legacy DCF that provides a prioritized level of QoS. The 802.11 legacy MAC does not support the concept of differentiating frames with different priorities. The DCF provides a channel access with equal probabilities to all STAs contending for the channel access in a distributed manner. However, equal access probabilities are not desirable among STAs with different priority frames. The emerging EDCA is designed to provide differentiated, distributed channel accesses for frames with eight different priorities (from 0 to 7) by enhancing the DCF. As distinct from the legacy DCF, the EDCA is not a separate coordination function. Rather, it is a part of a single coordination function, the HCF, of the 802.11e MAC.

Each frame arriving at the MAC from the higher layers carries a specific priority value. Each higher layer priority is mapped into an access category (AC) as shown in Table 5.4. Note the relative priority of 0 is placed between 2 and 3. This relative prioritization is rooted from IEEE 802.1d bridge specification

**Table 5.4** Priority to access category mappings

| Priority | Access category | AC # | AC designation | $CW_{min}$ | $CW_{max}$ | AIFSN | TXOP limit (msec) |
|----------|-----------------|------|----------------|------------|------------|-------|-------------------|
| 0,1,2 | AC_BK | 0 | Background | $aCW_{min}$ | $aCW_{max}$ | 2 | 0 |
| 3 | AC_BE | 1 | Best Effort | $aCW_{min}$ | $aCW_{max}$ | 1 | 0 |
| 4,5 | AC_VI | 2 | Video | $\dfrac{aCW_{min}+1}{2}-1$ | $aCW_{min}$ | 1 | 3 |
| 6,7 | AC_VO | 3 | Voice | $\dfrac{aCW_{min}+1}{4}-1$ | $\dfrac{aCW_{min}+1}{2}-1$ | 1 | 1.5 |

[8]. Then, each QoS data frame carries its priority value in the MAC frame header. An AC uses $AIFS[AC]$, $CW_{min}[AC]$, and $CW_{max}[AC]$ instead of $DIFS$, $CW_{min}$, and $CW_{max}$, of the DCF, respectively, for the contention to transmit a frame belonging to AC. $AIFS[AC]$ is determined by

$$AIFS[AC] = aSIFStime + AIFSN[AC] * aSlotTime \qquad (5.2)$$

where $AIFSN[AC]$ is an integer greater than zero. Moreover, the backoff counter is selected from $[1, 1 + CW[AC]]$, instead of $[0, CW]$ as in the DCF. Figure 5.7 shows the timing diagram of the EDCA channel access. The values of $AIFS[AC]$, $CW_{min}[AC]$, and $CW_{max}[AC]$, which are referred to as the EDCA parameters, are announced by the QoS Access Point (QAP) via beacon frames. The QAP can adapt these parameters dynamically depending on network conditions. Basically, the smaller $AIFS[AC]$ and $CW_{min}[AC]$, the shorter the channel access delay for the corresponding access category, and hence the more capacity share for a given traffic condition. However, the probability of collisions increases when operating with smaller $CW_{min}[AC]$. These parameters can be used in order to differentiate the channel access among different priority traffic.

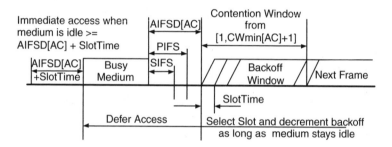

**Figure 5.7**   IEEE 802.11e EDCA channel access.

Figure 5.8 shows the 802.11e MAC with four transmission queues, where each queue behaves as a single enhanced DCF contending entity, i.e., an AC, where each queue has its own AIFS and maintains its own BC. When there is more than one AC finishing the backoff at the same time, the collision is handled in a virtual manner. That is, the highest priority frame among the colliding frames is chosen and transmitted, and the others perform a backoff with increased $CW$ values.

(1) EDCA Bursting.  The IEEE 802.11e defines a Transmission Opportunity (TXOP) as the interval of time when a particular QSTA has the right to initiate transmissions. Along with the EDCA parameters of $AIFS[AC]$, $CW_{min}[AC]$, and $CW_{max}[AC]$, the AP also determines and announces the TXOP limit of an EDCA TXOP interval for each AC, i.e., $TXOPlimit[AC]$, in beacon frames. During an EDCA TXOP, a QSTA is allowed to transmit multiple MSDU/MPDUs[d] from the same AC with a SIFS time gap between an ACK and the subsequent frame transmission [7,14]. Reference [7] refers to this multiple MSDU/MPDU transmissions as 'Contention-Free Burst (CFB)'. Figure 5.9 shows the transmission of two QoS data frames during a TXOP, where the whole transmission time for two data and ACK frames is less than the EDCA $TXOPlimit$ announced by the AP. As multiple MSDU transmission honors the $TXOPlimit$, the worst-case delay performance is not affected by allowing the CFB. We show in a later section that CFB increases the system throughput without

---

[d] MAC Physical Data Unit.

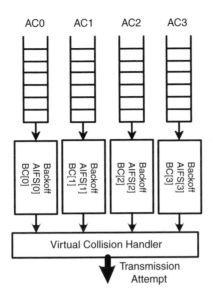

**Figure 5.8** Virtual transmission queues for EDCA. BC[AC] represents the backoff counter for the queue corresponding to AC.

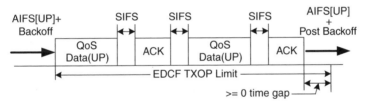

**Figure 5.9** IEEE 802.11e EDCA bursting timing structure.

degrading other system performance measures unacceptably as long as the EDCA *TXOPlimit* value is properly determined. A simple analytical model for the backoff procedure is outlined in the Appendix.

### 5.7.2 HCCA

The centrally controlled access mechanism of the IEEE 802.11e MAC, called the HCCA, adopts a poll and response protocol to control the access to the wireless medium and eliminate contention among wireless STAs. It makes use of the PIFS to seize and maintain control of the medium. Once the HC has control of the medium, it starts to deliver parameterized downlink traffic to STAs and issue QoS contention-free polls (QoS CF-Polls) frames to those STAs that have requested uplink or sidelink parameterized services. The QoS CF-Poll frames include the TXOP duration granted to the STA. If the STA being polled has traffic to send, it may transmit several frames for each QoS CF-Poll received, respecting the TXOP limit specified in the poll frame. Besides, in order to utilize the medium more efficiently, the STAs are allowed to piggyback both the acknowledgment (CF-ACK) and the CF-Poll onto data frames.

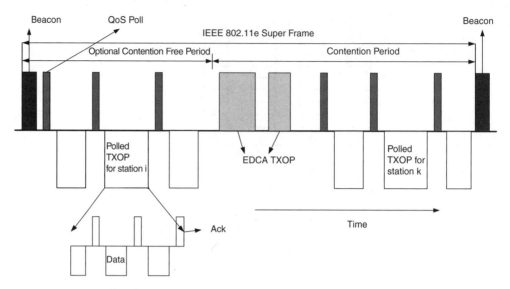

**Figure 5.10**   Operation of IEEE 802.11e HCCA in a Super Frame.

Differently from the PCF of the IEEE 802.11-99 standard [5], HCCA operates during both the CFP and CP (see Figure 5.10). During the CFP, the STAs cannot contend for the medium since their Network Allocation Vector (NAV), also know as virtual carrier sensing, is set, and therefore the HC enjoys free access to the medium. During the CP, the HC can also use free access to the medium once it becomes idle, in order to deliver downlink parameterized traffic or issue QoS CF-Polls. This is achieved by using the highest EDCA priority, i.e., $AIFS = PIFS$ and $CW_{min} = aCW_{max} = 0$. Note that the minimum time for any AC to access the medium is DIFS, which is longer than PIFS.

### 5.7.3 Support for Parameterized Traffic

An STA can request parameterized services using the Traffic Specification (TSPEC) element [7]. The TSPEC (see Figure 5.11) element contains the set of parameters that characterize the traffic stream that the STA wishes to establish with the HC. The parameters of the above field are explained in detail. The Nominal MSDU Size specifies the average size of the MSDU belonging to the Traffic Stream (TS). The Maximum MSDU Size field specifies the maximum size of an MSDU belonging to this TS. The Minimum Service Interval specifies the minimum interval between the start of two successive Service

| Octets: 1 | 1 | 3 | 2 | 2 | 4 | 4 | 4 | 4 |
|---|---|---|---|---|---|---|---|---|
| Element ID (13) | Length (55) | TS Info | Nominal MSDU Size | Maximum MSDU Size | Minimum Service Interval | Maximum Service Interval | Inactivity Interval | Suspension Interval |

| 4 | 4 | 4 | 4 | 4 | 4 | 4 | 2 | 2 |
|---|---|---|---|---|---|---|---|---|
| Service Start Time | Minimum Data Rate | Mean Data Rate | Peak Data Rate | Maximum Burst Size | Delay Bound | Minimum PHY Rate | Surplus Bandwidth Allowance | Medium Time |

**Figure 5.11**   TSPEC element as defined in IEEE 802.11e.

Periods (SPs). The Maximum Service Interval field specifies the maximum interval between the start of two successive SPs. The Inactivity Interval field specifies the maximum amount of time in units of microseconds that may elapse without arrival or transfer of an MSDU belonging to the TS before this TS is deleted by the MAC entity at the HC. The Suspension Interval field specifies the maximum amount of time in units of microseconds that may elapse without arrival or transfer of an MSDU belonging to the TS before the generation of successive QoS(+)CF-Poll is stopped for this TS. The Service Start Time field indicates the time, expressed in microseconds, when the SP starts. The Service Start Time indicates to QAP the time when a non-AP QSTA first expects to be ready to send frames and a power-saving non-AP QSTA will be awake to receive frames. This may help the QAP to schedule service so that the MSDUs encounter small delays in the MAC and the power-saving non-AP QSTAs to reduce power consumption. The Minimum Data Rate field specifies the lowest data rate that is required for proper deliver of this TS. The Mean Data Rate corresponds to the average rate of traffic arrivals from this TS. The Peak Data Rate specifies the maximum allowable data rate in units of bits/second from this TS. The Maximum Burst Size specifies the maximum burst of the MSDUs belonging to this TS that arrive at the MAC SAP at the peak data rate. The Delay Bound specifies the maximum amount of time allowed to transport an MSDU belonging to the TS. The Minimum PHY Rate specifies the desired minimum PHY rate to use for this TS. The Surplus Bandwidth Allowance field specifies the excess allocation of time (and bandwidth) over and above the stated application rates required to transport an MSDU belonging to the TS in this TSPEC. This should include retransmission which is a function of channel. The Medium Time specifies the amount of time admitted to access the medium, in units of 32 microsecond periods per second.

The User Priority, Minimum Data Rate, Mean Data Rate, Peak Data Rate, Maximum Burst Size, Minimum PHY Rate, and Delay Bound fields in a TSPEC express the QoS expectations requested by a application. Unspecified parameters are marked as zero to indicate that the non-AP QSTA does not have specific requirements for these parameters if the TSPEC request was issued.

Once the TSPEC request is received by the HC, it analyzes the TSPEC parameters and decides whether to admit the stream into the network. This is also known as the admission control process. If the stream is admitted, the HC schedules the delivery of downlink traffic and/or QoS CF- Polls in order to satisfy the QoS requirements of the stream as specified in the TSPEC. Several scheduling disciplines can be used in the HC. The performance of the HCCA is dependent on the choice of the admission control and the scheduling algorithm. Discussing the admission control and scheduling algorithms are beyond the scope of this chapter. In this chapter we use the simple scheduler [10] included as informative text in the IEEE 802.11e D7.0 [7], which is based on round robin scheduling. We will outline the admission control, scheduling and setting up of the retry-limit in the following subsections.

### 5.7.4 Simple Scheduler

This section describes the construction of a simple scheduler. This reference design was included in the draft [7] for implementors to show how to get a schedule from the TSPEC parameters. If both Maximum Service Interval and Delay Bound are specified by the non-AP QSTA in the TSPEC, the Scheduler uses the Maximum Service Interval for the calculation of the schedule. The schedule generated by any scheduler should meet the normative behavior. The normative behavior states the HC shall grant every flow the negotiated TXOP in a SI. This is calculated using the mandatory parameters from the TSPEC that shall be communicated by all non-AP QSTAs. The schedule for an admitted stream is calculated in two steps.

(1) The first step is the calculation of the scheduled Service Interval (SI).
(2) In the second step, the TXOP duration for a given SI, is calculated for the stream.

The calculation of the Service Interval is made as follows. First the scheduler calculates the minimum of all Maximum Service Intervals (MSI) for all admitted streams. Let this minimum be $m$. This is represented by:

$$m = \min\{MSI_1, MSI_2, ..., MSI_n\}. \tag{5.3}$$

Second, the scheduler chooses a number smaller than $m$ that is a submultiple of the beacon interval. This value is the Scheduled Service Interval ($SI$) for all non-AP QSTAs with admitted streams. For the calculation of the TXOP duration for an admitted stream, the Scheduler uses the following parameters: Mean Data Rate ($\rho$) and Nominal MSDU Size ($L$) from the negotiated TSPEC, the Scheduled Service Interval ($SI$) calculated above, Physical Transmission Rate ($R$), Maximum allowable Size of MSDU ($M = 2304$ bytes) and overheads in time units ($O$). The Physical Transmission Rate is the Minimum PHY Rate negotiated in the TSPEC. If Minimum PHY Rate is not committed in ADDTS response, the scheduler can use observed PHY rate as $R$. The Overheads in time includes interframe spaces, ACKs and CF-Polls. For simplicity, details for the overhead calculations are omitted in this description. The TXOP duration is calculated as follows. First the scheduler calculates the number of MSDUs that arrived at the Mean Data Rate during the SI.

$$N_i = \left\lceil \frac{SI \times \rho_i}{L_i} \right\rceil. \tag{5.4}$$

Then the scheduler calculates the TXOP duration as follows.

$$TXOP_i = \max\left( \frac{N_i L_i}{R_i}, \frac{M}{R_i} \right) + O. \tag{5.5}$$

Now any scheduler that is developed by the vendor must specify the two parameters: (1) the TXOP duration and the (2) the Service Interval. These two parameters are announced using a schedule element once a connection is admitted. The transmission of the respective stream shall not begin before this announcement.

### 5.7.4.1 Example of a Simple Scheduler Operation

Here, we give an example of the operation of the scheduler. A more detailed explanation can be found in [7]. An example is shown in Figure 5.12. A stream $i$ is admitted in the network. Therefore the HC allocates a TXOP for this stream. The beacon interval is 100 ms and the scheduler uses a Polling Interval, also known as Service Interval ($SI$), of 50 ms. In this example, the TXOP duration is derived from the Mean Data Rate and Nominal MSDU Size TSPEC Parameters, while the Polling Interval is calculated from the Delay Bound.

**Figure 5.12**   Schedule for stream $i$.

The same process is repeated continuously if more than one stream is in the network. Each stream is polled in round robin, and granted a specified TXOP duration according to the requirements of the stream. An example is shown in Figure 5.13: modifications can be implemented to improve the performance of the Simple Scheduler. For example, a scheduler may generate different polling intervals

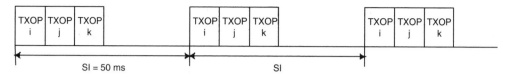

**Figure 5.13** Schedule for streams, $i$ to $k$.

for different streams/STAs, and/or a scheduler may consider a sandbag factor when calculating the TXOP durations to accommodate for retransmissions while allocating TXOP durations.

### 5.7.5 Admission Control at HC

An IEEE 802.11 network may use admission control to satisfy the QoS requests of the stream. As the HCF supports two access mechanisms, there are two distinct admission control mechanisms, one for contention-based access and another for controlled-access mechanism.

For the contention based admission control, the QAP uses the ACM (admission control mandatory) subfields advertised in the EDCA Parameter Set element to indicate whether admission control is required for each of the ACs. While the $CW_{min}$, $CW_{max}$, *AIFS, TXOP* limit parameters may be adjusted over time by the QAP, the ACM bit shall be static for the duration of the lifetime of the BSS. An ADDTS request shall be transmitted by a non-AP QSTA to the HC in order to request admission of traffic in any direction (uplink, downlink, direct or bidirectional) using an AC that requires admission control. The ADDTS request shall contain the user priority associated with the traffic and shall indicate EDCA as the access policy. The QAP shall associate the received user priority of the ADDTS request with the appropriate AC as per the UP to AC mappings [7]. The exact algorithm used for admission control is outside the scope of the standard.

In the case of HCCA based channel access, the draft [7] describes a reference design for an Admission Control Unit (ACU). The ACU uses the mandatory set of TSPEC parameters to make a decision to admit or reject a stream. When a new stream requests admission, the admission control process is performed in three steps.

(1) The ACU calculates the number of MSDUs that arrive at the Mean Data Rate during the Scheduled Service Interval. The Scheduled Service Interval (SI) is the one that the scheduler calculates for the stream as specified previous subsection. For the calculation of number of MSDUs the ACU uses the equation for $N_i$.
(2) The ACU calculates the required $TXOP_i$ for this stream.
(3) The ACU determines whether the $(k + 1)$th stream can be admitted or not using the following inequality:

$$\frac{TXOP_{k+1}}{SI} + \sum_{i=1}^{k} \frac{TXOP_i}{SI} \leq \frac{T - T_{CP}}{T}. \tag{5.6}$$

Here $k$ is the number of existing streams, $T$ indicates the beacon interval and $T_{CP}$ is the time used for EDCA traffic.

### 5.7.6 Power Management

A station may be in one of two power management modes: the Active mode or the Power Save (PS) mode. A station may switch from one power management mode to the other by changing the value of the PM bit. Frames destined for a station in PS mode will be buffered at the AP. Frames buffered at the

AP are delivered to a non-AP QSTA in PS mode only when it is in the Awake state. A non-AP QSTA may elect one of two methods for the delivery of the frames buffered at the AP while in the PS mode:

(1) by using PS Polls 28 or
(2) through APSD.

With PS Polls, the method available in the 802.11-1999 standard, a station transitions into the Awake state when AP sends the PS Poll and it can return to the Doze state after acknowledging receipt of a frame from the AP. With APSD, a non-AP QSTA is in the Awake state for the duration of the Service Period. To use APSD, a non-AP QSTA must submit a TSPEC request with the subfield APSD = 1. If APSD = 0, the non-AP QSTA may receive its frames sitting in the AP buffer through the use of PS Polls. Frames buffered at the AP while a station is in PS mode may also be received by the station switching to the Active mode. A non-AP QSTA in PS mode may use both delivery mechanisms at the same time for different types of traffic. The buffered frames may consist of a mix of traffic, such as a high-priority traffic stream for which an APSD TSPEC has been admitted and a low-priority data burst. All buffered frames associated with an admitted APSD TSPEC may be transmitted during a service period. Other lower-priority traffic that has been buffered at the AP, and is not associated with the admitted APSD TSPEC, may also be transmitted during the service period at the discretion of the AP, which should not favor the lower-priority traffic of an APSD station over higher-priority traffic of other stations. This provides an efficient way for a power-saving non-AP QSTA to receive its traffic. If traffic remains buffered at the AP at the end of a service period, which is signaled by the AP by setting the subfield EOSP=1, the subfield More Data is set to 1. The non-AP QSTA with frames remaining buffered at the AP at the end of a service period may receive its frames from the AP either through the use of PS Polls, or by switching to Active mode.

### 5.7.7 ACK Policies

IEEE 802.11e/D7.0 [7] specifies two ACK policies in addition to the Automatic Repeat Request (ARQ) mechanism used by legacy DCF. These are the No ACK and the Block ACK policy. We will briefly discuss both of them.

(1) No ACK Policy. If No ACK policy is used, the transmission of a frame is not accompanied by an ACK. This provides considerable throughput enhancement at the expense of potential performance degradation in high error rate conditions. When the error rate is close to zero, this mechanism saves considerable channel time and so the throughput of the channel is increased. A typical frame exchange using Polled TXOP and No ACK Policy is shown in Figure 5.14. In this frame exchange the AP is piggybacking data with the QoS CF-Poll.

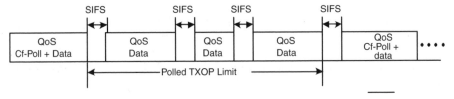

**Figure 5.14**   Frame exchange with No ACK policy.

(2) Block ACK policy. The Block Acknowledgement (Block Ack) mechanism allows a block of QoS Data frames to be transmitted successively, with each frame separated by a SIFS period, between two QSTAs. The mechanism is for improving the channel efficiency by aggregating several

acknowledgements into one frame. There are two types of Block ACK mechanisms: immediate and delayed. Immediate Block ACK is suitable for high-bandwidth, low latency traffic while the delayed Block ACK is suitable for applications that tolerate moderate latency.

The QSTA with data to send is referred to as the originator and the receiver of that data as the recipient. The Block ACK mechanism is initialized by an exchange of ADDBA request/response frames. The QSTA that sends ADDBA request frames is referred to as the initiator and the receiver of the frame as the responder. After initialization, blocks of QoS data type frames can be transmitted from the originator to the recipient. A block may be started within a polled TXOP or by winning EDCA contention. The MPDUs within this block usually fit within a single TXOP and are all separated by a SIFS. The number of frames in the block is limited, and the amount of states that must be kept by the recipient is bounded. The MPDUs within the block of frames are acknowledged by a BlockAck control frame, which is requested by a BlockAckReq control frame. The Block Ack mechanism does not require the setting up of a TS; however QSTAs using the TS facility may choose to signal their intention to use Block Ack mechanism for the scheduler's consideration in assigning TXOPs. Acknowledgements of frames belonging to the same TID, but transmitted during multiple TXOPs, may also be combined into a single Block Ack frame. This mechanism allows the originator flexibility regarding the transmission of Data MPDUs. The originator may split the block of frames across TXOPs, separate the data transfer and the block acknowledgement exchange and interleave blocks of MSDUs for different TIDs or RAs. Figure 5.15 illustrates the message sequence for the set up, data and block acknowledgement transfer and the tear down of Block Ack mechanism. In the next section we will explain the derivation of the error probability for a single frame.

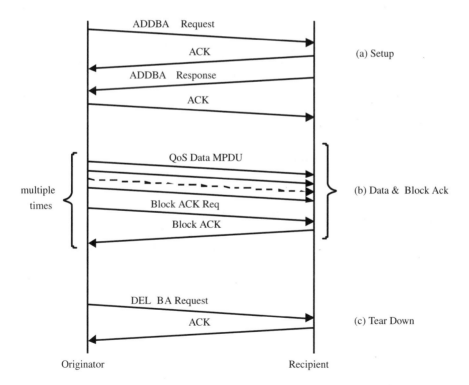

**Figure 5.15**   Frame exchange with Block ACK policy.

### 5.7.8 Direct Link Protocol (DLP)

In IEEE 802.11, STAs are not allowed to transmit frames directly to other STAs in a BSS and should always rely on the AP for the delivery of the frames. However, in IEEE 802.11e the QSTAs may transmit frames directly to another QSTA by setting up such data transfer using Direct Link Protocol (DLP). The need for this protocol is motivated by the fact that the intended recipient may be in Power Save Mode, in which case it can only be woken up by the QAP. The second feature of DLP is to exchange rate set and other information between the sender and the receiver. Finally, DLP messages can be used to attach security information elements. The security elements and the associated security handshake are not part of this standard. This protocol prohibits the stations going into power-save for the duration of the Direct Stream as long as there is an active direct link set up between the two stations. DLP does not apply in a QIBSS, where frames are always sent directly from one STA to another. A DLP set up is not required for sending MSDUs belonging to QoSLocalMulticast as the MAC cannot know the membership within a multicast/broadcast group. The DLP set up handshake is illustrated in Figure 5.16 and involves four steps.

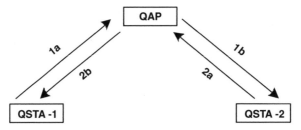

**Figure 5.16**    Operation of DLP.

(1) A station QSTA-1 that intends to exchange frames directly with another non-AP station QSTA-2 invokes DLP and sends a DLP Request frame to the AP (1a in Figure 5.16). This request contains the rate set, and capabilities of QSTA-1, as well as the MAC addresses of QSTA-1 and QSTA-2.
(2) If QSTA-2 is associated in the BSS, direct streams are allowed in the policy of the BSS and QSTA-2 is indeed a QSTA, the AP forwards the DLP-request to the recipient, QSTA-2 (1b in Figure 5.16).
(3) If QSTA-2 accepts the direct stream, it sends a DLP-response frame to the AP (2a in Figure 5.16), which contains the rate set, (extended) capabilities of QSTA-2 and the MAC addresses of QSTA-1 and QSTA-2.
(4) The AP forwards the DLP-response to QSTA-1 (2b in Figure 5.16), after which the direct link becomes active and frames can be sent from QSTA-1 to QSTA-2 and from QSTA-2 to QSTA-1.

## 5.8 Simulation Performance of IEEE 802.11

Event-driven simulation is used to evaluate the performance of the IEEE 802.11 MAC protocol. In this section, we present the simulation results obtained with our simulator BriarLAN. BriarLAN is based on OPNET 10.5.A PL1, with the May 2004 models. BriarLAN implements the 802.11e access mechanisms on top of the WLAN model provided by OPNET.

### 5.8.1 Throughput vs. Frame Size

In this scenario we use the following assumptions:

- there are two senders and two receivers running in the DCF mode with no interfering STAs nearby;
- the senders generate, at an infinite rate, $L$-byte long data frame (or MSDU);

- the MSDU is not fragmented;
- the propagation delays are neglected and
- the medium is error free.

One of the senders is the AP. The senders always have frames in the queues, and hence, the network is overloaded. The beacon interval is 100 ms. The PHY layer used in the simulation is 802.11a and 802.11b. In IEEE 802.11b (802.11a), the data transmission rate is 11 (54) Mb/sec, while the ACK and beacon transmission rates is 2 (6) Mb/s.

For EDCA simulation we use the default parameters shown in Table 5.2 for $AC\_VO$, with the exception of the TXOPlimit. The TXOPlimit is calculated so that five frames can always be transmitted during the TXOP independently of the frame size. This is shown in Figure 5.17 [27].

| Frame size | TXOP limit for 5 frames with Normal ACK (ms) | TXOP limit for 5 frames with No ACK (ms) |
|---|---|---|
| 100 | 2.765 | 1.475 |
| 200 | 3.129 | 1.839 |
| 400 | 3.856 | 2.566 |
| 600 | 4.583 | 3.293 |
| 800 | 5.310 | 4.020 |
| 1000 | 6.038 | 4.748 |
| 1200 | 6.765 | 5.475 |
| 1400 | 7.492 | 6.202 |
| 1600 | 8.220 | 6.930 |
| 1800 | 8.947 | 7.657 |
| 2000 | 9.674 | 8.384 |
| 2200 | 10.401 | 9.111 |
| 2304 | 10.780 | 9.490 |

**Figure 5.17**  TXOP duration for transmission of five frames in milliseconds.

Depending on the frame size of the STA, the TXOP limit is adjusted accordingly. For the HCCA simulations, the HC uses the Simple Scheduler as explained before. The HC piggybacks data with the CF-Poll, but does not piggyback data with the CF-ACK (when normal ACK policy is used). The CFP is 90 ms and the CP 10 ms. The wireless STA can use EDCA with $AC\_VO$ during the CP for the delivery of frames. The throughput and efficiency of the WLAN system is shown as a function of the frame size in Figure 5.18. This throughput is calculated at the MAC SAP, i.e., right above the MAC, and does not include the PHY or MAC overheads. When the frame sizes are smaller, the overheads like PLCP preamble, ACK or interframe spaces consume a significant amount of wireless channel capacity. When the frame size increases, these overheads become smaller and the throughput also increases. Figure 5.18 plots the throughput performance for the IEEE 802.11 legacy DCF and IEEE 802.11e access mechanisms with Normal and No ACK policies using 802.11b and 802.11a PHY respectively. Figure 5.18 also shows the throughput efficiency of the IEEE 802.11 DCF and IEEE 802.11e EDCA and HCCA as a function of the frame size. This efficiency measures the percentage of useful data bits carried over the 802.11 WLAN. For example, if the throughput at the MAC SAP, without PHY and MAC overheads, is 5.5 Mb/sec, and the PHY rate is 11 Mb/s, the efficiency is 50%. It is seen that, for the frame sizes of 2304, the HCCA with No ACK policy achieves around 88% efficiency.

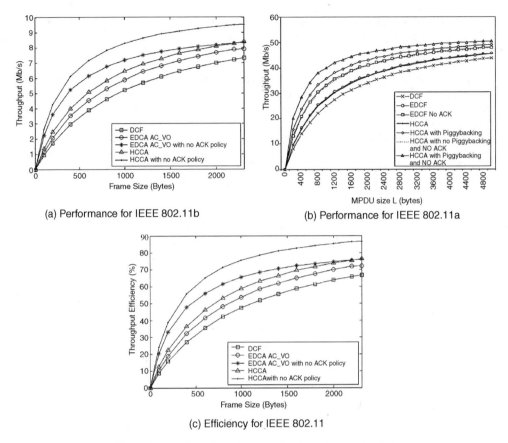

(a) Performance for IEEE 802.11b          (b) Performance for IEEE 802.11a

(c) Efficiency for IEEE 802.11

**Figure 5.18**   WLAN throughput as a function of frame size [27].

### 5.8.2 Throughput vs. Number of Stations

In this section, we run six separate simulation scenarios [27]: DCF, EDCA[AC_BK], EDCA[AC_BE], EDCA[AC_VI], EDCA[AC_VO], and HCCA. We evaluate the throughput performance as a function of the number of STAs in the network. The frame size is fixed to 1500 bytes. The beacon interval is 100 msec and the data transmission rate is 11 Mb/sec, while the ACK and beacon transmission rates is 2 Mb/sec. Each STA generates 2 Mb/sec of Constant Bit Rate (CBR) traffic.

For EDCA, each scenario consists of traffic belonging to the same and unique Access Category. The EDCA parameters in Table 5.2 are used. For HCCA the polled TXOP duration is fixed to 6 msec (therefore three frames of 1500 bytes can be transmitted during the TXOP with normal ACK policy and four frames with No ACK policy). The Simple Scheduler is used in this scenario as well, with the CFP of 90 ms. The results are shown in Figure 5.19.

In the case of EDCA[AC_VO] , the throughput drops dramatically after eight STAs start sending traffic in the network. As we are simulating using IEEE 802.11b PHY, with $aCW_{max} = 15$ for $AC\_VO$, the probability of collision in a slot increases considerably as the number of transmitting STAs increases. On the other hand DCF and EDCA[AC_BK] perform better as the $aCW_{max} = 1023$ and we would need more than 1000 STAs for the curve to drop faster towards 0. For the HCCA scenario, the throughput increases as a function of the number of STAs, and stabilizes when 20 or more STAs have traffic to send. It should be noted that the HC polls every STA, once during the CFP, even when the STA

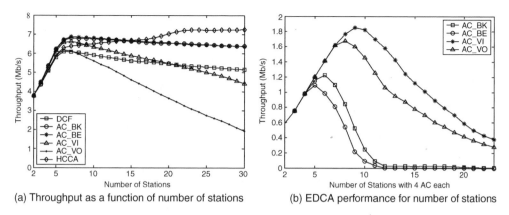

(a) Throughput as a function of number of stations          (b) EDCA performance for number of stations

**Figure 5.19**    WLAN throughput as a function of number of stations [27].

does not have traffic to send. Therefore, at the beginning of the curve, some time is spent on polling STAs that do not have any traffic. The throughput does not decrease for the HCCA scenario, due to the scheduling algorithm in the polling access mechanism.

### 5.8.3 EDCA ACs

In this scenario we consider EDCA STAs, each having traffic for all four ACs. Each AC has CBR traffic of 250 kb/sec with the frame size of 1500 bytes. Each STA will therefore generate a total of 1 Mb/s of traffic. The rest of simulation settings are the same as in the previous section. The results are shown in Figure 5.19(b). In this plot we find that, as the number of STAs is increased above five, the EDCA[AC_BK] starts dropping. EDCA[AC_BE] starts dropping with seven STAs in the network. The reason for this is that the higher priority ACs access the medium more aggressively than the lower priority ACs. As the number of STAs increases, the probability of collision increases, resulting in a steep fall in the throughput curves.

### 5.8.4 Effect of Bad Link

In this scenario we consider the effect of mobility on the overall performance of the network. Here we consider five STAs for each of the three different scenarios, namely: DCF, EDCA[AC_VO] and HCCA. We use the parameters specified in Table 5.2 for AC_VO and for the HCCA we use a TXOP limit of 5 msec. In all the above scenarios we allow one STA to move away from the AP, thus changing its modulation and reducing its physical transmission rate. Each scenario consists of four simulations with one step rate reduction in only one STA. We study the impact that this mobile STA can cause on the networkwide throughput performance, and in the individual throughput performances of other STAs. Figure 5.20 shows the network performance for each of the scenarios with four steps in rate reduction. DCF performs the worst. We observe that the introduction of a bad link also penalizes all other 'good link' STAs and their throughput is reduced to the throughput of the bad link STA (see also Figure 5.20(b)). This is because of the fair channel access provided by the CSMA/CA protocol. That is, DCF provides fair share per STA because the parameters like CWmin and DIFS are the same for all the STAs. Since all STAs have the same probability to access the channel, the resulting network throughput is reduced because of the increased time taken to transmit a frame by the mobile STA. The throughput of EDCA[AC_VO] is reduced but not that much. HCCA performs the best. In case of EDCA[AC_VO] and HCCA, the allocation of bandwidth is made by time or TXOPs, that is, EDCA and HCCA allocate the same time to the STAs irrespective of its modulation scheme. This is known as time fairness [11], so the

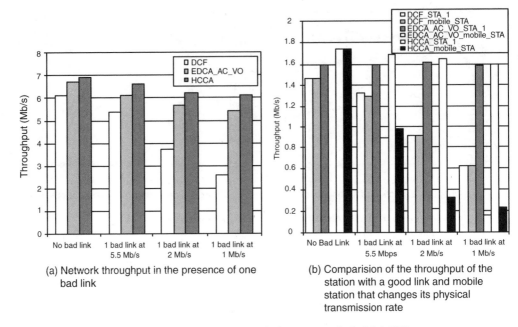

**Figure 5.20**  WLAN performance in the presence of a bad link [27].

STAs get their throughput share even in the presence of other bad link STAs. The 'bad link' STA must fragment the frame to fit within its fixed allocated TXOP limit. The overall throughput is slightly reduced due to the performance degradation in the 'bad link' STA. Figure 5.20(b) shows the comparison of throughput of a STA with a good link (STA1), and the STA with bad link (mobile STA). In the case of DCF, the 'bad link' STA also reduces the throughput of the 'good link' STAs as explained above, while in the case of EDCA[*AC_VO*] and HCCA the good link STA, denoted by STA1, is not affected. The EDCA[*AC_VO*] and HCCA's mobile STAs must use fragmentation in order to fit their frame sizes within their TXOP limit, so their throughput reduces considerably.

We also conclude that in the design of the EDCA as well as the HCCA parameters, one must use the TXOP limit other than zero. This is because a TXOP limit equal to zero, as specified in [7], allows an STA to transmit a single frame, irrespective of the length of the frame and/or physical transmission rate. Therefore mobility would have the same performance degradation as seen with DCF.

## 5.9 Support for VoIP in IEEE 802.11e

In this section, we provide simple formulae for calculating the maximum number of VoIP calls that can be admitted using AC3 in IEEE 802.11e EDCA [29]. In order to determine the maximum number of VoIP calls that can be supported by EDCA, we have to determine the maximum number of times a particular WSTA can wait before it successfully transmits its frame. This is represented by $N_{\text{max wait}}$ and is given by

$$N_{\text{max wait}} \leq \frac{D_{\text{end to end}}}{E[D]}. \tag{5.7}$$

Voice packets that cannot meet the delay bound or that have reached the maximum retry count should be dropped. Let $P_{\text{loss}}$ be the maximum loss probability as demanded by the application. Since all the VoIP stations have the same contention parameters, they access the medium with probability $1/N$. Assume that the tagged station has to wait $N_{\text{max wait}}$ before it can transmit. This is possible if the number of stations in the medium has reached the optimal point so that the frames are not dropped. This is represented by the equation below.

$$\sum_{k=1}^{N_{\text{max wait}}} \frac{1}{N}\left(1-\frac{1}{N}\right)^{k-1} \geq 1 - P_{\text{loss}}. \tag{5.8}$$

Simplifying the above equation we have

$$N \leq \frac{1}{1-(P_{\text{loss}})^{\frac{E[D]}{D_{\text{end to end}}}}}. \tag{5.9}$$

### 5.9.1 Comparison of Simulation and Analysis for VoIP Traffic

Simulations were carried out in OPNET by considering the entire IP stack implemented on top our IEEE 802.11e MAC and PHY. We modified the existing DCF code of OPNET and included HCF, IEEE 802.11a PHY and the channel errors. In order to compare the analysis presented for EDCA with the simulations, we consider the only VoIP scenario and calculate the maximum number of calls that can be admitted satisfying the QoS requirements of delay and packet loss. The packet loss was fixed at 1% and the Delay bound was fixed at 100 ms. The Table 5.5 summarizes the comparison between analysis and simulation. From the table V, there are few inferences that have to be noted. As the packet error rate increases, the number of VoIP calls admitted by the HCCA decreases proportionally to the PER. So we see for 10% PER, we have the capacity reduced to 58 calls from 65 calls and for 30% PER, we have the number of calls reduced to 45 from 68. This implies that the capacity of a VoIP over HCCA is proportional to the PER. Whereas the EDCA performs worst. This is due to the non-optimal and aggressive contention parameters set by the standard. Because of errors, there are lots of collisions due to retransmission attempts. More collisions reduce the throughput of the network. On the other hand, DCF performs well for VoIP in the presence of errors. This is because the contention parameters that are very large for successive retransmission attempts result in less collisions as the backoff stage increases. This does not mean that DCF is a good solution. It says that the probability of collision is low for DCF as the the backoff is doubled each time there is a collision or channel error. On the other hand the delay experienced by the VoIP call using EDCA is minimum because of the aggressive contention parameters, whereas the delays are larger in DCF because of larger contention windows. The second reason for DCF to perform better is because of the large Delay Bound of 100 ms. If we were to reduce the Delay Bound then the performance of DCF will begin to deteriorate. In the case of small Delay Bound, EDCA will still work well. Also, as we add more and more background traffic, the performance of DCF will deteriorate considerably, whereas the performance of EDCA will not.

**Table 5.5** Comparison of maximum number of VoIP calls that can be admitted by analysis and simulation [29]

| PER (%) | DCF | | EDCA | | HCCA | |
|---|---|---|---|---|---|---|
| | Analysis | Simulation | Analysis | Simulation | Analysis | Simulation |
| 0 | 45 | 45 | 35 | 35 | 65 | 65 |
| 10 | 42 | 45 | 33 | 34 | 58 | 58 |
| 30 | 40 | 42 | 11 | 13 | 46 | 45 |

**Table 5.6** Performance of DCF and EDCA in the presence of background traffic

| PER (%) | DCF | EDCA | HCCA |
| --- | --- | --- | --- |
| 0 | 41 | 14 | 65 |
| 10 | 40 | 11 | 58 |
| 30 | 24 | 4 | 45 |

Next, we consider the simulation scenario in the presence of background traffic. We have 1 HDTV source and four heavy FTP sources as background traffic. The HDTV source requires a bandwidth of 20 Mbps and the packet sizes of both FTP and HDTV are fixed at 1500 bytes. In HDTV uses AC2 in EDCA, which has the contention parameters given in Table 5.6. Table 5.6 summarizes the performance of DCF, EDCA and HCCA in the presence of background traffic. From the Table 5.6 we find that the HCCA is able to provide clear isolation for the admitted VoIP calls from the background traffic. In EDCA, we see that the VoIP frames experience more collisions but have least delay. For the DCF it is quite the opposite of EDCA. So far we have talked only about contention parameters like $aCW_{min}$ and $AIFS$. The $TXOPlimit$ selected by the draft standard, [7], for the EDCA $AC\_VO$ is optimal as it takes into account the effect of bad link [10]. Bad link is caused when the WSTAs start to roam and move to the edge of the cell and lower their PHY rate from 54 Mbps to 6 Mbps through link adaptation. Now the time required to transmit the same VoIP frame is nine times greater than that which would be required at 54 Mbps. The time to transmit a single VoIP frame is approximately 122 µsec. Even if the PHY transmission rate were to be reduce to 6 Mbps (this is required to maintain connection with the QAP), the TXOP time of 1.5 ms (from Table 5.4) is sufficient to transmit the VoIP frame at this reduced PHY rate. However, this results in less VoIP calls being packed. So a careful re-design of the contention parameters of EDCA is mandatory to make efficient use of EDCA for admitting more VoIP calls and still satisfying QoS constraints.

## 5.10 Video Transmission Over IEEE 802.11e

In this section, we compare the polling-based HCCA and the contention-based EDCA for their QoS support via simulations. We will emphasize the advantages of using the enhanced EDCA for QoS support as we discussed in [30] and to verify the effectiveness of the integrated airtime-based admission control and enhanced EDCA. The simulations are carried out in OPNET for two scenarios. In scenario 1, we compare the system efficiency, in terms of the number of streams being admitted into a wireless LAN under the EDCA and the HCCA. In scenario 2, we compare the two controlling methods, namely, controlling the TXOP limit and controlling medium accessing frequency, under the EDCA. We have modified the wireless LAN MAC of OPNET to include the admission control algorithm and the signaling procedures as explained above.

### 5.10.1 Scenario 1: System Efficiency

We assume that each station carries a single traffic stream which requests a DVD quality video using MPEG2 with guaranteed rate of 5 Mbps.[e] We also assume that all stations are required to transmit at 54 Mbps for QoS guarantees, and do not change their PHY rates. We increase the number of stations,

---

[e]The average bit rate of a DVD-quality (MPEG-2) video is about 5 Mbps.

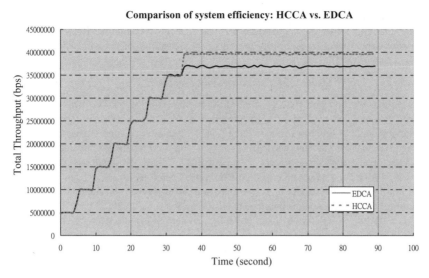

**Figure 5.21**   Comparison of system efficiency, in terms of the total throughput, between the HCCA and the EDCA. A new station carrying a single stream is added to the wireless LAN about every 5 seconds and transmits at 54 Mbps. The height of each 'stair' is equal to a stream's guaranteed rate = 5 Mbps.

starting from 1, until the wireless LAN cannot accommodate any more stations (or streams). For the EDCA case, we control the TXOP limit for airtime usage control. Since all streams have the same guaranteed rate ($g_i$ =5 Mbps) and minimum PHY rate ($R_i$ =54 Mbps), each station uses the same TXOP limit in this scenario. For the HCCA case, we follow the procedures in Section 5.5.7.

Figure 5.21 plots the total throughput under the HCCA and the EDCA. Since all stations request the same guaranteed rate, one can easily convert the total throughput to the total number of stations (i.e., streams) admitted into the wireless LAN. We increment the number of stations every 5 seconds in order to explicitly show the throughput received by individual streams. Prior to $t = 35$ second, every admitted stream gets exactly the 5-Mbps guaranteed rate under both the HCCA and the EDCA. It shows that using the enhanced EDCA can achieve the same QoS guarantees as using the polling-based HCCA.

After $t = 35$, the number of stations is increased to eight. The figure shows that using the EDCA cannot guarantee the streams' QoS any more because it needs a total throughput of 40 Mbps to support eight streams, but the wireless LAN can only provide about 37 Mbps. In our OPNET simulation trace, we observe a lot of frame drops under the EDCA (not shown but can be inferred from the figure), starting at $t = 37$ second. However, under the HCCA, all streams are still provided with the 5-Mbps guaranteed rate. This result is expected because the HCCA uses the polling-based channel access (in contrast to the contention-based EDCA), hence resulting in a higher efficiency. Based on the simulation results, one can also obtain the values of the effective airtime $EA$. Because all streams are transmitted at the same PHY rate, the value of $EA$ can be computed by

$$EA = \frac{System\ total\ throughput}{PHY\ rate} \tag{5.10}$$

Although using the HCCA achieves a better efficiency, it only generates $0.06 = 0.73 - 0.67$ second more data-transmission time (within a one-second period) or about 3 Mb more data frames when all stations transmit at 54 Mbps (the maximal PHY rate in the 802.11a PHY spec.). When stations use smaller PHY rates, the small difference between the EA values of the EDCA and the HCCA results in an even smaller throughput difference. Therefore, one can expect that using the EDCA and the HCCA will generate a similar performance, especially in terms of the total number of admissible streams.

### 5.10.2  Scenario 2: TXOP Limit vs. Medium Accessing Frequency

In this subsection, we compare the two controlling methods in the EDCA, namely, controlling the stations' TXOP limits and medium accessing frequency. We still assume that each stream requires a 5-Mbps guarantee rate. In order to emphasize the EDCA's quantitative control over the stations' diverse airtime usage, we assume that stations 1 and 2 carry a single traffic stream but stations 3 and 4 carry carry two streams. That is, there are six traffic streams in total. We again assume that all stations transmitted at 54 Mbps and do not change their PHY rate. Therefore, all streams are able to obtain their guaranteed rate based on the results in Scenario 1. In order to control the stations' medium accessing rate, we choose $CW_{min}$ as the control parameter. Therefore, we choose $CW_{min,1} = CW_{min,2} = 15(2^4 - 1)$ and $CW_{min,3} = CW_{min,4} = 31(2^5 - 1)$ based on [30], and set $CW_{max} = 63(2^6 - 1)$ for all stations. The TXOP limits are chosen according to [30].

Figure 5.22(a) plots the total throughput of using the two controlling methods. It shows that both methods generate identical results (in terms of throughput). One can observe that stations 1 and 2 both receive the 5-Mbps guaranteed rate after they join the wireless LAN at $t = 0$ and $t = 5$, while stations 3 and 4 both receive 10 Mbps (5 Mbps for each of their own two streams) after they join the wireless LAN at $t = 10$ and $t = 15$. The results show that both controlling methods can realize the distributed and quantitative control over stations' airtime usage. Here, the throughput is proportional to airtime usage, since all stations transmit at the same PHY rate.

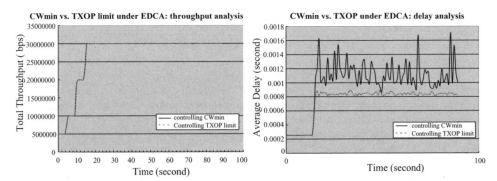

(a) Comparison of throughput between controlling stations' TXOP limits and $CW_{min}$ values. *The figures shows that in the EDCA, controlling stations' TXOP limits and $CW_{min}$ values result in the same performance in terms of streams' throughput

(b) Comparison of delay between controlling stations' TXOP limits and $CW_{min}$ values. *The figures shows that in the EDCA, controlling $CW_{min}$ values may result in a large delay variance but still satisfy all stream's delay bound.

**Figure 5.22**   Performance of tuning knobs for EDCA [30].

Figure 5.22(b) plots the delay under the two controlling methods. Once all four stations (all six streams) are admitted to the wireless LAN, the delay remains around 0.8 ms if using the TXOP Limit control, or fluctuates around 1.2 ms if using the $CW_{min}$ control. The reason why the delay fluctuates in the latter is that if stations using larger $CW_{min}$ (i.e., 31) collide with other stations, they use $CW_{max} = 63$ as the contention window size due to the exponential random backoff. Thus, these stations may wait much longer as compared to the case of controlling the TXOP Limit where stations (rarely) use

$CW_{max} = 63$ only when two consecutive collisions occur. In any case, the delay under both methods are well below the streams' delay bound, which is 200 ms in our simulation.

## 5.11  Comparison of HiperLAN/2 and IEEE 802.11e

In this section, we briefly compare the IEEE 802.11e and the HiperLAN/2 systems in terms of two different aspects: one is the protocol overhead, and the other is their features [28].

### 5.11.1  Protocol Overhead

Figure 5.23 shows the comparison between IEEE 802.11(e) and HiperLAN/2 in terms of the maximum achievable throughput for each PHY speed without considering the channel errors. While the throughput is not the measure to compare the QoS support capability, it is an important measure of the network performance. Note that the actual throughput will be lower than the maximum throughput depending on the channel condition, network topology, number of stations, frame sizes and higher layer protocols. The frame size from the higher layer used for 802.11(e) is assumed to be 1500 bytes.

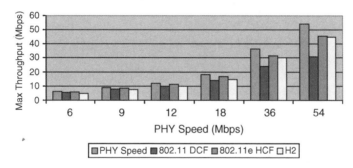

**Figure 5.23**   Network throughput of HiperLAN/2 and IEEE 802.11 at different physical rates [28].

First, we observe that 802.11e HCF achieves a significant improvement over the legacy 802.11 DCF. This is due to the reduced overheads achieved by HCF. This maximum throughput can be especially achieved thanks to (i) the piggybacking as was the case with PCF and (ii) the introduction of the block and No ACK mechanism. Second, we observe that the maximum throughput of HCF is better than that of HiperLAN/2 at all the PHY speeds. DCF is slight better than HiperLAN/2 for low PHY speeds, but significantly worse than HiperLAN/2 for high PHY speeds. The fact that HCF can be better than H2 in terms of the throughput is quite remarkable, since it has been known that 802.11 is worse than HiperLAN/2, which is more or less true if we consider the 802.11 DCF.

### 5.11.2  Comparison of Features

Looking at Figure 5.24, we can clearly see the advantages and implementation aspects of HiperLAN/2 and IEEE 802.11e. It is very important to note that both WLAN systems have the capability to support QoS. However, how a particular system works really depends on a number of algorithms implemented. Those algorithms include: (i) packet (or polling in case of HCF) scheduling algorithm; (ii) connection (or stream) admission control algorithm; (iii) link adaptation algorithm, i.e., to choose the best PHY speed; and (iv) DFS algorithm. Without having implemented such reasonably good algorithms, either system will fail to support the desirable QoS support. Since these algorithms are out of the scope of

| Features | 802.11e | | HiperLAN/2 |
|---|---|---|---|
| **MAC/DLC** | | | |
| QoS | Yes | | Yes |
| Centralized MAC | Yes (HCF) | | Yes |
| Distributed MAC | Yes   (EDCF) | | No |
| MAC FEC | Yes | | Yes |
| Direct Transmission | Yes | | Yes |
| **PHY** | 802.11b | 802.11a | |
| Frequency Bands | 2.4 GHz | 5 GHz | 5 GHz |
| Tx Speeds | 1–11 Mbps | 6–54 Mbps | 6–54 Mbps |
| Modulation | CCK | OFDM | OFDM |
| **Others** | | | |
| Dynamic Frequency Selection (DFS) | TGh working on it [8] | | Yes |
| Transmit Power Control (TPC) | TGh working on it [8] | | Yes |
| Wireless 1394 Support | TGe and 1394 WWG working on it | | Yes |
| Overlapping WLANs in one channel | Nearly fair channel sharing possible due to carrier sensing and NAV policy | | Not well supported; one WLAN should go to another channel via DFS |

**Figure 5.24**   Features of different standars [28].

standard specifications, and are implementation and vendor specific, one should be reminded that there will be a potential difference in performance among WLAN systems coming from different vendors.

## 5.12  Conclusions

In this chapter we gave a brief overview of HiperLAN and IEEE 802.11 protocols and brought out their essential features, showing how they can be used for granting QoS. It is clear that IEEE 802.11 has proliferated in the market more than HiperLAN because of its simplicity in hardware implementation. Also, we detailed how the IEEE standardization is addressing the shortfalls of the current IEEE 802.11n and how it is improving its efficiency. One major drawback of IEEE 802.11 is that it has overheads that are accompanied with the frame transmission that is clearly reducing its efficiency. So reducing the overheads is one of the key roles of the emerging IEEE 802.11n.

## 5.A  APPENDIX: Analysis of the Frame Error Rate and Backoff Process of EDCA Using One Dimensional Markov Chain

Assume that $n$ stations in a QBSS are operating under the EDCA mode. Each OFDM symbol is transmitted at the same power level. The frames are transmitted as a whole instead of fragments and we do not consider the hidden terminal effects. Also, we assume that the ACK frame is transmitted at the lowest rate, so that the probability of error in the ACK frame is zero. We calculate the frame error probability and include it in the analysis of the backoff process.

### 5.A.1  MAC/PHY Layer Overheads

In the IEEE 802.11e MAC, each data-type MPDU consists of the following basic components, as shown in Figure 5.25: a MAC header, which comprises a two-octet frame control field, a two-octet duration

| Frame Control (2) | Duration/ID (2) | Address 1 (6) | Address 2 (6) | Address 3 (6) | Sequence Control (2) | Address 4 (6) | QoS Control (2) | Frame Body (n) | FCS (4) |
|---|---|---|---|---|---|---|---|---|---|

**Figure 5.25**  Frame format of a data-type MSDU.

field, four six-octet address fields, and a two-octet sequence control information, a variable length information frame body, and a Frame Check Sequence (FCS), which contains an IEEE 32-bit cyclic redundancy code (CRC). All the fields except the frame body contribute to the MAC overhead for a data/fragment frame, which is 36 octets in total. Figure 5.26 illustrates the frame format of an ACK frame MPDU, which consists of a two-octet frame control field, a two-octet duration field, a six-octet address field, and a four-octet FCS field.

| Frame Control 2 Octets | Duration 2 Octets | RA 6 Octets | FCS 4 Octets |
|---|---|---|---|

**Figure 5.26**  Frame format of an ACK frame MSDU.

The PPDU format of the IEEE 802.11a PHY is shown in Figure 5.27, which includes PLCP preamble, PLCP header, MPDU, tail bits, and pad bits. The PLCP preamble field, with the duration of *tPLCPPreamble*, is composed of ten repetitions of a short training sequence (0.8 μs) and 2 repetitions of a long training sequence (4 μs). The purpose of the preamble is to enable the receiver to determine the channel status of each carrier and then take corrective actions on the bits received. The PLCP preamble is a fixed sequence and the receiver determines how the received signal is corrupt because of multi-path fading. In our case we don't consider the errors in the PLCP preamble. We assume that the preamble is transmitted correctly. Since the channel state can vary during the transmission time of a MAC frame, the PLCP does not give any information on the change of the channel status during the frame's transmission.

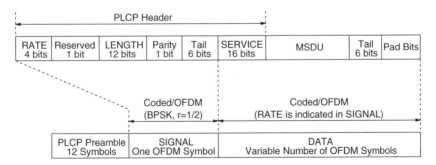

**Figure 5.27**  PPDU frame format of IEEE 802.11a OFDM PHY.

The PLCP header except the SERVICE field, with the duration of *tPLCPHeader*, constitutes a separate OFDM symbol, which is transmitted with BPSK modulation and the rate-1/2 convolutional coding. The six 'zero' tail bits are used to return the convolutional codec to the 'zero state', and the pad bits are used to make the resulting bit string a multiple of OFDM symbols. Each OFDM symbol interval, denoted by *tSymbol*, is 4 µs. The 16-bit SERVICE field of the PLCP header and the MPDU (along with six tail bits and pad bits), represented by DATA, are transmitted at the data rate specified in the RATE field.

### 5.A.2 Link Model

As indicated before, most of the studies, [16–19, 22, 23] assumed independent errors in modeling the channel state, which is not realistic in WLAN environment. We use the simple Gilbert–Elliot two state Markov model to account for the bursty nature of the wireless medium. The two states are 'GOOD' and 'BAD' states respectively. When the link state is 'GOOD', the probability of symbol error is negligible. (We consider symbol error instead of bit error.) When the link state is 'BAD' the symbol error probability is significant, resulting in a frame error. It is assumed that the sojourn times in 'GOOD' and 'BAD' states are geometric. We do the error analysis at the output of the decoder as the decoder uses symbol interleaving to correct some errors. So the error analysis is done at the output of the decoder which is appropriate to calculate the errors in the MAC frame.

As mentioned in the previous paragraph, the channel state process measured at the output of the decoder are described by the process $X(i)$. Let $p_{i,j}$ be the transition probability from state $i$ to state $j$ for $i, j \in \{g, b\}$. Here $g$ represents the 'GOOD' state and $b$ the 'BAD' state. The probability that the link state changes during the transmission time of one symbol is negligibly small. The transition probabilities of the link states of the Markov chain embedded at the beginning of the symbol are given by $u = p_{g,b}$ and $v = p_{b,g}$. This is shown in Figure 5.28.

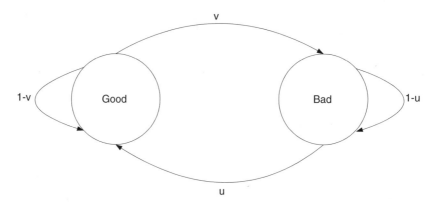

**Figure 5.28**   Two state Markov model.

For the channel state Markov chain described above, we have the following transition probabilities:

$$P_X(X(i) = g | X(i-1) = g) = 1 - v^{(m)}$$
$$P_X(X(i) = g | X(i-1) = b) = u^{(m)}$$
$$P_X(X(i) = b | X(i-1) = g) = v^{(m)}$$
$$P_X(X(i) = b | X(i-1) = b) = 1 - u^{(m)}. \qquad (5.11)$$

Here $m$ denotes the PHY modulation scheme used by the QSTA among the set of available modulations.

The second process is the status of the symbols transmitted over the channel. Let this be denoted by $Y(i)$, $i = 0, 1, 2, \ldots$. The symbol status is either 0, which means the symbol was transmitted successfully, or 1, meaning the symbol was in error. We have now the following relationship between the processes $X$ and $Y$:

$$P_Y(Y(i) = 0 | X(i) = g) = 1 - SER_g^{(m)}$$
$$P_Y(Y(i) = 1 | X(i) = g) = SER_g^{(m)}$$
$$P_Y(Y(i) = 0 | X(i) = b) = 1 - SER_b^{(m)}$$
$$P_Y(Y(i) = 1 | X(i) = b) = SER_b^{(m)}. \tag{5.12}$$

Here $SER_g$ and $SER_b$ are the symbol error probabilities when the channel state is in 'GOOD' and 'BAD' respectively.

The third process is the frame status. This process is described by $Z(j)$, $j = 0, 1, 2, \ldots$. The possible events of $Z$ are the frame being in error or error-free and are denoted by $B$ and $A$ respectively. Let the length of the MAC frame be $L$ bytes. The frame is composed of bits $1, 2, \ldots, L$. A frame is assumed to be correct if and only if all bits are transmitted correctly. For the $j$th frame, this is given by

$$P(Z(j) = A) = \prod_{i=1}^{\frac{L+30.75}{BpS}} P(Y(i) = 0). \tag{5.13}$$

Using the above three processes we can now evaluate the frame error probability for a modulation scheme $m$, if the channel state at the beginning first symbol being transmitted is known. The channel state is referred to the initial state of the frame. Remember that the channel state could change only at the edge of the symbol and not during the transition time of the symbol. For a frame of size $L$ bytes, the channel state can be $2^{\frac{L+30.75}{BpS}}$, where 30.75 represents the header overhead and $BpS$ represents the bytes per symbol. The probability of the channel state for each symbol can be calculated by the process $X$. Let us condition that the transmission of the first symbol of this frame happens when the channel state was 'GOOD'. So for the remaining $L + 30.75/BpS$ symbols the number of channel states is $2^{\frac{L+30.75}{BpS} - 1}$. Considering the error probability of each symbol, the total probability theorem is used to calculate the frame error probability. So the frame error probability given that the initial state is 'GOOD' is given by

$$P(Z(j) = A | X(1) = g) = \sum_{i=1,2,\ldots,\frac{L+30.75}{BpS};\; x_i \in \{g,b\}} \phi(g, x_1, x_2, \ldots, x_{\frac{L+30.75}{BpS}}), \tag{5.14}$$

where

$$\phi(g, x_1, x_2, \ldots, x_{\frac{L+30.75}{BpS}}) = P_Y(0/g) \prod_{i=2}^{\frac{L+30.75}{BpS}} P_Y(0|x_i) P_X(x_i | x_{i-1}).$$

A simple expression for $P(Z(j) = A | X(1) = g)$ is given by Equation 5.15.

$$[1\ 0] \times \begin{bmatrix} (1 - u^{(m)})(1 - SER_g^{(m)}) & u^{(m)}(1 - SER_g^{(m)}) \\ v^{(m)}(1 - SER_b^{(m)}) & (1 - v^{(m)})(1 - SER_b^{(m)}) \end{bmatrix}^{\frac{L+30.75}{BpS} - 1} \times \begin{bmatrix} 1 - SER_g^{(m)} \\ 1 - SER_b^{(m)} \end{bmatrix}. \tag{5.15}$$

Similarly the expression for $P(Z(j) = A | X(1) = b)$ is given by Equation 5.16.

$$[0\ 1] \times \begin{bmatrix} (1 - u^{(m)})(1 - SER_g^{(m)}) & u^{(m)}(1 - SER_g^{(m)}) \\ v^{(m)}(1 - SER_b^{(m)}) & (1 - v^{(m)})(1 - SER_b^{(m)}) \end{bmatrix}^{\frac{L+30.75}{BpS} - 1} \times \begin{bmatrix} 1 - SER_g^{(m)} \\ 1 - SER_b^{(m)} \end{bmatrix} \tag{5.16}$$

The PLCP header is transmitted using PHY mode 1. So in Equations (5.15) and (5.16) we will have one more matrix before the frame matrix, which is raised to the power 40. From the above equation we can get the unconditional probability of successful frame transmission as

$$
\begin{aligned}
1 - p_e^{(m)} = P(Z(j) = A) = P(Z(j)) \\
= A|X(1) = g)P(X(1) = g) + P(Z(j)) \\
= A|X(1) = b)P(X(1) = b),
\end{aligned}
\tag{5.17}
$$

where $p_e$ is the frame error probability and

$$
P(X(1) = g) = \frac{u^{(m)}}{u^{(m)} + v^{(m)}} \text{ and } P(X(1) = b) = \frac{v^{(m)}}{u^{(m)} + v^{(m)}}.
$$

The above equation can also be extended on the reverse channel where the ACK is transmitted. But we assume that this error is 0 for simplicity as the ACK frame is transmitted using the most robust modulation.

As indicated before, we use Rayleigh fading channel model and relate the parameters of the simple two state Markov chain to the Rayleigh fading model as explained in [24]. The Markov chain parameters are

$$
\varepsilon = 1 - e^{-1/F},
\tag{5.18}
$$

where $\epsilon$ is the steady state probability of symbol error and $F$ is the fading margin. The probability of error in symbol $i$ given that there was error in the $(i-1)$th symbol is given by $r$ in Eqn. (5.19).

$$
1 - v = r = P_X(X(i) = g|X(i-1) = g) = \frac{Q(\theta, \rho\theta) - Q(\rho\theta, \theta)}{e^{1/F-1}},
\tag{5.19}
$$

where

$$
Q(x, y) = \int_y^\infty e^{-\frac{x^2+w^2}{2}} I_0(xw) w \, dw
\tag{5.20}
$$

is the Marcum $Q$ function [25] and $I_0(x)$ is the modified Bessel function of the first kind. Also, $\rho = J_0(2\pi f_D t)$ is the Gaussian correlation coefficient of two samples of complex amplitude of a fading channel with Doppler frequency $f_D$, taken at a distance $T$. $J_0$ is the Bessel function of the first kind and zero order. $\theta$ is given by

$$
\theta = \sqrt{\frac{2}{F(1 - \rho^2)}}
\tag{5.21}
$$

The probability of no error in symbol $i$ given that there was error in $(i-1)$th symbol is given by

$$
u = P_X(X(i) = g|X(i-1) = b) = 1 - \frac{r\epsilon}{1 - \epsilon}.
\tag{5.22}
$$

$\varepsilon$ is related to the chain parameters in the following way:

$$
\varepsilon = P(X(1) = g)SER_g^{(m)} + P(X(1) = b)SER_b^{(m)}.
\tag{5.23}
$$

With the above equation and Equations 5.22 and 5.19 we can easily deduce the parameters of the Markov chain.

### 5.A.3 Backoff Procedure

As discussed in [18, 19], $b(t)$ is defined as a stochastic process that represents the value of the backoff counter for a given WSTA or queue at time slot $t$. Each WSTA or queue has a $AIFS[AC]$, $aCW_{min}[AC]$, $aCW_{max}[AC]$, $TXOP[AC]$ and $m[AC] + 1$ stages of backoff. Let $m'[AC]$ denote the maximum number of retries that is set for this access category. If the retry limit is greater than the maximum number of stages the contention window will not double after it has reached the maximum number of stages. On the other hand the retry limit can be less than the number of stages, also resulting in packet drop earlier. This is represented by the equation below (for the sake of convenience, we omit AC in the following equations and will add it appropriately).

$$W_i = \begin{cases} 2^i W_0 & : \text{ if } i \leq m, \ m < m' \\ 2^m W_0 & : \text{ if } i > m, \ m > m' \end{cases}.$$

Here $W_0 = aCW_{min}$ denotes the minimum contention window size for that particular access category. In fact we use $m'$ to show the short and long retry count as represented in Equation (5.24) in the form of two conditions. This is very important in determining the performance of the protocol. Before we proceed to analyze the backoff process of IEEE 802.11e EDCA, we need to understand the asynchronism introduced in IEEE 802.11e EDCA because of different $AIFS$. Figure 5.29 shows the channel access by stations in the EDCA mode. Since each station (or an EDCA entity) has a different AIFS, the station starts/resumes decrementing its backoff counter at a different time. Let $AIFS[i]$ be the value of station $i$'s AIFS. For example, station 1 may have a smaller AIFS than station 2 and let $AIFS[2] - AIFS[1] = 2$. Thus, every time station 2 starts to decrement its backoff counter, station 1 has already decremented its backoff counter by two slot times. Let $D$ be this 'decrementing lag' of station 2 with respect to station 1. $D$ is then a random variable with possible values 1 and 2. The reason why $D$ could be less than 2 is that if station 1 chooses a backoff counter value less than 2 (say 1), then station 2 will have no chance to start/resume its backoff process before station 1 finishes its transmission. In this case, $D = 1$. So when we are modelling the EDCA backoff process we must take into consideration this asynchronism introduced by different AIFS.

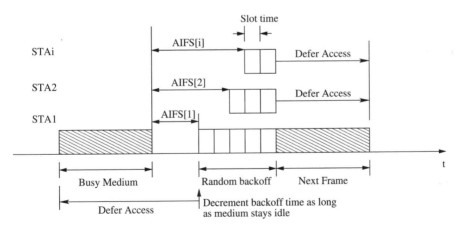

**Figure 5.29** Asynchronism due to different AIFS for different priorities.

Instead of using a two dimensional Markov chain, we use a single dimension Markov chain to characterize the backoff process. Let $b_{i,j,k}$ represent the backoff process with variable $i$ representing the backoff stage, variable $j$ representing the exact backoff value and variable $k$ representing the slot in the

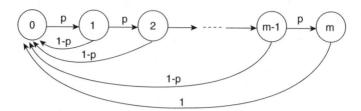

**Figure 5.30**   Representation of the backoff procedure by a Markov chain.

decrementing lag. The backoff stage alone can be represented by the Markov chain shown in Figure 5.30. From the Markov chain it is easily know on how the backoff process evolves. The node enters state $i + 1$ from $i$, if the frame was corrupted because of noise or collision and this is represented by $p = 1 - (1 - p_e)(1 - p_c)$. The following equation represents this behavior:

$$P_{i,i+1} = \Pr\{B_{t+1} = i + 1/B_t = i\} = p. \tag{5.24}$$

If the frame transmission was successful, then the backoff process is reset and will start the next backoff from stage 0 for the transmission of the new frame. This is represented by the following equation:

$$P_{i,0} = \Pr\{B_{t+1} = 0/B_t = i\} = 1 - p$$
$$P_{i,j} = 0, \ \forall j \neq i + 1. \tag{5.25}$$

Let $P_i$ be the steady state probability that the node is in state $i$. This is given by

$$P_i = \frac{1 - p}{1 - p^{m+1}} p^i. \tag{5.26}$$

Upon entering each state $i$, the node chooses a backoff value, $b_i$, uniformly distributed between $[1, W_i(= 2^i W_0)]$. The backoff value chosen by a node depends on the backoff stage it is currently in. So the average value of the backoff chosen by the node is given by

$$\overline{b}_i = \sum_{k=1}^{W_i} \frac{1}{W_i} k = \frac{W_i + 1}{2}. \tag{5.27}$$

Let $B_i$ be the probability that the node is in state $i$ at any given time. This is given by

$$B_i = \frac{P_i b_i}{\sum_{j=0}^{m} P_j b_j} = \frac{(1 - p)(1 - 2p)p^i(W_i + 1)}{W_0(1 - (2p)^{m+1})(1 - p) + (1 - 2p)(1 - p^{m+1})}. \tag{5.28}$$

From the above equation it is clear that the maximum value of $p$ is 0.5. In fact $p \leq 0.5$ is the necessary condition for the system to reach the steady state. So we have derived the first equation showing $B_i$ as the function of $W_0$ and $p$. But we now need to consider the other dynamics of EDCA, such as decrementing lag and the influence of the backoff process every slot time based on the medium being idle or busy.

In order to characterize the EDCA, let us now define the conditional probability of a node being in stage $i$. This is represented by the following equation:

$$B_i = \sum_{j=1}^{W_i+1} \sum_{k=1}^{q} \Pr\{t = j, k/i\} = \sum_{j=1}^{W_i+1} \sum_{k=1}^{q} B_{i,j,k}. \tag{5.29}$$

Here $B_{i,j,k}$ is the probability that the backoff counter has a value $j$ in stage $i$ and has a decrementing lag of $k$. This means that the node has to wait for another $k$ slots before it can start decrementing its backoff time. The backoff counter can be decremented only if there is no activity in the medium for the period of the decrementing lag. Once the medium is idle for the AIFS period, the backoff counter is decremented every slot time if the medium is idle and freezes with the current value if the medium is busy. Let $p_b$ be the probability that the medium is busy in a arbitrary slot after the node's AIFS and let $p_{ab}$ denote the probability that the slot is busy during the decrementing lag. So the backoff process of the decrementing lag portion as well as the normal backoff process can be represented by the Markov chain shown in Figures 5.31 and 5.32.

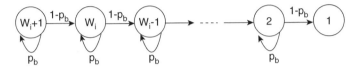

**Figure 5.31**   Representation of the backoff process by a Markov chain.

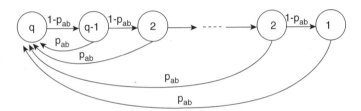

**Figure 5.32**   Representation of the decrementing lag process by a Markov chain.

For highest priority AC, this probability is 0 as there is no decrementing lag. Now we can express $B_{i,j,k}$ as follows:

$$B_{i,j,k} = B_{i,W_i,q} \frac{(W_i - j)}{(1 - p_b)} \frac{(q - k)}{(1 - p_{ab})}. \tag{5.30}$$

The denominator terms account for the average time it takes to decrement one slot time. The $q$ in the above equation denotes the decrementing lag between this node's AC and the highest priority AC in the network. So the Equation (5.28) can be re-written as

$$B_i = \sum_{j=1}^{W_i} \sum_{k=0}^{q} B_{i,j,k}. \tag{5.31}$$

Equation (5.39) is derived in the same way as we get Equation (5.29). Equation (5.39), after some algebra based on Markov chains in Figures 5.31 and 5.32 can be rewritten as:

$$B_i = \sum_{j=1}^{W_i} \sum_{k=0}^{q} B_{i,j,k} \cdot \frac{(W_i - j)}{(1 - p_b)} \cdot \frac{(q - k)}{(1 - p_{ab})}. \tag{5.32}$$

$B_{i,j,k}$ is expressed as a function of $B_{1,W_i,q}$ by the following equation:

$$B_{i,j,k} = B_{i,W_i,q}(q - k)(W_i - j). \tag{5.33}$$

Equation (5.32) can be simplified to yield

$$B_{1,W_i,q} = \frac{4B_i(1-p_b)(1-p_{ab})}{W_i q(q+1)}.$$ (5.34)

Using the above equation and Equation (5.33) we have

$$B_{i,j,k} = \frac{(q-k)(W_i-j)4(1-p)(1-2p)p^i(1-p_b)(1-p_{ab})}{\left[W_0(1-(2p)^{m+1})(1-p) + (1-2p)(1-p^{m+1})\right]W_i q(q+1)}.$$ (5.35)

Now making $j$ and $k$ equal to 0 in Equation (5.35) we get

$$B_{i,0,0} = \frac{4(1-p)(1-2p)p^i(1-p_b)(1-p_{ab})}{\left[W_0(1-(2p)^{m+1})(1-p) + (1-2p)(1-p^{m+1})\right](q+1)}.$$ (5.36)

From the above equation it is easier to determine the probability of transmission, $\tau$, in an arbitrary slot:

$$\tau = \sum_{i=0}^{m} B_{i,0,0} = \frac{4(1-2p)(1-p_b)(1-p_{ab})(1-p^{m+1})}{\left[W_0(1-(2p)^{m+1})(1-p) + (1-2p)(1-p^{m+1})\right](q+1)}.$$ (5.37)

The above equation considers the decrementing lab as the difference between the current node's AIFS and the highest priority's AIFS. But if we have a system with more than one priority, we have to include intermediate priorities and their access probabilities before the considered node can decrement its backoff value. For the sake of simplicity we assume the $1-p_{ab}$ is valid even if there are multiple priorities with different AIFS and hence the above equation is valid.

### Example 1
Consider that there are only two ACs in the system. Let their $CW_{\min}$, $CW_{\max}$ and $TXOP$ be equal. Let AC1 have $AIFS{=}DIFS$ and the second AC2 have $AIFS{=}PIFS$. From Equation (5.37), we get a following simple relation on the access probabilities:

$$\tau_2 = \frac{1-p_{ab}}{2}\tau_1.$$ (5.38)

This implies that the access probability is lowered by half and if the number of stations of priority 1 is very large then it further lowers the probability of priority 2's access.

Let us now introduce the four access categories as in EDCA and determine all the associated probabilities. The probability that the tagged station of priority $(AC = l(= 0, 1, 2, 3))$ transmits at slot $t$ is given by

$$\tau_t^l = \begin{cases} \sum_{i=0}^{m} B_{i,0,0}^l & : \quad \text{if } t > AIFS[l] \\ 0 & : \quad \text{if } t \leq AIFS[l]. \end{cases}$$ (5.39)

Equation (5.39) states that after the medium becomes idle following the busy period, the transmission probability of the node with priority $l$ is 0 if $AIFS[l]$ is not completed.

If the number of stations of each class is $N_l(l = 0, \ldots, 3)$, then the probability of the channel is busy at an offset slot $t$ is given by

$$p_{b,t}^l = 1 - (1-\tau_t^l)^{N_l-1}\prod_{h\neq l}(1-\tau_t^h)^{N_h}.$$ (5.40)

Equation (5.40) accounts for the fact that the tagged station of class $l$ sees the channel is busy only when at least one of the other station transmits. After calculating the busy probability, we go on to find the probability of successful transmission of priority $l$ in an offset slot $t$. This is given by

$$p^l_{succ,t} = \binom{N^l}{1} \tau^l_t (1 - \tau^l_t)^{N^l-1} \prod_{h \neq l} (1 - \tau^h_t)^{N^h}.$$  (5.41)

Similarly we the probability that the offset slot $t$ is idle is given by

$$p_{idle,t} = \prod_{h=0}^{3} (1 - \tau^h_t)^{N_h}.$$  (5.42)

Now we can easily evaluate the probability of collision at offset slot $t$ as

$$p_{coll,t} = 1 - p_{idle,t} - \sum_{h=0}^{3} p^h_{succ,t}.$$  (5.43)

Based on the above three equations, it is easy to calculate the throughput of the EDCA system. The transmission cycle under the EDCA of the IEEE 802.11e MAC consists of the following phases, which are executed repetitively: the $AIFS[AC]/SIFS$ deferral phase, the backoff/contention phase if necessary, the data/fragment transmission phase, the SIFS deferral phase, and the ACK transmission phase. The related characteristics for the IEEE 802.11a PHY are listed in Table 5.3. As indicated in [19,21], we assume that each transmission, whether successful or not, is a renewal process. Thus it is sufficient to calculate the throughput of the EDCA protocol during a single renewal interval between two successive transmissions. We extend the same philosophy for the EDCA bursting. The throughput of the protocol without bursting is given by Equation (5.44):

$$S^h = \frac{E[Time \; for \; successful \; transmission \; in \; an \; interval]}{E[Length \; between \; two \; consecutive \; transmissions]}$$

$$= \frac{\sum_t \sum_{h=0}^{3} \tau^h_t p^h_{succ,t} L^h}{\sum_t \tau^h_t (\sum_{h=0}^{3} T^h_s p^h_{succ,t} + T_c p_{coll,t} + aSlotTime \cdot p_{idle,t})}.$$  (5.44)

$T_s$ is the average time the channel is captured with successful transmission and $T_c$ is the average time the channel is captured by unsuccessful transmission. The values of $T_s$ and $T_c$ are given by

$$T_s = AIFS[AC] + \delta + T^m_{data}(L) + aSIFSTime + T^m_{ack} + \delta$$  (5.45)

$$T_c = AIFS[AC] + T^m_{data}(L) + aSIFSTime + T^m_{ack}.$$  (5.46)

The $\delta$ in the above equation represents the propagation delay. Also the $T_c$ is equal to the frame transmission time excluding the propagation delay because of Network Allocation Vector (NAV) set by the transmitting QSTA.

*5.A.4 Throughput Analysis for EDCA Bursting*

In the case of EDCA bursting, we need to know the maximum number of frames that can be transmitted during the EDCA TXOP limit. Let $T^l_{EDCA\_txop}$ represent the TXOP limit for this AC. Therefore the maximum number of frames of priority $l$, $N^l_{max}$, that can be transmitted by a specific queue when it gets to access the channel $T^l_{EDCA\_txop}$ is given by Equation (5.47):

$$N_{max} = \left\lfloor \frac{T^l_{EDCA\_txop}}{(AIFS[l] - aSIFSTime) + 2 \cdot (aSIFSTime + \delta) + T^m_{data}(L) + T^m_{ack}(L)} \right\rfloor.$$  (5.47)

In Equation (5.47) , the first term on the denominator comes from the fact that we have used *aSIFSTime* as the time between the transmission of the data frame as well as acknowledgment frame. In reality the first frame has deference given by *AIFS*[*l*]. For the throughput analysis, as we considered for single frame transmission, we consider the period between two transmissions. This assumption is valid as each WSTA that contends for the channel normally and if it gets the channel time, it transmits multiple frames instead of one. Once the WSTA wins the contention, the number of frames it transmits is upper bounded by Equation (5.47) . So on an average, the number of successful frame transmissions during and EDCA TXOP limit is given by:

$$N'_{max} = \left\lfloor \frac{T_{EDCA\_txop}[l]}{[(AIFS[l] - aSIFSTime) + 2 \cdot (aSIFSTime + \delta) + T^m_{data}(L) + T^m_{ack}(L)]N_{Transmissions}} \right\rfloor. \quad (5.48)$$

The throughput is the same as discussed in the previous subsection.

# References

[1] ETSI, HiperLAN Functional Specification, ETSI Draft Standard, July 1995.
[2] G. Anastasi, L. Lenzini and E. Mingozzi, Stability and Performance Analysis of HiperLAN, *IEEE JSAC*, **30**(90), 1787–1798, 2000.
[3] K. Pahlavan and P. Krishnamurthy, Principles of Wireless Networks, Prentice Hall, 2002.
[4] B. Walke, N. Esseling, J. Habetha, A. Hettich, A. Kadelka, S. Mangold, J. Peetz, and U. Vornefeld, IP over Wireless Mobile ATM – Guaranteed Wireless QoS by HiperLAN/2, in *Proceedings of the IEEE*, **89**, pp. 21–40, January 2001.
[5] IEEE Std 802.11-1999, Part 11: Wireless LAN Medium Access Control (MAC) and Physical Layer (PHY) specifications, Reference number ISO/IEC 8802-11:1999(E), IEEE Std 802.11, 1999.
[6] IEEE Std 802.11a, Wireless LAN Medium Access Control (MAC) and Physical Layer (PHY) specifications: Higher-speed Physical Layer Extension in the 5 GHz Band, Supplement to Part 11, IEEE Std 802.11a-1999, 1999.
[7] IEEE 802.11e/D7.0, Draft Supplement to Part 11: Wireless Medium Access Control (MAC) and physical layer (PHY) specifications: Medium Access Control (MAC) Enhancements for Quality of Service (QoS), June 2003.
[8] IEEE 802.1d-1998, Part 3: Media Access Control (MAC) bridges, ANSI/IEEE Std. 802.1D, 1998.
[9] Sunghyun Choi, Javier del Prado, Sai Shankar N and Stefan Mangold, IEEE 802.11e Contention-Based Channel Access (EDCA) Performance Evaluation, in *Proc. IEEE ICC'03*, Anchorage, Alaska, USA, May 2003
[10] Javier del Prado and Sai Shankar et al. Mandatory TSPEC Parameters and Reference Design of a Simple Scheduler, IEEE 802.11-02/705r0, November 2002.
[11] C.T. Chou, Sai Shankar N and K.G. Shin, Distributed control of airtime usage in multi-rate wireless LANs, submitted to *IEEE Transactions on Networking*.
[12] Maarten Hoeben and Menzo Wentink, Enhanced D-QoS through Virtual DCF, IEEE 802.11-00/351, October 2000.
[13] Stefan Mangold, Sunghyun Choi, Peter May, Ole Klein, Guido Hiertz and Lothar Stibor, IEEE 802.11e Wireless LAN for Quality of Service, in *Proc. European Wireless '02*, Florence, Italy, February 2002.
[14] Sunghyun Choi, Javier del Prado, Atul Garg, Maarten Hoeben, Stefan Mangold, Sai Shankar and Menzo Wentink, Multiple Frame Exchanges during EDCA TXOP, IEEE 802.11-01/566r3, January 2002.
[15] J. G. Proakis, *Digital Communications, 3rd ed*, McGraw Hill, New York, NY, 1995.
[16] M. B. Pursley and D. J. Taipale, Error probabilities for spread spectrum packet radio with convolutional codes and viterbi decoding, *IEEE Trans. Commun.*, **35**(1), pp. 1–12, Jan. 1987.
[17] D. Haccoun and G. Begin, High-rate punctured convolutional codes for Viterbi and sequential decoding, *IEEE Trans. Commun.*, **37**(11), pp. 1113–1125, November 1989.
[18] F. Cali, M. Conti and E. Gregori, Dynamic Tuning of the IEEE 802.11 Protocol to achieve a theoretical throughput limit, *IEEE/ACM Trans. Netw.*, **8**(6), December 2000.
[19] G. Bianchi, Performance Analysis of the IEEE 802.11 Distributed Coordination Function, *IEEE Journal on Selected Areas in Communications*, **18**(3), March 2000.

[20] H. S. Chhaya and S. Gupta, Performance modeling of asynchronous data transfer methods of IEEE 802.11 MAC protocol, *Wireless Networks*, **3**, pp. 217–234, 1997.

[21] H. Wu *et al.*, Performance of reliable transport protocol over IEEE 802.11 Wireless LAN: Analysis and enhancement, *Proc. IEEE INFOCOM'02*, New York, June 2002.

[22] Daji Qiao and Sunghyun Choi, Goodput enhancement of IEEE 802.11a wireless LAN via link adaptation, in *Proc. IEEE ICC'01*, Helsinki, Finland, June 2001.

[23] Daji Qiao, Sunghyun Choi, Amjad Soomro and Kang G. Shin, Energy-efficient PCF operation of IEEE 802.11a wireless LAN, in *Proc. IEEE INFOCOM'02*, New York, June 2002.

[24] M. Zorzi, Ramesh R. Rao and L. B. Milstein, On the accuracy of a first-order Markov model for data transmission on fading channels, in *Proc. IEEE ICUPC'95*, pp. 211–215, November 1995.

[25] J. I. Marcum, A Statistical theory of target detection by pulsed radar: mathematical appendix, *IEEE Trans. Info. Theory*, pp. 59-267, Apr. 1960.

[26] M. Heusse, Franck Rousseau, Gilles Berger-Sabbatel and Andrzej Duda, Performance anomaly of IEEE 802.11b, in *IEEE INFOCOM 2003*, San Francisco, USA.

[27] Delprado, J. and Sai Shankar N., Impact of frame size, number of stations and mobility on the throughput performance of IEEE 802.11e WLAN, in *IEEE WCNC 2004*, Atlanta, USA.

[28] Sunghyun Choi, Chiu Ngo, and Atul Garg, Comparative Overview on QoS Support via IEEE 802.11e and HIPERLAN/2 WLANs, Philips Research USA Internal Document, June 2000.

[29] Sai Shankar N., Javier Delprado, and Patrick Wienert, Optimal packing of VoIP calls in an IEEE 802.11a/e WLAN in the presence of QoS Constraints and Channel Errors, to appear in *IEEE Globecom 2004*, Dallas, USA.

[30] Chou, C. T., Sai Shankar N, and Shin K. G. Per-stream QoS in the IEEE 802.11e Wireless LAN: An integrated airtime-based admission control and distributed airtime allocation, submitted to *IEEE INFOCOM 2005*, Miami, USA.

# 6

# Wireless Multimedia Personal Area Networks: An Overview

Minal Mishra, Aniruddha Rangnekar and Krishna M. Sivalingam

## 6.1 Introduction

The era of computing has now shifted from traditional desktop and laptop computers to small, handheld personal devices that have substantial computing, storage and communications capabilities. Such devices include handheld computers, cellular phones, personal digital assistants and digital cameras. It is necessary to interconnect these devices and also connect them to desktop and laptop systems in order to fully utilize the capabilities of the devices. For instance, most of these devices have personal information management (PIM) databases that need to be synchronized periodically. Such a network of devices is defined as a Wireless Personal Area Network (WPAN). A WPAN is defined as a network of wireless devices that are located within a short distance of each other, typically 3–10 meters. The IEEE 802.15 standards suite aims at providing wireless connectivity solutions for such networks without having any significant impact on their form factor, weight, power requirements, cost, ease of use or other traits [1]. In this chapter, we will explore the various network protocol standards that are part of the IEEE 802.15 group. In particular, we describe IEEE 802.15.1 (Bluetooth®) offering 1–2 Mbps at 2.4 GHz, IEEE 802.15.3 (WiMedia) offering up to 55 Mbps at 2.4 GHz, IEEE 802.15.3a offering several hundred Mbps using Ultra-wide-band transmissions, and IEEE 802.15.4, which is defined for low-bit rate wireless sensor networks.

The IEEE 802.15 group adopted the existing Bluetooth® standard [2] as part of its initial efforts in creating the 802.15.1 specifications. This standard uses 2.4 GHz RF transmissions to provide data rates of up to 1 Mbps for distances of up to 10 m. However, this data rate is not adequate for several multimedia and bulk data-transfer applications. The term 'multimedia' is used to indicate that the information/data being transferred over the network may be composed of one or more of the following media types: text, images, audio (stored and live) and video (stored and streaming). For instance, transferring all the contents of a digital camera with a 128 MB flash card will require a significant amount of time. Other high-bandwidth demanding applications include digital video transfer from a camcorder, music transfer from a personal music device such as the Apple iPod™. Therefore, the 802.15 group is examining newer technologies and protocols to support such applications.

*Emerging Wireless Multimedia: Services and Technologies*   Edited by A. Salkintzis and N. Passas
© 2005 John Wiley & Sons, Ltd

There are two new types of Wireless Personal Area Networks (WPAN) that are being considered: the first is for supporting low speed, long life-time and low cost sensor network at speeds of a few tens of kbps and the other is for supporting the multimedia applications with higher data rates of the order of several Mbps with better support for Quality of Service (QoS). Our focus, in this chapter, is on the second type of WPAN dealing with multimedia communication. In an effort to take personal networking to the next level, a consortium of technology firms has been established, called the WiMedia Alliance[3]. The WiMedia Alliance develops and adopts standards-based specifications for connecting wireless multimedia devices, including: application, transport, and control profiles; test suites; and a certification program to accelerate wide-spread consumer adoption of 'wire-free' imaging and multi-media solutions.

Even though the operations of the WPAN may resemble that of WLAN (Wireless Local Area Networks), the interconnection of personal devices is different from that of computing devices. A WLAN connectivity solution for a notebook computer associates the user of the device with the data services available on, for instance, a corporate Ethernet-based intranet. A WPAN can be viewed as a personal communications bubble around a person, which moves as the person moves around. Also, to extend the WLAN as much as possible, a WLAN installation is often optimized for coverage. In contrast to a WLAN, a WPAN trades coverage for power consumption.

The rest of this chapter is organized as follows. The following section gives a brief overview of the multimedia data formats and application requirements. In Section 6.3, we present the Bluetooth protocols as described in the IEEE 802.15.1 standard. In Section 6.4, we discuss issues related to coexistence of Bluetooth networks with other unlicensed networks operating in the same frequency region. The IEEE 802.15.3 protocol suite for multimedia networks is considered in Section 6.5. In addition, we also describe ultra-wide-band (UWB) based networks that offer data rates of several hundred Mbps. In order to complete the discussions of the entire IEEE 802.15 group of standards, we also present the IEEE 802.15.4 standard for low-rate Wireless Personal Area Networks.

## 6.2 Multimedia Information Representation

In general, the term 'multimedia traffic' denotes a set of various traffic types with differing service requirements. The classical set of multimedia traffic include audio, video (stored or streaming), data and images [4,5]. The different types of media have been summarized in the Figure 6.1. Some applications generate only one type of media, while others generate multiple media types. The representation and compression of multimedia data has been a vast area of research. In this section, we present an overview of multimedia information representation. We will consider an example scenario that consists of a desktop computer, a laptop computer, and several digital peripheral devices such as digital camera, digital camcorder, MP3 player, Personal Music Storage device (e.g. iPod$^{TM}$), laser printer, photo printer, fax machine, etc.

The applications involving multimedia information comprise blocks of digital data. For example, in the case of textual information consisting of strings of characters entered at a keyboard, each character is represented by a unique combination of fixed number of bits known as a codeword. There are three types of text that are used to produce pages of documents: unformatted or plain text, formatted text and hypertext. Formatted text refers to text rich documents that are produced by typical word processing packages. Hypertext is a form of formatted text that uses hyperlinks to interconnect a related set of documents, with HTML, SGML and XML serving as popular examples.

A display screen of any computing device can be considered to be made of a two dimensional matrix of individual picture elements (pixels), where each pixel can have a range of colors associated with it. The simplest way to represent a digitized image is using a set of pixels, where each pixel uses 8 bits of data allowing 256 different colors per pixel. Thus, a $600 \times 300$ picture will require approximately 175 kb of storage. Compression techniques can be used to further reduce the image size. An alternate representation is to describe each object in an image in terms of the object attributes. These include

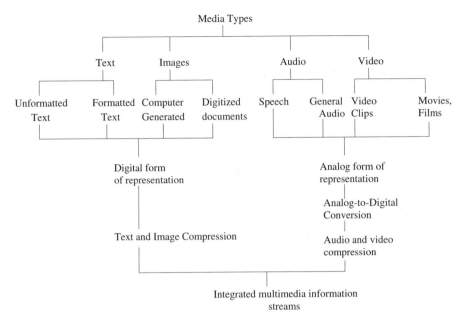

**Figure 6.1** Different types of media used in multimedia applications.

its shape, size (in terms of pixel positions of its border coordinates), color of the border, and shadow. Hence the computer graphic can be represented in two different ways: a high level version (specifying the attributes of the objects) and an actual pixel image of the graphic, also referred to as the bitmap format. It is evident that the high level version is more compact and requires less memory. When the graphic is transmitted to another host, the receiver should be aware of high-level commands to render the image. Hence, bitmaps or compressed images are used more often.

The commonly used image formats are GIF (graphic interchange format), TIFF (tagged image file format), JPEG (Joint Photographers Experts Group) and PNG (Portable Network Graphics). Compressed data formats also exist for transferring fax images (from the main computer to the fax machine). In order to understand the data requirements, let us consider a 2 Mega-Pixel (2 MP) digital camera, where the size of each image typically varies from 1 Mb to 2 Mb, depending on the resolution set by the user. A 256 Mb memory card can store approximately 200 photos. There is always a need to periodically transfer these digital files to a central repository such as a PC or a laptop. This is often done using the USB interface, which can provide data rates of up to 12 Mbps for USB 1.1 and up to 480 Mbps for USB 2.0. However, our intention is to use wireless networking for interconnecting such multimedia devices and the computer. The Bluetooth® standard provides data rates of 1 Mbps which is inadequate compared with the USB speeds. For instance, 128 Mb worth of multimedia files would take at least 18 minutes to transfer from a camera to PC. This is the reason for the development of the higher bit-rate IEEE 802.15.3 wireless PAN standard.

For audio traffic, we are concerned with two types of data: (i) speech data used in inter-personal applications including telephony and video-conferencing and (ii) high-quality music data. Audio signals can be produced either naturally using a microphone or electronically using some form of synthesizer [5]. The analog signals are then converted to digital signals for storage and transmission purposes. Let us consider the data requirements for audio traffic. Audio is typically sampled at 44100 samples per second (for each component of the stereo output) with 1 byte per second to result in a total of approximately 705 kbps. This can be compressed using various algorithms, with MP3 (from the Motion Picture Experts

Group) [6] being one of the most popular standards that can compress music to as around 112–118 kbps for CD-quality audio. Thus, streaming audio between a single source-destination pair is possible even with Bluetooth®. However, if there are several users in a WPAN, each having different audio streams in parallel, then higher bandwidths are necessary.

However, to store a CD-quality 4–5 minute song requires approximately 32 Mb of disk space. Hence, bulk transfer of audio files between a computer and a personal music device (such as the Apple iPod™) requires a large bandwidth for transmission. There are several different ways to compress this data before transmission and decompress it at the receiver's end. The available bandwidth for transmission decides the type of audio/video compression technique to be used.

Real-time video streaming with regular monitor-sized picture frames is still one of the holy grails of multimedia networking. Video has the highest bandwidth requirement. For instance, a movie with 30 frames per second (fps), with $800 \times 600$ pixels per frame and 8 bits per pixel requires an uncompressed bandwidth of 115 Mbps. There have been several compression standards for video storage. The MPEG-1 standard used on Video-CDs requires bandwidth of approximately 1.5 Mbps for a $352 \times 288$ pixel frame. The MPEG-2 standard used on DVDs today supports up to $720 \times 576$ pixel-frame with 25 fps for the PAL standard and $720 \times 480$ pixel-frame with 30 fps for the NTSC standard. The effective bandwidth required ranges from 4 Mbps to 15 Mbps. The MPEG-4 standard, approved in 1998, provides scalable quality, not only for high resolution, but also for lower resolution and lower bandwidth applications. The bandwidth requirements of MPEG-4 are very flexible due to the versatility of the coding algorithms and range from a few kbps to several Mbps. It is clear that higher bandwidth WPANs such as IEEE 802.15.3 are necessary to handle video traffic. Other video standards such as High-Definition Television (HDTV) can require bandwidths of around 80–100 Mbps, depending upon the picture quality, compression standards, aspect ratios, etc.

In the following sections, we describe the various WPAN networking protocols and architectures.

## 6.3 Bluetooth® (IEEE 802.15.1)

Bluetooth® is a short-range radio technology that enabled wireless connectivity between mobile devices. Its key features are robustness, low complexity, low power and low cost. The IEEE 802.15.1 standard is aimed at achieving global acceptance such that any Bluetooth® device, anywhere in the world, can connect to other Bluetooth® devices in their proximity. A Bluetooth® WPAN supports both synchronous communication channels for telephony-grade voice communication and asynchronous communications channels for data communications. A Bluetooth® WPAN is created in an ad hoc manner when devices desire to exchange data. The WPAN may cease to exist when the applications involved have completed their tasks and no longer need to continue exchanging data.

The Bluetooth® radio works in the 2.4 GHz unlicensed ISM band. A fast frequency hop (1600 hops per second) transceiver is used to combat interference and fading in this band. Bluetooth® belongs to the contention-free, token-based multi-access networks. Bluetooth® connections are typically ad hoc, which means that the network will be established for a current task and then dismantled after the data transfer has been completed. The basic unit of a Bluetooth® system is a piconet, which consists of a master node and up to seven active slave nodes within a radius of 10 meters. A piconet has a gross capacity of 1 Mbps without considering the overhead introduced by the adopted protocols and polling scheme. Several such basic units having overlapping areas may form a larger network called a scatternet. A slave can be a part of a different piconet only in a time-multiplexing mode. This indicates that, for any time instant, the node can only transmit or receive on the single piconet to which its clock is synchronized and to be able to transmit in another piconet it should change its synchronization parameters. Figure 6.2 illustrates this with an example. A device can be a master in only one piconet, but it can be a slave in multiple piconets simultaneously. A device can assume the role of a master in one piconet and a slave in other piconets. Each piconet is assigned a frequency-hopping channel based on the address of the master of that piconet.

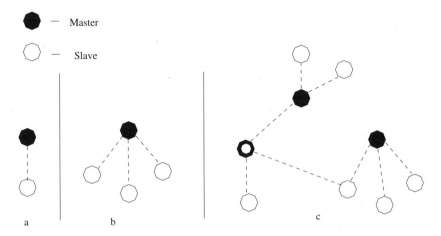

**Figure 6.2** (a) Point-to-point connection between two devices; (b) point-to-multi-point connection between master and three slaves and (c) scatternet that consists of three piconets.

### 6.3.1 The Bluetooth® Protocol Stack

The complete protocol stack contains a Bluetooth® core of certain Bluetooth® specific protocols: Bluetooth® radio, baseband, link manager protocol (LMP), logical link control and adaptation protocol (L2CAP) and service discovery protocol (SDP) as shown in Figure 6.3. In addition, non-Bluetooth specific protocols can also be implemented on top of the Bluetooth® technology.

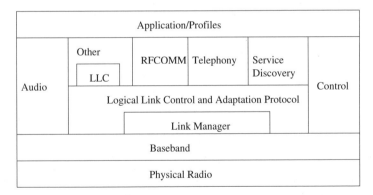

**Figure 6.3** Bluetooth® protocol stack.

The bottom layer is the physical radio layer that deals with radio transmission and modulation. It corresponds fairly well to the physical layer in the OSI and 802 models. The baseband layer is somewhat analogous to the MAC (media access control) sublayer but also includes elements of the physical layer. It deals with how the master controls the time slots and how these slots are grouped into frames. The physical and the baseband layer together provides a transport service of packets on the physical links.

Next comes a layer of somewhat related protocols. The link manager handles the setup of physical links between devices, including power management, authentication and quality of service. The logical link control and adaptation protocol (often termed L2CAP) shields the higher layers from the details of transmission. The main features supported by L2CAP are: protocol multiplexing and segmentation and reassembly. The latter feature is required because the baseband packet size is much smaller than the usual size of packets used by higher-layer protocols. The SDP protocol is used to find the type of services that are available in the network. Unlike the legacy wireless LANs, there is no system administrator who can manually configure the client devices. In the following sections the lower layers of the Bluetooth® protocol stack have been examined in detail.

### 6.3.2 Physical Layer Details

Bluetooth® radio modules use Gaussian Frequency Shift Keying (GFSK) for modulation. A binary system is used where a '1' is represented by a positive frequency deviation and a '0' is represented by a negative frequency deviation. The channel is defined by a pseudo-random hopping sequence hopping through 79 RF (radio frequency) channels 1 MHz wide. There is also a 23 channel radio defined for countries with special radio frequency regulations. The hopping sequence is determined by the Bluetooth® device address (a 48 bit address compliant with IEEE 802 standard addressing scheme) of the master and hence it is unique to the piconet. The phase or the numbering of the hopping sequence is determined by the bluetooth clock of the piconet master. The numbering ranges from 0 to $2^{27} - 1$ and is cyclic with a cycle length of $2^{27}$ since the clock is implemented as a 28-bit counter. Therefore, all devices using the same hopping sequence with the same phase form a piconet. With a fast hop rate, good interference protection is achieved. The channel is divided into time slots (625 microseconds in length) where each slot corresponds to particular RF hop frequency. The consecutive hops correspond to different RF hop frequencies. The nominal hop rate is 1600 hops/s. The benefit of the hopping scheme is evident when some other device is jamming the transmission of a packet. In this scenario, the packet is resent on another frequency determined by the frequency scheme of the master [2].

Bluetooth® provides three different classes of power management. Class 1 devices, the highest power devices operate at 100 milliwatt (mW) and have an operating range of up to 100 meters (m). Class 2 devices operate at 2.5 mW and have an operating range of up to 10 m. Class 3, the lowest power devices, operate at 1 mW and have an operating range varying from 0.1 to 1 m. The three levels of operating power is summarized in the Table 6.1.

**Table 6.1** Device classes based on power management

| Type | Power | Power level | Operating range |
| --- | --- | --- | --- |
| Class 1 Devices | High | 100 mW (20 dBm) | Up to 100 meters (300 feet) |
| Class 2 Devices | Medium | 2.5 mW (4 dBm) | Up to 10 meters (30 feet) |
| Class 3 Devices | Low | 1 mW (0 dBm) | 0.1–1 (less than 3 feet) |

A time division duplex (TDD) is used where the master and slave transmit alternately. The transmission of the master shall start at the beginning of the even numbered slots and that of the slave shall start in the odd numbered time slots only. Figure 6.4 depicts the transmission when a packet covers a single slot.

In multi-slot packets, the frequency remains the same until the entire packet is sent and frequency is derived from the Bluetooth® clock value in the first slot of the packet. While using multi-slot packets, the data rate is higher because the header and the switching time are needed only once in each packet [7]. Figure 6.5 shows how three and five slot packets are used at the same frequency throughout the transmission of the packets.

### 6.3.3 Description of Bluetooth® Links and Packets

Bluetooth® offers two different types of services: a synchronous connection-oriented (SCO) link and an asynchronous connectionless link (ACL). The first type is a point-to-point, symmetric connection

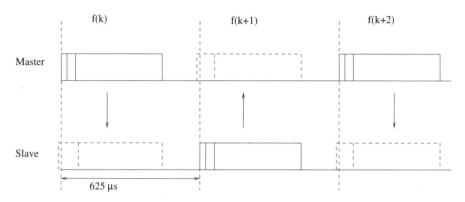

**Figure 6.4** Single slot packets depicting time division duplexing. ($f(k)$ represents frequency at time-slot $k$).

between a master and a specific slave. It is used to deliver time-bounded traffic, mainly voice. The SCO link rate is maintained at 64 kbit/s and the SCO packets are not retransmitted. The SCO link typically reserves a couple of consecutive slots, i.e. the master will transmit SCO packets at regular intervals and the SCO slave will always respond with a SCO packet in the following slave-to-master slot. Therefore, a SCO link can be considered as a circuit switched connection between the master and the slave.

The other physical link, ACL, is a connection in which the master can exchange packets with any slave on a per-slot basis. It can be considered a packet switched connection between the Bluetooth® devices and can support the reliable delivery of data. To assure data integrity, a fast automatic repeat request scheme is adopted. A slave is permitted to return an ACL packet in the slave-master slot if and only if it has been addressed in the preceding master-to-slave slot. An ACL channel supports point-to–multipoint transmissions from the master to the slaves.

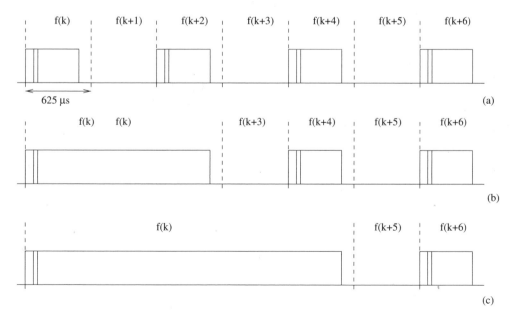

**Figure 6.5** (a) Single-slot packets; (b) three-slot packet; (c) five-slot packet. Three-slot and five-slot long packets reduce overhead compared with one-slot packets. 220 s switching time after the packet is needed for changing the frequency.

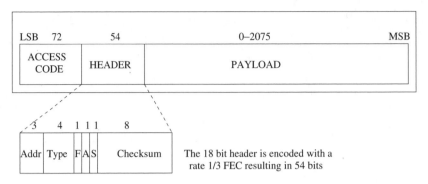

**Figure 6.6** Standard packet format.

The general packet format transmitted in one slot is illustrated in Figure 6.6. Each packet consist of three entities: the access code, the header and the payload. The access code and the header are of fixed size 72 and 54 bits respectively, but the payload can range from 0 to 2075 bits. The bit ordering when defining packets and messages in the Baseband Specification, follows the Little Endian format, i.e. the following rules apply.

- The LSB is the first bit sent over the air.
- In Figure 6.6, the LSB is shown on the left-hand side.

The access code is derived from the master device's identity, which is unique for the channel. The access code identifies all the packets exchanged on a piconet's channel, i.e. all packets sent on a piconet's channel are preceded by the same channel access code (CAC). The access code is also used to synchronize the communication and for paging and inquiry procedures. In such a situation, the access code is considered as a signaling message and neither header nor payload is included. To indicate that it is a signaling message only the first 68 bits of access code are sent. The packets can be classified into sixteen different types, using the four TYPE bits in the header of the packets. The interpretation of the TYPE code depends on the physical link type associated with the packet, i.e. whether the packet is using SCO or an ACL link. Once that is done, it can be determined which type of SCO or ACL packet has been received. Four control packets are common to all the link types. Hence, twelve different types of packet can be defined for each of the links. Apart from the type, the header also contains a 3-bit active member's address, 1-bit sequence number (S), 1-bit (F) for flow control of packets on the ACL links and a 1-bit acknowledge indication. To enhance the reliable delivery of the packets, forward error correction (FEC) and cyclic redundancy check (CRC) algorithms may be used. The possible presence of FEC, CRC and multi-slot transmission results in different payload lengths. As the SCO packets are never retransmitted, the payload is never protected by a CRC. The presence or absence of FEC also provides two types of ACL packets: DMx (medium speed data) or DHx (high speed data) respectively where x corresponds to the slots occupied by the packets. All ACL packets have a CRC field to check the payload integrity.

### 6.3.4 Link Manager

The Link Manager Protocol (LMP) provides means for setting up secure and power efficient links for both data and voice. It has the ability to update the link properties to obtain optimum performance. The Link Manager also terminates connections, either on higher layers request or because of various failures. Apart from these services, the LMP also handles different low-power modes.

- Sniff mode. The duty cycle of the slave is reduced, the slave listens for transmissions only at sniff-designated time slots. The master's link manager issues a command to the slave to enter the sniff mode.
- Hold mode. A slave in this mode does not receive any synchronous packets and listens only to determine if it should become active again. The master and slave agree upon the duration of the hold interval, after which the slave comes out of the Hold mode. During Hold mode, the device is still considered an active member of the piconet and it maintains its active member address.
- Park mode. This mode provides the highest power savings, as the slave has to only stay synchronized and not participate on the channel. It wakes up at regular intervals to listen to the channel in order to re-synchronize with the rest of the piconet, and to check for page messages. The master may remove the device from the list of active members and may assign the active member address to another device.

The services to upper layers in the complete protocol are provided by the Bluetooth® Logical Link Control and Adaptation Protocol (L2CAP), which can be thought to work in parallel with LMP. L2CAP must support protocol multiplexing because the Baseband protocol does not support any 'type' field identifying the higher layer protocol being multiplexed above it. L2CAP must be able to distinguish between upper layer protocols such as the Service Discovery Protocol, RFCOMM and Telephony Control. The other important functionality supported by L2CAP is segmentation and reassembly of packets larger than those supported by the baseband. If the upper layers were to export a maximum transmission unit (MTU) associated with the largest baseband payload, then it would lead to an inefficient use of bandwidth for higher layer protocols (as they are designed to use larger packets).

L2CAP provides both a Connection-Oriented and a Connectionless service. For the Connectionless L2CAP channel, no Quality of Service (QoS) is defined and data are sent to the members of the group in a best effort manner. The Connectionless L2CAP channel is unreliable, i.e. there is no guarantee that each member of the group receives the L2CAP packets correctly. For the Connection-Oriented channel, quality of service is defined and the reliability of the underlying Baseband layer is used to provide reliability. For example, delay sensitive traffic would be transmitted over an ACL link between the two communicating devices. Between any two Bluetooth® devices there is at most one ACL link. Therefore, the traffic flows generated by each application on the same device compete for resources over the ACL link. These traffic flows, however, may have different QoS requirements in terms of bandwidth and delay. Hence, when there is contention for resources, the QoS functions enable service differentiation. Service differentiation improves the ability to provide a 'better' service for one traffic flow at the expense of the service offered to another traffic flow. Service differentiation is only applicable when there is a mix of traffic. The service level is specified by means of QoS parameters such as bandwidth and delay. In many cases there is a trade-off between QoS parameters, e.g. higher bandwidth provides lower delay. It should be noted that service differentiation does not improve the capacity of the system. It only gives control over the limited amount of resources that satisfy the needs of the different traffic flows [8].

### 6.3.5 Service Discovery and Connection Establishment

The channel in the piconet is characterized entirely by the master of the piconet. The channel access code and frequency hopping sequence is determined by the master's Bluetooth® device address. The master is defined as the device that initiates communication to one or more slave units. The names master and slave refer only to the protocol on the channel and not the devices. Therefore, any device can be a master or a slave in the piconet.

There are two major states that a bluetooth device can be in: Standby and Connection. In addition there are seven sub-states: page, page scan, inquiry, inquiry scan, master response, slave response, inquiry response. The sub-states are interim states that are used to add new slaves to the piconet. Internal signals from the link controller are used to move from one state to another.

In order to set up a connection, a device must detect which other devices are in range. This is the goal of the inquiry procedure. The inquiry procedure must overcome the initial frequency discrepancy between the devices. Therefore, inquiry uses only 32 of the 79 hop frequency. A device in inquiry state broadcasts ID packets, containing the 68-bit inquiry access code, on the 32 frequencies of the inquiry hopping sequence. The inquiry hopping sequence is derived from the General Inquiry Access Code (GIAC) that is common for all devices and hence is the same for all devices. A device wishing to be found by inquiring units periodically enters inquiry scan sub-state. The device listens at a single frequency of the inquiry hopping sequence for a period determined by the inquiry scan window. Upon reception of an ID packet, a device in an inquiry scan sub-state will leave the inquiry scan sub-state for a random backoff time, this is done to reduce the probability that multiple devices would response to the same ID packet, thus colliding. After the random backoff, the device enters the inquiry response period to listen for a second ID packet. Upon reception of this, the device responds with a packet containing its device information, i.e. its device address and its current clock.

The connection establishment is handled by the page process, which requires the knowledge of the device address with which the connection is to be established. The page hopping sequence consist of 32 frequencies, derived from the device address which is being paged. Furthermore, the device being paged must be in the page scan sub-state, i.e. listening for page messages. When a unit in the page scan mode receives an ID packet containing its own device access code (DAC), it acknowledges the page message with an ID packet and enters the slave response state. After receiving the ACK from the paged device, the paging device enters the master response sub-state and sends a frequency hop selection (FHS) packet containing its native clock, which will be the piconet clock and the active member address (AMADDR) that the paged device shall use. The paged device acknowledges the FHS packet and switches to the hopping sequence of the master. The two units are now connected and they switch to connection state. Figure 6.7 presents the various state transitions during the inquiry and paging process.

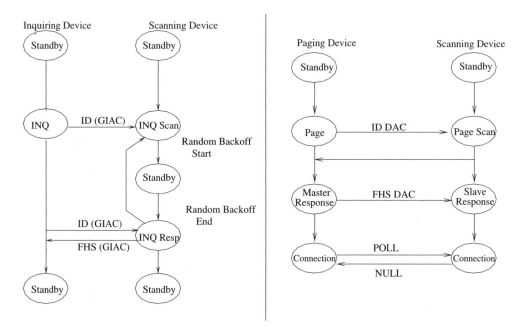

**Figure 6.7** State transitions during the inquiry and paging processes.

*6.3.6 Bluetooth® Security*

The Bluetooth® specification includes security features at the link level. The three basic services defined by the specification are the following.

(1) Authentication. This service aims at providing identity verification of the communicating Bluetooth® devices. It addresses the question 'Do I know with whom I am communicating?'. If the device is unable to authenticate itself, then the connection is aborted.
(2) Confidentiality. Confidentiality, or privacy, is yet another security goal of Bluetooth®. The information compromise caused by passive attacks is prevented by encrypting the data being transmitted.
(3) Authorization. This service aims at achieving control over the available resources, thus preventing devices that do not have access permissions from misusing network resources.

These services are based on a secret link key that is shared by two or more devices. To generate this key a pairing procedure is used when the devices are communicating for the first time. For Bluetooth® devices to communicate, an initialization process uses a Personal Identification Number (PIN), which may be entered by the user or can be stored in the non-volatile memory of the device. The link key is a 128-bit random number generated by using the PIN code, Bluetooth® Device address, and a 128-bit random number generated by the other device as inputs. The link key forms a base for all security transactions between the communicating devices. It is used in the authentication routine and also as one of the parameters in deriving the encryption key. The Bluetooth® authentication scheme uses a challenge-response protocol to determine whether the other party knows the secret key. The scheme is illustrated in Figure 6.8. To authenticate, the verifier first challenges the claimant with a 128-bit random number. The verifier, simultaneously, computes the authentication response by using the Bluetooth® device address, link key and random challenge as the inputs. The claimant returns the computed response, SRES, to the verifier. The verifier then matches the response from the claimant with that computed by the verifier. Depending on the application, there can be either one-way or two-way authentication.

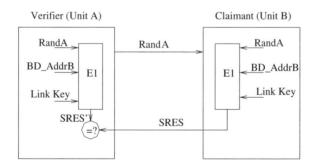

**Figure 6.8** Challenge–Response mechanism for Bluetooth authentication scheme [9].

The Bluetooth® encryption scheme encrypts the payloads of the packets. When the link manager activates encryption, the encryption key is generated and it is automatically changed every time the Bluetooth® device enters encryption mode. The encryption key size may vary from 8 to 128 bits and is negotiated between the communicating devices. The Bluetooth® encryption procedure is based on stream cipher algorithm. A key stream output is exclusive-OR-ed with the payload bits and sent to the receiving device, which then decrypts the payload. The Bluetooth® security architecture, though relatively secure, is not without weaknesses. References [9, 10–12] have identified flaws in Bluetooth® security protocol architecture.

*6.3.7 Application Areas*

The ad hoc method of untethered communication makes Bluetooth® a very attractive technology and can result in increased efficiency and reduced costs. The efficiencies and cost savings can lure both the home user and the enterprise business user. Many different user scenarios can be imagined for Bluetooth® wireless networks as outlined below.

- Cable replacement. Today, most of the devices are connected to the computer via wires (e.g., keyboard, mouse, joystick, headset, speakers, printers, scanners, faxes, etc.). There are several disadvantages associated with this, as each device uses a different type of cable, has different sockets for it and may hinder smooth passage. The freedom of these devices can be increased by connecting them wirelessly to the CPU.
- File sharing. Imagine several people coming together, discussing issues and exchanging data. For example, in meetings and conferences you can transfer selected documents instantly with selected participants, and exchange electronic business cards automatically, without any wired connections.
- Wireless synchronization. Bluetooth® provides automatic synchronization with other Bluetooth® enabled devices. For instance, as soon as you enter your office the address list and calendar in your notebook will automatically be updated to agree with the one in your desktop, or vice versa.
- Bridging of networks. Bluetooth® is supported by a variety of devices and applications. Some of these devices include mobile phones, PDAs, laptops, desktops and fixed telephones. Bridging of networks is possible, when these devices and technologies join together to use each others capabilities. For example, a Bluetooth®-compatible mobile phone can act as a wireless modem for laptops. Using Bluetooth®, the laptop interfaces with the cell phone, which in turn connects to a network, thus giving the laptop a full range of networking capabilities without the need of an electrical interface for the laptop-to-mobile phone connection.
- Miscellaneous. There are several other potential applications for the Bluetooth® enabled devices. For example, composing emails on the portable PC while on an airplane. As soon as the plane lands and switches on the mobile phone, all messages are immediately sent. Upon arriving home, the door automatically unlocks, the entry way lights come on, and the heat is adjusted to pre-set preferences.

When comparing Bluetooth® with the wireless LAN technologies, we have to realize that one of the goals of Bluetooth® was to provide local wireless access at low costs. The WLAN technologies have been designed for higher bandwidth and larger range and are, thus, much more expensive.

## 6.4 Coexistence with Wireless LANs (IEEE 802.15.2)

The global availability of the 2.4 GHz industrial, scientific, medical (ISM) unlicensed band, is the reason for its strong growth. Fuelling this growth are the two emerging wireless technologies: wireless personal area networks (WPAN) and wireless local area networks (WLAN). Bluetooth®, as the frontrunner of personal area networking is predicted to flood the markets by the end of this decade. Designed principally for cable replacement applications, Bluetooth® has been explained in significant detail in Section 3.

The WLAN has several technologies combating for dominance; but looking at the current market trends, it is apparent that Wi-Fi (IEEE 802.11b) has been the most successful of them all. With WLANs, applications such as Internet access, email and file sharing can be done within a building, supported by the technology. Wi-Fi offers speed upto 11 Mbps and a range of up to 100 m. The other WLAN technologies include the 802.11a and 802.11g standard. The 802.11a was developed at the same time as 802.11b but, due its higher costs, 802.11a fits predominately in the business market. 802.11a supports bandwidth up to 54 Mbps and signals in a regulated 5 GHz range. Compared with 802.11b, this higher frequency limits the range of 802.11a. The higher frequency also means that 802.11a signals have more difficulty penetrating walls and other obstructions. In 2002, a new standard called 802.11g began to appear on the scene. 802.11g attempts to combine the best of both 802.11a and 802.11b. 802.11g supports bandwidth up to 54 Mbps, and it uses the 2.4 GHz frequency for greater range. 802.11g is backward compatible with 802.11b, meaning that 802.11g access points will work with 802.11b wireless network adapters and vice versa.

The wireless local area networking and the wireless personal area networking are not competing technologies; they complement each other. There are many devices where different radio technologies can be built into the same platform (e.g., Bluetooth® in a cellular phone), collocation of Wi-Fi and Bluetooth® is of special significance because both occupy the 2.4 GHz frequency band. This sharing of spectrum among various wireless devices that can operate in the same environment may lead to severe interference and result in performance degradation. Owing to the tremendous popularity of Wi-Fi and Bluetooth® enabled devices the interference problem would spiral out of proportions. To prevent this, there have been a number of industry led activities focused on coexistence in the 2.4 GHz band. One such effort was the formation of the IEEE 802.15.2 Coexistence Task Group [13]. It was formed to evaluate the performance of Bluetooth® devices interfering with WLAN devices and to develop a model for coexistence that will consist of a set of recommended practices and possibly modifications to the Bluetooth® and the IEEE 802.11 [14] standard specifications that allow proper operation of these protocols in a cooperating way. The Bluetooth® SIG (Special Interest Group) has also created a Coexistence Working Group, in order to achieve the same goals as the 802.15.2 Task Group.

### 6.4.1 Overview of 802.11 Standard

The IEEE 802.11 [14] standard defines both physical (PHY) and medium access control (MAC) layer protocols for WLANs. The standard defines three different PHY specifications: direct sequence spread spectrum (DSSS), frequency hopping spread spectrum (FHSS) and infrared (IR). Our focus would be on the 802.11b standard (also called Wi-Fi), as it works in the same frequency band as Bluetooth®. Data rates up to 11 Mbps can be achieved using techniques combining quadrature phase shift keying and complementary code keying (CCK).

The MAC layer specifications coordinate the communication between stations and control the behavior of users who want access to the network. The MAC specifications are independent of all PHY layer implementations and data rates. The Distributed Coordination function (DCF) is the basic access mechanism of IEEE 802.11. It uses a Carrier Sense Multiple Access with Collision Avoidance (CSMA/CA) algorithm to mediate the access to the shared medium. Prior to sending a data frame, the station senses the medium. If the medium is found to be idle for at least DIFS (DCF interframe space) period of time, the frame is transmitted. If not, a backoff time is chosen randomly in the interval [0, CW], where CW is the Contention Window. After the medium is detected idle for at least DIFS, the backoff timer is decremented by one for each time slot the medium remains idle. If the medium becomes 'busy' during the backoff process, the backoff timer is paused. It is restarted once again after the medium is sensed idle for a DIFS period. When the backoff timer reaches zero, the frame is transmitted. Figure 6.9 depicts the basic access procedure of the 802.11 MAC. A virtual carrier sense mechanism is also incorporated at the MAC layer. It uses the request-to-send (RTS) and clear-to-send (CTS) message exchange to make predictions of future traffic on the medium and updates the network allocation vector (NAV) available in all stations that can overhear the transmissions. Communication is established when one of the stations sends an RTS frame and the receiving station sends the CTS frame that echoes the sender's address. If the CTS frame is not received by the sender then it is assumed that a collision has occurred and the process is repeated. Upon successful reception of the data frame, the destination returns an ACK frame. The absence of ACK frame indicates a collision has taken place. The contention window is doubled, a new backoff time is then chosen, and the backoff procedure starts over. After a successful transmission, the contention window is reset to $CW_{min}$.

### 6.4.2 802.11b and Bluetooth® Interference Basics

Both Bluetooth® and Wi-Fi share the same 2.4 GHz band, which extends from 2.402 to 2.483 GHz. The ISM band under the regulations of Federal Communications Commission (FCC) is free of tariffs, but

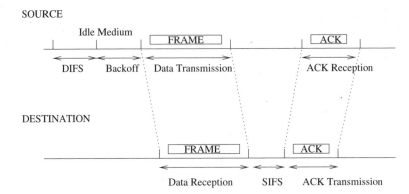

**Figure 6.9** 802.11 frame transmission scheme.

must follow some rules related to total radiated power and the use of spread spectrum modulation schemes. These constraints are imposed to enable multiple systems to coexist in time and space. A system can use one of the two spread spectrum (SS) techniques to transmit in this band. The first is the Frequency Hopping Spread Spectrum (FHSS), where a device can transmit at high power in a relatively narrow band but for a limited time. The second is the Direct Sequence Spread Spectrum (DSSS), where a device occupies a wider band with relatively low energy in a given segment of the band and, importantly, it does not hop frequencies [15].

As outlined in the preceding sections, Bluetooth® selected FHSS, using 1 MHz width and a hop rate of 1600 times/s (i.e. 625 microseconds in every frequency channel) and Wi-Fi picked DSSS, using 22 MHz of bandwidth to transmit data at speeds of up to 11 Mbps. An IEEE 802.11b system can use any of the eleven 22 MHz wide sub-channels across the acceptable 83.5 MHz of the 2.4 GHz frequency band, which obviously results in overlapping channels. A maximum of three Wi-Fi networks can coexist without interfering with one another, since only three of the 22 MHz channels can fit within the allocated bandwidth [14].

A wireless communication system consist of at least two nodes, one transmitting the data and the other receiving it. Successful operation of the system depends upon whether the receiver is able to distinctly identify the desired signal. Further, this depends upon the ratio of the desired signal and the total noise at the receiver's antenna. This ratio is commonly referred to as the signal-to-noise ratio (SNR). The main characteristic of the system is defined as the minimum SNR at which the receiver can successfully decode the desired signal. A lower value of SNR increases the probability of an undesired signal corrupting the data packets and forcing retransmission. Noise at the receiver's antenna can be classified into two categories: in-band and out-of-band noise. The in-band noise, which is the undesired energy in frequencies the transmitter uses to transmit the desired signal, is much more problematic. The noise generated outside the bandwidth transmission signal is called out-of-band noise and its effect can be minimized by using efficient band-pass filters. Noise can be further classified as white or colored. The white noise generally describes wideband interference, with its energy distributed evenly across the band. It can be modeled as a Gaussian random process where successive samples of the process are statistically uncorrelated. Colored noise is usually narrowband interference, relative to the desired signal, transmitted by intentional radiators. The term intentional radiator is used to differentiate signals deliberately emitted to communicate from those that are spurious emissions. Figure 6.10 illustrates white and colored noise [16].

When two intentional radiators, Bluetooth® and IEEE 802.11b, share the same frequency band, receivers also experience in-band colored noise. The interference problem is characterized by a time and frequency overlap, as depicted in Figure 6.11. In this case, a Bluetooth® frequency hopping signal is shown to overlap with a Wi-Fi direct sequence spread spectrum signal.

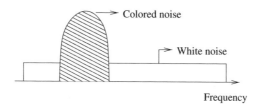

**Figure 6.10** White and Colored noise.

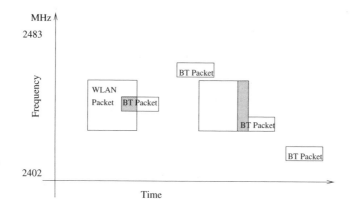

**Figure 6.11** Bluetooth® and 802.11b packet collisions in the 2.4 GHz band.

### 6.4.3 Coexistence Framework

As the awareness of the coexistence issues has grown, groups within the industry have begun to address the problem. The IEEE 802.15.2 and Bluetooth® Coexistence Working Group are the most active groups. The proposals considered by the groups range from collaborative schemes intended for Bluetooth® and IEEE 802.11 protocols to be implemented in the same device to fully independent solutions that rely on interference detection and estimation. Collaborative mechanisms proposed by the IEEE 802.15.2 working group [13] are based on a MAC time domain solution that alternates the transmission of Bluetooth® and WLAN packets (assuming both protocols are implemented on the same device and use a common transmitter) [17]. Bluetooth® is given priority access while transmitting voice packets, while WLAN is given priority for transmitting data. The non-collaborative mechanism uses techniques for detecting the presence of other devices in the band like measuring the bit or frame error rate, the signal to interference ratio, etc. For example, all devices can maintain a bit error rate measurement per frequency used. The frequency hopping devices can then detect which frequencies are occupied by other users and adaptively change the hopping pattern to exclude them. Another way is to let the MAC layer abort transmission at the particular frequencies where users have been detected. The latter case is easily adaptable to the existing systems, as the MAC layer implementation is vendor specific and hence, the Bluetooth® chip set need not be modified. Though, the adaptive frequency hopping scheme requires changes to the Bluetooth® hopping pattern, and therefore a new chip set design, its adoption can increase the Bluetooth® throughput by maximizing the spectrum usage. The other alternative would be migration to the 5 GHz ISM band. This will come at the cost of higher power consumption and expensive components since the range decreases with the increase in frequency.

## 6.5 High-Rate WPANs (IEEE 802.15.3)

This section presents the details of the IEEE 802.15.3 standard being considered for high datarate wireless personal area networks.

### 6.5.1 Physical Layer

The 802.15.3 PHY layer operates in the unlicensed frequency band between 2.4 GHz and 2.4835 GHz, and is designed to achieve data rates of 11–55 Mbps, which are required for the distribution of high definition video and high-fidelity audio. Operating at a symbol rate of 11 Mbaud, five distinct modulation schemes have been specified, namely, uncoded Quadrature Phase Shift Keying (QPSK) modulation at 22 Mbps and trellis coded QPSK, 16/32/64-Quadrature Amplitude Modulation (QAM) at 11, 33, 44, 55 Mbps respectively [18]. With higher speeds, even a small amount of noise in the detected amplitude or phase can result in error and, potentially, many corrupted bits. To reduce the chance of an error, standards incorporating high data rate modulation schemes do error correction by adding extra bits to each sample. The schemes are referred to as Trellis Coded Modulation (TCM) schemes. For instance, a modulation scheme can transmit 5 bits per symbol, of which, with trellis coding, 4 bits would be used for data and 1 bit would be used for parity check. The base modulation format for 802.15.3 standard is QPSK (differentially encoded). The higher data rates of 33–55 Mbps are achieved by using 16, 32, 64-QAM schemes with 8-state 2D trellis coding, which depends on the capabilities of devices at both ends. The 802.15.3 signals occupy a bandwidth of 15 MHz, which allows for up to four fixed channels in the unlicensed 2.4 GHz band. The transmit power level complies with the FCC rules with a target value of 0 dBm.

Each WPAN frame contains four segments: a preamble, a header, data and a trailer. The preamble is used to perform gain adjustment, symbol and carrier timing compensation, equalization, etc. The header is used to convey physical layer data necessary to process the data segment, such as modulation type and frame length. The tail is used to force the trellis code to a known state at the end of the frame to achieve better distance properties at the end of the frame. The preamble and header will utilize the base QPSK modulation while the data and tail will use one of the four defined modulations.

### 6.5.2 Network Architecture Basics

WPANs are not created a priori. They are created when an application on a particular device wishes to communicate with similar applications on other devices. This network, created in an ad hoc fashion, is torn down when the communication ends. The network is based on a master–slave concept, similar to the Bluetooth® network formation. A piconet is a collection of devices such that one device is the master and the other devices are slaves in that piconet. The master is also referred to as the piconet controller (PNC). The master is responsible for synchronization and scheduling the communication between different slaves of its piconet.

In the 802.15.3 WPAN, there can be one master and up to 255 slaves. The master is responsible for synchronization and scheduling of data transmissions. Once the scheduling has been done, the slaves can communicate with each other on a peer-to-peer basis. This is contrary to Bluetooth® PAN, where devices can only communicate with the master in a point to point fashion. In Bluetooth, if device $d_1$ wants to communicate with $d_2$, $d_1$ will send the data to the master and the master will forward the data to $d_2$. The two slave devices cannot communicate on peer basis. A scatternet is a collection of one or more piconets such that they overlap each other. Thus, devices belonging to different piconets can communicate over multiple hops. The piconet can be integrated with the wired network (802.11/ Ethernet) by using a IEEE 802 LAN attachment gateway. This gateway conditions MAC data packet units to be transported over Bluetooth® PAN.

The IEEE 802.15.3 standard defines three types of piconets. The independent piconet is a piconet with no dependent piconet and with no parent piconet. A parent piconet has one or more

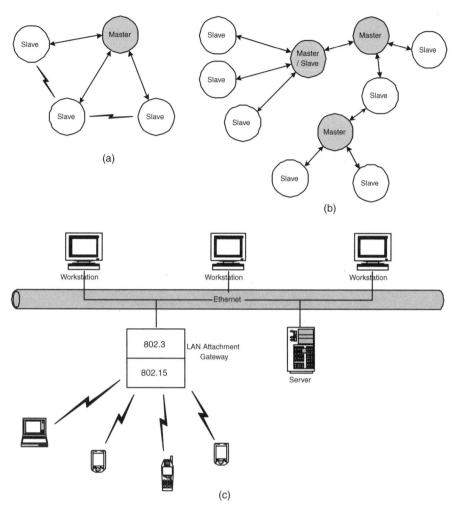

**Figure 6.12** (a) Piconet communication in IEEE 802.15.3 WPAN; (b) scatternet formation; (c) IEEE 802.3 LAN attachment gateway.

dependent piconets. A dependent piconet is synchronized with the parent piconets timing and needs time allocation in the parent piconet. There are two types of dependent piconets: child piconet and neighbor piconet. A child piconet is a dependent piconet where the PNC is a member of the parent piconet. The PNC is not a member of the parent piconet in the case of neighbor piconet.

All devices within a piconet are synchronized to the PNC's clock. The PNCs of the dependent piconets synchronize their operation to the parent PNC's clock and time slot allocated to it. Periodically, the PNC sends the information needed for synchronization of the devices.

The functions discussed by IEEE 802.15.3 standard include: formation and termination of piconets, data transport between devices, authentication of devices, power management of devices, synchronization, fragmentation/defragmentation, piconet management functions like electing a new PNC and formation of child and neighbor piconets

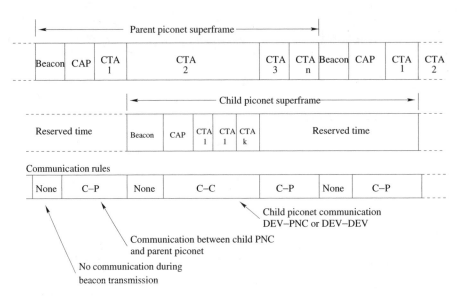

**Figure 6.13** Timing diagram of parent piconet and child piconet.

### 6.5.3 Piconet Formation and Maintenance

A piconet is initiated when a device, capable of assuming the role of PNC, starts transmitting beacons. Before transmitting the beacon, the potential PNC device scans all the available channels to find a clear channel. The device uses passive scan to detect whether a channel is being used by any other piconet. The device listens in receive mode for a fixed amount of time for beacon signals from the other piconet's PNC. The device may search for piconets by traversing the available channels in any order as long as all the channels are searched. Once a free channel is found, the piconet is started by sending a beacon on the channel, after it has been confirmed that the channel has remained idle for a sufficient time period. If a channel cannot be found, the device may form a dependent piconet. The formation of dependent piconets is explained later. Once the piconet has been formed, other devices may join this piconet on receiving the beacon. Since the PNC may not be the most 'capable' device, a handover process has been defined so that the role of the PNC may be transferred to a more capable device.

A child piconet is formed under an existing piconet, and the existing piconet becomes the parent piconet. A parent piconet may have more than one child piconet. A child piconet may also have other child piconets of its own. The child piconet is an autonomous entity and the association, authentication and other functionality of the piconet are managed by its PNC without the involvement of the parent PNC. The child PNC is a member of the parent piconet and hence can communicate with any device within the parent piconet. When a PNC-capable device in the parent piconet wants to form a child piconet, it requests the parent PNC to allocate a time slot for the operation of the child piconet. Once the time slot is allocated, the device, now the child PNC, starts sending beacons and operates the piconet autonomously.

The standard does not allow direct communication between a device in the child piconet and a device in the parent piconet. But since the child PNC is a member of both the child and parent piconets, it can act as an intermediary for such a communication.

A neighbor piconet is formed when a PNC capable device cannot find an empty channel to start a new piconet. The device communicates with the PNC of an active piconet and asks for a time slot for the operation of the new piconet on the same channel. If there is sufficient time available, the PNC of the parent piconet will allocate a time slot for the neighbor piconet. The neighbor piconet operates in this

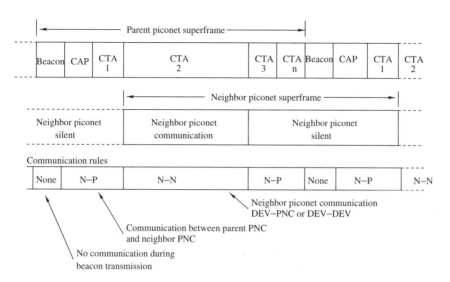

**Figure 6.14** Timing diagram of parent piconet and neighbor piconet. CAP - Contention Access Period; CTA - Channel Time-slot Assignment.

time slot only. The neighbor PNC is not a member of the parent piconet and hence cannot communicate with any device in the parent piconet. The neighbor piconet is autonomous except that it is dependent on the time slot allocated by the parent piconet.

If the PNC wishes to leave the piconet and there is no other device in the piconet that is capable of being the new PNC, then the PNC will initiate procedure to shutdown the piconet. The PNC sends the shutdown command to the devices through the beacon. If the piconet is itself not a dependent piconet but has one or more dependent piconets, then it chooses one of the dependent piconets (the one with the lowest value of device id) to continue its operation. The chosen dependent piconet is granted control over the channel and is no longer a dependent piconet. All other piconets have to either shutdown, change channels or join other piconets as dependent piconets.

If the PNC that requests a shutdown is a dependent PNC, it will stop the operations for its piconet and then inform its parent PNC that it no longer needs the time period allocated to it. The parent PNC may then allocate this time slot to any other device.

A parent PNC can also terminate the operation of a dependent piconet. If it wishes to stop a dependent piconet, the parent PNC sends a command to the dependent PNC. On receiving this command, the dependent PNC must either initiate its shutdown procedure, change channels or join some other piconet as a dependent piconet. The parent PNC listens for the shutdown behavior of the dependent piconet before reclaiming the time slots allocated to the dependent piconet. If the dependent piconet does not cease operations within a predetermined time limit, the parent PNC will remove the time slots irrespective of the state of the dependent piconet.

Some of the devices may not have the capabilities to become the PNC and such devices join piconets formed by other devices. Membership of a piconet depends on the security mode of the piconet. If the piconet does implement security, then the joining device has to go through the authentication process before gaining membership of the piconet.

A device initiates the association process by sending an association request to the PNC of the piconet. When the PNC receives this request, it acknowledges it by sending an immediate acknowledge to the device. The acknowledgment message is not an indication of acceptance, it just implies that the request has been received and is being processed. The acknowledgment message is needed since the PNC requires time to process the association request. The PNC checks the availability of resources to support an additional device. The PNC may maintain a list of devices that are allowed to join the piconet. If such a list is maintained, then the PNC needs to compare the address of the requesting device with the list to determine its response. If the association fails for any of the above mentioned reasons, the PNC sends a response to the device with an explanation of the reason for failure. On the other hand, if the PNC decides to accept the association request, it assigns a device id to the device and sends a successful response to the device. The time slot allocation for the new device will be sent during the subsequent beacon messages.

Once the device joins the piconet, the PNC broadcasts information about the new device to all other devices. This enables any device in the piconet to keep track of all other devices in the piconet, their properties and any services that they may provide. Once the device has successfully joined the piconet, it may initiate or receive data communication from any other device in the piconet. If the PNC decides to remove a device from the piconet, it sends a disassociation request to the device with the appropriate reason. Similarly, if the device decides to leave the piconet, it sends a disassociation request to the PNC with the reason. On receiving this request, the PNC reclaims the time slots assigned to the device and informs the other devices about the departure of the node by setting appropriate bits in the following beacon messages.

When the PNC leaves the piconet it transfers its functionality to another device in the piconet that is capable of being a PNC. Each device maintains information of its capabilities such as maximum transmit power, number of devices it can handle, etc. The departing PNC initiates PNC handover by querying for the capabilities of each device in the piconet. Based on the information collected, it chooses the best possible PNC capable device and sends a PNC handover command to this device. If the piconet is not a dependent piconet, the device will accept the new role and receive information about the piconet from the old PNC. If the device is already a PNC of some dependent piconet, it may reject the handover by sending a refuse command back to the PNC. If both the PNC and the chosen device are part of the same dependent piconet, then the device will accept the handover only if it is able to join the parent piconet as either a regular device or a dependent PNC. If it is unable to join the parent piconet, it will refuse the PNC handover. Once a device accepts the role of the PNC, the old PNC will transfer all information necessary for the operation of the piconet. Once the new PNC has all the required information and is capable of taking over the piconet, it will send a handover response to the old PNC and start sending beacon messages that will allow other devices in the piconet to recognize the new PNC. On receiving the response from the new PNC, the old PNC will stop sending beacons and give up its control over the piconet. The PNC handover maintains the association of all other devices in the piconet and hence they do not need to re-associate themselves with the new PNC.

### 6.5.4 Channel Access

Channel access in the 802.15.3 MAC is based on superframes, where the channel time is divided into variable size superframes, as illustrated in Figure 6.15. Each superframe begins with a beacon that is sent by the PNC and is composed of three main entities: the beacon, the contention access period (CAP) and the contention free period (CFP). The beacon and the CAP are mainly used for synchronization and

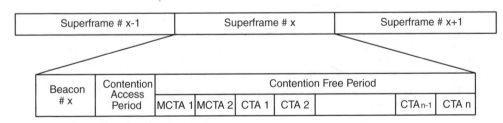

**Figure 6.15** IEEE 802.15.3 superframe format.

control information whereas the contention free period is used for data communication. During the CAP, the devices access the channel in a distributed manner using CSMA/CA with a specified backoff procedure. The CFP is regulated by the PNC, which allocated time slots to various devices based on their demand and availability.

The beacon is used to send the timing information and any piconet management information that the PNC needs to send to the devices. The beacon consists of a beacon frame and any commands sent by the PNC as beacon extensions. The beacon information contains details about the superframe duration, CAP end time, maximum transmit power level and piconet mode. The superframe duration specifies the size of the current superframe and is used along with the CAP end time to find the duration of the CAP. The resolution of superframe duration and CAP end time is 1 μs and the range is [0–65535] μs. The value of the maximum transmit level is specified in dBm and may vary for each superframe. The piconet mode field describes some of the characteristics of the piconet and the superframe. It specifies whether the CAP may contain data, command or association traffic and may be used to disallow a certain type of traffic to be sent during the CAP. The piconet mode field specifies if the management time slots are being used in the current superframe. It also defines the security mode of the piconet.

All the devices in the piconet reset their superframe clock on receiving the beacon preamble. All times in the superframe are measured relative to the beacon preamble. Each device in the piconet calculates its transmission time based on the information contained in the beacon.

The contention access period (CAP) is used to communicate commands and asynchronous data traffic, if any. Carrier sense multiple access with collision avoidance (CSMA/CA) is the basic medium access technique used in the CAP. The type of data or commands that a device may send during the CAP is governed by the PNC by setting appropriate bits in the piconet mode field of the beacon. Before transmitting each frame, the device senses the medium for a random period of time and transmits only if the medium is idle. Otherwise, it will perform a backoff procedure that is maintained across superframes and is not reset at the start of a new superframe. That is, if the backoff interval has not expired and there is not enough time left in the CAP, then the backoff interval is suspended and restarted at the begin of the next superframe's CAP. When the device gains control of the medium, it checks if there is sufficient time in the CAP for the transmission of the whole frame. CAP traffic is not allowed to intrude in the contention free period and the device must back off until the beginning of the next superframe's CAP.

The contention free period (CFP) consists of channel time allocations (CTAs). CTAs are used for management commands as well as synchronous and asynchronous data streams. The PNC divides the CFP into channel time allocations that are allocated to individual devices. A device may or may not fully utilize the CTA allocated to it, with no other device being allowed to transmit during this period. The order of transmission of the frames is decided locally by the device without the knowledge of the PNC. Depending on its position in the superframe, there are two type of CTAs: dynamic CTA and pseudo-static CTA. The devices in the piconet have the choice for requesting either of the CTAs. The position of a dynamic CTA, within a superframe, can be moved on a superframe to superframe basic. This allows the PNC the flexibility to rearrange the CTAs to obtain the most efficient schedule. The scheduling mechanism for the CTAs is not specified by the draft standard and is left to the discretion of the implementer. Pseudo-static CTAs maintain their position within the superframe and are allocated for isochronous streams only. The PNC is allowed to move the location of these CTAs as long as the old location is not allocated to any other stream for a pre-defined constant period. The CFP may also contain Management CTAs (MCTA) that are allocated just after the contention access period. MCTAs are used to send command frames that have the PNC either as the source or the destination. The PNC is responsible for determining the number of MCTAs for each superframe.

Whenever a device needs to send data to another device in the piconet, it sends a request to the PNC. The PNC allocates the CTAs based on the current outstanding requests of all the devices and the available channel time. When a source device has a frame to be sent to a destination, it may send it during any CTA allocated for that source destination pair. If such a CTA does not exist, the source may send the frame in any CTA assigned to that source as long as the source device can determine that the destination device will be receiving during that period. A device may not extend its transmission, started in the CTA, beyond the end of that CTA. The device must check whether there is enough time for transmission of the frame during the current CTA to accommodate the frame. If a device receives the beacon in error, it shall not transmit during the CAP or during any management or dynamic CTA during that superframe. The device is allowed to use the pseudo-static CTAs until the number of consecutive lost beacons exceeds a constant value. Any device that misses a beacon may also listen for the entire superframe to receive frames for which it is the destination.

There are three acknowledgement schemes defined by the standard: no acknowledgement (no-ACK), immediate acknowledgement (Imm-ACK) and delayed acknowledgement (Dly-ACK). A transmitted frame with ACK policy set to no-ACK shall not be acknowledged by the receiver. The transmitting device assumes that the frame was received successfully and proceeds to the next frame. When the acknowledgement policy is set to Imm-ACK, the receiver sends the ACK frame after every successfully received data frame. Delayed acknowledgement is used only for isochronous data streams where the acknowledgement mechanism is negotiated by the sender and receiver. The receiver notifies the sender of the number of frames that the sender may send in one burst. This number depends on the receiver's buffer size and the frame size. The acknowledgement policy of a data stream may be changed from a Dly-ACK to Imm-ACK or no-ACK by sending a frame specifying the new acknowledgement policy.

There are four inter frame spacings defined by the standard: short interframe spacing (SIFS), minimum interframe spacing (MIFS), backoff interframe spacing (BIFS) and retransmission interframe spacing (RIFS). The actual value of these parameters depends on the physical layer. The acknowledgment frames in the Imm-ACK and Dly-ACK case shall be transmitted a SIFS duration after the transmission of the frame that requested the ACK. Consecutive frames within a CTA are sent a MIFS duration apart if the acknowledgement policy is no-ACK or Dly-ACK. If a device needs to retransmit some frame, the frame has to wait for a RIFS amount of time. This retransmission scheme is used only during the contention free period. The CAP follows its own retransmission scheme which uses BIFS to backoff while gaining access to the medium.

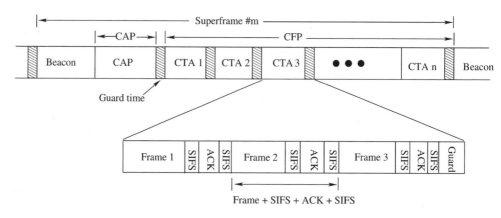

**Figure 6.16** Interframe spacing (IFS) for a connection with Imm-ACK policy.

### 6.5.5 Power Management

The standard defines three power management modes to enable devices to turn off or reduce power: device synchronized power save (DSPS) mode, piconet synchronized power save (PSPS) mode and asynchronous power save (APS) mode. Any device in a piconet operates in one of the following four power management (PM) modes: ACTIVE mode, DSPS mode, PSPS mode or APS mode. In each PM mode, a device may be in either of two power states: AWAKE or SLEEP. The device is in AWAKE state when it is either transmitting or receiving. SLEEP state is defined as the state where the device is neither transmitting nor receiving.

Piconet synchronized power save (PSPS) mode allows the devices to sleep for intervals defined by the PNC. When the devices want to enter the PSPS mode, they send a request to the PNC. The PNC conveys this information to the piconet by setting appropriate bits in the beacon. The PNC chooses some of the beacons to be the system wake beacons and notifies the next wake beacon to the PSPS set. All devices in the PSPS mode have to listen to the system wake beacons.

Device synchronized power save (DSPS) mode enables groups of devices to sleep for multiple superframes but still be able to wake up at the same superframe. Devices that want to synchronize their sleep patterns, join a DSPS set. The DSPS set determines the sleep interval and the next time that the devices will be awake. This mode not only allows devices to sleep at the same time, but it also allows nodes in the piconet to determine the time that the devices in the DSPS mode will be awake to receive data. A device may be in multiple DSPS sets at a time and hence may have multiple wake beacons.

Asynchronous power save (APS) mode allows a device to save power for long periods of time by switching off the transceiver until it decides to receive a beacon. The wake beacon for a device in APS mode is decided by the device, independent of the PNC or any other device in the piconet. Unlike DSPS and PSPS, the wake beacon in APS is not periodic but occurs only once.

### 6.5.6 Security

The 802.15.3 standard supports two different modes of security, no security (mode 0) and use of strong cryptography (mode 1).

- Mode 0–No security. Neither secure membership nor payload protection is provided. The system does not provide either data integrity or data encryption. The only way a PNC may restrict admission to the piconet is through the use of admission control list.
- Mode 1–Secure membership and payload protection. Devices need to establish a secure membership with the PNC before they can be a part of the piconet. The data communication in the piconet may be protected by using both data integrity and data encryption techniques. The control traffic in the piconet also needs to be protected.

The security mechanisms provided in mode 1 allow the devices to be authenticated before admission in the piconet. The standard provides a symmetric cryptographic mechanism to assist in providing security services. The establishment and management of security keys is done at the higher layers and is out of the scope of this standard. Data encryption is done by using symmetric key cryptography. The data may be encrypted by using either the key shared by all the piconet or by using the key shared by the two communicating devices. Similarly, data integrity is maintained by using an integrity code to protect the data from being modified. The integrity code may either be the key shared by all the devices in the piconet or the key shared by the two communicating devices. Data integrity is also provided for the beacons and the commands exchanged between the PNC and the devices.

### 6.5.7 802.15.3a–Ultra-Wideband

An alternative PHY layer, based on Ultra-wideband radio transmission, has been proposed for the 2.4 GHz PHY layer described in IEEE 802.15.3. The IEEE TG802.15.3a is in the process of developing an alternative high-speed (greater 110 Mbps) link layer design conformal with the 802.15.3 multiple access (MAC). Ultra-Wideband transmissions can achieve data rates that are significantly higher than devices based on the 802.15.3 PHY specification and the currently available WLAN devices. The UWB technology itself has been in use in military applications since the 1960s, based on exploiting the wideband property of UWB signals to extract precise timing / ranging information. However, recent

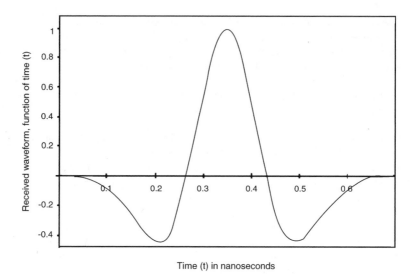

Figure 6.17 Sample UWB waveform in time domain.

Federal Communication Commission (FCC) regulations has paved way for the development of commercial wireless communication networks based on UWB in the 3.1–10.6 GHz unlicensed band [23]. The following two modes of operation are supported in the standard's specification: 110 and 200 Mbps and higher bit rates, such as 480 Mbps.

Ultra Wideband technology is defined as any transmission scheme whose fractional bandwidth, $\eta$, is greater than 20% of a center frequency [23, 24], where the fractional bandwidth is defined as:

$$\eta = \frac{\{f_H - f_L\}}{\left\{\frac{f_H + f_L}{2}\right\}} \times 100\%,$$

where $f_H$ and $f_L$ are respectively defined as the highest and lowest frequencies in the transmission band. The data transmission over a wide range of frequency bands results in a high data rate. The UWB devices have low power utilization and adhere to the rules of FCC's part 15 regulations for emission control in an unlicensed spectrum. Due to this, simultaneous operation of UWB devices with existing narrowband systems is enabled, thereby increasing spectrum reuse. This also helps in reducing the probability of detection and interception. Unlike continuous wave technology that use sine waves to encode information, UWB technology uses very short (sub-nanosecond), low power pulses (mono-cycles) with a sharp rise and fall time, resulting in a waveform that occupies several GHz of bandwidth [24]. Another important property of UWB signal is the high immunity to multipath fading. Multipath fading is a phenomenon observed in continuous wave signals. It occurs due to the reflection of the signals off objects resulting in destructive cancelation and constructive addition. Since UWB is not a continuous wave technology, it is not affected as the reflections can be resolved in time.

The two most common methods of generating signals with UWB characteristics are Time-modulated UWB (TM-UWB) and Direct Sequence Phase Coded UWB (DSC-UWB). TM-UWB is a pulse position modulation (PPM) scheme. PPM is based on the principle of encoding information with two or more positions in time, referred to the nominal pulse position. A pulse transmitted at the nominal position represents a '0', and a pulse transmitted after the nominal position represents a '1'. Additional positions can be used to provide more bits per symbol. The time delay between positions is typically a fraction of a nanosecond, while the time between nominal positions is typically much longer to avoid interference between impulses. Furthermore, pseudo-random time hopping codes are used to allow several users to share the transmission range. The pseudo-random nature of the codes eliminates catastrophic collisions in multiple accessing. DSC-UWB uses M-ary Pulse Amplitude Modulation where multiple pulses or 'chips' are sent per bit. The chip pulse sequence corresponds to a short binary pseudo-random code sequence, analogous to CDMA (Code Division Multiple Access). With a matched filter receiver, DSC-UWB multiple access is more suited for higher rates as it can accommodate more users than the TH-PPM for a given BER. At lower data rates, the multiple accessing capacity of the two systems are approximately the same; in such cases, TH-PPM may be preferred to DSC-UWB since it is potentially less susceptible to the near-far effect.

A variety of UWB system can be designed to use the 7500 MHz (3.1–10.6 GHz) available UWB spectrum. A signal's envelope can be so shaped to occupy the entire spectrum or to occupy only 500 MHz of bandwidth. In the latter case, 15 such signals can be used to cover the entire spectrum. Bandwidth and number of available bands generate different performance tradeoffs and design challenges. The single-band system, for example, is more sensitive to intersymbol interference (ISI) generated by delay spread. The two systems also present different implementation challenges. The single-band system's transmit signal is much higher bandwidth than the multiband, requiring very fast switching circuits. The multiband system, on the other hand, requires a signal generator that is able to switch quickly between frequencies.

As of November 2004, the IEEE task group TG802.15.3a has not choosen the physical layer design but is considering two proposals. The first proposal, promoted by the UWB Forum, is based on the principles of direct sequencing. DS-UWB provides support for data rates of 28M, 55M, 110M, 220M, 500M, 660M and 1320M bits/sec. The other proposal, developed by the Multiband OFDM Alliance (MBOA [26]), is based on the concept of Multi-band Orthogonal Frequency Division Multiplexing (Multi-band OFDM) and supports data rates of 55M, 110M, 200M, 400M, 480M bits/sec. Multi-band OFDM is a transmission technique where the available spectrum is divided into multiple bands. Information is transmitted on each band using OFDM modulation. The information bits are interleaved across all the bands to provide robustness against interference. Multi-band OFDM divides the available spectrum (3.1–10.6 GHz) into 13 bands of 528 MHz each. These bands are grouped into four groups, as shown in Figure 6.18, to enable multiple modes of operation for multi-band OFDM devices. Currently two modes of operation have been specified. Mode 1 is mandatory and operates in frequency bands 1–3, i.e. Group A. Mode 2 is optional and uses seven frequency bands, three bands from group A and four bands from group C. Groups B and D are reserved for future use. Channelization in multiband OFDM is achieved by using different time-frequency codes, each of which is a repetition of an ordered group of channel indexes. An example of time-frequency codes is given in Table 6.2. The beacon frames are transmitted using a pre-determined time-frequency code. This facilitates reception of beacon frames by devices that have not been synchronized. There are still many technological challenges ahead, mostly

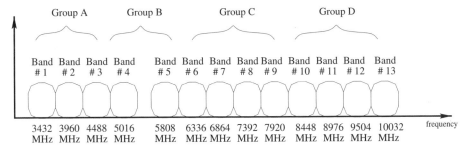

**Figure 6.18** Frequency spectrum allocation for multiband OFDM.

**Table 6.2**  Time frequency codes for multiband OFDM devices

| Ch. no. | Time frequency codes (Mode 1) | | | | | | Time frequency codes (Mode 2) | | | | | | |
|---|---|---|---|---|---|---|---|---|---|---|---|---|---|
| 1 | 1 | 2 | 3 | 1 | 2 | 3 | 1 | 2 | 3 | 4 | 5 | 6 | 7 |
| 2 | 1 | 3 | 2 | 1 | 3 | 2 | 1 | 7 | 6 | 5 | 4 | 3 | 2 |
| 3 | 1 | 1 | 2 | 2 | 3 | 3 | 1 | 4 | 7 | 3 | 6 | 2 | 5 |
| 4 | 1 | 1 | 3 | 3 | 2 | 2 | 1 | 3 | 5 | 7 | 2 | 4 | 6 |
| 5 | — | — | — | — | — | — | 1 | 5 | 2 | 6 | 3 | 7 | 4 |
| 6 | — | — | — | — | — | — | 1 | 6 | 4 | 2 | 7 | 5 | 3 |

around the high level of integration that UWB products require: they need to be developed at low cost and low power to meet the vision of integrated connectivity for PANs.

## 6.6  Low-Rate WPANs (IEEE 802.15.4)

With the explosive growth of the Internet, the major focus to date has been insatisfying the need for shared high-speed connectivity. the wireless networking community has been in focused on enhancing WLAN capabilities and developing new approaches to meet the needs of the growing pool of applications requiring wireless devices. On the other side of the spectrum, applications such as home automation, security and wireless sensor networks have low throughput requirements. Apart from this, such applications cannot handle the complexity of a heavy protocol stack that impacts the power consumption and utilize too many computational resources. For example, let us consider an array of temperature sensors in a room. The sensors may require to report the temperature only a few times every hour, have a small form factor and be extremely cheap, as they would be deployed in large numbers. It would be an impractical option to connect sensors with wires because the installation of wires would cost more than the sensors. Also, frequent battery replacement is impractical, hence low power consumption is another essential characteristic for such an application. Bluetooth® was originally conceived as an option but it has followed a high complexity trend, making it unsuitable for low-power consumption applications. Fueled by the drive to run such applications, in December 2000 Task Group 4, under the IEEE 802 Working Group 15, was formed to begin the development of a LR-WPAN (Low Rate-Wireless Personal Area Network) standard IEEE 802.15.4. The goal of Task Group 4 is to provide a standard that has the characteristics of ultra-low complexity, low-cost and extremely low-power for wireless connectivity among inexpensive, fixed, portable and moving devices[19]. Yet another standards group, ZigBee [25] (a HomeRF [21] spinoff), has been working along with the IEEE 802.15.4 working group to achieve the above mentioned goals. Table 6.3 highlights the main characteristics of the standard.

**Table 6.3**  Main characteristics of LR-WPAN

| Property | Range |
|---|---|
| Frequency bands | 868 MHz (Europe), 915 MHz (USA) and 2.4 GHz |
| Raw data rate | 868 MHz: 20 kbps; 915 MHz: 40 kbps; 2.4 GHz: 250 kbps |
| Range | 10–20 m |
| Latency | around 15 ms |
| Channels | 868 MHz: 1 channel; 915 MHz: 10 channels, 2.4 GHz: 16 channels |
| Addressing | Short 16 bit or 64 bit IEEE |
| Channel access | Carrier Sense Multiple Access with Collision Avoidance (CSMA/CA) |

### 6.6.1 Applications

The intent of IEEE 802.15.4 is to address applications where existing WPAN solutions are too expensive and the performance of a technology such as Bluetooth® is not required. The application areas are numerous: industrial control and monitoring; sensing and location determination at disaster hit sites; precision agriculture, such as sensing soil moisture and pH levels; automotive sensing, such as tire pressure monitoring, smart badges and tags. Most of these applications require LR-WPAN devices to be used in conjunction with the sensors so that data will be gathered, processed and analyzed to determine if or when user interaction is required. Sensor applications could range from detecting emergency situations, such as hazardous chemical levels and fires to the monitoring and maintenance of rotating machinery. From the multimedia application point of view, the LR-WPAN radios can work in conjunction with high rate devices for providing faster access to signalling and control information. A LR-WPAN would definitely reduce the installation cost of a new small devices network and simplify expansion of existing network installations. The initial implementation will most likely occur in monitoring applications with non-critical data where longer latencies would be acceptable. The emphasis would be placed on low power consumption in order to maximize battery lifetime. Apart from these applications, the consumer and home automation market presents a significant potential because of its size. The different sectors of the market would be: PC peripherals, including wireless mice, keyboards and joysticks; consumer electronics, including radios, televisions, VCRs, CDs and remote controls; home automation, including heating, ventilation, security and lighting; health monitoring, including sensors, monitoring and diagnostics; and toys and games [22]. The maximum data rate that would be required by these applications would not be more than 200 kbps with a minimum latency of 10–15 ms, which can easily be provided by LR-WPAN.

### 6.6.2 Network Topologies

The implementation of the higher layer protocol is left to the discretion of the individual applications and has not been specified in the standard. Like any other 802 standard, the 802.15.4 draft encompasses only those layers up to and including the Data Link Layer.

However, the standard suggests two network topologies that can used depending on the applications. They are the star topology and the peer-to-peer topology. Both are illustrated in Figure 6.19. In the star topology, the communication is established between devices and a single network controller called the PAN coordinator. The PAN coordinator is mainly used to establish, terminate and route communications. The PAN coordinator can have a continuous supply of power while the remaining devices are battery powered.

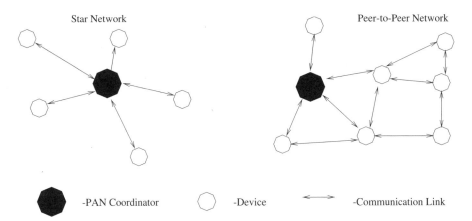

**Figure 6.19** Star and peer-to-peer networks.

The peer-to-peer topology also has a PAN coordinator; however it differs from the star network as any device can communicate with any other device as long as they are within the range of one another, similar to 802.15.3 network architecture. The selection of the network topology is an application design choice; some applications, such as PC peripherals and home automation, require the low-latency connection of a star network while others, such as industrial control and monitoring and wireless sensor networks would benefit from a peer-to-peer topology. All devices shall have unique 64 bit extended addresses. This address can be used for direct communication within the PAN, or it can be exchanged for a short address allocated by the PAN coordinator when the device associates.

### 6.6.3 Overview of the 802.15.4 standard

The Data Link Layer can be divided into two sublayers: the MAC and the logical link control (LLC) sublayers. The 802 standard has a common LLC sublayer for all the standards under it. The MAC sublayer is closer to the hardware and hence largely depends on the underlying physical layer (PHY) implementation. The MAC frame format is kept very flexible to accommodate the needs of different applications. A generic frame format is illustrated in Figure 6.20. The MAC protocol data unit is composed of the MAC header, MAC service data unit and the MAC footer. The MAC header has three fields: the first being frame control, which indicates the type of MAC frame being transmitted, format of the address field and controls the acknowledgment; the second being the sequence number to uniquely identify the transmitted packet and also to match the acknowledgment with the data unit sent; and the third being the address field. The size of the address field may vary depending on the type of frame. For example, a data frame may contain both the source and the destination information while the return acknowledgment frame does not contain any address information at all. In addition, either a short 16-bit or 64-bit addresses may be used. The payload field is flexible too, but the entire MAC protocol data unit cannot exceed 127 bytes in length. The MAC footer contains the frame check sequence to verify the integrity of the frame.

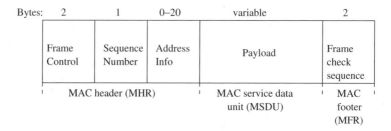

**Figure 6.20** Generic MAC frame format.

The medium access specified in the standard is Carrier Sense Multiple Access with Collision Avoidance (CSMA-CA). To enable transmissions with low latencies the 802.15.4 can operate in optional superframe mode. The PAN coordinator transmits a beacon at the start of every superframe. The time between two beacons is variable and is divided into 16 equal slots independent of the duration. A device can transmit at any instant but must complete its transaction before the next beacon. The channel access in the time slots is contention based, however, the PAN coordinator may assign slots to devices requiring dedicated bandwidth. The assigned time slots are referred to as the guaranteed time slots (GTS) and form a contention free period (CFP) located immediately before the next beacon as shown in the Figure 6.21. When GTS are employed, all devices must complete their contention-based transaction before the CFP begins. The beacon contains the details associated with a superframe, i.e. the duration of the superframe and instant when the contention free period would begin. In a non-beacon

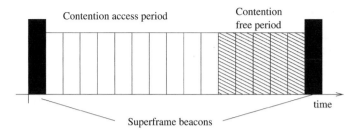

**Figure 6.21** Superframe structure with GTSs (guaranteed time slots).

enabled network, a device first checks if another device is currently transmitting on the same channel and, if so, it may then backoff for a random period. Another important function of the MAC layer is to confirm the successful reception of a received frame. This is done by sending an acknowledgment frame immediately after receiving the previous frame. The acknowledgment frame does not make use of the CSMA mechanism.

The IEEE 802.15.4 standard provides two physical layer (PHY) options that can combine with the above described MAC layer. The fundamental difference between the two PHYs is the frequency band. The 2.4 GHz PHY specifies operation in the unlicensed 2.4 GHz ISM band, which is available universally. While the 868/915 MHz PHY specifies operation in the 868 MHz band in Europe and 915 MHz band in the USA [19]. Obviously, moving from one geographic region to another is not anticipated with home networking appliances, but the international availability of the 2.4 GHz band does offer the advantages of larger markets and lower manufacturing costs. Another point of difference is that the 2.4 GHz band provides data rate of 250 kbps, while the 868/915 MHz offers rates of 20 and 40 kbps respectively. The 868/915 MHz PHY have the advantage of better sensitivity and larger coverage area when compared with 2.4 GHz PHY, which achieves better throughput and lower latencies. It is expected that the two PHYs will find applications for which its strengths are best suited.

## 6.7 Summary

In this chapter, we discussed various protocols related to multimedia wireless personal area networks. We first discussed the Bluetooth® standard that supports 1–2 Mbps. This was followed by a discussion of the coexistence of Bluetooth® and Wireless LANs that operate using the same frequency band. We also discussed the multi-megabit 802.15.3 and 802.15.3a (UWB) based standard for WPANs and finally concluded with the low bit-rate IEEE 802.15.4 standard that is being considered for networks such as wireless sensor networks.

## References

[1]  IEEE 802.15 Working Group for WPAN, Part 15.1: Wireless Medium Access Control (MAC) and Physical Layer (PHY) Specifications for Wireless Personal Area Networks (WPANs), IEEE Std 802.15.1, 2002.
[2]  Specification of the Bluetooth System, Core v1.2, Bluetooth SIG, http://www.bluetooth.com, November 2003.
[3]  WiMedia Alliance, www.wimedia.org
[4]  Ralf Steinmetz and Klara Nahrstedt, *Multimedia Systems*, Springer-Verlag, 2004.
[5]  Fred Halsall, *Multimedia Communications – Applications, Networks, Protocols and Standards*, Addison-Wesley, 2000.
[6]  Motion Picture Experts Group, http://www.chiarogonline.org/mpeg/, 2004.
[7]  Raffaele Bruno, Marco Conti and Enrico Gregori, Traffic integration in personal, local and geographical wireless networks, in *Handbook of Wireless Networks and Mobile Computing*, Ivan Stojmenovic, ed., pp. 145–168, 2002.

[8] Martin van der Zee and Geert Heijenk, Quality of Service in Bluetooth Networking Part I, citeseer.ist.psu.edu/vanderzee01quality.html

[9] D.L. Lough, A taxonomy of computer attacks with applications to wireless networks, Ph.D Dissertation, Virginia Polytechnic Institute and State University, April 2001.

[10] Tom Karygiannis and Les Owens, Wireless network security 802.11, Bluetooth and handheld devices, NIST Special Publication, November 2002.

[11] Creighton T. Hager and Scott F. Midkiff, An Analysis of Bluetooth Security Vulnerabilities, *Wireless Communications and Networking Conference, WCNC 2003*, Vol. 3, pp. 1825–1831, March 2003.

[12] M. Jakobsson and S. Wetzel, Security weaknesses in Bluetooth, in *Crytographers Track at RSA Conference*, San Franscisco, CA, 2001.

[13] IEEE 802.15 Working Group for WPAN, Part 15.2: Coexistence of Wireless Personal Area Networks with Other Wireless Devices Operating in Unlicensed Frequency Bands, IEEE Std 802.15.2, 2003.

[14] IEEE, Wireless LAN medium access control (MAC) and physical layer (PHY) Spec, IEEE 802.11 standard, 1998.

[15] N. Golmie, Interference in the 2.4 Ghz ism band: Challenges and Solutions, in http://w3.antd.nist.gov/pubs/golmie.pdf, Gaithersburg, Maryland, National Institute of Standards and Technology.

[16] Ron Nevo, Jim Lansford, Adrian Stephens, Wi-fi(802.11b) and Bluetooth: Enabling coexistence, *Network, IEEE*, **15**(5), 20–27, October 2001.

[17] E. Zehavi, Jim Lansford and Ron Nevo, Metha: A method for coexistence between co-located 802.11b and Bluetooth systems, IEEE P802.11 Working Group Contribution, vol. IEEE P802.15-00/360r0, November 2000.

[18] G. Ungerboeck, Trellis Coded Modulation with redundant signal sets part 1: Introduction, *IEEE Communication Magazine*, **25**(2), February 1987.

[19] IEEE 802.15 Working Group for WPAN, Part 15.4: Wireless Medium Access Control (MAC) and Physical Layer (PHY) Specifications for Low-Rate Wireless Personal Area Network (LR-WPANs), IEEE Std 802.15.4, 2003.

[20] I. Poole, What exactly is ... zigbee? *Communications Engineer*, **2**(4), September 2004.

[21] K. P. Negus, A. P. Stephens and J. Lansford, HomeRF: wireless networking for the connected home, *Personal Communications, IEEE*, **7**(1), February 2000.

[22] E. Callaway, P. Gorday, L. Hester, J. A. Gutierrez, M. Naeve, B. Heile and V. Bahl, Home Networking with IEEE 802.15.4: a developing standard for low-rate Wireless Personal Area Networks, *Communications Magazine, IEEE*, **40**(8), 70–77, August 2002.

[23] FCC First Report & Order, *Revision of part 15 of the commission's rules regarding UWB transmission systems (FCC 02-48)*, ET-Docket 98-153, 2002.

[24] M. Z. Win and R. A. Scholtz, *Impulse Radio: How it Works*, IEEE Communications Letters, vol. 2, pp. 36–38, February 2002.

[25] Zigbee Alliance, *Zigbee Alliance: Wireless control that simply works*, www.zigbee.org

[26] Multiband OFDM Alliance, www.multibandofdm.org, May 2005.

# 7

# QoS Provision in Wireless Multimedia Networks

Nikos Passas and Apostolis K. Salkintzis

## 7.1 Introduction

This chapter addresses several issues related to the Quality of Service (QoS) provision in wireless multimedia networks. First, we focus on some special topics of QoS in Wireless Local Area Network (WLANs) that serve as an expansion to the WLAN QoS topics discussed in Chapter 5. We start with the problem of maintaining QoS during handover and traffic scheduling at the wireless medium. Concerning handover, we explain the weaknesses in combining RSVP and mobility and present some of the major proposed solutions. To provide more information to the interested reader, we describe in detail a solution that extends the Inter-Access Point Protocol (IAPP) [12] and use it to reduce handover latency. Simulation results are provided to show the performance of the described solution. Next, we focus on IEEE 802.11e [17], the extension of the IEEE 802.11 standard for QoS provision, and emphasize the need for an efficient traffic-scheduling algorithm. To show how such an algorithm can play a key role to the performance of the protocol, we describe in detail a proposed solution that aims at maintaining the required QoS parameters of different kinds of traffic over the wireless medium. Again, we provide specific simulation results that show the considerable improvement attained with the use of the algorithm, compared with conventional solutions, such as the Simple Scheduler included in IEEE 802.11e.

Subsequently, we study the problem of RSVP over wireless networks. Designed for fixed networks, RSVP assumes a stable given overall bandwidth per link, that can be distributed to different active flows, based on their requirements. Clearly, this is not the case in wireless networks, where bandwidth can experience sudden and unpredictable fluctuations due to factors such as interference, fading, attenuation, etc. Several solutions have been proposed to this problem, with dynamic RSVP (dRSVP) being one of the major ones. dRSVP extends the standard RSVP protocol to use ranges of traffic specifications, instead of absolute values. The benefit is that intermediate nodes can adjust allocations, as long as these are within the specified ranges, depending on changes of the available bandwidth in the output links. After presenting the basic features of dRSVP, we focus on a component that is essential for its performance, i.e., the flow rejection algorithm. This algorithm has to detect the instances where the

---

*Emerging Wireless Multimedia: Services and Technologies*   Edited by A. Salkintzis and N. Passas
© 2005 John Wiley & Sons, Ltd

overall bandwidth is reduced below the required limit, and reject a number of flows in order to allow the rest to operate as required. The question is how many and which in particular should be rejected, while causing as little inconvenience to the users and the network as possible. We present a specific flow rejection algorithm proposal that aims at minimizing the number of rejected flows, while at the same time preventing system under-utilization. The illustrated simulation results show that the described algorithm can actually improve overall performance, compared with random rejection.

Finally, we address the challenges related to QoS in hybrid 3G/WLAN networks and in particular we focus on the provision of consistent QoS across these network technologies. We discuss a simple example as a way of easily explaining and understanding the main QoS topics in this area. This example considers a real-time video session, which needs to be handed over from a 3G environment[a] to a WLAN environment. Our main objective is to evaluate the conditions and restrictions under which the video session can seamlessly (i.e. without any noticeable QoS differences) continue across the two networks. For this purpose, we formulate a number of practical interworking scenarios, where UMTS subscribers with ongoing real-time video sessions handover to WLAN, and we study the feasibility of seamless continuity by means of simulation. We particularly quantify the maximum number of UMTS subscribers that can be admitted to the WLAN, subject to maintaining the same level of UMTS QoS and respecting the WLAN policies. Our results indicate that the WLAN can support seamless continuity of video sessions for only a limited number of UMTS subscribers, which depends on the applied WLAN policy, access parameters and QoS requirements. In addition to this study, we address several other issues that are equally important to 3G/WLAN seamless session continuity, such as the QoS discrepancies across UMTS and WLAN, the vertical handover details, and various means for access control and differentiation between regular WLAN data users and UMTS subscribers. The framework for discussing these issues is created by considering a practical UMTS/WLAN interworking architecture.

## 7.2 QoS in WLANs

### 7.2.1 QoS During Handover

The demand for multimedia services in mobile networks has raised several technical challenges, such as the minimization of handover latency. In this context, soft and softer handover techniques have played a key role and provided the means for eliminating the handover latency and thus enabling the provision of mobile multimedia services. However, not all radio access networks support soft handover techniques. For example, the notorious IEEE 802.11 WLANs [26] support only hard handovers and consequently the use of multimedia services over such WLANs raises considerable concerns. Another major challenge in both fixed and wireless networks today is QoS provision. For wireless access networks, such as WLANs, the Integrated Services (IntServ) framework [1], standardized by the Internet Engineering Task Force (IETF), provides the necessary means for requesting and obtaining QoS per traffic flow. IntServ uses the Resource reSerVation Protocol (RSVP) [2] for implementing the required QoS signaling. RSVP is problematic in wireless networks, basically due to the need for re-establishing resource reservations every time a Mobile Node (MN) changes its point of attachment and the IP route with its Corresponding Node (CN) has to be updated end-to-end, resulting in increased handover latency. This problem and major solutions are described below in more detail.

RSVP was designed to enable hosts and routers to communicate with each other in order to setup the necessary states for Integrated Services support. It defines a communication *session* to be a data flow with a particular destination and transport layer protocol, identified by the triplet (destination address, transport-layer protocol type, destination port number). Its operation applies only to packets of a

---

[a]In the context of 3G/WLAN QoS interworking we consider UMTS as a typical realization of a 3G environment. However, most of the topics discussed can be easily extended to other 3G environments.

particular session, and therefore every RSVP message must include details of the session to which it applies. Usually, the term *flow* is used instead of *RSVP session*. The RSVP protocol defines seven types of messages, of which the PATH and RESV messages carry out the basic operation of resource allocation. The PATH message is initiated by the sender and travels towards the receiver to provide characteristics of the anticipated traffic, as well as measurements for the end-to-end path properties. The RESV message is initiated by the receiver, upon receipt of the PATH message, and carries reservation requests to the routers along the communication path between the receiver and the sender. After the path establishment, PATH and RESV messages should be issued periodically to maintain the so-called soft states that describe the reservations along the path.

RSVP assumes fixed end-points and for that reason its performance is problematic in mobile networks. When an active MN changes its point of attachment with the network (e.g., upon handover), it has to re-establish reservations with all its CNs along the new paths. For an outgoing flow, the MN has to issue a PATH message immediately after the routing change, and wait for the corresponding RESV message, before starting data transmission through the new attachment point. Depending on the hops between the sender and the receiver, this can cause considerable delays, resulting in temporary service disruption. The effects of handover are even more annoying in an incoming flow because the MN has no power to immediately invoke the path re-establishment procedure. Instead, it has to wait for a new PATH message, issued by the sender, before responding with a RESV message in order to complete the path re-establishment. Simply decreasing the period of the soft state timers is not an efficient solution because this could increase signaling overhead significantly. A number of proposals can be found in the literature, extending RSVP for either inter-subnet or intra-subnet scenarios. For intra-subnet scenarios, proposals that combine RSVP with micro-mobility solutions, such as Cellular IP [3], can reduce the effects of handover on RSVP, since only the last part of the virtual circuit has to be re-established. For inter-subnet scenarios, the existing proposals include advance reservations, multicasting, RSVP tunnelling, etc. We look at some of them in more detail below.

Talukdar *et al.* [4] proposed Mobile RSVP (MRSVP), an extension of RSVP that allows the MN to pre-establish paths to all the neighboring cells. All reservations to these cells are referred to as *passive* reservations, in contrast to the *active* reservations in the cell that the MN actually is. To achieve this, proxy agents are introduced and a distinction is made between active and passive reservations. Although this proposal solves the timing delay for QoS re-establishment with over-reservation, it has several disadvantages. First, RSVP has to be enhanced to support passive reservations. Furthermore, the introduction of several proxy agents together with their communication protocol augments the complexity of the network. Finally, a drawback of MRSVP is that it relies on the MN to supply its mobility specification (i.e., a list of care-of addresses in the foreign subnets it may visit). Das *et al.* [5] attempt to tackle this last issue by introducing two new protocols called Neighbor Mobility Agent Discovery Protocol (NMADP) and Mobile Reservation Update Protocol (MRUP).

Tseng *et al.* [6] proposed the Hierarchical MRSVP (HMRSVP), in an attempt to reduce the number of required passive reservations. According to HMRSVP, passive reservations are performed only when an MN is moving in the overlapping area of two or more cells. Although this proposal outperforms MRSVP, in terms of reservation blocking, forced termination and session completion probabilities while achieving the same QoS, it does not cater for the other disadvantages introduced by MRSVP.

Another proposal by Kuo and Ko [7] extends RSVP with two additional processes, a resource clear and a resource re-reservation, in order not to release and re-allocate reservations in the common routers of the old and the new path. This solution performs well in reducing the path re-establishment time, but modifies the RSVP protocol significantly.

Chen *et al.* [8] proposed an RSVP extension based on IP multicast to support MNs. RSVP messages and actual IP datagrams are delivered to an MN using IP multicast routing. The multicast tree is modified dynamically every time an MH is roaming to a neighboring cell. This method can minimize service disruption due to rerouting of the data path during handovers, but it introduces extra overhead for the dynamic multicast tree management and requires for multiple reservations in every multicast tree.

Hadjiefthymiades *et al.* [9] proposed a Path extension scheme. The RSVP protocol is modified in such a way that the existing reservation is preserved and an *extension* to the reservation is performed locally from the old to the new Access Router. To deploy such a solution several modifications are required in the network components and the related protocols.

All these approaches, while trying to improve the performance of RSVP in mobile networks, either result in low resource utilization, due to advance reservations, or require considerable modifications of protocols and network components operation. In micro-mobility environments, only a small part of the path is changed, while the remaining path can be re-used. Accordingly, a scheme for partial path re-establishment can be considered, which handles discovery and setup of the new part between the crossover router and the MN. Paskalis *et al.* [10] have proposed a scheme that reduces the delay in data path re-establishment without reserving extra resources, while it requires modifications only in the crossover router between the old and the new path. According to this scheme, an MN may acquire different *local* care-of addresses, while moving inside an access network, but is always reachable by a *global* care-of address through tunneling, address translation, host routing, or any other routing variety, as suggested in various hierarchical mobility management schemes. The crossover router, referred to as *RSVP Mobility Proxy*, handles resource reservations in the last part of the path and performs appropriate mappings of the global care-of address to the appropriate local care-of address.

A similar approach for partial path re-establishment has been proposed by Moon and Aghvami in [11]. According to this scheme, if an MN is a sender, it initiates an RSVP PATH message after the route update is completed. When the RSVP daemon on the crossover router, which is determined by the route update message, receives an RSVP PATH message after a handover, it immediately sends an RSVP RESV message to the MN. If an MN is a receiver, the RSVP daemon on the crossover router can trigger an RSVP PATH message immediately after detecting any changes to the stored PATH state or receiving a notification from the underlying routing daemon. Both [10] and [11] reduce the handover delay, but their performance depends on the distance between the MN and the crossover router. Additionally, they do not handle the delay introduced by the transmission of the required RSVP signaling through the radio interface that is usually subject to collisions and increased delays, compared with fixed links.

Below, a solution that combines RSVP with the Inter-Access Point Protocol IAPP [12] is presented, referred to as IAPP+, destined mainly for intra-subnet scenarios applicable to WLANs. More information on this solution can be found in [12]. Its main advantage is that it requires minimum enhancements at the APs, while it can be combined either with the standard RSVP operations, or partial path-re-establishment solutions, such as [10] and [11].

IAPP is a recommended practice recently standardized by the IEEE Working Group 802.11f, that provides the necessary capabilities for transferring context from one AP to another, through the fixed distribution system, in order to facilitate fast handovers across multi-vendor APs. The operation of the protocol is depicted in Figure 7.1. The standard describes a service access point (SAP), service primitives, a set of functions and a protocol that will allow conformant APs to interoperate on a common distribution system, using the Transmission Control Protocol over IP (TCP/IP) to exchange IAPP packets. The IAPP message exchange during handover is initiated by the *reassociate-request* message, which carries the Basic Service Set Identifier (BSSID) of the old AP (message 1 in Figure 7.1). In order to communicate with the old AP through the fixed distribution system, the new AP needs its IP address. The mappings between BSSIDs and IP addresses can be either stored inside all APs or obtained from a Remote Authentication Dial-In User Service (RADIUS) server [14] (messages 2 and 3). If communication between APs needs to be encrypted, security blocks are exchanged before actual communication (messages 4 and 5). The two most significant messages of the IAPP are the *MOVE-Notify* (message 6), issued by the new AP to inform the old AP that the MN has moved to its own area, and the *MOVE-Response* (message 7), issued by the old AP, containing a *Context Block*. The MOVE-notify and MOVE-response are IP packets carried in a TCP session between the two APs. Although initially intended to contain authentication information, to allow the new AP to accept the MN without re-authentication, the Context Block has a flexible structure, able to support any type of information exchange. More specifically, it can consist of a variable number of *Information Elements* (IEs) of the form (Element ID,

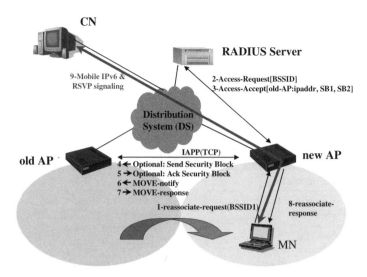

**Figure 7.1**   Standard IAPP and RSVP signaling.

Length, Information). In this way, every IE can contain variable length information, whose type is specified by the Element ID. Processing of the information transferred inside the IEs is beyond the scope of the IAPP, and depends on the functionality of the APs.

During handover of an MN from one AP (old AP) to another (new AP), the aim of the IAPP is to avoid time-consuming procedures that in different case would be needed before transmitting through the new link. The two most significant messages of IAPP are the *MOVE-Notify*, issued by the new AP to inform the old AP that a MN has moved to its area, and the *MOVE-Response*, issued by the old AP, containing the Context Block (see Figure 7.1). Although initially intended to contain authentication information, to allow the new AP to accept the MN without re-authentication and reduce handover latency, the Context Block has a flexible structure able to transfer any kind of information.

The key concept of the IAPP+ scheme is briefly summarized below and illustrated with the aid of Figures 7.2 and 7.3. Standard IAPP and RSVP messages are marked with solid lines, while new messages are marked with dashed lines. When an MN decides to handover, it switches to the frequency of the new AP, transmits the *reassociate-request* message and immediately switches back to the old AP to continue data transmission. By means of conventional IAPP signaling (*MOVE-Response* message), the new AP receives context information pertaining to the MN. In general, both authentication and RSVP context needs to be transferred to the new AP. RSVP context contains information about the traffic characteristics and QoS requirements of all the active sessions of the MN. Since the new AP might be running out of resources, an admission control algorithm decides which sessions can be maintained. The results of the admission control are transmitted to the old AP through the *MOVE-Accept* message and to the MN through the *ACCEPTED-Request* message. If the results of the admission control are acceptable to the MN, it responds back to the old AP with a positive *ACCEPTED-Response* message, and then switches to the new AP waiting for the handover procedure to finish. Otherwise, the answer is negative and the MN should try to handover to another AP, provided there is one available. In the case of a positive answer, the old AP issues a positive *MN-Response* message to the new AP, in order to switch the data path (through a *Layer-2 Update Frame*) and start the RSVP signaling with the CNs of the accepted sessions. For an outgoing session, the MN can start transmitting data as soon as it receives the RSVP RESV message from the CN while, for an incoming session, the MN can simply wait to receive data packets. In this way, the MN does not have to cease transmission until the IAPP signaling is over, but it is allowed to use the old AP during this time.

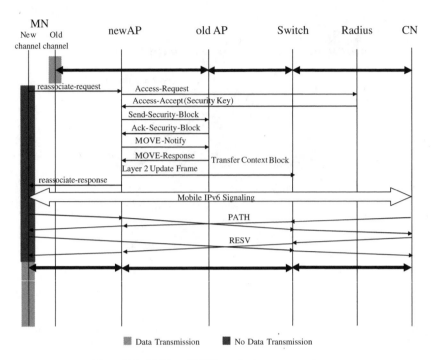

**Figure 7.2**    IAPP and RSVP standard message exchange.

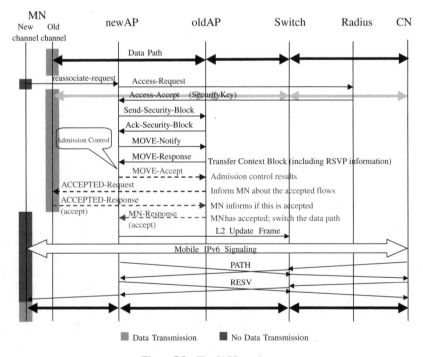

**Figure 7.3**    The IAPP+ scheme.

Additionally, the new AP acts as a proxy of the MN for the required RSVP signaling, reducing the path re-establishment time.

PATH and RESV messages contain a large set of objects, in order to cover all network cases. For example, they contain objects to indicate non-RSVP nodes, or nodes without policy control, objects for authentication or confirmation. Most of these objects can be omitted from the respective IAPP information elements. More specifically, the PATH IE should contain the objects that describe the traffic that the sender intends to enter to the network, i.e., the SESSION, SENDER_TEMPLATE and SENDER_TSPEC objects, while the RESV IE should contain the objects describing the anticipated offered bandwidth and QoS, i.e., the SESSION, STYLE and flow descriptor list (consisting of FLOWSPEC and FILTER_SPEC objects). We assume IPv4 addressing, Fixed-Filter style and Guaranteed Service. For IPv6, the only difference is that source and destination addresses occupy 16 bytes instead of 4. Guaranteed Service is considered for the FLOWSPEC, in order to offer bounded end-to-end delays and bandwidth. For Controlled-Load Service the FLOWSPEC should not include the *Rate* and *Slack Term* parameters. Details for the specific fields can be found in [1].

It is clear that many fields inside the objects could be omitted or forced to consume less space. Nevertheless, the objects are included as defined for RSVP in order to be compatible with future versions of the RSVP protocol that use more fields. Additionally, the above IE structure is flexible in including more objects if needed. The only thing that is needed in order to include a new object is to add the object at the end of the respective IE, as defined in RSVP together with its object header, and update the IE Length field accordingly. A typical length of the PATH IE is 64 bytes, while the length of the RESV IE is 84 bytes. Note that, using the solution described above, this information is transmitted through a high-speed distribution system connecting the two APs. In different case, the same information would have to be transmitted using standard RSVP signaling through the wireless medium, after re-association of the MN with the new AP, resulting in considerable delays and signaling overhead.

To obtain numerical results and quantify the improvement attained with the use of the IAPP+ scheme, a simulation model was developed focusing on modeling the message exchange between the involved nodes during handover. The aim was to compare IAPP+ with the conventional path re-establishment procedure during handover (referred to as the NO IAPP scheme). Comparison was made in terms of path re-establishment delays, under different traffic conditions in the radio interface. The simulation model included:

- the '*observed*' Mobile Node (MN), which represents the origination of signaling for outgoing RSVP sessions;
- the Corresponding Node (CN), which connects to the distribution system and represents the origination of signaling for incoming RSVP sessions;
- two Access Points (APs), between which consecutive handovers of the MN were performed, and
- a variable number of '*background*' MNs in the cell of each AP, generating the background (or interfering) traffic on the radio interface.

The AP and MNs in each cell were operating based on IEEE 802.11b in Distributed Coordination Function mode, with a bit rate of 11 Mbps, while the two APs were connected through a fixed link of 100 Mbps. Depending on the scheme under consideration, the execution of the corresponding message exchange was modeled separately. Three different levels of mean background traffic were considered in every cell, corresponding to 10%, 50% and 80% of the maximum WLAN loading. For each level of background traffic load, several simulation runs were performed for different numbers of active sessions in the observed MN. In every run, the sessions were equally divided between incoming and outgoing, while a large number of handovers were executed in order to obtain accurate mean values.

The mean total path re-establishment delay for different numbers of sessions at the observed MN and traffic levels of 10% (light) and 80% (heavy) is illustrated in Figures 7.4 and 7.5, for incoming and outgoing sessions respectively. As can be observed, the improvement attained with IAPP+ is greater for incoming sessions, since both PATH and RESV messages are avoided over the radio interface. On the

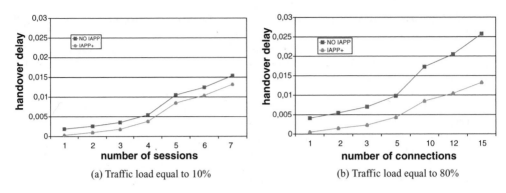

(a) Traffic load equal to 10%                    (b) Traffic load equal to 80%

**Figure 7.4**   Path re-establishment delay for incoming sessions.

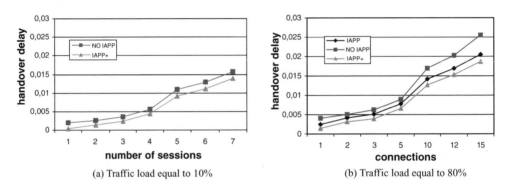

(a) Traffic load equal to 10%                    (b) Traffic load equal to 80%

**Figure 7.5**   Path re-establishment delay for outgoing sessions.

other hand, only the PATH message is avoided for outgoing sessions. In all cases, the delay improvement of IAPP+ increases with the number of active sessions at the MN, as more messages are not transmitted on the radio interface. With the improvement added in the IAPP+ scheme, the data blocking time due to handover is reduced, leading to lower handover delay. As expected, the improvement from the use of the IAPP+ scheme is relatively the same, independently of the traffic load and the number of active sessions at the observed MN. This is because, as shown in Figure 7.2 and 7.3, the part of blocking time that is saved (IAPP signaling transmission at the fixed network) is not affected significantly by the traffic load.

### 7.2.2 Traffic Scheduling in 802.11e

As described in Chapter 5, IEEE 802.11e [17] introduces two new access modes, namely the Enhanced Distributed Coordination Access (EDCA) and the HCF Controlled Channel Access (HCCA). A STA that supports these access modes is referred to as QoS STA (QSTA). Access to the wireless channel is composed of Contention-Free Periods (CFPs) and Contention Periods (CPs), controlled by the Hybrid Coordinator (HC), located at the Access Point (AP). EDCA is used in CPs only, while HCCA is used in both periods. EDCA is an enhanced version of the legacy DCF and it is used to provide the prioritized QoS service. HCCA on the other hand, provides a parameterized QoS service, based on specific traffic and QoS characteristics, as expressed through TSPECs. A TXOP is defined as a period of time during which the STA can send multiple frames. One of the main components of the HCCA is the traffic scheduler, an algorithm that decides on the allocation of TXOPs. The *Simple Scheduler*, included in the draft amendment, is not expected to work sufficiently well under all kinds of conditions. Owing to its operation to assign static TXOPs at static intervals for all QSTAs, it is expected to perform relatively well in scenarios with uniform Constant Bit Rate (CBR) traffic, where the amount of data to be transmitted in every constant interval is stable, and the QoS requirements (mainly packet loss and delay) are similar for all transmitted traffic. For Variable Bit Rate (VBR) traffic, the Simple Scheduler has no means to adjust TXOPs based on time-varying transmission requirements.

Identifying the weaknesses of the Simple Scheduler, the scheduling algorithm proposed in [15] provides improved flexibility by allowing the HC to poll each QSTA at variable intervals, assigning variable length TXOPs. The algorithm is referred to as *Scheduling based on Estimated Transmission Times – Earliest Due Date* (SETT-EDD), indicating that TXOP assignments are based on earliest deadlines, to reduce transmission delay and packet losses due to expiration. SETT-EDD is a flexible and dynamic scheduler, but it lacks an efficient mechanism for calculating the exact required TXOP duration for each QSTA transmission. TXOP duration is calculated based on estimations derived from the mean data rate of each Traffic Stream (TS) and the time interval between two successive transmissions, a scheme that can be very inefficient for bursty traffic.

In both Simple and SETT-EDD, TXOP durations are calculated by using some kind of estimation of the amount of data waiting to be transmitted by every QSTA. The scheduling algorithm described below is referred to as '*Adaptive Resource Reservation Over WLANs*' (ARROW) [16], as it adapts TXOP reservations based on real-time requirements declared by the QSTAs. To achieve this, ARROW utilizes the *Queue Size (QS)* field, introduced by 802.11e as part of the new *QoS Data* frames, not supported by legacy 802.11 systems. The QS field can be used by the QSTAs to indicate the amount of buffered traffic for their TSs, i.e., their transmission requirements. A scheduler that utilizes this information should allocate TXOPs to QSTAs in such a way that satisfies these transmission requirements, as long as they comply with the traffic specifications declared through the TSPECs.

Before we proceed with the description of ARROW, it is essential to refer to some parameters that are utilized by the HC in order to calculate an aggregate service schedule for a $QSTA_i$ having $n_i$ active TSs. These parameters can be derived from the individual TSPEC parameters as follows.

- Minimum TXOP duration ($mTD$). This is the minimum TXOP duration that can be assigned to a QSTA and equals the maximum time required to transmit a packet of maximum size for any of the QSTA's TSs. Thus, $mTD_i$ of $QSTA_i$ is calculated as:

$$mTD_i = \max\left(\frac{M_{ij}}{R_{ij}}\right), j = 1 \ldots n_i. \tag{7.1}$$

- Maximum TXOP duration ($MTD$). This is the maximum TXOP duration that can be assigned to a QSTA. It should be less than or equal to the transmission time of the Aggregate.
- Maximum Burst Size ($AMBS$) of a QSTA. The $AMBS$ is the sum of the maximum burst sizes (MBSs) of all TSs of a QSTA. Thus for $QSTA_i$ it holds:

$$AMBS_i = \sum_{j=1}^{n_i} MBS_{ij} \tag{7.2}$$

and

$$MTD_i \leq \frac{AMBS_i}{R_i}, \tag{7.3}$$

where $R_i$ is the minimum physical bit rate assumed for $QSTA_i$ ($R_i = \min(R_{ij}), j = 1 \ldots n_i$).

- Minimum Service Interval (*mSI*). The minimum time gap required between the start of two successive TXOPs assigned to a specific QSTA. It is calculated as the minimum of the *mSI*s of all the QSTA's TSs:

$$mSI_i = \min(mSI_{ij}), j = 1 \ldots n_i \qquad (7.4)$$

If not specified in the TSPEC, $mSI_{ij}$ of $TS_{ij}$ is set equal to the average interval between the generation of two successive MSDUs, i.e., $mSI_{ij} = L_{ij}/\rho_{ij}$.
- Maximum Service Interval (*MSI*). The maximum time interval allowed between the start of two successive TXOPs assigned to a QSTA. Although no specific guidelines for calculating *MSI* are provided, an upper limit exists to allow an MSDU generated right after a TXOP assignment to be transmitted at the next TXOP. Accordingly:

$$MSI_i \leq D_i - MTD_i, \qquad (7.5)$$

where $D_j$ is defined as the minimum delay bound of all TSs of $QSTA_i$ $(D_i = \min(D_{ij}), j = 1 \ldots n_i)$. This is an upper limit that ensures that successive TXOPs will be assigned close enough to preserve delay constraints [17]. As shown later, the maximum allowed value of *MSI* may be lower, depending on the specific operation of the scheduler.

An example of the use of QS in ARROW is depicted in Figure 7.6. The allocation procedure will be described later in this section. For simplicity, one TS per QSTA is assumed. At time $t_i(x)$, $QSTA_i$ is assigned $TXOP_i(x)$, according to requirements declared earlier through the QS field. Using a *QoS Data* frame, $QSTA_i$ transmits its data together with the current size of its queue in the QS field $(QS_i(x))$. At time $t_i(x+1)$ the scheduler assigns $TXOP_i(x+1)$ to $QSTA_i$, in order to accommodate part or all of $QS_i(x)$. During the interval $[t_i(x), t_i(x+1)]$ new data are generated in $QSTA_i$, therefore $QSTA_i$ uses the *QoS Data* frame transmitted at $TXOP_i(x+1)$ to indicate the new queue size $(QS_i(x+1))$. In the same manner, at $t_i(x+2)$ the scheduler assigns $TXOP_i(x+2)$ to $QSTA_i$, accommodating part or all of the data declared in $QS_i(x+1)$ and gets the new queue size from $QSTA_i$ $(QS_i(x+2))$. As clearly shown, by utilizing the QS field, ARROW has very accurate information about the time varying properties of each TS, and is able to adapt the TXOP duration accordingly. This is considered essential, especially in the case of bursty and VBR traffic, where transmission requirements are time varying.

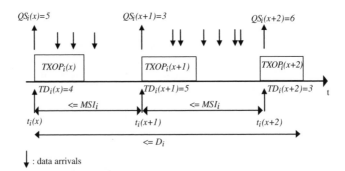

**Figure 7.6**  TXOP assignment with ARROW.

As can be observed in Figure 7.6, for every $QSTA_i$, data arriving within the interval $[t_i(x), t_i(x+1)]$ can be transmitted no earlier than $TXOP_i(x+2)$ starting at $t_i(x+2)$. Therefore, in order not to exceed the delay deadline of MSDUs, assuming the worst case that service intervals are equal to $MSI_i$ and

$TXOP_i(x+2) = MTD_i$, it should hold that:

$$D_i \geq 2MSI_i + MTD_i \Leftrightarrow$$
$$\Leftrightarrow MSI_i \leq \frac{D_i - MTD_i}{2}. \tag{7.6}$$

If the scheduler should also take into account possible retransmissions, relation (6) becomes:

$$MSI_i \leq \frac{D_i - MTD_i}{2 + m}, \tag{7.7}$$

where $m$ is the number of allowed retransmission attempts.

Equations (7.6) and (7.7) show that $MSI$ used in ARROW should be less than half the upper limit calculated with (7.5). This means that, for the same traffic, ARROW should schedule shorter TXOPs closer to each other, compared with schedulers such as Simple and SETT-EDD. Although this results in increased overhead, it is compensated for by the accurate TXOP assignment, as shown in the simulation results given later.

Additionally, ARROW incorporates a traffic policing mechanism to ensure that the transmission requirements expressed through the QSs do not violate traffic characteristics expressed through the TSPECs. For that purpose, a $TXOP$ $timer$ is used, that implements the operation of a leaky bucket of time units. The TXOP timer value $T_i$ for a $QSTA_i$ having $n_i$ active TSs, increases with rate $r(T_i)$:

$$r(T_i) = \sum_{j=1}^{n_i} \left( \left( \frac{L_{ij}}{R_{ij}} + O \right) \Big/ \frac{L_{ij}}{\rho_{ij}} \right). \tag{7.8}$$

Equation (7.8) means that during the time interval needed for the generation of an MSDU of Nominal Size at mean data rate, the TXOP Timer should be increased by the time required for the transmission of this MSDU. The maximum TXOP Timer value, max($T_i$), equals the time required for the transmission of all maximum bursts:

$$\max(T_i) = \sum_{j=1}^{n_i} \left( \frac{MBS_{ij}}{R_{ij}} + O \right). \tag{7.9}$$

According to the operation of ARROW described below, no TXOP longer than the current value of $T_i$ can be assigned to $QSTA_i$ at any time. After each TXOP assignment, the value of the respective TXOP timer is reduced accordingly.

Based on the above discussion, the operation of ARROW is as follows:

(1) The scheduler waits for the channel to become idle.
(2) When the channel becomes idle at a given moment $t$, the scheduler checks for QSTAs that:
    (a) can be polled without violating $mSI$ and $MSI$, i.e., for a $QSTA_i$ that was last polled at time $t_i$, it should hold that:

$$t_i + mSI_i \leq t \leq t_i + MSI_i \tag{7.10}$$

    and
    (b) their TXOP timer value $T$ is greater than the value of their $mTD$, to ensure enough time for the minimum TXOP duration.
(3) If no QSTAs are found, the scheduler proceeds to the next time slot and returns to step 1.
(4) In a different case, the scheduler polls the QSTA with the earliest deadline. The deadline for a $QSTA_i$ is the latest time that this QSTA should be polled, i.e., $t_i + MSI_i$, where $t_i$ is the time of the last poll for $QSTA_i$.

(5) Assuming a $QSTA_i$ having $n_i$ active TSs is selected for polling, the scheduler calculates the TXOP duration $TD_i$, in three steps:

    (a) First, for every $TS_{ij}$ of $QSTA_i$ ($j = 1 \ldots n_i$), the scheduler calculates $TD_{ij}$, as the maximum of (i) the time required to accommodate the pending traffic, as indicated by the queue size of that TS ($QS_{ij}$), plus any overheads and, (ii) $mTD_{ij}$, to ensure that the assigned TXOP will have at least the minimum duration:

$$TD_{ij} = \max\left(\frac{QS_{ij}}{R_{ij}} + O, mTD_{ij}\right). \tag{7.11}$$

    In the special case where $QS_{ij}$ is equal to zero, $TD_{ij}$ is set equal to the time for the transmission of a Null-Data MSDU, to allow $QSTA_i$ to update the queue size information for $TS_{ij}$.

    (b) $TD_i$ for $QSTA_i$ is calculated as the sum of all $TD_{ij}$:

$$TD_i = \sum_{j=1}^{n_i} TD_{ij}. \tag{7.12}$$

    (c) Finally $TD_i$ obtained from (7.12) is compared with the current TXOP Timer value $T_i$, to ensure conforming traffic:

$$TD_i = \min(TD_i, T_i). \tag{7.13}$$

(6) After the scheduler assigns the TXOP, it reduces the respective TXOP timer value accordingly and returns to step 1:

$$T_i = T_i - TD_i. \tag{7.14}$$

Although the above calculations are considered simple enough to be executed at the beginning of an idle slot, it is possible in most cases to execute steps 2–5 in parallel with step 1, while waiting for the channel to become idle, provided that the QSTA currently transmitting is not eligible for the next poll (e.g., it does not satisfy (7.10)).

The exploitation of queue size information for calculating accurate TXOP durations is particularly effective for both Constant Bit Rate (CBR) and Variable Bit Rate (VBR) traffic, but introduces some extra delay and increases the transmission overhead percentage. For VBR traffic this seems to be unavoidable, since its behaviour cannot be accurately predicted. On the other hand, due to its periodic nature, CBR traffic has a much more predictable behaviour. An enhanced version of ARROW scheduler can take advantage of this characteristic to reduce overhead and delays. To differentiate the two versions of the algorithm, we refer to '*basic ARROW*' and '*enhanced ARROW*'. The idea behind the enhancement is that, instead of waiting for the QS information or use Null-Data TXOPs to get the current queue size, the scheduler can estimate the current queue size of a CBR TS. Every time a TXOP is assigned to a QSTA with CBR TSs, the scheduler calculates the TXOP duration for each of these TSs by adding the queue size value, indicated by the previous MSDU transmission of the same TS, and the estimated (using Mean Data Rate) generated traffic in the time interval between the previous and the current transmission. Accordingly, for CBR TSs Equation (7.11) can be replaced by:

$$TD_{ij} = \max\left(\frac{QS_{ij} + \left(\frac{\rho_{ij}}{t - t'}\right)}{R_{ij}} + O, mTD_{ij}\right). \tag{7.15}$$

In a fully synchronized system, the use of $QS_{ij}$ in Equation (7.15) would not be required, as it should always be equal to zero (the queue would be emptied in every transmission). Nevertheless, it is used at no expense to cover cases when, for example, a MSDU generation is delayed for some reason (e.g., high computing load), and misses the scheduled TXOP. Assuming a CBR TS in Figure 7.6, the duration of $TXOP_i(x + 1)$ is calculated based on the $QS_i(x)$ and the estimation for the generated MSDUs within the interval $[t(x), t(x + 1)]$. This estimation is very accurate for CBR TSs, leading to considerably lower transmission delays, since the MSDUs generated in the interval $[t(x), t(x + 1)]$ are transmitted at $t(x + 1)$, instead of $t(x + 2)$ with basic ARROW. This strategy also reduces transmission overheads and leads to lower average channel occupancy (i.e., better channel utilization) since, by picking a suitable value for $mSI_i$, adequate for the generation of at least one MSDU, no Null-Data TXOPs for CBR TSs are required.

To show how traffic scheduling can affect the performance of 802.11e, simulation results from the comparison of Simple, SETT-EDD, basic ARROW and enhanced ARROW are presented below. The simulation scenarios considered an increasing number of QSTAs attached to an AP. All QSTAs and the AP were supporting the extended MAC layer specified in IEEE 802.11e [17] and the PHY layer specified in IEEE 802.11g [18], with a transmission rate of 12 Mbps. An ideal, error-free wireless channel was assumed, as the focus was on the scheduling procedure. In order to investigate the limits and the maximum scheduling capability of each algorithm under heavy traffic conditions, no admission control was applied. Additionally, no minimum fraction of time for the operation of EDCA was reserved (i.e., the whole Beacon Interval could be utilized by HCCA) to focus solely on the performance of the HCCA mechanism. Each QSTA had two active sessions:

(1) a bi-directional G.711 voice session (CBR traffic), mapped into two TSs (one per direction), and
(2) an uplink (from QSTA to AP) H.261 video session at 256 Kbps (VBR traffic), mapped into one uplink TS.

Figure 7.7 depicts throughput of non-delayed MSDUs for voice and video traffic. For voice traffic (Figure 7.7(a)), basic ARROW accommodates up to 18 QSTAs, while SETT-EDD can manage up to 14 QSTAs and Simple up to only 7 QSTAs. Using the enhancement, the number of QSTAs can be increased to 19 with enhanced ARROW. For video traffic (Figure 7.7(b)), basic and enhanced ARROW outperform both SETT-EDD and Simple, accommodating up to 19 QSTAs, as opposed to 13 with SETT-EDD and 6 with Simple. The main reason for the considerably improved performance of basic ARROW is the accurate TXOP assignment it performs, as a result of the accurate queue size information. This is also shown in more detail later in this section. As for the enhanced ARROW, it appears that the admission capacity limit of the Scheduler compared with the Standard ARROW for Voice traffic is somewhat increased since, with Enhanced ARROW up to 19 G.711 TSs can be accommodated comfortably.

It is interesting to observe that throughput of SETT-EDD and ARROW (both basic and enhanced) reduces rapidly immediately after reaching its maximum value. The reason is that, due to the dynamic TXOP assignment performed by these algorithms, new TSs entering the system can participate equally in the channel assignment. Thus, when the scheduler exceeds its maximum scheduling capability, service for all TSs is degraded abruptly. The Simple Scheduler on the other hand, manages to provide a stable throughput regardless of the offered load, because static allocations for existing TSs are not affected as the traffic load increases. This effect highlights the need for an effective admission control scheme for SETT-EDD and ARROW, which would prevent the offered load from exceeding the maximum scheduling capability.

In Figure 7.8, the mean delay for non-delayed voice and video MSDUs is shown. For voice traffic (Figure 7.8(a)), both Simple and SETT-EDD perform better than basic ARROW, since they are able to perform a fairly accurate estimation of expected CBR traffic. But enhanced ARROW achieves considerable improved delays, comparable to both Simple and SETT-EDD, as a direct effect of the incorporation of the traffic load prediction for CBR traffic. For video traffic (Figure 7.8(b)), Simple achieves a low and almost stable mean delay, but suffers from the considerable low number of accommodated QSTAs, as shown earlier in Figure 7.7(b). The improved delay of basic ARROW

**G.711 Non-Delayed MSDU Throughput**

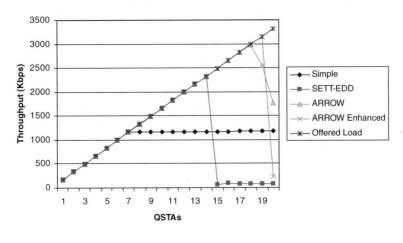

(a) G.711 Voice

**H.261 Non-Delayed MSDU Throughput**

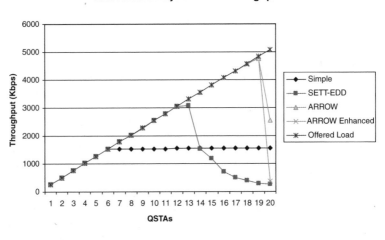

(b) H.261 video

**Figure 7.7**    Throughput of non-delayed MSDUs.

compared with SETT-EDD shows that the extra scheduling delay introduced by ARROW is balanced by the accurate TXOP assignment. With the enhanced version of ARROW, the mean delay for video traffic is the same as with basic ARROW for low traffic loads (up to seven QSTAs), but for higher traffic loads the delay is somewhat increased. This occurs because of the use of the estimation mechanism that eliminates the need for Null-Data TXOPs for CBR traffic and results in the allocation of larger TXOPs for voice traffic, compared with basic ARROW. This delays the TXOP allocation for video MSDUs and increases their mean delay in heavy traffic conditions. But, in any case, the transmissions are within the delay bounds of video traffic.

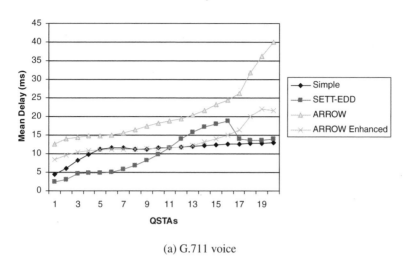

(a) G.711 voice

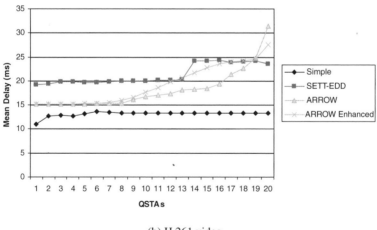

(b) H.261 video

**Figure 7.8** Mean delay of non-delayed MSDUs.

## 7.3 RSVP over Wireless Networks

QoS provision is considered today as one of the basic requirements for next generation networks. These networks are expected to integrate a large number of access systems, both wired and wireless, in one common core, and rely on the Internet Protocol (IP) for data exchange. The IP protocol, up to its current version 4, was designed for applications with low network requirements, such as e-mail and ftp, and accordingly offers an unreliable service that is subject to packet loss, reordering, packet duplication and unbounded delays. The expected new version 6 of IP provides some means for QoS support, but it still needs supporting mechanisms to provide guaranteed service. To treat the problem of QoS in IP networks, the Internet Engineering Task Force (IETF) has introduced two main frameworks, namely the Integrated Services (IntServ) [1] and the Differentiated Services (DiffServ) [18]. DiffServ classifies and possibly conditions the traffic, in order to ensure similar behaviour throughout the network. It performs well in core networks, due to its scalability to support large numbers of flows. IntServ on the other hand, is targeted mainly for the access systems, by providing means to request and obtain end-to-end QoS per flow. As already mentioned, the Resource Reservation Protocol (RSVP) [2] is considered the most popular signaling protocol in IntServ for requesting QoS per flow, and setting up reservations end-to-end upon admission.

A static resource reservation based approach (such as RSVP) performs well in fixed networks where links are stable, but exhibits low performance in a variable bandwidth environment. RSVP assumes a stable given overall bandwidth per link, that can be distributed to different active flows, based on their requirements. When a new flow requests admission, RSVP is used to check if the overall bandwidth per link is sufficient for all (active and new) flows. If the result of this check is positive, the new flow can be accepted. But if the available bandwidth is reduced after admission control (as can be the case in wireless links), the network may not be able to meet commitments for flows that have been successfully admitted. In this case, none of the flows operates according to its requirements, while the QoS observed by the user can be poor.

A number of solutions can be found in the literature, addressing this problem. For example, dynamic RSVP (dRSVP) [19] uses ranges of traffic flows specifications, instead of absolute values. The benefit is that intermediate nodes can adjust allocations, as long as these are within the specified ranges, depending on changes of the available bandwidth in the output links. As expected, dRSVP introduces a number of extensions/modifications of dRSVP, compared with the standard RSVP as listed below.

- An additional flow specification object (FLOWSPEC) in RESV messages and an additional traffic specification object (SENDER TSPEC) in PATH messages have been introduced, so as to describe ranges of traffic flows.
- A measurement specification object (MSPEC) has been added in the RESV messages, used to allow nodes to learn about downstream resource bottlenecks.
- A measurement specification object (SENDER MSPEC) has been introduced in the PATH messages, used to allow nodes to learn about upstream resource bottlenecks.
- An admission control process has been added, able to handle ranges of required bandwidth.
- A bandwidth allocation algorithm has been introduced that divides the available bandwidth among admitted flows, taking into account the desired range for each flow, as well as any upstream or downstream bottlenecks.

Measurements presented in [19] have shown that dRSVP can significantly improve the performance of RSVP over wireless. Nevertheless, as listed above, it introduces a number of enhancements to the standard protocol. Another solution that does not introduce changes to the protocol operation is to detect the instances where the overall bandwidth is reduced below the required limit and reject a number of flows in order to allow the rest to operate as required. The question is how many and which in particular should be rejected, to cause as little inconvenience to the users and the network as possible. Below, we present such a mechanism for fair and efficient flow rejection. More information on this mechanism can be found in [20].

The presented flow rejection mechanism can operate at a central point of the wireless link (e.g., the Access Point of a cell), where it can observe the overall provided bandwidth. When the provided bandwidth falls to a level that is insufficient for maintaining the reserved bandwidth values for all flows, one or more flows must be rejected, in order to maintain committed values for the rest. This can occur in two cases.

(1) *When a new flow requests admission.* If bandwidth is insufficient, the admission control will not allow the new flow to be accepted.

(2) *When the overall available bandwidth is decreased to a point that the requested bandwidth per flow cannot be maintained for all flows.* In this case, one or more of the flows have to be rejected, in order to maintain the reserved bandwidth for the rest of the flows within limits.

Flow rejection can be performed through standard RSVP signaling (RESV_Tear, PATH_Tear). A simple mechanism can be used that rejects flows randomly until the total requested bandwidth is reduced below the available capacity. This can lead to low efficiency, in terms of flow rejection probability and bandwidth utilization, as it might tear down:

- *more flows than necessary* (this can happen when a large number of flows with low bandwidth requirements is randomly selected for rejection, instead of a smaller number of high bandwidth flows),
- *high priority flows* (if the mechanism does not differentiate the flows, based on their importance, it can reject high priority instead of low priority flows), or
- *flows that utilize a large portion of the available bandwidth* (this could lead to low bandwidth utilization).

Below, a more efficient flow rejection mechanism is presented, which attains a considerably improved flow rejection probability. In this, three criteria are considered for providing fair and efficient rejection.

(1) Reject the lower priority flows first. In order to achieve this, a scheme is needed that classifies flows into a number of priority classes. The priority classes may be defined according to a number of criteria, such as the flow type, traffic characteristics, QoS requirements, etc. In its simplest form, the classification scheme can use the transport-layer protocol type included in every session identification triplet to classify flows. For example, UDP flows can be considered as high priority, as they usually carry real time data, while TCP flows can be considered as low priority.

(2) Minimize the number of rejected flows. If there are more that one set of flows that can be rejected, then the set with the fewer members (i.e., flows) should be selected for rejection. This criterion prevents one from rejecting a large number of low bandwidth flows.

(3) Prevent underutilization. This could happen if the mechanism chose to reject flows with large bandwidth requirements. Accordingly, the mechanism should reject a set of flows that leaves a total required bandwidth lower than but as close to the available capacity as possible.

More specifically, the mechanism takes effect when the available bandwidth of the wireless link falls below the total required bandwidth of all flows. Assuming an ascending list of flows $\{f_{i,1}, f_{i,2}, \ldots, f_{i,n}\}$ in each priority class $C_i$, according to their bandwidth requirements, the mechanism starts with the first flow $f_{1,1}$ of the lowest priority list $C_1$, and checks if, by rejecting it, the total required bandwidth falls below the available. It continues traversing the list until:

(1) either such a flow is found, or,
(2) the end of the list is reached.

In case (1), it rejects the flow and stops, since the total required bandwidth is below the available. In case (2), it rejects the last flow of the list $f_{1,n}$ (the one with the maximum bandwidth reservations in the list) and starts over from the beginning of the list. If all of the flows in the list are rejected (i.e., all the flows of the particular priority class), it continues with the flows of the next higher priority class $C_2$, and so on, until the total required bandwidth falls below the available, or until all flows have been rejected.

Below, the results of a comparison between the previously presented mechanism and a random rejection mechanism are presented, based on a simulation model with the following characteristics.

(1) Flows belong to two different priority classes, $C_1$ = Low and $C_2$ = High, although the mechanism can work with any number. The mechanism starts dropping flows belonging to priority class Low, as previously described in detail.
(2) Each priority class includes a mix of flow types with different bandwidth requirements, assuming that the type of a flow is an independent discrete random variable.
(3) The number of active flows is not constant. In each class, flows arrive according to an independent Poisson process. Flow durations are assumed to be exponential independent random variables.
(4) The channel can be found in two states, 'good' or 'bad', offering two levels of finite bandwidth. The time the channel stays in each state is an exponential random variable, thus the channel can be modeled as a two state Markov chain.
(5) New flows are not allowed to enter the system when bandwidth is insufficient to satisfy the requirements of all (existing and new) flows.

Two simulation scenarios were considered, where flows in each class were generated as in Table 7.1. As can be observed, the flow mix included a large number of 'light' flows and a smaller number of 'heavy' flows. To experiment with different conditions, we used different channel capacities and flow interarrival times per class. In both scenarios, the rejection probability per class was measured, defined as the average number of flows dropped divided by the total number of flows in the class.

**Table 7.1**   Different bandwidth requirements per class

| Flow types | Bandwidth requirements | Probability |
|------------|:----------------------:|:-----------:|
| 1 | 0.5 | 0.6 |
| 2 | 2 | 0.25 |
| 3 | 5 | 0.1 |
| 4 | 10 | 0.05 |

*Scenario 1*
Here, a low capacity channel was considered, with bandwidth equal to 600 and 400 while in the 'good' and 'bad' state, respectively. The mean dwell times in each state were 8 and 1, respectively. The mean flow duration was equal to 4 for all flows, while six experiments were performed with different interarrival times per class in the range [0.012, 0.040]. As can be seen in Figure 7.9, the proposed mechanism attains lower overall rejection probability in all experiments, while for the high priority class the performance is significantly improved. The increased rejection probability for the low priority class is considered acceptable, since this class is destined mostly for best-effort flows.

*Scenario 2*
In this scenario, the channel capacity was increased by a factor of 4 (Good = 2400, Bad = 1600), while the range of interarrival times was decreased by the same factor ([0,003, 0,009]), resulting in a considerably larger number of active flows. The mean time per channel state and the mean flow duration were as in Scenario 1. Six experiments were performed with different interarrival times per class, and the results are presented in Figure 7.10. Again, the overall rejection probability is improved significantly, especially for large values of the interarrival time. The rejection probability for the high priority class is much lower than the corresponding mean value of the randomly rejection mechanism.

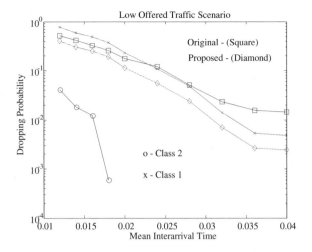

**Figure 7.9**   Rejection probability vs. mean interarrival time (low offered traffic scenario).

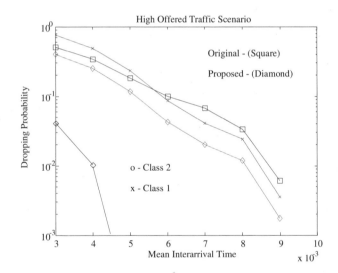

**Figure 7.10**   Rejection probability vs. mean interarrival time (high offered traffic scenario).

In both scenarios, the channel utilization of the proposed mechanism was equal or slightly improved than that for the random rejection mechanism, indicating that the proposed mechanism manages to reduce the flow rejection probability without decreasing the bandwidth utilization.

## 7.4  QoS in Hybrid 3G/WLAN Networks[b]

The interworking between 3G cellular and Wireless Local Area Networks (WLANs) has been considered as a suitable and viable evolution path towards the next generation of wireless networks.

---

[b]Parts of the following sections have been published in [34].

Yet this interworking raises considerable challenges, especially when we demand seamless continuity of multimedia sessions across the two networks. To deal with these challenges several 3G/WLAN interworking requirements need to be identified and fulfilled.

Typically, the 3G/WLAN interworking requirements are specified and categorized in terms of several usage scenarios [22, 23]. For example, a common usage scenario is when a 3G subscriber is admitted to a WLAN environment by re-using his regular 3G credentials, and then obtains an IP connectivity service (e.g., access to the Internet). In this case, the interworking requirements include support of 3G-based access control, signaling between the WLAN and the 3G network for Authentication, Authorization and Accounting (AAA) purposes, etc. Other scenarios can call for more demanding interworking requirements. We may envisage, for instance, a scenario in which a 3G subscriber initiates a video session in his home 3G network and subsequently transits to a WLAN environment, wherein the video session is continued *seamlessly*, i.e., without any noticeable change to the quality of service (QoS). In this case, not only 3G-based access control is required, but also access to 3G-based services is needed over the WLAN network, which in turn calls for appropriate routing enforcement mechanisms. More importantly, however, there is need for QoS consistency across 3G and WLAN, which does not appear to be very straightforward given the different QoS features offered by these networks. Indeed, WLANs have initially been specified without much attention being paid to QoS aspects and aimed primarily to satisfy simple and cost-effective designs. Even with the recent IEEE 802.11e [17] developments, WLAN QoS still exhibits several deficiencies with respect to the 3G QoS (this is further discussed later). On the contrary, 3G cellular networks were built with the multimedia applications in mind and trade simplicity and cost for inherently providing enhanced QoS in wide-area environments.

Our main interest in the following sections of this chapter is to examine the challenges of seamless session continuity across UMTS and WLAN, and to evaluate the conditions and restrictions under which seamless continuity is feasible. To this purpose, we formulate a number of practical interworking scenarios, where UMTS subscribers with ongoing real-time video sessions handover to WLAN, and we consider the capability of WLAN to provide seamless session continuity under several policy rules and WLAN traffic loads. One measure that we are particularly interested to quantify is the maximum number of UMTS subscribers[c] that can be admitted to the WLAN, subject to maintaining the level of UMTS QoS and respecting the WLAN policies. Although our study focuses primarily on QoS consistency, we do address several other issues that are equally important for enabling seamless session continuity, such as routing enforcement, access control and differentiation between the traffic of regular WLAN data users and UMTS roamers. The framework for discussing these issues is created by considering a practical UMTS/WLAN interworking architecture that conforms to the 3GPP specifications [23, 24] and other interworking proposals found in the technical literature (e.g., [22]).

## 7.5 UMTS/WLAN Interworking Architecture

The end-to-end interworking architecture we are considering is illustrated in Figure 7.11 and is compliant with the proposals in [22] and [23]. Below we briefly discuss the main characteristics of this architecture. Although we do not provide a comprehensive description, we set the ground for the next sections and define the real-life environment to which the subsequent study applies. This creates the right context for the following discussion and makes it easy to assess the importance of the provided results. For more detailed information on the considered architecture the interested reader is referred to [22–26].

As shown in Figure 7.11, the 3G network supports access to a variety of IP multimedia services via two radio access technologies: UMTS Terrestrial Radio Access (UTRA) and WLAN access. Access

---

[c] The UMTS subscribers admitted to the WLAN are also referred to as *UMTS roamers*.

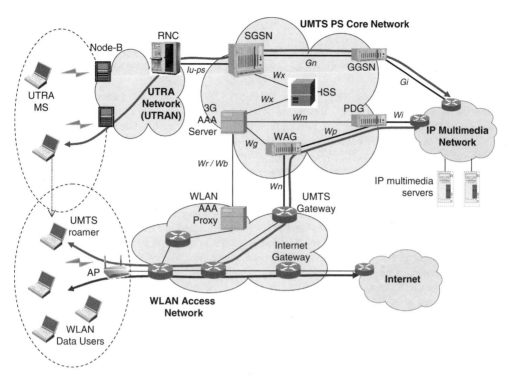

**Figure 7.11** The considered end-to-end interworking architecture for seamless multimedia session continuity.

control and traffic routing for 3G subscribers in UTRA is handled entirely by the UMTS Packet-Switched (PS) network elements, which encompass the Serving GPRS Support Node (SGSN) and the Gateway GPRS Support Node (GGSN) (see [25] for more details). On the other hand, access control and traffic routing for 3G subscribers in WLAN (*UMTS roamers*) is shared among the WLAN and the UMTS network elements as discussed below. The important assumption we make, as shown in Figure 7.11, is that 3G subscribers can change radio access technology and keep using their ongoing multimedia sessions in a seamless fashion. Thus, we assume that *seamless service continuity* is provided. This assumption raises considerable challenges and, as noted above, we are interested in investigating the capability of WLAN to support this seamless service continuity under specific scenarios. Note that by the term *seamless service continuity* we to refer to the continuation of a session (that has been initiated in the UMTS environment) in the WLAN environment *without* any noticeable QoS degradation or any other discrepancies whatsoever. We do not take into account however, the handover latency. In other words, we focus on 'how well' a session is continued in the WLAN and not 'how fast' the session is handed over from UMTS.

The WLAN access network may be owned either by the UMTS operator or by any other party (e.g., a public WLAN operator, or an airport authority), in which case the interworking is enabled and governed by appropriate business and roaming agreements. As shown in Figure 7.11, in a typical deployment scenario the WLAN network supports various user classes, e.g., UMTS roamers and regular WLAN data users (i.e., no 3G subscribers). Differentiation between these user classes and enforcement of corresponding policies is typically enabled by employing several Service Set Identifiers (SSIDs). For example, the regular WLAN data users may associate with the SSID that is periodically broadcast by the Access Point (AP), whereas the UMTS roamers may associate with another SSID that is also configured in the AP, but not broadcast (see [26] about the usage of SSIDs). In this case, the WLAN can apply distinct access control and routing policies for the two user classes and can forward the traffic of WLAN data users, e.g. to the Internet and the traffic of UMTS roamers to the UMTS PS core network (as shown in Figure 7.11). Such routing enforcement is vital for supporting seamless service continuity and can be implemented as discussed in [22]. Moreover, different AAA mechanisms could be used for the different user classes.

For enabling interworking with WLANs, the UMTS PS core network incorporates three new functional elements: the *3G AAA Server*, the *WLAN Access Gateway* (WAG) and the *Packet Data Gateway* (PDG). The WLAN needs to also support similar interworking functionality to meet the access control and routing enforcement requirements. The 3G AAA Server in the UMTS domain terminates all AAA signaling originated in the WLAN that pertains to UMTS roamers. This signaling is securely transferred across the *Wr/Wb* interface, which is typically based on *Radius* [14] or *Diameter* [27] protocols. The 3G AAA Server interfaces with other 3G components, such as the PDG, the Home Subscriber Server (HSS), the Home Location Register (HLR), the Charging Gateway/Charging Collection Function (CGw/CCF) and the Online Charging System (OCS). Note that, for simplicity, not all these elements are shown in Figure 7.11. Both the HLR and the HSS are basically subscription databases, used by the 3G AAA Server for acquiring subscription information for particular WLAN MSs. Typically, if an HSS is available, the HLR need not be used. The CGw/CCF and the OCS are 3G functional elements used to provide off-line and on-line charging services respectively. The 3G AAA Server can also route AAA signaling to/from another 3G networks, in which case it serves as a proxy and it is referred to as *3G AAA Proxy* (see [22]).

As shown in Figure 7.11, traffic from UMTS roamers is routed to the WAG across the *Wn* interface and finally to the PDG across the *Wp* interface. This routing is enforced by establishing appropriate traffic tunnels after a successful access control procedure. The PDG functions much like a GGSN in a UMTS PS core network. It routes the user data traffic between the MS and an external *Packet Data Network* (PDN) (in our case, the IP Multimedia Network) and serves as an anchor point that hides the mobility of the MS within the WLAN domain. The WAG functions mainly as a route policy element, i.e., it ensures that user data traffic from authorized MSs is routed to the appropriate PDGs, located either in the same or in a foreign UMTS network.

### 7.5.1 Reference Points

Some of the reference points (also referred to as *interfaces*) illustrated in the architecture diagram of Figure 7.11, are briefly discussed below. A more thorough discussion of all reference points can be found in [23]. A brief description of the 3G internal interfaces, such as *Gr*, *Iu-ps*, *Gi*, *Gn*, can be found in [25].

- **Wr/Wb**. This interface carries AAA signaling between the WLAN and the 3G visited or home PLMN in a *secure manner*. The proposed protocol across this interface is *Diameter* [27], which is used for providing AAA functions and for carrying the Extensible Authentication Protocol (EAP) messages exchanged between the MS and the 3G AAA Server. However, since legacy WLANs need to be supported, the Radius [14] protocol must also be supported across *Wr/Wb* and this calls for Radius–Diameter interworking functions across the legacy WLAN and the 3G AAA Proxy/Server. Diameter is the preferred AAA protocol since it can provide enhanced functionality compared with Radius. Note that the term *Wr* is used to refer specifically to the interface that carries authentication and authorization information, whilst the *Wb* is used to refer to the interface that carries accounting information.
- **Wf**. This is located between the 3G AAA server and the 3G CGw/CCF (not shown in Figure 7.11). The prime purpose of the protocols on this interface is to transport charging information towards the 3G operator's CGw/CCF located in the visited or home PLMN. The application protocol on this interface is Diameter based.

- **Wo**. This interface is used by the 3G AAA server to communicate with the 3G OCS and exchange online charging information so as to perform credit control for the online charged subscribers. The application protocol across this interface is again Diameter based.
- **Wx**. This interface is located between the 3G AAA Server and HSS, and is used primarily for accessing the WLAN subscription profiles of the users, retrieving authentication vectors, etc. As noted before, if Wx is implemented, then the interface to the HLR is not required. The protocol across Wx is either Mobile Application Part (MAP) or Diameter-based.

### 7.5.2 Signaling During UMTS-to-WLAN Handover

Although Figure 7.11 shows the architecture that can support seamless session continuity, it does not address the dynamics of handover procedure, which is especially important for the provision of seamless continuity. To elaborate further on this key procedure, we depict in Figure 7.12 a typical signaling diagram that pertains to a situation where a mobile station (MS) hands over from UMTS to WLAN in the middle of an ongoing packet-switched video session. The establishment of the video session is triggered at instant *A* and, in response, the MS starts the Packet Data Protocol (PDP) context establishment procedure for requesting the appropriate QoS resources (described by the 'Req. QoS' Information Element (IE), see [33]). The UMTS network acknowledges the request and indicates the negotiated QoS resources (specified by the 'Neg. QoS' IE) that could be provided. After that, video traffic on the user plane commences and the video session gets in progress. At some point the MS enters a WLAN coverage area and it starts receiving Beacons[d] from the nearby Access Points (APs). We

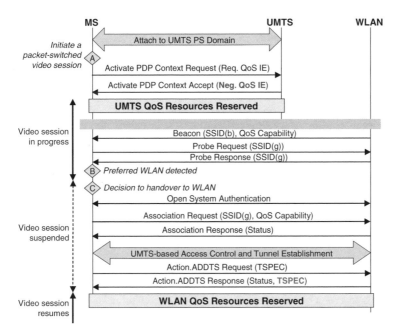

**Figure 7.12** Typical signaling during handover of a video session from UMTS to WLAN (HCCA availability is assumed).

---

[d]From the Beacons the MS discovers what particular QoS features the WLAN supports, if any.

assume that this can happen concurrently with the ongoing video session because, although the MS has one transceiver available, it can periodically decode signals on other frequency channels for inter-system handover purposes. The MS may need to check if the detected WLAN supports one of its preferred *Service Set Identifiers* (SSIDs) before considering it valid for inter-system change. For this purpose, the MS *probes* for a preferred SSID, denoted as SSID(g), according to the applicable procedures in [26].

At instant *C* the MS takes the decision to handover to the detected WLAN and thus suspends the ongoing video session. This may demand further signaling with the UMTS but here we skip this for simplicity. After switching to the WLAN channel, the normal 802.11 authentication and association procedures [26] are carried out. Subsequently, the UMTS-based access control procedure is executed in which the MS is authenticated and authorized by means of its regular 3G credentials (see [22], [23] for more details). At this stage, a tunnel will also be established for routing further MS traffic to a UMTS entry point (the WAG according to Figure 7.11). Next, the MS uses 802.11e QoS signaling (assuming it is supported by the WLAN) to reserve the appropriate resources for its suspended video session. The Traffic Specification (TSPEC) element carries a specification of the requested QoS resources. For the objectives of seamless continuity, it is apparently that TSPEC needs to be set consistently with the QoS negotiated in the UMTS system. After this point, the video session is finally resumed in the WLAN, possibly after some high-layer mobility management procedures (e.g., Mobile IP or SIP).

From the above discussion it becomes evident that vertical handovers from UMTS to WLAN (and vice versa) present several challenges, especially for minimizing the associated latencies and the interruption of ongoing multimedia sessions. Apart from that however, the maintenance of consistent QoS across the UMTS and WLAN networks is equally challenging and is the focus of our discussion below.

## 7.6 Interworking QoS Considerations

One vital component for the provision of seamless multimedia session continuity is the *QoS consistency* across the WLAN and UMTS networks. This is indeed vital because without QoS consistency the multimedia sessions would experience different QoS levels in the two network domains and hence seamless continuity would not be doable. It is unfortunate, however, that the UMTS and WLAN specifications were based on rather different set of requirements and they ended up supporting rather different set of QoS features. Consequently, the QoS consistency turns out to be a quite challenging issue. To provide more insight on this issue, we discuss below a list of WLAN QoS deficiencies with respect to UMTS QoS. Apparently, when we target multimedia session continuity across UMTS and WLAN, we should carefully take these deficiencies into consideration and understand their impact. The discussion is based on the assumption that the WLAN MAC layer complies with IEEE 802.11 [26] plus the amendments of IEEE 802.11e [17] and the physical layer complies with IEEE 802.11g [18]. For a good introduction to the QoS aspects of IEEE 802.11e the reader is referred to Chapter 5 as well as to [28].

(1) In a WLAN we cannot support unequal error protection across different media streams. For instance, an AMR payload, an H.263 payload and a HTTP payload will all be subject to the same channel coding (for a given transmission rate) and therefore all media streams will be equally protected against transmission errors, no matter what their different bit error rate requirements are. On the contrary, in UTRAN different radio transport channels (each with their own channel coding) can be established for streams with different error rate requirements. Thus, error rate can be controlled on a per stream basis.

(2) Unequal error protection in a WLAN cannot also be supported between different flows in the same media stream. For instance, the class A and class B bits of an AMR payload [29] will be equally protected against transmission errors, although class B bits can tolerate a higher bit error rate. In

fact, the WLAN layers are unable to distinguish the different flows of the AMR stream. On the contrary, UTRAN typically employs different radio transport channels for the different AMR flows.

(3) Although different media streams may tolerate different residual bit error rates, in a WLAN there is no way to control the residual bit error rate. This is because the WLAN has been optimized to support data streams and therefore enforces a very small residual bit error rate (by using 32-bit long CRC codes). In practice, this may result in increased MAC Service Data Unit (MSDU) loss rate (especially when acknowledges are not used) since all erroneous packets will be dropped even if some of them could be tolerated by the application. Moreover, the channel utilization will be decreased because, according to 802.11 [26], when a station receives an erroneous MSDU it cannot access the channel until after an Extended Inter-Frame Space (EIFS) period.

(4) WLAN stations cannot have dedicated radio channels as in UTRAN and therefore the queuing delay and jitter figures could be increased. Note that in 802.11e Hybrid Coordinator Channel Access (HCCA) mode the stations transmit with a polling discipline and hence delay and jitter will depend on the overall number of stations requesting resources and on the scheduler characteristics. In 802.11e Enhanced Distributed Channel Access (EDCA) mode the stations transmit with a random access discipline tailored to support several different traffic classes.

(5) In the WLAN there is no adaptive mechanism for controlling the MSDU loss rate in real-time. The typical way to control MSDU loss rate is via link adaptation with transmission rate change. However, this adaptation is not mandatory in all WLAN stations, it is implementation dependent and, more importantly, it is typically based on some *predefined* loss rate thresholds that do not correlate directly with the loss rate requirements of the transmitted media streams. The IEEE 802.11g standard allows the transmit power level to vary (the same holds true for 802.11b and 802.11a) but in practice all WLAN stations tend to use the maximum power level at all times, since no fast power control mechanism exists. Note also that in 802.11e EDCA loss rate is even harder to control since collisions are unavoidable.

(6) There are no soft handovers in WLAN. Handovers are typically hard in nature, i.e., follow the break-then-make approach, and hence considerable transmission disruptions may exist that result in QoS degradation. Moreover, handovers in 802.11 are solely controlled by the WLAN stations, so the WLAN infrastructure cannot provide tight control of the QoS provisioning. If a WLAN station tends to 'ping-pong' between two APs the QoS will be severely affected and the WLAN infrastructure has no means to prevent that.

One of our main conclusions is that the 'Service Data Unit (SDU) Error Ratio' and the 'Residual Bit Error Ratio' attributes used in UMTS QoS profile (see [30]) cannot be negotiated and controlled in an 802.11 WLAN, mainly due to physical layer restrictions. Also, the WLAN infrastructure is nearly impossible to guarantee a strict QoS level[e] given that there is no standardized mechanism for soft handovers. Of course, these deficiencies represent the price we pay for facilitating simple and cost-efficient WLAN designs.

Nevertheless, it is important to stress that the above QoS deficiencies of WLANs do not necessarily mean that seamless session continuity from UMTS to WLAN cannot be supported. Under certain conditions, seamless session continuity can be provided (but not guaranteed) even with inefficient utilization of WLAN radio resources (but these are cheap anyway!). To validate our point, in the next sessions we carry out a performance study and evaluate the number of UMTS roamers that can be admitted to the WLAN under certain restrictions, e.g., maintain the QoS level that was experienced in UMTS, respect WLAN policies, etc.

---

[e]It is interesting to observe that IEEE 802.11e carefully refers to the provided QoS as 'differentiated' and 'parameterized' QoS, not as 'guaranteed' QoS.

## 7.7 Performance Evaluation

Given the aforementioned QoS discrepancies across UMTS and WLAN and, in particular, the limited QoS features of WLAN with respect to UMTS, it is interesting to investigate the feasibility and constraints of seamless QoS provision. In this context, we are interested in investigating the capability of WLAN to support seamless QoS provision for multimedia sessions that have previously been initiated in the UMTS environment. Say, for example, that the WLAN starts accepting UMTS roamers, each one with a video session in progress. Will these video sessions experience a QoS level consistent with the QoS level provided before in the UMTS environment? Under what conditions is this possible? And how many such video sessions can be admitted into the WLAN without compromising the requirements for seamless QoS provision and comply with the possible interworking policy of the WLAN? These are the important questions that we are dealing with in this section.

To derive realistic answers to the above questions we consider the interworking scenario illustrated in Figure 7.13. In this scenario, there is only one AP, which provides access services to two classes of users: (*a*) the WLAN data users and (*b*) the UMTS roamers, who are UMTS subscribers handed over from UMTS. All UMTS roamers are considered statistically identical and the same holds true for the WLAN data users. In addition, the AP as well as the UMTS roamers comply with the procedures and elements specified in IEEE 802.11i (or Wi-Fi Protected Access) for enhanced security provision. This is required for supporting UMTS-based access control and having the UMTS roamers authenticated and authorized by their UMTS home environment, as specified in [22, 23]. All WLAN stations, including the AP, support the extended MAC layer specified in IEEE 802.11e and the physical layer specified in IEEE 802.11g. We assume that the transmission rate of WLAN stations is 24 Mbps and that no transmission errors occur (the channel is ideal).

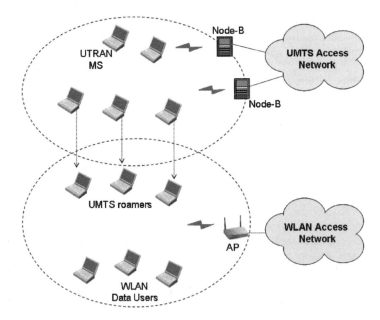

**Figure 7.13**    The interworking scenario considered for performance evaluation.

Each WLAN data user has a number of ongoing non-real-time data sessions (e.g., web browsing, e-mail, ftp, etc.), which generate an aggregate user traffic described as a Poisson process with 256 kbps mean rate. On the other hand, each UMTS roamer has a uni-directional (uplink) real-time video session in progress, which has been initiated in the UMTS domain and granted the negotiated QoS parameters

**Table 7.2** The QoS values negotiated in UMTS for the H.263 video sessions

| UMTS QoS parameter | Parameter value | Comments |
|---|---|---|
| Traffic class | Conversational | |
| Residual BER | $10^{-5}$ | Corresponds to 16-bit CRC. This cannot be controlled in the WLAN since the supported CRC is always 32 bits long. |
| SDU error ratio | $10^{-3}$ | |
| Maximum SDU size | 1500 bytes | |
| Transfer delay | 40 msec | |
| Guaranteed bit rate for uplink | 64 kbps | |
| Maximum bit rate for uplink | 128 kbps | |
| Guaranteed bit rate for downlink | 2 kbps | To support RTCP traffic. |
| Maximum bit rate for downlink | 2 kbps | |
| Delivery order | No | Re-ordering should be taken care by application in order to minimize the transfer delay. |
| Delivery of erroneous SDUs | No | |
| Traffic handling priority | Subscribed | Not relevant |
| SDU format information | Not used | Unequal error protection within the bits of the same SDU is not required. |
| Allocation/retention priority | Subscribed | Not relevant |
| Source statistics descriptor | Unknown | Only used for voice and allows the network to assess the utilization of trunk resources. |

shown in Table 7.2. The selection of these parameters is based on [31] and the assumption that H.263 video coding is used with a target bit rate of 64 kbps. In our simulations the video packet traffic is generated with the aid of the video trace files found in http://trace.eas.asu.edu/TRACE/ltvt.html and the use of Real Time Protocol (RTP) and Real Time Control Protocol (RTCP).

For the interworking scenario discussed above, our main goal is to evaluate how many UMTS roamers the WLAN network can support under the following two constraints.

(1) The video streams of all UMTS roamers admitted to the WLAN must experience at least the same QoS level as the one negotiated in the UMTS network (see Table 7.2). For example, the MAC SDU (MSDU) loss rate in the WLAN must not exceed the corresponding UMTS SDU error ratio, i.e., $10^{-3}$. This constraint is required for satisfying the seamless service continuity requirements.

(2) The bandwidth available to WLAN data users must not diminish below a predefined threshold. This constraint makes it possible to enforce a bandwidth reservation policy consistent with the WLAN operator's interworking requirements. For example, the WLAN operator may need to ensure that the WLAN data users will have at least 7 Mbps of bandwidth available no matter how many UMTS roamers are admitted into the WLAN. Therefore, UMTS roamers can be served with possibly higher priority than WLAN data users (in order to meet the seamless continuity requirements) but they cannot consume all the available bandwidth. For this purpose, the WLAN needs to apply an admission control function that would reject further association requests from UMTS roamers should the bandwidth reservation limit be reached. In our study we assume that such an admission control function is implemented and we take it into account for calculating the maximum number of UMTS roamers that can be admitted into the WLAN.

## 7.8 Performance Results

For performing our evaluations we consider two practical WLAN deployment scenarios: (i) a Contention-based Scenario, where the WLAN AP does not support HCCA access mode and thus both UMTS roamers and WLAN data users employ contention-based channel access, and (ii) a Contention-Free Scenario, where the WLAN AP supports HCCA access mode and all UMTS roamers are serviced in this mode (i.e., contention free).

### 7.8.1 Contention-Based Scenario

In this scenario we assume that both UMTS roamers and WLAN data users use contention-based channel access. A key policy applied by the WLAN indicates that at least $L$ Mbps must be available to the WLAN data users. Hence, UMTS roamers can be admitted to the system as long as (i) the WLAN can support $L$ Mbps of data traffic and (ii) the QoS experienced by the video streams meets or exceeds the QoS negotiated in the UMTS environment (according to Table 7.1).

First, we consider a typical case that is expected during the early deployment of UMTS/WLAN interworking, where the WLAN data users use legacy terminals with no 802.11e support. These users access the channel by employing the Distributed Coordination Function (DCF) with the following access parameters (see [26]): DCF Inter-Frame Space (DIFS) = 2 slots, minimum Contention Window (CWmin) = 15, maximum Contention Window (CWmax) = 1023 and Persistence Factor (PF) = 2. On the contrary, UMTS roamers use 802.11e aware terminals and employ an EDCA access class with the following access parameters (see [17] and Chapter 5): Arbitration Inter-Frame Space Number (AIFSN) = 3, CWmin = 15, CWmax = 1023 and PF = 2. All terminals maintain uplink buffers that can hold up to eight maximum sized MSDUs. The terminals of UMTS roamers make every effort to transmit all video packets within their delay bound, which is considered equal to 40 ms for consistency with UMTS. If a video packet, however, is delayed for more than 40 ms, it is dropped. This policy guarantees that the delay experienced by all successfully transmitted video packets will be less than 40 ms. Hence, the key parameter affecting the QoS of video streams will be the loss rate, which should be kept below the corresponding UMTS limit ($10^{-3}$).

Our simulation results for the above case reveal that the limiting factor for the maximum number of UMTS roamers in the WLAN is not the bandwidth reservation constraints but rather the loss rate of video streams. As illustrated in Figure 7.14, the MSDU loss rate for video traffic reaches the UMTS

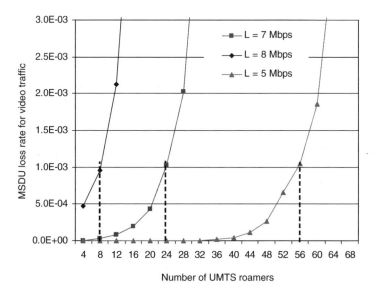

**Figure 7.14** The MSDU loss rate for video traffic vs. the number of UMTS roamers admitted in the WLAN.

negotiated value ($10^{-3}$) when there are 56 UMTS roamers (or equivalently 56 video streams) for $L = 5$ Mbps, or 24 UMTS roamers for $L = 7$ Mbps, or eight UMTS roamers for $L = 8$ Mbps. Apparently, when $L$ increases (i.e., when there are more WLAN data users in the system), the transmission delay of video packets is increased and the probability to reach the 40 msec delay bound is increased as well. Hence, the video packet loss rate rises accordingly.

In Figure 7.15 we display the maximum delay of 99% of the successfully delivered packets for both UMTS roamers and WLAN data users. As expected, the delay of video packets is always larger than the delay of WLAN data traffic, since the latter employs a smaller inter-frame space and thus gains priority over video traffic. We also note that the maximum delay experienced by the video packets in the WLAN domain is far less than the limit of 40 msec that was negotiated in the UMTS domain. For example, as shown in Figure 7.15, when $L = 7$ Mbps and there are 24 UMTS roamers (thus the loss rate is within the UMTS negotiated limit) the maximum delay is about 15 msec. This leads to a significant conclusion: The WLAN can meet or exceed the QoS negotiated in the UMTS domain but in a very inefficient way. Indeed, we can readily derive that, when the transmission rate is 24 Mbps and the WLAN data users offer 7 Mbps aggregate data traffic, there will be about 4.3 Mbps available for UMTS roamers (this corresponds to 47% max channel utilization). However, only up to 1.5 Mbps ($24 \times 64$ kbps) can be utilized for meeting the loss rate requirements.

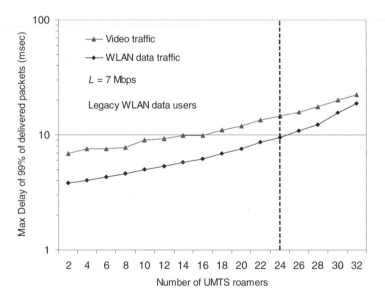

**Figure 7.15** The max delay of 99% of the delivered MSDUs for both the video and the data traffic vs. the number of UMTS roamers admitted in the WLAN.

We consider another case where both UMTS roamers and WLAN data users are 802.11e aware. The two user classes are mapped however to different EDCA access classes with the UMTS roamers given higher access priority for meeting their demanding QoS needs. In particular, we assume that the UMTS roamers are mapped to an EDCA access class with the following access parameters: AIFSN = 3, CWmin = 6, CWmax = 511 and PF = 2. Similarly, the WLAN data users are mapped to an EDCA access class with the following access parameters: AIFSN = 6, CWmin = 31, CWmax = 1023 and PF = 2.

In contrast to the previous case (legacy WLAN data users), our simulation now indicates that the loss rate experienced by video packets is almost negligible since the UMTS roamers are given preferential access to the wireless medium. Therefore, the limiting factor for the maximum number of UMTS roamers in the WLAN is not the loss rate of video streams but rather the bandwidth reservation constraints. Indeed, as displayed in Figure 7.16, when $L = 7$ Mbps, the MSDU loss rate for data traffic is equal to zero for up to 40 UMTS roamers. Up to this number of roamers, the capacity offered to the WLAN data users is indeed 7 Mbps and hence the bandwidth reservation policy is respected. However, when more than 40 UMTS roamers are admitted to the WLAN, this policy cannot be satisfied as the bandwidth utilized by WLAN data users is quickly diminished.

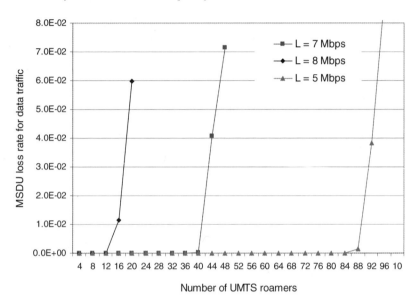

**Figure 7.16**   The MSDU loss rate for data traffic vs. the number of UMTS roamers.

From the above, it can easily be deduced that we can experience significant capacity benefits when the WLAN data users become 802.11e aware and the appropriate access parameters are used; e.g., from 24 UMTS roamers with legacy WLAN data users we can climb to 40 UMTS roamers with 802.11e aware WLAN data users. Similar observations can be made in the case of the other two scenarios, namely when $L = 5$ and $L = 8$, where the maximum number of UMTS users that can be supported increases to 84 and 12 respectively. Yet, this gain is achieved with a cost on the delay performance of WLAN data users. For example, as illustrated in Figure 7.17, when $L = 7$ the maximum delay experienced by the WLAN data users, when there are 40 UMTS roamers in the system is about 140 msec. In the case of legacy WLAN data users, however, the delay is about 10 msec when there are 24 UMTS roamers (see Figure 7.15). By comparing Figures 7.15 and 7.17, it becomes evident that when the WLAN data users become 802.11e aware, they can still use the same total bandwidth (7 Mbps) but they experience considerably increased delay, which accounts for the considerably more UMTS roamers in the WLAN. Moreover, not only can we have more UMTS roamers admitted in the WLAN but each one will experience reduced delay, i.e., 2–4 msec in Figure 7.17 as compared with 7–15 msec in Figure 7.15.

### 7.8.2 Contention-Free Scenario

In the contention-free scenario it is assumed that all UMTS roamers operate in HCCA mode and therefore they do not contend with the WLAN data users. HCCA implements a polling mechanism which allows the Hybrid Coordinator (HC) entity of 802.11e (normally implemented in the AP) to

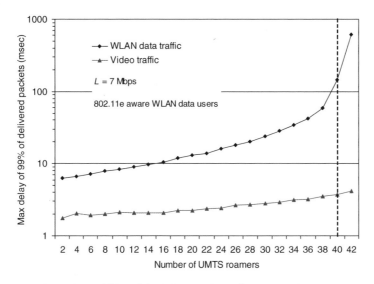

**Figure 7.17**   The maximum delay of 99% of the delivered MSDUs for both the video and the data traffic vs. the number of UMTS roamers admitted in the WLAN.

control the access to the wireless channel by assigning Transmission Opportunities (TXOPs) to the requesting WLAN terminals. Since the access to the channel is centrally controlled and there is no contention or collisions, HCCA is appropriate for providing parameterised QoS services with specific bounds. In our simulation, each handover of an H.263 video session from UMTS triggers the establishment of a new 802.11e Traffic Stream (TS) with the signaling messages illustrated at the end of Figure 7.12. The traffic characteristics and QoS requirements of each TS are described by the Traffic Specification (TSPEC) element shown in Table 7.3 (see Chapter 5 and [17] for details on these parameters).

**Table 7.3**   The TSPEC values negotiated in WLAN for the H.263 video sessions

| 802.11e TSPEC parameter | Parameter value | Comments |
| --- | --- | --- |
| Nominal MSDU Size | 320 bytes | Calculated from the H.263 trace file |
| Maximum MSDU Size | 1917 bytes | Calculated from the H.263 trace file |
| Minimum service interval | 0 msec | 'Don't care' |
| Maximum service interval | 40 msec | Considered equal to delay bound |
| Inactivity interval | | Not used |
| Suspension interval | | Not used |
| Service start time | | Not used |
| Minimum data rate | | Not used |
| Mean data rate | 64 kbps | Calculated from the H.263 trace file |
| Peak data rate | 383.4 kbps | Calculated from the H.263 trace file |
| Maximum burst size | 1917 bytes | Calculated from the H.263 trace file |
| Delay bound | 40 msec | For consistency with UMTS (see Table 7.1) |
| Minimum PHY rate | 24 Mbps | All stations can transmit with 24 Mbps |
| Surplus bandwidth allowance | | Not used |
| Medium time | | Not used |

The allocation of TXOPs to UMTS roamers is performed by the Scheduler (implemented by the HC) and is based on the QoS requirements defined by the corresponding TSPEC. The scheduler decides for both the polling time of a UMTS roamer as well as the duration of the allocated TXOP. For the purposes of our simulation, it is assumed that the WLAN implements the reference scheduler described in [17], which is also referred to as the Simple Scheduler. Based on the negotiated Mean Data Rate and Delay Bound of each video session, the Simple Scheduler calculates a fixed TXOP length for each UMTS roamer and a Service Interval (SI). The result is that the scheduler implements a periodic service pattern by sequentially serving all UMTS roamers every SI time units.

The impact of the WLAN data users, which still operate in contention mode, is considered by limiting the percentage of channel time available to the Simple Scheduler. In particular, assuming again the same bandwidth reservation policy, i.e., allowing 7 Mbps minimum bandwidth for the WLAN data users, we readily calculate that about 40% of channel time is left for HCCA operation. Given that constraint, our main objective is to assess the performance of the Simple Scheduler in terms of channel utilization and maximum number of UMTS roamers that can be supported.

Figure 7.18 depicts the Channel Occupancy (i.e., the fraction of the total channel time spent in HCCA mode) versus the UMTS roamers admitted in the WLAN. Observe that the occupancy increases linearly and that the 40% of channel capacity available for HCCA is reached for 17 UMTS roamers. This exposes a fairly inefficient channel utilization. Indeed, with 40% of the available radio resources up to approximately 4.8 Mbps can be accommodated, yet the Simple Scheduler manages to accommodate only 1.06 Mbps (17 × 64 kbps).

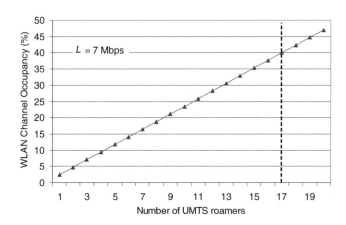

**Figure 7.18** Percentage of WLAN channel time spent in HCCA mode vs. the number of UMTS roamers.

In contrast to the contention-based scenario, video packets transmitted in HCCA mode are not lost because the scheduler tries to respect the negotiated delay bounds and allocate enough TXOPs to accommodate all traffic load offered by the UMTS roamers. Therefore, the QoS experienced by the UMTS roamers in HCCA mode is affected only by the delay characteristics. Figure 7.19 illustrates the delay experienced by the delivered video packets, which is nearly constant throughout the considered range of UMTS roamers and remains below the delay bound negotiated in the UMTS domain (40 msec). It is interesting to note, however, that this delay is larger than the corresponding delay in contention-based scenario. Comparing for example Figure 7.19 with Figure 7.17, we identify that the delay in contention-free scenario increases by nearly four times. This is a result of the inefficient TXOP allocation of the Simple Scheduler.

As already pointed out, the main issue in HCCA operation is the inefficient resources utilization and hence the support of a relatively small number of users (17) as compared with the contention-based

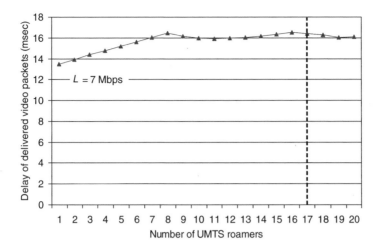

**Figure 7.19**   Delay of delivered video packets vs. the number of UMTS roamers.

scenario. However this should not be attributed to the HCCA mechanism itself but rather to the inherent characteristics of the Simple Scheduler, which employs a greedy, over-provisioning strategy for TXOP allocation in order to meet the QoS requirements. Our simulation results denote that the TXOP Loss Factor, i.e., the percentage of unused time allocated to a UMTS roamer (because there were no video packets for transmission), is a bit more than 80%. This basically indicates that the Simple Scheduler allocates four times more radio resources to a UMTS roamer than what is required. This happens because when the Simple Scheduler calculates the TXOP duration per UMTS roamer, it makes a worst-case estimation so as to be able to accommodate the largest MSDU size. With Constant Bit Rate (CBR) traffic, where the MSDU size features nearly no variation, this sounds like a reasonably simple and acceptable approach. However, with Variable Bit Rate (VBR) traffic like the considered H.263 video streams, this strategy leads to excessive inefficiency and thus the performance of the contention-free scenario does not compare favorably with the performance of the contention-based scenario. To improve the scheduling efficiency for VBR traffic, more sophisticated schedulers have been proposed in the technical literature; these adapt the TXOP lengths to the actual size of buffered MSDUs. Examples of such schedulers are presented in [15,32].

## 7.9 Conclusions

This chapter addressed some of the major issues in QoS provision over wireless networks. As WLANs become more and more important in integrated network architectures, we focused on their weaknesses in QoS provision and described specific solutions. We started with the problem of QoS during handover and, more specifically, we studied the performance of RSVP signaling combined with mobility. A specific solution was described, accompanied by simulation results.

We moved on to emphasize the importance of traffic scheduling in WLANs, especially in IEEE 802.11e networks. To prove our concept, we described a proposed scheduling algorithm and compared it with a simple solution included in the IEEE 802.11e specification. An interesting challenge for future research in the area of QoS over WLANs is the combination of different techniques and mechanisms that can support both efficient transmission over the wireless medium, and fast but yet simple handover execution, with strict QoS guarantees for different kinds of traffic. This will allow for the use of a wide range of multimedia applications over WLANs and extend the set of environments that such technologies can operate.

Another major problem in providing QoS for wireless multimedia is the low performance of RSVP in wireless networks, as a result of its design for fixed environments with stable links. Towards solving this problem, we presented one of the most important proposals, namely, the dynamic RSVP (dRSVP). dRSVP extends the standard RSVP protocol to use ranges of traffic specifications, instead of absolute values. The benefit is that intermediate nodes can adjust allocations, as long as these are within the specified ranges, depending on changes of the available bandwidth in the output links. After presenting the basic features of dRSVP, we focused on a component that is essential for its performance, i.e., the flow rejection algorithm. The presented algorithm was found to considerably improve the protocol's performance, especially in environments with heavy bandwidth fluctuations. The main characteristic of this algorithm is that it rejects the minimum number of flows required to reduce the overall bandwidth requirements below the offered capacity at any time. As for the future, research at international level is moving towards designing new QoS provision protocols, assuming both fixed and wireless environments. This means that the resulting protocols should be able to handle both user mobility and unstable links.

Finally, we focused on the problem of WLANs interworking with UMTS. Undoubtedly, there are still several challenges to be addressed for enabling the seamless interworking of wireless LAN and UMTS networks. As we discussed in this chapter, the QoS discrepancies between these networks raise some of these challenges. Although the WLAN QoS capabilities have recently been extended considerably with the introduction of IEEE 802.11e, the WLAN is still incapable of supporting all the QoS features provided by UMTS (e.g., to dynamically control the MSDU error rate, the residual bit error ratio, unequal error protection, etc.). This is partially attributed to the characteristics of the WLAN physical layer, which has been kept relatively simple in order to enable low-cost designs. As a result, multimedia transmission over WLANs turns out to be less efficient than the UMTS (but definitely less expensive). Vertical handovers also present another challenge for seamless UMTS/WLAN interworking. Since they are usually implemented as mobile-controlled, hard handovers, they bring up considerable QoS concerns, which severely affect the provision of seamless interworking.

In our study of various interworking scenarios, where UMTS subscribers with ongoing real-time video sessions handed over to WLAN, we mainly quantified the maximum number of UMTS subscribers that can be admitted to the WLAN, subject to maintaining the same level of UMTS QoS and respecting the WLAN bandwidth reservation policies. Our results suggest that the WLAN can support seamless continuity of video sessions for only a limited number of UMTS subscribers, depending on the bandwidth reservations, on the WLAN access parameters and on the QoS requirements of video sessions. The operation of the Simple Scheduler was proved inefficient for video traffic and consequently the contention-based scenario gave a better performance, even though the video traffic had to content with data traffic for channel access.

## Acknowledgments

The authors would like to give special thanks to Professor F-N. Pavlidou, G. Dimitriadis and D. Skyrianoglou for their valuable contributions to this chapter.

## References

[1]  R. Braden, D. Clark and S. Shenker, Integrated Services in the Internet Architecture: an Overview, RFC 1633, June 1994.
[2]  R. Braden, L. Zhang, S. Berson, S. Herzog and S. Jamin, Resource ReSerVation Protocol (RSVP) – Version 1 Functional Specification, RFC2205, September 1997.

[3]  A. G. Valko, Cellular IP – A new approach to Internet host mobility, *ACM Computer Communication Review*, January 1999.

[4]  A. K. Talukdar, B. R. Badrinath and A. Acharya, MRSVP: A resource reservation protocol for an integrated services network with mobile hosts, *The Journal of Wireless Networks*, **7**(1), 2001.

[5]  S. Das, R. Jayaram, N. Kakani and S. Sen, A resource reservation mechanism for mobile nodes in the Internet, in *Proc. IEEE Vehicular Technology Conference (VTC)*, Spring 1999, Houston, Texas, May 1999.

[6]  C.-C. Tseng, G.-C. Lee and R.-S. Liu, HMRSVP: A hierarchical mobile RSVP protocol, in *Proc. International Workshop on Wireless Networks and Mobile Computing (WNMC) 2001*, Valencia, Spain, April 2001.

[7]  G.-S. Kuo and P.-C. Ko, Dynamic RSVP for Mobile IPv6 in Wireless Networks, in *Proc. IEEE Vehicular Technology Conference (VTC) 2000*, Tokyo, Japan, May 2000.

[8]  W.-T. Chen and L.-C. Huang, RSVP mobility support: A signaling protocol for integrated services Internet with mobile hosts, in *Proc. INFOCOM 2000*, Tel Aviv, Israel, March 2000.

[9]  S. Hadjiefthymiades, S. Paskalis, G. Fankhauser and L. Merakos, Mobility management in an IP-based wireless ATM network, in: *Proc. of ACTS Mobile Summit*, June 1998.

[10] S. Paskalis, A. Kaloxylos, E. Zervas and L. Merakos, An efficient RSVP/mobile IP interworking scheme, *ACM Mobile Networks Journal*, **8**(3), June 2003.

[11] B. Moon and A. H. Aghvami, Reliable RSVP Path Reservation for Multimedia Communications under IP Micro Mobility Scenario, *IEEE Wireless Communications Magazine*, October 2002.

[12] IEEE Std. 802.11f/Draft 6.0, Draft Recommended Practice for Multi-Vendor Access Point Interoperability via an Inter-Access Point Protocol Across Distribution Systems Supporting IEEE 802.11 Operation, November 2003.

[13] N. Passas, A. Salkintzis, G. Nikolaidis and M. Katsamani, WLAN signaling enhancements for improved handover performance, *Proc. Wireless Personal Multimedia Communications (WPMC)*, Abano Terme, Italy, September 2004.

[14] C. Rigney, S. Willens, A. Rubens and W. Simpson, Remote Authentication Dial In User Service (RADIUS), RFC 2865, June 2000.

[15] Grilo, M. Macedo and M. Nunes, A scheduling algorithm for QoS support in IEEE 802.11e networks, *IEEE Wireless Communications*, 36–43, June 2003.

[16] D. Skyrianoglou, N. Passas and A. Salkintzis, Traffic scheduling for multimedia QoS over WLANs, *Proc. IEEE ICC*, Seoul, Korea, May 2005.

[17] IEEE draft standard 802.11e/D9.0, Medium Access Control (MAC) Quality of Service (QoS) Enhancements, August 2004.

[18] IEEE 802.11g, Supplement to IEEE Standard for Telecommunications and Information Exchange Between Systems – LAN/MAN Specific Requirements – Part 11: Wireless LAN Medium Access Control (MAC) and Physical Layer (PHY) Specifications – Amendment 4: Further Higher-Speed Physical Layer Extension in the 2.4 GHz Band, June 2003.

[19] S. Blake, D. Black, M. Carlson, E. Davies, Z. Wang and W. Weiss, An Architecture for Differentiated Services, RFC 2475, December 1998.

[20] M. Mirhakkak, N. Schult and D. Thomson, Dynamic bandwidth management and adaptive applications for a variable bandwidth wireless environment, *IEEE Journal on Selected Areas in Communications*, **19**(10), October 2001.

[21] N. Passas, E. Zervas, G. Hortopan and L. Merakos, A Flow Rejection Algorithm for QoS Maintenance in a Variable Bandwidth Wireless IP Environment, *IEEE Journal on Selected Areas in Communications (JSAC)*, special issue on All-IP Wireless Networks, **22**(4), May 2004.

[22] A. K. Salkintzis, Interworking techniques and architectures for WLAN/3G integration towards 4G mobile data networks, *IEEE Wireless Communications*, **11**(3), 50–61, June 2004.

[23] 3GPP TS 23.234 v6.0.0, 3GPP system to WLAN Interworking; System Description (Release 6), March 2004.

[24] 3GPP TR 22.934 v6.2.0, Feasibility study on 3GPP system to WLAN interworking (Release 6), September 2003.

[25] 3GPP TS 23.060 v5.6.0, General Packet Radio Service (GPRS); Service description; Stage2 (Release 5), June 2003.

[26] IEEE standard 802.11, Wireless LAN Medium Access Control (MAC) and Physical Layer (PHY) Specifications, 1999.

[27] P. Calhoun, J. Loughney, E. Guttman, G. Zorn and J. Arkko, Diameter Base Protocol, *IETF RFC 3588*, September 2003.

[28] Stefan Mangold, Sunghyun Choi, Guido R. Hiertz, Ole Klein and Bernhard Walke, Analysis of IEEE 802.11e for QoS support in Wireless LANs, *IEEE Wireless Communications*, **6**, 40–50, December 2003.

[29] 3GPP TS 26.071 v5.0.0, AMR speech Codec; General description (Release 5), December 2002.

[30] 3GPP TS 23.107 v5.12.0, Quality of Service (QoS) concept and architecture (Release 5), March 2004.

[31] 3GPP TS 26.236 v5.6.0, Transparent end-to-end Packet-switched Streaming Service (PSS), Protocols and codecs (Release 5), September 2003.

[32] Pierre Ansel, Qiang Ni and Thierry Turletti, An efficient scheduling scheme for 802.11e, in *Proc. WiOpt (Modeling and Optimization in Mobile, Ad Hoc and Wireless Networks)*, Cambridge, UK, March 2004.

[33] 3GPP TS 24.008 v5.12.0, Mobile radio interface Layer 3 specification; Core network protocols; Stage 3 (Release 5), June 2004.

[34] A. K. Salkintzis, G. Dimitriadis, D. Skyrianoglou, N. Passas and N. Pavlidou, seamless continuity of real-time video across UMTS and WLAN networks: Challenges and performance evaluation, *IEEE Wireless Communications*, June 2005.

# 8

# Wireless Multimedia in 3G Networks

George Xylomenos and Vasilis Vogkas

## 8.1 Introduction

This chapter describes the support for IP based multimedia services on Third Generation (3G) wireless cellular networks. While earlier cellular networks offered only basic IP connectivity on a best-effort basis, 3G networks provide explicit support for real time multimedia services with guaranteed Quality of Service (QoS). As a result, they can offer traditional voice telephony, rich telephony services and multimedia streaming over IP, both inside a 3G network and in conjunction with external networks such as the Internet, the Public Switched Telephone Network (PSTN) and the Integrated Services Digital Network (ISDN). In fact, 3G networks will deploy for the first time in large scale technologies such as IP version 6 (IPv6) and policy based IP QoS, spearheading the introduction of such technologies into the Internet.

The two most important aspects of 3G networks with respect to IP based multimedia services are the IP Multimedia Subsystem (IMS) and the Multimedia Broadcast / Multicast Service (MBMS). The IMS enables complex IP based multimedia sessions to be created with guaranteed QoS for each media component. Example applications include voice telephony and video conferencing. The IMS interoperates with both traditional telephony services and external IP based multimedia services. The MBMS provides native IP broadcast and multicast support in 3G networks, allowing high bandwidth services to be economically offered to multiple users. Example applications include video streaming via multicast and location based services via broadcast. The MBMS interoperates directly with IP multicasting. While both the IMS and the MBMS are IP based, their standardization is proceeding independently. It is however clear that their combination would allow numerous new services to be provided.

The outline of this chapter is as follows. In Section 8.2 we give an overview of cellular networks and their evolution, while in Section 8.3 we describe in detail a specific 3G network, the Universal Mobile Telecommunications System (UMTS). Section 8.4 presents an introductory description of the features and services provided by the IMS and the MBMS. In Section 8.5 we provide the details of the IMS, including service architecture, session setup and control and interworking issues. Section 8.6 describes the MBMS in the same manner as for the IMS. Finally, Section 8.7 discusses the QoS issues for IP based multimedia services, describing the overall QoS concept, the policy based QoS control scheme and its

*Emerging Wireless Multimedia: Services and Technologies*   Edited by A. Salkintzis and N. Passas
© 2005 John Wiley & Sons, Ltd

application to IMS sessions. At the end of the chapter, a glossary lists the acronyms used. An extensive UMTS vocabulary is provided in [1].

## 8.2 Cellular Networks

This section provides an overview of cellular network evolution, in terms of both technology and services. First generation (1G) systems used analog transmission and provided only circuit switched voice telephony. Second generation (2G) systems were fully digital and initially offered voice and circuit switched data services. The increasing popularity of the Internet led to the addition of packet switched data services to 2G networks, turning them into 2.5G networks. Finally, 3G networks are planning to provide all services over packet switching, including voice telephony. This focus on packet switching enables new services to be offered, including IP based multimedia ones.

### 8.2.1 First Generation

All wireless systems face the problem that all transmissions must share the frequency range allocated to the system, thereby limiting the number of simultaneous users. One means to increase the number of users for a given frequency range is the cellular concept, illustrated in Figure 8.1. The area to be covered is divided into hexagonal cells, with a Base Transceiver Station (BTS) located at the center of each cell. The BTS transmits with just enough power to reach the outer limits of its cell. Each user has a Mobile Station (MS), i.e. a cellular telephone. Depending on the cell where the MS is located, it communicates with the network via the corresponding BTS. To avoid interference between transmissions in neighboring cells, the available frequency range is divided into non-overlapping frequency bands and different bands are assigned to adjacent cells. However, the same band can be reused in a non-adjacent cell, thereby increasing the total number of users that can be simultaneously served in the area of coverage.

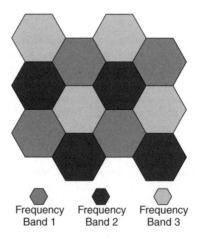

| Frequency Band 1 | Frequency Band 2 | Frequency Band 3 |

**Figure 8.1** The cellular concept.

The cellular network infrastructure consists of a BTS in each cell plus a network interconnecting them. The cellular network must also provide connectivity to external networks so that an MS may place calls to or receive calls from fixed telephony networks. When users move to another cell while making a call, a handover occurs. That is, the cellular network ensures that the MS transparently switches to the frequency band used by the BTS of its new cell without interrupting the call. When multiple cellular operators exist in the same area, they are allocated different frequency ranges from the local licensing authority. Each operator can deploy a completely different cell structure and frequency band allocation, using its own network to interconnect these cells.

The oldest cellular system, the Advanced Mobile Phone Service (AMPS), started operation in the USA during the 1970s. It is based on the cellular concept described above. Active calls within the same cell share the frequency band via Frequency Division Multiple Access (FDMA), that is, each call between an MS and the BTS is allocated a separate part of the frequency band. In addition, the transmissions from the MS to the BTS, called the uplink, and from the BTS to the MS, called the downlink, use separate frequency ranges, in a technique called Frequency Division Duplexing (FDD). Note the distinction between FDMA and FDD: the former refers to frequency sharing between different calls, while the latter refers to frequency sharing between the two directions of the same call. The only service offered by AMPS is voice telephony. Voice is transmitted in analog mode, that is, a carrier of the appropriate frequency is modulated by the voice signal. It is also possible to transmit data using an analog modem to modulate the carrier. Since each voice call is allocated a frequency band of 30 kHz, the resulting data rate is limited.

During the 1980s many other analog cellular systems were deployed in various countries, especially in Europe, with each system used in at most a few countries. All these systems have practically vanished due to their low market penetration and the emergence of much more popular 2G systems. In contrast, AMPS is still in operation in the USA, and it actually remains quite popular, since it is the only system providing coverage in the entire country.

### 8.2.2 Second Generation

The increasing use of digital transmission and switching inside the fixed telephony network, the PSTN, led, during the 1980s, to the introduction of its fully digital successor, the ISDN. Similarly, 2G digital cellular systems were introduced to replace their analog predecessors. Realizing that their national markets were too small to sustain a 2G cellular system able to compete with AMPS, numerous European manufactures and operators collaborated to create a single pan-European 2G standard. The outcome was the Global System for Mobile Communications (GSM), a fully digital cellular system.

While GSM inherits the cellular concept from AMPS and uses FDD to separate the frequencies used for the uplink and downlink transmissions of a call, it differs from AMPS in nearly every other aspect. Voice is digitized and compressed before transmission, therefore each call requires much less bandwidth. Rather than allocating a different frequency range to each call, GSM first divides the frequency band of each cell into smaller bands using FDMA. In each band Time Division Multiple Access (TDMA) is used, that is, time is divided into slots of equal duration and each call gets to use a single slot periodically. As a result, multiple calls share the same band by transmitting in different time slots. The advantage of TDMA is that it can support calls that require more bandwidth by allocating them more slots per period.

The original version of GSM, called Phase 1, was purely circuit switched, like the PSTN and ISDN, therefore only circuit switched data services were available. In contrast to the PSTN, and similarly to the ISDN, since GSM is digital there is no need to use a modem; data are transmitted directly in the digital GSM format. Since compressed digitized voice requires lower bandwidth than uncompressed analog voice, the bandwidth available to GSM circuits was low however; circuit switched data calls only supported speeds of up to 9.6 kbps.

In order for GSM to establish voice calls with the PSTN, an InterWorking Function (IWF) is needed at the edge of the GSM network to convert between analog and digital voice and signaling. For data calls, the IWF uses a modem at the PSTN side and relays data from (to) the digital GSM circuit to (from) the analog PSTN circuit. While the ISDN is also digital, the voice and data formats it uses are different from those of GSM, therefore an IWF is also used in this case to convert between the two digital formats.

In parallel with GSM in Europe, other digital 2G systems were deployed in the USA and other parts of the world. The Digital Advanced Mobile Phone Service (D-AMPS) adds digital transmission to the AMPS system. Since D-AMPS needs 10 kHz per call, it can multiplex three digital calls in the frequency band consumed by one analog call, thus tripling system capacity. By using the same frequency allocations as AMPS, D-AMPS can coexist with it, thus allowing gradual migration from analog to digital.

A more radical departure from AMPS is CDMAone, which uses the Code Division Multiple Access (CDMA) scheme to share the available frequency band between users. With CDMA each call employs the complete frequency range all the time, regardless of the cell that it takes place in. However, each call uses a different code to scramble its data at a very high speed. By the choice of appropriate codes and coding and decoding techniques, CDMA systems have the remarkable property that each call distinguishes its own data, treating all other transmissions as random noise. Each cell is assigned a different primary code and all calls in that cell employ secondary codes derived from the primary one. CDMA can also support calls that require higher bandwidth by allocating them additional secondary codes.

While GSM was the only 2G system in the entire European market, in the USA three 2G systems competed, D-AMPS, CDMAone and GSM. In addition, while GSM did not have any serious 1G competitors in Europe, AMPS was a well entrenched analog competitor in the USA. As a result, regardless of the technical merits of each 2G system, GSM became the most popular 2G system due to its dominance in Europe, and it was deployed in numerous countries around the world. In the following we will focus on the evolution of GSM.

The increased importance of data services, and especially connectivity to the Internet, has led to the introduction of additional services in the latest version of GSM, called Phase 2+. The High Speed Circuit Switched Data (HSCSD) service allows circuit switched calls to use more bandwidth by allocating them more TDMA slots per period. This is roughly equivalent to combining multiple circuits to increase the data rate. As a result, circuit switched data calls can reach speeds of up to 64 kbps. Unfortunately, circuit switching means that these resources are tied up whether the call has any data to transmit or not.

In contrast, the General Packet Radio Service (GPRS) offers packet switched data calls. When a GPRS call has data to transmit, the system dynamically allocates it some TDMA slots, but only for the duration of the transmission. As a result, many GPRS calls can dynamically share the available bandwidth without being charged when they are idle. By allocating multiple TDMA slots per period for each transmission, GPRS provides speeds of up to 171 kbps. GPRS packets are transported inside the cellular network using a packet switched backbone, separate from the circuit switched backbone used for voice calls. This packet switching backbone originally provided support for many data protocols, but eventually only IP survived.

### 8.2.3 Third Generation

While the success of the various 2G systems has been spectacular, the fierce competition between them and their technological limitations have led to an effort for the standardization of a worldwide digital 3G cellular system. This system, besides using the most advanced technology available, should place emphasis on packet switched data services, especially IP based multimedia. This effort began in the International Telecommunications Union (ITU) with the name International Mobile Telecommunications 2000 (IMT-2000). The IMT-2000 system would replace all 2G systems, providing a way for operators to gradually migrate from 2G to 3G. Even though all parties agreed that this system should be based on CDMA, there was disagreement on the details and on the evolutionary path from 2G to 3G.

The result was the formation of two separate groups, comprising operators, manufacturers and regulators. The 3G Partnership Project (3GPP) is designing a system based on Wideband CDMA (W-CDMA) for the radio part and GSM for the network backbone, while the 3G Partnership Project 2 (3GPP2) is designing a system based on CDMA2000, an evolution of the CDMAone system [2]. While both projects are moving in the same general direction and are co-ordinated to a large extent, 3GPP is one step ahead in the area of support for IP based multimedia. In the following we will concentrate on the 3GPP system, usually referred to as the Universal Mobile Telecommunications System (UMTS).

A UMTS network consists of two parts: the Radio Access Network (RAN) and the Core Network (CN). The User Equipment (UE) connects to the network via the RAN. The RAN is composed of all the network elements providing users with radio access to the UMTS; it is inherently tied to a specific wireless technology. The CN is composed of all the network elements providing UMTS services to the users; it is independent of the wireless technology used. As shown in Figure 8.2, this separation allows different RAN and CN elements to be combined, thus providing multiple evolutionary paths from 2G to 3G.

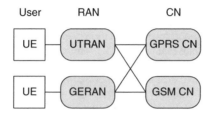

**Figure 8.2** Componets of a UMTS network.

For the CN the options are the GSM based circuit switched network and the GPRS based packet switched network, both already provided by 2G networks. For the RAN the options are the GSM EDGE RAN (GERAN) and the Universal Terrestrial RAN (UTRAN). The GERAN is based on the Enhanced Data rates for GSM Evolution (EDGE) technology, which reuses the frequency allocations of GSM and provides higher bandwidths by using more advanced modulation and coding schemes [3]. The UTRAN is based on the new W-CDMA technology. An operator may migrate to 3G by first upgrading the CN components to the UMTS specifications and using the GERAN for radio access. The UTRAN can be introduced later, when its high deployment cost is justified.

The UMTS specifications were originally released yearly by the European Telecommunications Standards Institute (ETSI), and they were named accordingly, e.g. Release 97. Starting from Release 99, the 3GPP assumed responsibility for UMTS. This release incorporates GSM Phase 2+, which includes GPRS, and the UTRAN. The numbering scheme was changed after Release 99. The next step, Release 4, introduces modifications to the CN and more radio access options. Release 4 also allows IP to be used inside the CN to transport both voice and data. Release 5 introduces the IMS and allows IP to be used inside the RAN to transport both voice and data. Release 6 is currently under development, introducing the MBMS and further extending the IMS [4].

## 8.3 UMTS Networks

This section presents an overview of a UMTS network, based on the 3GPP Release 6 specifications. We first discuss the overall service architecture and its impact on the UE and then separately describe the CN and the RAN in detail. We omit the IMS and MBMS specific functionality, as well as QoS issues, which are covered in later sections.

### 8.3.1 Services and Service Capabilities

Unlike 1G networks that provided only a single service, i.e. voice telephony, 2G networks provide multiple services, divided into three types. A bearer service is a signal transmission facility inside the network; the signal may be voice. A teleservice combines a bearer service with functionality in the terminals to provide a service to the user, such as voice telephony. Finally, a supplementary service provides additional facilities for another service, such as call forwarding.

While 3G networks provide all 2G teleservices and supplementary services, for compatibility, the UMTS architecture emphasizes service capabilities, that is, parametric bearer services and mechanisms required to implement services. This allows a UMTS network to offer new services based on the same service capabilities, without requiring changes to the network architecture. For this reason, Release 6 specifications define only those capabilities sufficient to implement the required services, without standardizing the services themselves [5]. This emphasis on service capabilities is reflected in the choice of IP for transport: packet switched IP bearers can be used to provide both voice and data services, unlike circuit switched bearers which are more suitable for voice.

Another area where this separation is clear is on the UMTS definition of the UE. Unlike earlier networks where a cellular phone provided both communication and application functionality, i.e. wireless access and voice telephony, the UE in UMTS is split into the Mobile Terminal (MT), providing communication, and the Terminal Equipment (TE), providing application functionality. A cellular phone can thus either include application functionality, or it may interface to an external device providing application functionality, such as a notebook computer or a personal digital assistant [6].

### 8.3.2 Core Network

The main function of the CN is to provide routing and switching for user and control traffic. The CN is divided into the circuit switched (CS) and packet switched (PS) domains, as shown in Figure 8.3. The CS domain is an evolution of GSM, while the PS domain is an evolution of GPRS, modified in both cases so as to handle the new UMTS services. Both domains employ the Home Subscriber Server (HSS) that contains all information related to a user, including user preferences, authentication data and location management information.

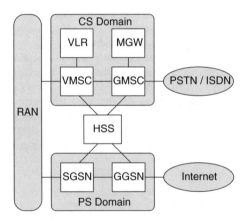

**Figure 8.3** Core Network architecture.

In the CS domain, calls are handled by two Mobile services Switching Centers (MSC): the Gateway MSC (GMSC) is located at the user's home network and the Visitor MSC (VMSC) is located at the network the user is currently visiting. When the user enters a new network, its VMSC informs the local Visitor Location Register (VLR) about the user, and the VLR in turn informs the appropriate HSS that the user is currently located there. Incoming calls are first directed to the GMSC, which asks the HSS about the user's current location and then directs the call to the appropriate VMSC. Outgoing calls use the reverse path, from the VMSC to the GMSC in the user's home network. For ISDN or PSTN originated or terminated calls, a Media GateWay (MGW) is employed by the GMSC for the appropriate translations.

In the PS domain, calls are similarly handled by two GPRS Support Nodes (GSN): the Gateway GSN (GGSN) is located at the user's home network and the Serving GSN (SGSN) is located at the network the user is currently visiting. While the HSS provides all information related to the user, the GGSN always knows the current SGSN handling the user, therefore there is no need for a VLR in the PS domain. For Internet originated or terminated calls, the GGSN acts as an IP gateway router between the UMTS network and the Internet.

In Release 99, user and control data in both the CS and PS domains were transferred over Asynchronous Transfer Mode (ATM) virtual circuits, with ATM Adaptation Layer 2 (AAL2) used for the CS domain and ATM Adaptation Layer 5 (AAL5) used for the PS domain. However, starting with Release 4, the CN can employ IP for data transfer in both the CS and PS domains. This reflects the increased importance of IP in UMTS networks.

Since the PS domain is based on GPRS, some additional details on GPRS are provided below. The GPRS was originally designed to transport any type of packet data efficiently. In UMTS only IP is supported, in unicast and multicast modes. In order for a UE to gain IP connectivity, it must first attach to the GPRS. During this procedure the user is authenticated by the HSS and its local SGSN creates a mobility management context to handle the UE. At this point the UE is known to the UMTS network but it is not yet able to communicate. The next step is for the UE to create a Packet Data Protocol (PDP) context at the GGSN in its home network. The PDP contains the IP address assigned to the user, which is forwarded to the UE and the SGSN. At this point the user can send and receive IP packets via the SGSN and the GGSN [7].

### 8.3.3 Radio Access Network

The main function of the RAN is to provide connectivity between the UE and the CN. The first RAN option supported by UMTS is the GERAN, as shown in Figure 8.4. A GERAN covering a large area is called a Base Station Subsystem (BSS). The BSS is divided into smaller regions, with each region controlled by a Base Station Controller (BSC). Each BSC region is divided into cells, with each cell served by a BTS. The GERAN is basically an evolution of the GSM radio network with GPRS support. Each circuit switched channel can either be allocated to a single voice call or to multiple packet data

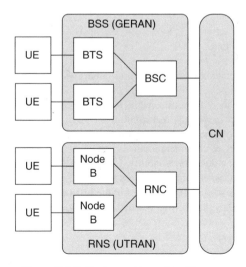

**Figure 8.4** Radio Access Network architecture.

calls. For channels allocated to packet data, the system dynamically allocates one or more TDMA slots in each period to each UE that needs to transmit or receive packets, but only for the duration required. Uplink and downlink TDMA slots are allocated separately, thus efficiently supporting asymmetric services, such as file downloads. Depending on the number of TDMA slots allocated to a transmission and the coding scheme used over the channel, GPRS can support data rates between 9 kbps and 171 kbps.

The EDGE version of GSM offered by the GERAN provides more advanced modulation and coding schemes, supporting data rates of at least 384 kbps for urban environments and 144 kbps for rural environments, without changing the FDMA or TDMA structures of GSM. As a result, the GERAN can be used to support many of the services offered by UMTS networks. However, compatibility with GSM means that the system is fundamentally limited, therefore services requiring higher data rates must resort to a different technology.

This technology is the UTRAN, the second RAN option supported by the UMTS, also shown in Figure 8.4. In the UTRAN one or more cells are served by a Node-B. Multiple Node-Bs are connected to a Radio Network Controller (RNC), and multiple RNCs form a Radio Network Subsystem (RNS). The UTRAN uses W-CDMA to share the available frequency range between multiple simultaneous calls, with each call assigned one or more codes depending on the required bandwidth. The W-CDMA system can operate in two modes: in FDD mode uplink and downlink transmissions take place in different frequency bands, while in TDD mode uplink and downlink transmissions use the same frequency band but are multiplexed in time. The TDD mode is more flexible in terms of spectrum allocation but also more complicated in terms of synchronization.

The UTRAN, besides matching the speeds provided by the GERAN, can also support speeds of more than 2048 kbps in small cells or indoor areas. To achieve these high data rates, besides exploiting the flexible bandwidth allocation scheme of W-CDMA, the UMTS network also employs power control. This means that the transmissions between a UE and a Node-B use the minimum amount of power required to achieve the required level of reliability, depending on the distance between them. As a result, the interference among different calls is minimized, and the system provides higher capacity.

Another area where the UTRAN is superior to the GERAN is in the handling of handovers. A handover is performed when a UE making a call moves to a new cell. Since the GERAN uses different frequency bands in each cell, it only supports hard handover in which the radio link in the old cell is disconnected before establishing a radio link in the new cell. In contrast, the UTRAN uses the same frequency band everywhere, with cells differentiated only by the primary code used in each cell. The UTRAN therefore supports soft handover in which the radio link to the new cell is established before removing the radio link to the old cell. As a result, the user never loses connectivity during a handover.

## 8.4 Multimedia Services

Besides combining multiple media components, multimedia services include at least one continuous, i.e. time sensitive, media component, such as audio or video, and they require all media components to be synchronized with each other. Therefore, while simple IP connectivity allows diverse media to be transmitted over IP in cellular networks, real multimedia services require additional support from the network, at least in the area of QoS provisioning for the continuous media components. Despite many efforts to provide such support, the Internet remains a best effort network, providing no guarantees about end-to-end packet transmission delay or reliability. In contrast, UMTS networks have made significant progress in this direction. One aspect of this progress is the addition of the IMS and MBMS components that will be introduced in this section and analyzed in the two following sections. Another aspect is the provisioning of guaranteed QoS that will be discussed in the last section.

### 8.4.1 IP Multimedia Subsystem

The IP Multimedia Subsystem (IMS) enhances the basic IP connectivity of UMTS by adding network entities that handle multimedia session setup and control and QoS provisioning. Note that the IMS uses the term session in place of call: a session includes the senders, receivers and data streams participating in an application. The new IMS entities ensure that multimedia sessions will be able to reserve the resources they need in order to perform satisfactorily. A session is able to request different QoS levels for each of its media components and modify these levels during its lifetime.

Following the general philosophy of UMTS, the IMS does not standardize any applications, only the service capabilities required to build various services. As a result, real-time and non real-time multimedia services can easily be integrated over a common IP based transport. These services can directly interwork with all the services available over the Internet. Eventually, even legacy circuit switched services, like voice, may be replaced inside the UMTS network by real-time IP based packet switched services, making the CS domain obsolete.

Some of the services that can be provided over IMS are voice and video telephony, rich telephone calls, presence services, instant messaging, chat rooms, voice and video conferencing and multiparty gaming. While the possible services are numerous, they are all based on a small set of capabilities [8,9]:

- endpoint identities, including telephone numbers and Internet names;
- media description capabilities, including coding formats and data rates;
- person-to-person real-time multimedia services, including voice telephony;
- machine-to-person streaming multimedia services, including TV channels;
- generic group management, enabling chat rooms and messaging;
- generic group communication, enabling voice and video conferencing.

These basic services can be controlled by an external Application Server (AS) in order to provide actual applications. For example, the IMS does not offer a conferencing or chat room service, but it provides (i) point-to-point and point-to-multipoint transmission facilities, (ii) group management facilities, and (iii) the ability for an external AS to control the group communication. Depending on the functionality of the AS, the application built on top of these capabilities may be a video conference or a chat room.

To maximize flexibility, the IMS organizes its functionality in three layers, as shown in Figure 8.5 (see Section 8.5 for the acronyms). The transport and endpoint layer initiates and terminates the signaling needed to set up and control sessions and provides bearer services and support for media conversions. The session control layer provides functionality that allows endpoints to be registered with the network, sessions to be set up between them and media conversions to be controlled. In this layer multiple transport services may be combined in a single session. The application server layer allows sessions to interact with various AS entities. In this layer multiple sessions may be combined in a single application.

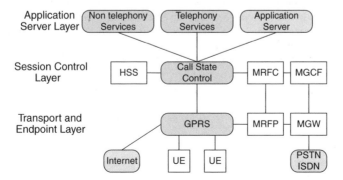

**Figure 8.5** IMS layered service architecture.

### 8.4.2 Multimedia Broadcast/Multicast Service

When packets must be transmitted to many users in a cell, it is more economical to transmit them only once over a common channel received by all users. Since Release 4, UMTS networks have provided a low bandwidth Cell Broadcast Service (CBS) that can transmit short messages to all users in a particular region, called a cell broadcast area [10]. CBS messages are unacknowledged, and each one may be directed to a different cell broadcast area, i.e. set of cells. These messages may originate from a number of information providers, which transmit them through a Cell Broadcast Center (CBC). The CBC broadcasts each message periodically, at a frequency and duration arranged between the CBC and the information provider. The frequency normally depends on message content. For example, volatile content such as road traffic reports will probably require more frequent transmissions than weather reports.

The CBS is targeted to text messages, therefore it is unsuitable for multimedia services, due to the high bandwidth these services require. Since Release 99, UMTS networks have also supported IP multicasting, in which IP packets are forwarded to all users belonging to a multicast group. The multicast group is identified by a class D multicast IP address. Unfortunately, as shown in Figure 8.6, this service is implemented by separately sending packets from the GGSN to each UE. Since multicast packets are sent separately to each receiver in the same cell, no sharing gain is achieved and high bandwidth multimedia services cannot be provided [11].

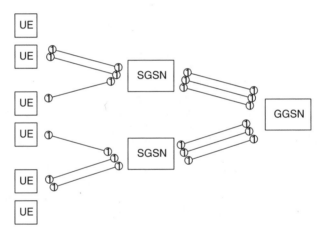

**Figure 8.6** Multicasting without MBMS.

In order to overcome the limitations of these options, the Release 6 specifications include the Multimedia Broadcast/Multicast Service (MBMS) that supports native multicasting and broadcasting over UMTS networks. MBMS is interoperable with IP multicasting, that is, IP multicast packets can be transmitted over MBMS. As shown in Figure 8.7, the GGSN and SGSN send multicast packets only once to each downstream node. More importantly, these packets are transmitted only once over the wireless link, regardless of the number of receivers in a cell. Note that the services provided by MBMS are unidirectional, that is, data are transmitted only from the GGSN to the UE [12].

The services that may be provided over MBMS are classified as follows [13].

- Streaming. Continuous media flows, such as audio and video, plus supplementary text or images. These services are similar to TV channels, but enhanced with multimedia content.
- File download. Reliable binary data transfers, without any delay constraints. These services are similar to conventional file transfers, but with multiple receivers.
- Carousel. A combination of streaming and file download; static media are sent, but with synchronization constraints. These services are similar to stock quote ticker tapes.

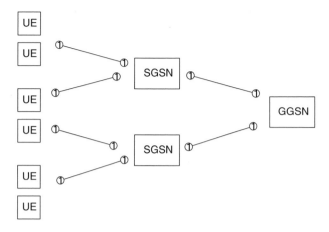

**Figure 8.7** Multicasting with MBMS.

All services can be offered in two modes: in broadcast mode all users that have activated the service receive all data transmitted; in multicast mode only those users that choose to join a multicast group receive the data transmitted to that group. In broadcast mode only the sender can be charged, not the receivers, therefore this mode is suitable for advertising supported services. In multicast mode both the sender and the receivers can be charged, therefore this mode is suitable for either pay per view or advertising supported services. It should also be noted that each service covers a specific area of the network; this allows a provider to transmit different content in each area.

## 8.5 IMS Architecture and Implementation

### 8.5.1 Service Architecture

The relationship of the IMS to the PS and CS domains of a UMTS network is shown in Figure 8.8. From the figure is should be clear that the IMS is indeed a subsystem of the CN that depends on the PS

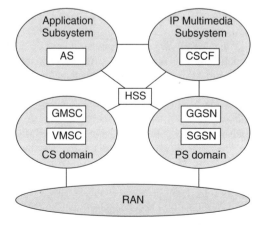

**Figure 8.8** IMS within a UMTS core network.

domain. The actual data transport services are provided by the existing IP based mechanisms offered by GPRS, as enhanced for UMTS networks. What the IMS does is to provide flexible multimedia session management using these IP bearer services. For complete application functionality to be provided, the IMS may have to rely on services provided by an external AS, for example, a conferencing server. The IMS itself only provides session setup and control functions, media processing functions, and media and signaling interworking functions [14]. It also mandates, however, that all media should be transported using the Real Time Protocol (RTP) over UDP/IP, and, most importantly, that the IMS should use exclusively IPv6.

The general architecture of the IMS is outlined in Figure 8.9. Multimedia sessions are set up and controlled via various types of Call Session Control Functions (CSCF): the Proxy CSCF (P-CSCF) is the local contact point of the UE in the network it is visiting, analogous to the SGSN in GPRS; the Serving CSCF (S-CSCF) is controlling the session at the user's home network, analogous to the GGSN in GPRS. Since a network may contain many S-CSCFs for load balancing, an Interrogating CSCF (I-CSCF) may be provided at the entry point to an operator's network so as to direct sessions to the appropriate S-CSCF. The I-CSCF and the S-CSCF rely on the HSS for user related information. Networks with multiple HSSs also provide a Subscription Locator Function (SLF) that locates the HSS handling a given user.

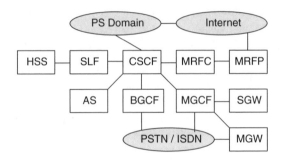

**Figure 8.9** IMS architecture.

While the IMS does not standardize any applications, it provides a Media Resource Function Processor (MRFP) that is able to mix, generate and process media streams under the control of a Media Resource Function Controller (MRFC). These entities can be used in conjunction with an appropriate AS to support applications such as voice conferencing, with the MRFP mixing voice streams, or announcement services, with the MRFP generating announcements. The MRFP can also provide transcoding to allow IMS applications to interoperate with other IP based applications employing different encoding schemes. By providing only this basic functionality inside the IMS, many types of applications can be supported by an AS, without having to transfer the actual media streams to the AS. By separating the MRFP from the MRFC a single control function can oversee many processing functions to achieve scalability.

The IMS also provides Media GateWay (MGW) functions to allow IMS sessions to interwork with circuit switched networks, including the PSTN and the ISDN, and even the CS domain of UMTS. The MGW simply transcodes the data streams to and from the format used in the external network. It is controlled by a Media Gateway Control Function (MGCF) that also handles the signaling to and from the circuit switched network. For some types of circuit switched networks, the MGCF is supported by a separate Signaling GateWay (SGW). Finally, a Breakout Gateway Control Function (BGCF) determines where breakout should occur, i.e. where an outgoing session should exit the IMS and enter the circuit switched network.

### 8.5.2 *Session Setup and Control*

The various CSCFs provide session setup and control for users accessing the IMS services. All signaling between the UE and the CSCFs, as well as between the CSCFs themselves, is based on the Session Initiation Protocol (SIP) [15], with some extensions specific to UMTS [16]. Since SIP is described in detail in a separate chapter, we will omit the SIP signaling details below, concentrating on the concepts. All SIP messages that are used to set up a session or modify its parameters contain the requirements of its media components using the Session Description Protocol (SDP) [17].

The CSCF may take on various roles, as shown in Figure 8.10. In order to understand these roles, we will explain how a UE registers with a SIP server so that it may start originating and terminating IMS sessions. The UE must first attach to the GPRS network and activate at least one PDP context for IPv6, to be used for signaling within the IMS, as explained in the previous section. Having established IP connectivity, the UE must then contact a SIP registry in its home network and register one or more SIP identities. These identities may be either telephone number style or Internet name style SIP identities. All IMS entities that are involved with SIP signaling also have SIP identities.

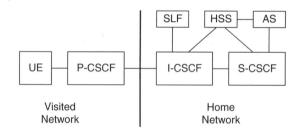

**Figure 8.10** The various call state control functions.

The UE discovers its P-CSCF either during the initial PDP context activation or by using some other mechanism such as querying the Domain Name System (DNS). The UE sends a SIP register message to the P-CSCF, including its SIP identity and the name of its home network. The P-CSCF maps the home network name to an I-CSCF address for that network and forwards the register message there. If the home network contains many HSSs, the SLF is used by the I-CSCF to find the HSS containing the user's information. The I-CSCF contacts this HSS to ask which S-CSCF should handle the user. The HSS checks the user's profile to see which applications the user has enabled, and replies with the address of an S-CSCF that has sufficient resources and capabilities to handle the user. The I-CSCF then forwards the registration to that S-CSCF. The S-CSCF queries the HSS for subscriber information, registers the user, informs the HSS that it is handling the user and returns a SIP confirmation via the I-CSCF and the P-CSCF to the UE. At this point, the user can originate and terminate SIP sessions.

An example of session setup between two UEs is shown in Figure 8.11 [18]. The caller UE sends a SIP invite message to its P-CSCF, including an SDP description of the media components of the session and the SIP identity of the called UE. The P-CSCF may accept or deny the request, depending on resource availability and the QoS policy of the visited network. If accepted, the invite message will be forwarded to the user's S-CSCF, possibly via the I-CSCF of the user's home network. The S-CSCF will then contact the HSS and accept or deny the request depending on the user's profile. If accepted, the invite message will be forwarded to the S-CSCF handling the called UE at its own home network, again possibly via an I-CSCF. The S-CSCF will contact the appropriate HSS to check whether the call can be accepted and, if so, it will forward the message to the called UE via the corresponding P-CSCF.

After the SIP invite is received by the called UE, its SDP payload will be examined, and the called UE will return a counter proposal in a SIP reply. This reply will follow the reverse path through the CSCFs to reach the caller UE, which will then respond with a final mutually acceptable SDP specification to the called UE. At this point the usual SIP signaling continues, always via the CSCFs, until the session is

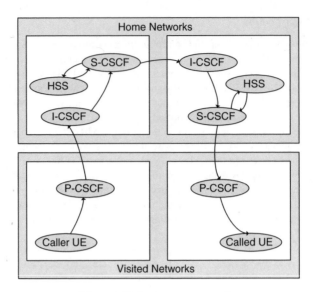

**Figure 8.11** Session setup example.

established. The actual data of the session can then be exchanged directly between the two UEs, without passing any more through the CSCFs. It should be noted that before the actual data exchange begins, each UE must activate secondary PDP contexts for each media component, using the same IP address as for the already established primary PDP context, but different QoS parameters, as appropriate for each media component of the session. This procedure will be described in the last section.

In order to allow an external AS to provide services, an AS can install a set of filters in the HSS handling a user. A filter describes an event that should cause the AS to be notified. For example, a filter could specify that the AS should be notified when the user sends or receives a SIP invite message with specific parameters in the SDP description. When a UE registers with the IMS, the S-CSCF handling the user receives from the HSS the filters associated with the user and stores them in its database. Then, when SIP messages from or to the user are received by the S-CSCF, they are matched against the filter database and the corresponding ASs are notified by SIP messages. For example, when a UE sends a SIP invite message indicating that it wants to participate in a video conference, the AS controlling the video conference may be notified so as to handle the user.

### 8.5.3 Interworking Functionality

Since the GGSN is essentially an IP router, basic connectivity between the IMS and external IP based networks is simple. One complication is that while UMTS networks support both IP version 4 (IPv4) and IPv6, the IMS uses exclusively IPv6, and most other IP networks use exclusively IPv4. Therefore, IPv4 to IPv6 translation gateways are needed at the edge of an IMS network to convert between the different header formats and addresses. Note that this is an issue for all IPv6 networks, not just IMS.

Another complication is that while SIP is an Internet standard, it has been extended to better handle the requirements of the IMS. As a result, when SIP requests are received from or sent to external IP networks, the S-CSCF will discover that one side does not support the IMS specific extensions. Depending on operator policy, the S-CSCF may either refuse to set up sessions with non IMS conformant SIP endpoints, or accept them and translate between IMS and non IMS specific SIP semantics [19]. If media transcoding is also needed, it may be performed by the MRFP under the control of the MRFC.

Since voice telephony is so important in circuit switched networks, the IMS must interwork as seamlessly as possible with the PSTN, the ISDN and the CS domain of UMTS networks. As explained above, this is achieved by using an MGW to perform the required media translations under the control of an MGCF, which also performs the required signaling translations, possibly aided by an SGW. Even though IMS applications can use any codec they desire, each UE and MGW must support at least some standardized codecs so as to provide a minimal guaranteed level of interoperation with other networks. For example, for sessions to or from the PSTN the MGW must convert the RTP packets containing Adaptive Multi-Rate (AMR) encoded voice to bit streams containing Pulse Code Modulation (PCM) encoded voice, and vice versa. The MGW may also have to generate in band signaling for the PSTN, such as dialing tones. Similarly, the SGW and MGCF must convert the SIP signaling of IMS to PSTN signaling, and vice versa [20].

For sessions originating from the UMTS, the S-CSCF handling the caller UE forwards the SIP invite to the BGCF of the home network. If the BGCF determines that the breakout is to occur locally, it selects the MGCF that will be responsible for the interworking and forwards the SIP invite there. The MGCF will then complete the signaling with the external network in order to set up and control the session. If the breakout is to occur in another network, the BGCF forwards the SIP invite to a BGCF in that network. For sessions originating from external circuit switched networks, the signaling is handled by the MGCF in the home network of the called UE, since the session setup request will arrive there. The MGCF acts then as a SIP proxy server in order to complete the session setup inside the UMTS network.

## 8.6 MBMS Architecture and Implementation

### 8.6.1 Service Architecture

While the MBMS affects both the CN and the RAN, the emphasis is on the conservation of resources over the air interface, i.e. the wireless link between the RAN and the UE. The MBMS provides various multicast and broadcast services, with each service offering some content and covering a possibly different set of cells. The actual transmission of data within a service is a session. A service can only have one active session at any given time, but it may use multiple sessions over its lifetime [12]. In order to support MBMS services and sessions, a new functional entity, the Broadcast/Multicast Service Center (BM-SC), is added to the PS domain of the CN, as shown in Figure 8.12. Changes must also be made to the GGSN, SGSN, RAN and UE to provide MBMS support [21].

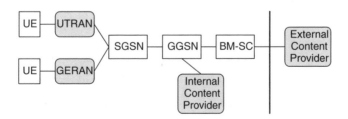

**Figure 8.12** MBMS architecture.

The BM-SC is the core MBMS component, handling all MBMS services and sessions. With MBMS data can only be transmitted towards the UEs, using IP multicast packets for transport. The content transmitted may originate from either external or internal content providers. External content providers must first be authenticated by the BM-SC and they must then transmit only via the BM-SC. The BM-SC may encode the data to provide additional error resilience. In order for a session to begin, the BM-SC instructs the GGSN to set up MBMS bearers in the required service area at the appropriate QoS level. The session data are always transmitted via the GGSN, which may receive them either directly or indirectly, via the BM-SC, from the content provider.

The data are delivered via the GGSN and SGSN to the RAN. The GGSN is responsible for creating appropriate MBMS bearers, based on instructions from the BM-SC, and routing data to the appropriate SGSNs, i.e. all of them in broadcast mode or only those serving interested UEs in multicast mode. The GGSN generates charging data for each receiver of a multicast session, while the BM-SC generates charging data for the content provider of both multicast and broadcast sessions. The SGSN is responsible for routing the data to the appropriate BSCs or RNCs, depending on the type of RAN used.

The main issue for MBMS multicast data delivery inside the RAN is whether a single point-to-multipoint or multiple point-to-point physical channels should be used over the air interface. Since point-to-multipoint channels are more expensive in terms of resources, each operator must set a threshold: when the number of recipient UEs in the cell exceeds the threshold, a point-to-multipoint channel is established; when the number of recipient UEs drops below the threshold, separate point-to-point channels are established. For MBMS broadcast data delivery, a broadcast physical channel is always used.

Finally, the UE must be modified so as to handle activation and deactivation of broadcast services and joining and leaving of multicast services. The former is a purely local operation, since broadcast data are always transmitted anyway. The latter requires signaling between the UE and the network. Two options are being considered for signaling, either creating an MBMS specific scheme or using the standardized group management protocols of the Internet, i.e. the Internet Group Management Protocol (IGMP) for IPv4 or the Multicast Listener Discovery (MLD) protocol for IPv6. An MBMS specific solution may be more efficient, but it will require a new protocol to be designed and multiple network elements to be modified. In the following subsections we assume that the IGMP/MLD option is used.

### 8.6.2 Service Setup and Control

The MBMS services follow the phases shown in Figure 8.13. The broadcast service is simpler, as there is no interaction between the network and the users, so we will describe it first. In the service announcement phase the users are informed about the service availability in their area. The BM-SC can generate service announcements and distribute them over MBMS using the Session Announcement Protocol (SAP) [22]. Announcements can include descriptions of the media that will be included (using SDP), the scheduled time of transmission and other relevant details. The users may also discover the services available in their area by other means, such as web pages or CBS messages. The users may choose to either activate or deactivate reception for each broadcast service locally, i.e. in the UE.

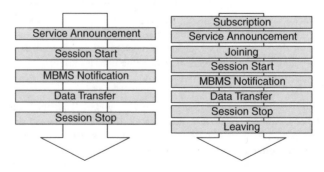

**Figure 8.13** Multicast and broadcast mode phases.

When data are about to be delivered, the BM-SC controlling the service initiates the session start phase, during which the GGSN is instructed to set up MBMS bearers. In broadcast mode, this requires the creation of bearers towards all cells in the service area. In the MBMS notification phase the users are notified that a session for a specific service is about to begin transmission. In the data transfer phase the actual data are transmitted to all cells in the service area and are received by all UEs that have not deactivated the service. Finally, in the session stop phase the MBMS bearers of the session are released.

The multicast mode adds three more phases, performed separately by each user that wants to participate in a multicast session, unlike the phases already described that are performed only once per session. In the subscription phase the interested users subscribe to specific multicast services via external means, for example, via a web page. This informs the BM-SC that the user has agreed to potentially receive a multicast service. The service announcement phase is the same as in broadcast mode, but the announcement must also include the multicast address that will be used by the service. Either before or after the session start phase, users interested in receiving data enter the joining phase, i.e. they send an IGMP/MLD join message to the GGSN indicating the multicast group corresponding to the desired service. These messages are forwarded to the BM-SC, which verifies that the user has already subscribed to the service.

During the session start phase, the BM-SC instructs the GGSN to create the appropriate MBMS bearers. In multicast mode bearers are created only towards cells that already contain UEs that have joined the multicast group. During the MBMS notification phase the BTS or Node-B in each cell, depending on the type of RAN used, counts the number of UEs that are interested in this service and establishes either a single point-to-multipoint or multiple point-to-point physical channels to the UEs. In the data transfer phase the actual data are transmitted to all UEs that have joined the service, and in the session stop phase the MBMS bearers are released. Either before or after the session stop phase, users no longer interested in receiving data enter the leaving phase, i.e. they send an IGMP/MLD leave message to the GGSN, indicating the multicast group that they wish to leave.

We will now explain how multicast distribution trees are constructed. When a UE wants to join a multicast group corresponding to a service, it sends an IGMP/MLD message to its GGSN using a signaling PDP context. The GGSN checks with the BM-SC if the user has subscribed to the service. If so, the GGSN informs the UE, via the SGSN, that the join is approved. The UE then asks the SGSN to create an MBMS UE context, indicating the multicast service and the UE. In turn, the SGSN asks the GGSN to create a similar MBMS UE context. If these are the first MBMS UE contexts created for a multicast service, the SGSN and the GGSN also create an MBMS bearer context indicating the QoS level for the service, based on information from the BM-SC. The MBMS bearer context also lists the downstream nodes that should receive data. In the session start phase, when the BM-SC instructs the GGSN to create multicast MBMS bearers, by consulting the MBMS UE and bearer contexts the GGSN knows which SGSNs are interested and each SGSN knows which UEs are interested in this multicast service, so bearers are created only towards the appropriate nodes, using the proper QoS level. If some UEs later join or leave the group, the MBMS UE and bearer contexts are updated and the multicast distribution tree is modified accordingly.

## 8.6.3 Interworking Functionality

The MBMS is a UMTS specific service, therefore it cannot directly interwork with services provided by external networks. The exception is IP networks, since the MBMS multicast mode is based on IP multicasting. Even in this case however, MBMS multicast groups are controlled for charging purposes and transmissions are strictly unidirectional towards the UEs, therefore the MBMS and IP multicasting service models are quite different. In general, when the content to be delivered via MBMS originates outside the UMTS network, it must be authorized and transmitted by the BM-SC, therefore it must enter the UMTS network in unicast mode and then be replicated as needed.

If an external IP multicast group includes receivers inside the UMTS network, IP multicast routing will terminate at the GGSN; from there on multicast group management will take over, unlike in regular

IP networks where multicast routing extends all the way to the local networks, i.e. the cells. However, even though regular multicast routing is not performed inside the UMTS network, the MBMS ensures that data are delivered in the most efficient manner to the UEs, i.e. not as separate unicasts, as was the case before MBMS.

## 8.7 Quality of Service

### 8.7.1 Quality of Service Architecture

The 3GPP has adopted the layered QoS architecture shown in Figure 8.14 for UMTS networks. The figure shows a user connected to a UMTS network communicating with a user connected to another network, which could be the PSTN, the ISDN or the Internet. The end-to-end Quality of Service (QoS) depends on the QoS offered by the bearer services supported along the path. The Release 6 specifications do not standardize the end-to-end service, since it partly lies outside the UMTS network; they deal only with the UMTS bearer service.

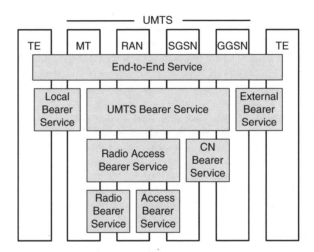

**Figure 8.14** The UMTS QoS architecture.

The local bearer service is offered over a local connection, or internally in UEs integrating both connectivity and terminal functionalities, i.e. both the MT and the TE; it deals with the signaling required to set up an end-to-end service. The UMTS bearer service, the actual service provided by the network, consists of two parts: the radio access bearer service provided by the RAN and the core network bearer service provided by the CN. These services are separated since a UMTS network may include different RANs, i.e. the UTRAN or the GERAN, and different CNs, i.e. the PS and CS domains and the IMS. The radio access bearer service in turn consists of a radio bearer service provided over the wireless link and an access bearer service provided inside the RAN. Finally, between the edge of the CN and the TE at the other end an external bearer service is provided by another network; this could be a PSTN voice call or a best-effort IP session. While these services are not part of the UMTS specifications, some guidelines are proposed to translate between them and UMTS services [23].

Internally, a UMTS network provides four different QoS classes [24]:

(1)  The conversational class is suitable for real-time applications with stringent end-to-end delay and delay variation requirements, such as telephony.
(2)  The streaming class is suitable for real-time applications with stringent delay variation requirements, such as media streaming.
(3)  The interactive class is suitable for request/reply applications that have high reliability but moderate delay requirements, such as web browsing.
(4)  The background class is suitable for bulk transfer applications that have high reliability requirements, such as file transfer.

**Table 8.1**  QoS parameters per class

| Traffic class | Conv. | Str. | Int. | Black. |
|---|---|---|---|---|
| Maximum bit rate | × | × | × | × |
| Maximum packet size | × | × | × | × |
| Packet error ratio | × | × | × | × |
| Residual bit error ratio | × | × | × | × |
| Delivery order | × | × | × | × |
| Delivery of erroneous packets | × | × | × | × |
| Allocation/Retention priority | × | × | × | × |
| Guaranteed bit rate | × | × | | |
| Transfer delay | × | × | | |
| Packet format information | × | × | | |
| Source statistics descriptor | × | × | | |
| Traffic handling priority | | | × | |
| Signaling indication | | | × | |

While both the conversational and streaming classes are real-time, only the conversational class requires low end-to-end delay. Similarly, while both the interactive and background classes are non real-time, only the interactive class requires moderate end-to-end delay. Besides its QoS class, a bearer is characterized by a number of additional QoS parameters, summarized in Table 8.1.

All service classes can specify the maximum bit rate, maximum packet size and packet error ratio for the bearer, all of which are self-explanatory. The residual error ratio refers to the bit errors that go undetected. The delivery order flag is set if the bearer should deliver packets in the order they were sent; this is important for protocols such as TCP. The delivery of erroneous packets flag is set if the endpoint wants to inspect these packets; this is useful for protocols performing error recovery. The allocation/retention priority indicates how important it is to establish/retain the bearer during resource shortages.

The real-time classes can also specify a guaranteed bit rate and a transfer delay limit; the latter should cover at least 95% of the packets. The packet format information indicates that the bearer will only carry specially formatted packets that can be optimally encoded by the network. The source statistics descriptor indicates if the bearer will transfer voice, for which statistical traffic models exist. Finally, the interactive class can specify a traffic handling priority, which is used to prioritize the interactive bearers, and a signaling indication flag that is set if the bearer will be used for signaling. Note that background bearers have lower priority than interactive bearers [23].

When IMS sessions are established, they may involve many media components, as specified by the SDP descriptions carried in the SIP messages. Therefore, in addition to the PDP context used for signaling, each UE must establish secondary PDP contexts with the same IP address but different QoS parameters for each media component. Each PDP context is associated with a filter that shows which IP packets should be handled by it; one PDP context may have no associated filter, serving as the default context. It should also be noted that MBMS sessions may only use the streaming or background QoS classes. This is justified by the fact that they are unidirectional and therefore unsuitable for the conversational and interactive classes. Another MBMS limitation is that the entire multicast or broadcast distribution tree must use the same QoS parameters, which are specified by the BM-SC during session start; these parameters cannot be changed during the session, thus simplifying tree management.

### 8.7.2 Policy-based Quality of Service

While the UMTS specifications describe the QoS classes and their parameters in detail, they do not indicate how the corresponding QoS must be provided. It is up to each implementation, i.e. to the manufacturer and operator of a UMTS network, to choose appropriate mechanisms and their parameters to provide the required QoS for the bearers. A clean way to separate the QoS policy from its implementation on various devices is to use a policy-based architecture. A UMTS network can optionally implement the policy-based architecture outlined in Figure 8.15; we will concentrate on this approach for the remainder of this section. In policy-based QoS the operator selects a set of policies to be enforced by the UMTS network, based on the services that should be provided to the users and on interworking requirements with external networks. These policies are translated into actual QoS mechanisms based on the devices used in the UMTS network [25].

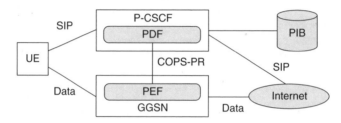

**Figure 8.15** Policy-based QoS architecture.

This approach splits QoS handling between two entities: the Policy Decision Function (PDF) and the Policy Enforcement Function (PEF). The PDF intercepts session setup messages specifying a particular QoS level, retrieves the operator's policy rules matching the request from a Policy Information Base (PIB) and decides whether the session can be accepted, based on current resource availability. If the session is accepted, the PDF translates the policy rules to appropriate configuration actions and sends them to the PEF, which actually implements the QoS provisioning mechanism. The PEF informs the PDF when the policy has been applied, since the PDF must always be aware of PEF resource availability in order to make decisions.

Even though the PDF and the PEF may be separate entities, it is more economical to combine them with existing UMTS entities [26]. The PDF must intercept session setup signaling, so it can be combined with the P-CSCF serving the UE. The PEF on the other hand must intercept actual data transmissions, so it can be combined with the GGSN serving the UE. The PEF and the PDF communicate via the COPS for Policy Provisioning (COPS-PR) protocol [27], a variant of the Common Open Policy Service (COPS) protocol [28]. The COPS-PR messages are transmitted over TCP. The policy rules defined by the operator are stored in a UMTS specific PIB [29].

The model for policy-based QoS provisioning in UMTS assumes that the PEF implements a gate for each bearer service. The gate is basically a QoS specification and a filter that matches the corresponding IP packets based on various header fields. The gate can be opened or closed based on policy decisions made by the PDF. When a SIP session setup request is intercepted by the PDF, the PDF examines the SDP description of its media components and decides whether the session should be established. If so, the PDF sends binding information to the UE; this includes an authorization token uniquely identifying the authorized request and filters matching the approved media components.

When the UE receives the binding information, it can start creating PDP contexts for the various media components of the session. These requests, containing a QoS specification and the binding information returned by the PDF, are sent to the GGSN. The PEF at the GGSN intercepts these messages, extracts the authorization token from the binding information and queries the PDF about the authorized resources. The PDF returns the filters and QoS specifications authorized; if these equal or exceed those in the UE's request, the PDP context is activated, the appropriate filters and QoS specifications are installed by the PEF and the PDF is notified to modify its records. Finally, when the session is established, the PDF intercepting the corresponding SIP message instructs the PEF to open the gate, and the session data may start flowing through the GGSN [26].

### 8.7.3 Session Setup and Control

To clarify how the policy-based scheme operates, we will present the procedures used for QoS provisioning using as an example a session between two UMTS UEs. There are eight procedures used for QoS control [30].

(1) Authorize QoS resources. During session setup, when the called UE returns to the caller UE a SIP message containing a modified SDP description, the PDF at the P-CSCF determines if the session can be accepted. If yes, it includes appropriate binding information in the SIP messages sent to both UEs. There is no need to authorize QoS resources earlier, since the called UE may not accept the session.

(2) Resource reservation. When the UE attempts to activate PDP contexts for the media components of the session, it sends a request to the GGSN containing the binding information returned to it by the PDF. The PEF at the GGSN asks the PDF what resources have been authorized for that authorization token. If the resources authorized equal or exceed the requested resources, the PEF installs the packet filters and notifies the PDF that the resources have been reserved.

(3) Approval of QoS commit. When the SIP setup signaling concludes and the session is established, the PDF at the P-CSCF updates its database of available resources and instructs the PEF at the GGSN to open the gate and allow data to flow.

(4) Removal of QoS commit. This procedure reverses the approval procedure, i.e. closes the gate. The filters and reservations are not removed, therefore this procedure is useful when the session is temporarily suspended.

(5) Revoke authorization for QoS resources. This procedure reverses both the reservation and authorization procedures. It is invoked when an established session is normally released using SIP signaling.

(6) Indication of PDP context release. This is similar to the revoke authorization procedure, but it is used when a PDP context is released without previous SIP signaling, i.e. in an abnormal termination.

(7) Authorization of PDP context modification. This procedure is used when the UE wishes to modify the session by requesting additional resources.

(8) Indication of PDP context modification. This procedure is used when a PDP context modification indicates that there is no need for the corresponding resources any more.

A full session setup example is shown in Figure 8.16. We assume that both UEs have already registered with their S-CSCFs. When the caller UE sends an SIP invite message, this is routed to the caller UE via the intermediate CSCFs. When the called UE responds with an updated SDP description of the media components, the PDF at the P-CSCF examines the request and decides if it can be authorized. If so, the PDF sends binding information to the called UE and forwards the SIP message to the PDF at the P-CSCF of the caller UE, which repeats the above procedure; if it also approves the request, it also sends binding information to the caller UE. At this point the authorization of resources is complete, but resources have not yet been reserved.

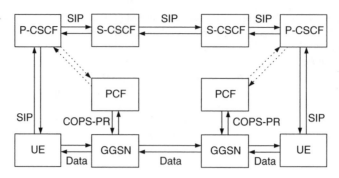

**Figure 8.16** QoS setup example.

The next step is for both UEs to independently activate appropriate PDP contexts at their GGSNs, using the binding information returned by the PDFs. The included authorization tokens are passed by each PEF at the GGSN to its PDF to determine the amount of resources that have been authorized; if these equal or supersede those requested, each PEF installs the corresponding filters, reports it to the PDF and informs the UE that the PDP contexts have been activated. At this point the reservation of resources is complete but the gates are closed, since the session has not been established.

Eventually the session is established, as indicated by a final SIP message sent by the called UE to the caller UE via the CSCFs. As this message passes through the PCF at each P-CSCF, the PCF instructs the PEF at the GGSN to open the gates, and the PEF confirms that the gates have indeed been opened. When the final SIP message reaches the caller UE, the gates are already open at both GGSNs, therefore session data may start flowing in both directions.

## 8.8 Summary

This chapter provided an introduction to the support for IP based multimedia services on 3G wireless cellular networks, focusing on the IP Multimedia Subsystem (IMS) and the Multimedia Broadcast/ Multicast Service (MBMS). An overview of cellular networks in general and UMTS networks in particular was first presented to lay the groundwork for the following discussion. Then the features and services of the IMS and the MBMS were introduced, followed by detailed descriptions of both. Finally, the QoS issues for IP based multimedia services were discussed, emphasizing the policy based QoS control scheme of UMTS and its application to the IMS.

## 8.9 Glossary of Acronyms

| | |
|---|---|
| 1G/2G/3G | First/Second/Third Generation |
| 3GPP | 3rd Generation Partnership Project |
| 3GPP2 | 3rd Generation Partnership Project 2 |
| AAL2/5 | ATM Adaptation Layer 2/5 |
| AMPS | Advanced Mobile Phone Service |
| AMR | Adaptive Multi-Rate |
| AS | Application Server |
| ATM | Asynchronous Transfer Mode |
| BGCF | Breakout Gateway Control Function |
| BM-SC | Broadcast / Multicast Service Center |
| BSC | Base Station Controller |

| | |
|---|---|
| BSS | Base Station Subsystem |
| BTS | Base Transceiver Station |
| CBC | Cell Broadcast Center |
| CBS | Cell Broadcast Service |
| CDMA | Code Division Multiple Access |
| CN | Core Network |
| COPS | Common Open Policy Service |
| COPS-PR | COPS for Policy Provisioning |
| CS | Circuit Switched |
| D-AMPS | Digital Advanced Mobile Phone Service |
| DNS | Domain Name System |
| EDGE | Enhanced Data Rates for GSM Evolution |
| ETSI | European Telecommunications Standards Institute |
| FDD | Frequency Division Duplexing |
| FDMA | Frequency Division Multiple Access |
| GERAN | GSM EDGE Radio Access Network |
| GGSN | Gateway GPRS Support Node |
| GMSC | Gateway Mobile services Switching Center |
| GPRS | General Packet Radio Service |
| GSM | Global System for Mobile Communications |
| GSN | GPRS Support Node |
| HSCSD | High Speed Circuit Switched Data |
| HSS | Home Subscriber Server |
| I-CSCF | Interrogating Call State Control Function |
| IGMP | Internet Group Management Protocol |
| IMS | IP Multimedia Subsystem |
| IMT-2000 | International Mobile Telecommunications 2000 |
| IPv4/6 | IP version 4/6 |
| ISDN | Integrated Services Digital Network |
| ITU | International Telecommunications Union |
| IWF | InterWorking Function |
| MBMS | Multimedia Broadcast/Multicast Service |
| MGCF | Media Gateway Control Function |
| MGW | Media GateWay |
| MLD | Multicast Listener Discovery |
| MRFC | Multimedia Resource Function Controller |
| MRFP | Multimedia Resource Function Processor |
| MS | Mobile Station |
| MSC | Mobile services Switching Center |
| MT | Mobile Terminal |
| P-CSCF | Proxy Call Session Control Function |
| PCM | Pulse Code Modulation |
| PDF | Policy Control Function |
| PDP | Packet Data Protocol |
| PEF | Policy Enforcement Function |
| PIB | Policy Information Base |
| PS | Packet Switched |
| PSTN | Public Switched Telephone Network |
| QoS | Quality of Service |
| RAN | Radio Access Network |
| RNC | Radio Network Controller |

| RNS | Radio Network Subsystem |
| RTP | Real Time Protocol |
| SIP | Session Initiation Protocol |
| S-CSCF | Serving Call State Control Function |
| SGSN | Serving GPRS Support Node |
| SGW | Signaling GateWay |
| SLF | Subscription Locator Function |
| TDD | Time Division Duplexing |
| TDMA | Time Division Multiple Access |
| TE | Terminal Equipment |
| UE | User Equipment |
| UMTS | Universal Mobile Telecommunications System |
| UTRAN | Universal Terrestrial Radio Access Network |
| VLR | Visitor Location Register |
| VMSC | Visitor Mobile services Switching Center |
| W-CDMA | Wideband Code Division Multiple Access |

# References

[1] 3GPP, Vocabulary for 3GPP specifications, TR 21.905, V6.5.0, January 2004.

[2] M. Zeng, A. Annamalai and V.K. Bhargava, Harmonization of global third-generation mobile systems, *IEEE Communications Magazine*, December 2000, 94–104.

[3] A. Furuskär, S. Mazur, F. Müller and H. Olofsson, EDGE: Enhanced Data Rates for GSM and TDMA/136 Evolution, *IEEE Personal Communications*, June 1999, 56–66.

[4] 3GPP, Evolution of 3GPP system, TR 21.902, V6.0.0, September 2003.

[5] 3GPP, Services and service capabilities, TS 22.105, V6.2.0, June 2003.

[6] 3GPP, Network architecture, TS 23.002, V6.3.0, December 2003.

[7] 3GPP General Packet Radio Service (GPRS); Service description; Stage 2, TS 23.060, V6.3.0, December 2003.

[8] 3GPP Service requirements for the Internet Protocol (IP) multimedia core network subsystem; Stage 1, TS 22.228, V6.5.0, January 2004.

[9] 3GPP IP Multimedia Subsystem (IMS) group management; Stage 1, TS 22.250, V6.0.0, December 2002.

[10] 3GPP, Technical realization of Cell Broadcast Service (CBS), TS 23.041, V6.2.0, December 2003.

[11] M. Hauge and Ø. Kure, Multicast in 3G networks: Employment of existing IP multicast protocols in UMTS, *ACM WoWMoM*, September 2002, 96–103.

[12] 3GPP Multimedia Broadcast/Multicast Service (MBMS); Stage 1, TS 22.146, V6.3.0, January 2004.

[13] 3GPP Multimedia Broadcast/Multicast Service (MBMS) user services; Stage 1, TS 22.246, V6.0.0, January 2004.

[14] 3GPP IP Multimedia Subsystem (IMS); Stage 2, TS 23.228, V6.4.1, January 2004.

[15] J. Rosenberg, H. Schulzrinne, G. Camarillo, A. Johnston, J. Peterson, R. Sparks, M. Handley and E. Schooler, SIP: Session Initiation Protocol, June 2002, RFC 3261.

[16] M. Garcia-Martin, E. Henrikson and D. Mills, Private Header (P-Header) extensions to the Session Initiation Protocol (SIP) for the 3rd-Generation Partnership Project (3GPP), January 2003, RFC 3455.

[17] M. Handley and V. Jacobson. SDP: Session Description Protocol, April 1998, RFC 2327.

[18] K.D. Wong and V.K. Varma, Supporting real-time IP multimedia services in UMTS, *IEEE Communications Magazine*, November 2003, 148–155.

[19] 3GPP Interworking between the IM CN subsystem and IP networks, TS 29.162, V1.0.0, March 2002.

[20] 3GPP Interworking between the IP Multimedia (IM) Core Network (CN) subsystem and Circuit Switched (CS) networks, TS 29.163, V6.1.0, December 2003.

[21] 3GPP Multimedia Broadcast/Multicast Service (MBMS) user services; Architecture and functional description, TS 23.246, V6.1.0, December 2003.

[22] M. Handley, C. Perkins and E. Whelan, Session Announcement Protocol, October 2000, RFC 2974.

[23] 3GPP Quality of Service (QoS) concept and architecture, TS 23.107, V6.0.0, December 2003.

[24] R. Koodli and M. Puuskari, Supporting packet-data QoS in next-generation cellular networks, *IEEE Communications Magazine*, February 2001, 180–188.

[25] W. Zhuang, Y.S. Gan, K.J. Loh and K.C. Chua, Policy-based QoS architecture in the IP multimedia subsystem of UMTS, *IEEE Network*, May/June 2003, 51–57.

[26] 3GPP End-to-end Quality of Service (QoS) concept and architecture, TS 23.207, V5.3.0, March 2002.

[27] K. Chan, J. Seligson, D. Durham, S. Gai, K . McCloghrie, S. Herzog, F. Reichmeyer, R. Yavatkar and A. Smith, COPS usage for Policy Provisioning (COPS-PR), RFC 3084, March 2001.

[28] D. Durham, J. Boyle, R. Cohen, S. Herzog, R. Rajan and A. Sastry, The COPS (Common Open Policy Service) Protocol, RFC 2748, January 2000.

[29] 3GPP Policy control over Go interface, TS 29.207, V5.7.0, March 2004.

[30] 3GPP End-to-end Quality of Service (QoS) signaling flows, TS 29.208, V5.7.0, March 2004.

# Part Two

# Wireless Multimedia Applications and Services

# 9

# Wireless Application Protocol (WAP)

Alessandro Andreadis and Giovanni Giambene

## 9.1 Introduction to the WAP Protocol and Architecture

WAP Forum was formed when a USA network operator, Omnipoint, issued a tender for the provision of mobile information services in early 1997. It received several responses from different suppliers using proprietary techniques such as Smart Messaging from Nokia and Handheld Device Markup Language (HDML) from Phone.com. These different approaches were not so different, thus implying that they could be combined and extended to form a powerful standard. Hence, Omnipoint informed the tender responders that it would not accept a proprietary approach and recommended that various vendors got together to explore the definition of a common standard. These events triggered the development of WAP, a standard for delivering Internet contents to wireless devices. Ericsson, Motorola, Nokia and Phone.com founded the Wireless Application Protocol Forum (WAP Forum) in 1997. Hundreds of members joined the WAP forum that, at present, has been consolidated into the Open Mobile Alliance (OMA) [1].

The explosive growth of the Internet has fuelled the creation of new and exciting information services. Most of the original technology developed for Internet services has been designed for large computers with medium-to-high bit-rate transmission capabilities, large displays, a keyboard and a mouse as input devices. Whereas, mobile devices have small displays and are constrained in terms of CPU processing capacity, available memory, energy consumption, displays size and input methods (i.e., there is not a mouse). An interesting approach for allowing the mobile access to the Internet is provided by WAP, a comprehensive and scalable protocol stack designed for use with:

- diverse mobile phones: from those with a one-line display to Personal Digital Assistants (PDAs), smart-phones and pagers;
- several network bearers, such as: Short Message Service (SMS), Circuit Switched Data (CSD), Unstructured Supplementary Services Data (USSD), General Packet Radio Service (GPRS), etc.;
- many mobile network standards, like: GSM 900, 1800 and 1900 MHz; Interim Standard (IS)-136; Digital European Cordless communication (DECT); Trans European Trunked RAdio (TETRA); third-generation (3G) cellular systems;

*Emerging Wireless Multimedia: Services and Technologies*   Edited by A. Salkintzis and N. Passas
© 2005 John Wiley & Sons, Ltd

- different operating systems, like: Windows CE, PalmOS, EPOC, Pocket PC, FLEXOS, OS/9, Linux and JavaOS.

Two main releases of the WAP protocol standard have been issued, namely WAP 1.X (with many variants, being WAP 1.2 the most significant one among them [2]) and WAP 2.0.

The WAP 1.X network architecture envisages WAP servers, hosting pages designed in the Wireless Markup Language (WML), and WAP gateways between the wireless network domain and the wireline Internet. The WML language (an eXtensible Markup Language[a] [3]) is specifically conceived for small screens and one-hand navigation without a keyboard. WML is scalable from two-line text displays up to graphic screens of smart phones and communicators. WAP also defines a markup script language, WMLScript, similar to JavaScript, but making minimal demands on memory and CPU power. At the application layer, WAP specifications define a Wireless Application Environment (WAE) aimed at enabling operators, manufacturers, and content developers to built advanced services and applications including a micro-browser (to display WML pages), scripting facilities, e-mail, World Wide Web (WWW)-to-mobile-handset messaging, and mobile-to-telefax access. Pages in WML are called decks. Decks are constructed according to a set of cards.

WAP contents are transported by using a set of standard communication protocols based on the Internet protocol suite. Moreover, WAP adopts a proxy approach (Performance-Enhancing Proxy, PEP) to improve the inter-connection between the wireless domain and the Internet. In particular, the proxy appears like a server towards the mobile user and like a client towards the Internet. The WAP proxy supports the following functionalities.

- Gateway. It performs the adaptation from the WAP protocol stack to the WWW protocol stack and vice versa.
- Coding and decoding. A binary encoding process is employed to make the contents more compact and suitable for transmission through the wireless link.
- Caching proxy. A caching proxy can improve both the access performance and the network utilization by maintaining a cache of frequently accessed resources.

WAP puts the intelligence in the WAP proxy&gateway, whilst adding just a micro-browser to the mobile phones, requiring only limited resources on the mobile phone. The following two cases are possible for the contents retrieved from the Internet.

(1) The Web server provides contents in the HyperText Markup Language (HTML) format: the proxy has to translate this information in WML format.
(2) The Web server directly provides WAP contents in WML format: the proxy doesn't have to perform any filtering action.

The WML compact format is further encoded by the proxy in a binary compressed representation, optimized for the transmission on the low-bandwidth links of the radio mobile network.

WAP version 1.2 has introduced new features such as: push services (proactive delivery of information from a WAP gateway to a WAP terminal), user profiles, WMLScript, CryptoLibrary, Wireless Telephony Application (WTA), WAE enhancements and other features.

On the basis of the WAP architecture described in Figure 9.1, WAP operates as follows.

(1) The user requests a Web page with a given Uniform Resource Locator (URL).
(2) The user agent sends a URL request to a WAP gateway by means of the WAP protocol (i.e., through WSP/WTP protocols that are described in the following sub-Sections 9.2.2 and 9.2.3).

---

[a]XML is a language proposed by the World Wide Web Consortium (W3C) in 1998 to allow data exchanges between heterogeneous systems. An XML document can be used to store and to transfer information in a form completely independent of both the platform and the device.

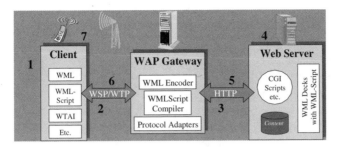

**Figure 9.1**  WAP basic architecture.

(3)  The WAP gateway creates a conventional HTTP request for the specified URL and sends it to the Web server.

(4)  The HTTP request is processed by the Web server. The URL may refer to a static file or to a script application. In the first case, the Web server fetches the file and adds an HTTP header to it. If the URL specifies a script application, the Web server runs the application.

(5)  The Web server returns the WML page (i.e., a deck; see subsection 9.3.1) with the added HTTP header or the WML output from the script application.

(6)  The WAP gateway verifies the HTTP header and the WML content and encodes them into a binary format that is delivered to the user agent.

(7)  The user agent receives the WAP response. It processes the WML response and displays the first card of the WML deck to the user.

In 2002, WAP Forum released the 2.0 version of the WAP Protocol. WAP 2.0 gets the wireless world closer to the Internet with a new suite of specifications. The access to the Internet through wireless devices is more similar to the access by means of a fixed terminal [4]. WAP 2.0 is based on the latest Internet standards: the WML2 markup language derived from the eXtensible HyperText Markup Language (XHTML), a wirelessly-optimized TCP/IP suite and the HyperText Transfer Protocol (HTTP/1.1). XHTML provides the ability to make graphical Web pages, similarly to common Web pages, but in a smaller size [5]. The wirelessly-profiled TCP and HTTP are optimized versions of the Internet TCP and HTTP protocols for a more efficient delivery of the contents over wireless links.

In the previous versions of the WAP standard a WAP gateway was needed in order to establish a connection between a mobile client and a server in the Internet (the WAP protocol was employed for the dialogue between the mobile client and the gateway; standard Internet protocols were used for the dialogue between the gateway and the server). Whereas, WAP 2.0 does not need a gateway since the communication between the client and the server is direct by means of the HTTP/1.1 protocol. However, a WAP proxy is still useful to cache frequently accessed Web pages and to provide the mobile terminal with localization, user privacy, push services and service personalization (the proxy communicates mobile phone capabilities to the application so that contents can be customized for a particular device).

In addition to the classical WAP architecture with Web server, WAP proxy and WAP client, other configurations can also be supported by release 2.0. In particular, we can consider supporting servers that provide useful functions for devices, proxies, and application servers. Examples of supporting servers are as follows.

- PKI (Public Key Infrastructure) Portal. The PKI portal allows devices to initiate the creation of new public key certificates [6].
- UAProf Server. The UAProf server allows applications to retrieve client capabilities and profiles of user agents and individual users [7]. CC/PP 1.0 is a standard for expressing device capabilities and user preferences by means of the Resource Description Framework (RDF) [7].
- Provisioning Server. The provisioning server is trusted by the WAP device to give its provisioning information [8].

WAP clients support the proxy selection mechanism that permits them to utilize the most appropriate proxy for a given service or to connect directly to a service, if necessary. In some cases, the device might make direct connections to application servers, for example to provide a secure connection between the device and the application server.

A complete description of the WAP 2.0 network architecture is shown in Figure 9.2.

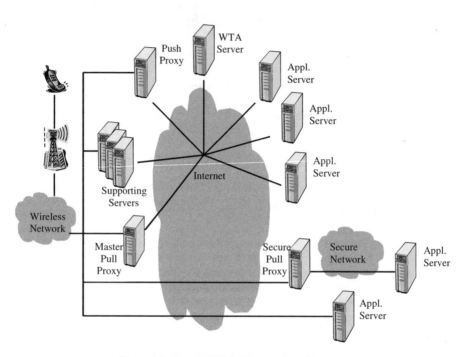

**Figure 9.2**   Detailed WAP 2.0 network architecture.

### 9.1.1 WAP-based Multimedia Services: Potentials and Limitations

It is expected that within few years the number of mobile devices accessing the Internet will exceed the number of Personal Computers (PCs). The use of mobile terminals is becoming attractive, since they allow the access on the move and require a shorter set-up time than the initial power on of a PC.

The benefits of WAP can be envisaged for different aspects as detailed below.

- Operators. For wireless network operators, WAP promises to cut costs and to increase the subscriber base both by improving existing services, such as interfaces to voice-mail and prepaid systems, and by facilitating an unlimited range of new value-added services and applications, such as account management and billing inquiries. New applications can be introduced quickly and easily without the need for additional infrastructures or modifications to the mobile phone.
- Content providers. By means of WAP, applications will be written by using the same model of the Internet.
- End users. End users of WAP will benefit from an easy mobile access to relevant Internet information and services, such as unified messaging, banking, and entertainment. Intranet information such as corporate databases can also be accessed via WAP technology. Moreover, WAP push capability will

enable weather and travel information to be provided to mobile users related to their positions. This push mechanism represents both a significant advantage with respect to WWW users and a tremendous potential for information providers and mobile operators.

WAP has dealt with many constraints that are typical of mobile communications such as limited capacity, frequent errors on wireless links, small displays, and lack of mouse for input. Even with these constraints WAP provides users on the move with appealing services such as:

- news;
- meteorological information;
- flight and train schedules;
- stock exchange news;
- browsing of e-mail;
- remote access to a bank account;
- location-aware information services, and
- Multimedia Messaging Services (MMS).

However, WAP has some problems that limit its potentialities. The first WAP releases, in particular, suffered from a lack of compatibility (in terms of both presentations on the display and supported functionalities) between the browsers of different mobile phone manufacturers.

Moreover, with the WAP 1.X standard, the translation operated by the WAP gateway (from HTML to WML) did not provide good quality results. Hence, the navigation was commonly limited to WAP servers containing pages in WML format. With WAP 2.0, pages are in the new WML2 (XHTML-based) language that is independent of the platform, thus solving the above problems.

Other limitations were related to the reduced dimensions of the displays, which did not allow presentation of icons and pictures of adequate dimensions.

Finally, another limitation that contributed to users' negative perception of WAP-based services was the low transmission capacity on wireless links. For instance, a circuit-switched data connection through GSM allows only 9.6 kbit/s, which appears to be insufficient for browsing pages in short times. The use of GPRS can increase the download capacity to the theoretical maximum of 36 kbit/s (or 53.6) if the CS-1 (CS-2) coding scheme is used with four slots assigned to a connection. However, the available capacity again limits the WAP potentialities and allows interactions at reduced speeds.

## 9.2 WAP Protocol Stack

The WAP protocol is structured according to a layered architecture that inherited many characteristics of the ISO/OSI reference model and has strong similarities to the Web protocol stack. However, it has been designed to take into account a different usage scenario, where mobile devices are connected through low bandwidth and high latency networks. Hence, lightness and compactness constitute the main peculiarities of the WAP protocol stack that is composed of five layers, as illustrated in Figure 9.3. The following subsections provide a detailed description of each layer.

### 9.2.1 Wireless Application Environment

The application layer in WAP is named the Wireless Application Environment (WAE) and offers operators and service providers an environment for the development and execution of applications and services targeted at a wide variety of wireless platforms. The main goals of WAE are to provide an application framework that is neutral to the network and is particularly suitable to narrow-band wireless devices, permitting a high degree of device independence. It adopts a Web programming model, leveraging on Internet standard technologies.

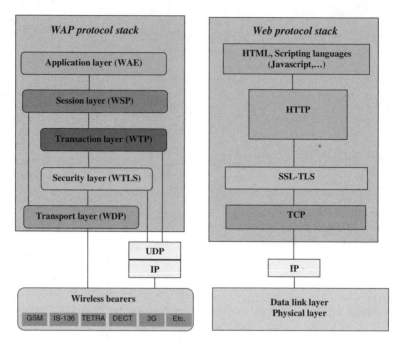

**Figure 9.3**  WAP and Web protocol stacks.

WAE enables WAP devices, generally assumed to have restricted input capabilities, small display size and limited bandwidth requirements, to access services and applications in an efficient manner. WAE allows the creation of systems and services that are specifically tailored for wireless devices; it is sensitive to the limited bandwidth resources of a wide variety of wireless devices and networks and it represents a flexible way to realize user interfaces that can enhance the user experience.

WAE provides tools to operate, through a secure interface, with additional hardware (e.g., smart cards) or software that may extend the functionality of a basic browser. Moreover, it includes different tools and formats that allow the creation and optimization of content presentation and interaction with devices of limited capability. WAE elements are described in Figure 9.4.

The WAE framework is then extended by WTA, which provides telephony services.

User agents and interchange formats are the building blocks of WAE and are described below.

The WAE user agent is a software implemented in the client, typically a micro-browser (textual browser, voice browser), a message editor or a phonebook, providing content visualization capabilities and other specific functionalities. WAE user agents must be compliant with the formats specified by WAE. One type of WAE user agent is the WML browser. It is able to interpret WML, WMLscript or other contents and it represents the user interface to access WAP pages; its behavior can be compared to an Internet browser for HTML pages. The WTA user agent is another type of WAE user agent; it extends the WML user agent capabilities and handles the telephony functions integrated in a WAP device. It adopts an interface (WTAI) in order to create call-control and call-handling applications and to interact with device-specific features (e.g., phone book interface).

WAE user agents may support different Multipurpose Internet Mail Extensions (MIME) media types, according to specific content interchange formats. They include WML, WMLscript, XHTML, electronic business cards (Internet Mail Consortium – IMC vCards), calendar events (IMC vCalendar), pictograms and images (Wireless BitMap – WBMP), WAP Cascading StyleSheets (CSS), Wireless Binary XML (WBXML).

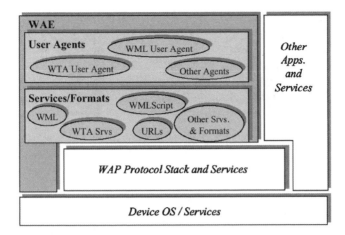

**Figure 9.4**   WAE client components.

The two most important formats defined in WAE are WML and WMLScript. A WML encoder at the WAP gateway, or 'tokenizer', converts a WML deck into its binary representation [9] and a WMLScript compiler transforms a script into byte-code. This process allows a significant compression of the data to be sent to the user on the wireless link, for an efficient utilization of air interface resources.

### 9.2.2 Wireless Session Protocol

Wireless Session Protocol (WSP) provides the application layer of WAP (i.e., WAE) with a consistent interface for data exchange, in an organized way between co-operating client/server applications. WSP offers two session services that allow a client and a proxy or a server to set up their connection: a connection-mode session service, operating over the reliable WTP, and a connectionless session service, operating over a secure or non-secure unreliable WDP service.

The connection-mode session service provides the following functionalities:

- to establish, manage and close a reliable session between a client and a server; a session can be suspended and resumed later, in case of data transmission problems (due to roaming or network disconnection);
- Capability negotiation – it allows client and server to agree on a common acceptable level of service. A set of capabilities are defined in WSP specifications [10];
- to exchange contents between client and server using compact encoding;
- Push modality – the server can send unsolicited data to the client. Push operations can be confirmed or not. Confirmed data push envisages notification by the client upon reception of pushed data. In unconfirmed data push, no acknowledgment is sent to the server.

Push technology allows trusted application servers to send proactively personalized contents to the end-user, such as a sale offer for a product, a new e-mail notification, or a location-dependent promotion. The push technology complements the traditional request–response mechanism ('pull' model) of the Internet where users request specific information from a Web site (see Figure 9.5).

The connectionless session service provides the functionality of non-confirmed content exchange between WSP users. It supports only unreliable method invocation and unconfirmed data push operation; it can be used without establishing a session.

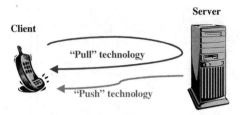

**Figure 9.5**  Push and pull applications.

WSP services are delivered through the use of primitives and related parameters. Service primitives are borrowed from ISO [11]. They enable communications between adjacent layers and between layers within the session layer. Primitives consist of commands and responses associated with a requested service facility. For example, S-connect.req(SA, CA, CH, RQ) is a WSP primitive requesting the establishment of a session with the server; the specified parameters have the following meanings: SA = Server Address, CA = Client Address, CH = Client Header, RQ = ReQuested capability [10]. The client generates a WSP Protocol Data Unit (PDU) with the above parameters, in order to send its connection request to the remote server. The capability functionalities can be modified and finally the server may accept or refuse the session establishment.

WSP semantics and mechanisms are based on HTTP/1.1, with enhancements and optimizations for wireless networks. WSP can be considered as a binary form of HTTP: server and client transmit data in a binary form and this allows one to overcome some limits related to narrow-band mobile networks and to connection-loss sensitivity due to coverage problems or cell overload.

WSP is a transaction-oriented protocol, based on the request/reply method and it supports all HTTP/1.1 methods. However, HTTP is not used because it does not optimize data sent over the network, it does not support push modality and it has inefficient capability negotiations. The main enhancements beyond HTTP offered by WSP are:

- binary header encoding;
- exchange of client and server session headers;
- confirmed and non-confirmed data push from server to client;
- capability negotiation;
- suspend and resume of the session;
- connectionless service.

Capability refers to information about the operation of the session service provider. The main defined capabilities are: client and server message size, protocol options (to enable push/confirmed push facility, session resume facility, acknowledgment headers), maximum outstanding requests that can be contemporarily active during the session, set of extended methods (beyond HTTP) to be used during the session, set of header code pages (beyond HTTP) to be used during the session. The peer entity that starts capability negotiation is called the initiator, the other peer is called the responder. The responder may reply with different capability requirements, provided that these do not imply a higher level of functionality than the one proposed by the initiator.

The suspend and resume feature allows interruption of the session and maintainance of the state of the session on both the client and the server, so that the same session can be resumed later. There can be different reasons for suspending a session: typically it is requested by the client before terminating a connection with an underlying bearer network, but also the service provider can do it due to poor network coverage or unavailability.

WSP inserts service requests and responses inside PDUs. Referring to Figure 9.6, A WSP PDU is composed of a payload, containing WML, WMLScript or images, and a header, containing information on the payload (content type, character set, languages, etc.) and on the transaction. WSP adopts a compact binary encoding of headers in order to reduce protocol overhead. Each PDU is passed to the transport level where it is further encapsulated. WSP PDU common fields are TID, Type and Type-Specific contents.

**Figure 9.6**   WSP Protocol Data Unit (WSP PDU).

The TID (Transaction Identifier) field has a length of 8 bits; it is present only in the connectionless service (it must not be included in connection-mode PDUs). It is used to associate requests with replies.

The Type field has a length of 8 bits; it specifies the type and the function of the PDU, according to a reference table of the WSP assigned numbers (e.g., connect $= 0 \times 01$, reply $= 0 \times 04$, push $= 0 \times 06$).

The Type-Specific contents field has a variable length and contains information generated by a WSP service primitive.

### 9.2.3 Wireless Transaction Protocol

Wireless Transaction Protocol (WTP) provides an efficient transport service, handling requests and replies between a user agent (e.g., a WAP browser) and an application server. It tries to optimize the transactions between the client and server by eliminating much of the TCP overhead. WTP represents a lightweight protocol, optimized for thin clients and for the adoption in networks with low-to-medium bandwidth. Owing to the reduced bandwidth utilization, WTP allows an operator to load more subscribers on the same network and allows individual users to experience an improved performance at reduced costs.

WTP is a connection-oriented protocol that makes no use of explicit connection setup or tear down phases. It simplifies the three-way handshake procedure at the beginning of a TCP session by carrying data in the first packet of the protocol exchange. Its main functionalities are [12]:

- three classes of transaction services;
- optional reliable data transfer (the WTP user confirms every received message);
- concatenation of multiple PDUs in a single Service Data Unit (SDU) at the datagram level and delayed acknowledgment in order to reduce the number of exchanged messages;
- out-of-band optional data (e.g., performance measurements for evaluating user perceived quality of service), inserted in the last acknowledgment of the transaction;
- support of asynchronous transactions: the results of a transaction are sent back by the responder, as data become available;
- transaction abort: an outstanding transaction can be aborted by the WTP user due to a human input or to a protocol error.

Three service primitives are provided by WTP to the upper layer:

(1)  TR-Invoke, used to initiate a new transaction;
(2)  TR-Result, used to return a result of an active transaction;
(3)  TR-Abort, used to abort an existing transaction.

The following classes of transaction services are supplied by WTP to the upper WSP layer.

- Class 0 provides an unreliable datagram service with no result message. This transaction class can be used for unreliable push services. The 'initiator' (the client, i.e., in this case a content server) sends a request (i.e., a PDU invoke message) to the 'responder' (the user agent), which does not reply with an acknowledgment.
- Class 1 provides a reliable datagram service with no result message. It can be used for reliable push services. The initiator sends a request to the responder, which replies with an acknowledgment; the responder maintains the transaction state information for some time, in order to handle possible re-transmissions of the acknowledgment in case this is requested again by the initiator.
- Class 2 provides a reliable transaction service with one reliable result message. It supports several transactions running at the same time during one WSP session. The initiator sends a request to the responder, which implicitly acknowledges it with one result message. The result message is then acknowledged by the initiator, which also maintains the transaction state information for some time, in case the acknowledgment fails to arrive at the responder.

Referring to classes 1 and 2 above, WTP provides WSP with a reliable transaction service, though running on top of an unreliable datagram transport service; hence, it requires re-transmissions of lost packets and acknowledgments. Moreover, it supports selective retransmission mechanisms. A packet is re-transmitted when its re-transmission timer has expired without getting a response. The number of re-transmissions of the same packet is limited by a maximum value; when this maximum value is reached, the WTP user is informed that the transaction is being terminated. The mechanism of implicit acknowledgment is adopted in order to reduce the number of packets sent over the air; for example, a result message implicitly acknowledges a request sent by the initiator.

Multiple PDUs, even belonging to different transactions, can be concatenated in a single SDU of the datagram layer, provided that they have the same source and destination ports and the same source and destination device addresses. Separation is then needed at the destination node, where PDUs are extracted from one SDU, before being passed to transactions. The process of concatenation and separation allows having fewer wireless transmissions, thus improving efficiency.

An important field contained in all PDUs is the Transaction Identifier (TID). Its function is to identify messages generated by the same transaction. When a message is re-transmitted, its TID is reused in the re-transmitted PDU. TID is 16 bit long, but the high order bit is used to indicate direction; hence, the maximum number of outstanding transactions at any time is $2^{15}$.

Multiple transactions can be activated at the same time, even before the initiator receives a response to the first one. Transactions should be handled asynchronously, hence each result should be sent back as soon as it is ready, independently of other active transactions. However, the maximum number of Transaction Identifiers (TID) limits the number of outstanding transactions at any moment.

WTP operates with two types of messages: data messages, carrying user data, and control messages, carrying control information such as error reports and acknowledgments. A PDU is composed of the header, with a fixed and a variable part, and data. The different PDU types defined by WTP are [12]:

- Invoke PDU (4 bytes), which initializes the transaction;
- Result PDU (3 bytes), which returns a result message to the invoking entity (initiator); it is used in class 2 transaction services;
- Ack PDU (3 bytes), which carries an acknowledgment to an Invoke or Result PDU; it is used in transaction services of classes 1 and 2;
- Abort PDU (4 bytes), which aborts an existing transaction; the reason for aborting the transaction is included in the PDU;
- Segmented Invoke PDU (4 bytes); which is used for invoking Segmentation And Re-assembly (SAR) operations, when implemented, it is optionally used when the message length exceeds the Maximum Transfer Unit (MTU) for the current bearer;

- Segmented Result PDU (4 bytes), which carries a result message to a class 2 SAR message; it is optionally used;
- Negative Ack PDU (4 + N bytes), which indicates that $N$ packets of a sequence have been lost. It applies only in SAR transactions.

### 9.2.4 Wireless Transport Layer Security

Wireless Transport Layer Security (WTLS) is an optional layer that provides security services between the WAP client (mobile device) and the WAP gateway. It is based on the Internet standard technology, in particular, on a modified version of the Secure Sockets Layer (SSL) protocol, that is the Transport Layer Security (TLS) protocol [13]. TLS is the security protocol adopted between browsers and Web servers. WTLS is modular and its use depends on the security level required by the given application. In the WAP architecture it is located above the transport protocol layer (WDP/UDP). WTLS is more efficient than TLS, since it is designed to minimize the number of exchanged messages; it is a lightweight protocol that adapts to high latency, narrow bandwidth, and the limited memory and processing power of mobile devices.

WTLS can be used between the client and the gateway, while TLS is adopted between the gateway and the origin server. Protocol conversion between WTLS and TLS is performed inside the WAP gateway. This is the reason why the WAP gateway is a crucial node that should be carefully protected from attacks. No unencrypted information is stored in the gateway, since this would be a weak point defeating all the security measures. The translation is performed in the memory of the WAP gateway (see Figure 9.7).

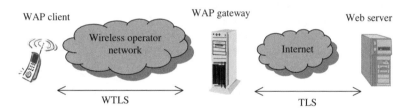

**Figure 9.7** WTLS-TLS translation.

WTLS provides the following main functionalities:

- privacy, through the adoption of encryption algorithms; data exchanged between client and gateway cannot be read by anyone else;
- data integrity, through the adoption of message authentication mechanisms; no one is allowed to alter the transmitted content without being noticed;
- authentication, the two communicating parts are authenticated through the use of digital certificates (asymmetric and symmetric algorithms).

WTLS has an internal architecture that is composed of two protocol layers, as detailed in Figure 9.8. The upper layer comprises the Handshake protocol, the Alert protocol and the Change Cipher Spec protocol; the lower layer is the Record protocol.

WTLS Record Protocol is a layered protocol that receives data from higher layers and encapsulates them into a WTLS PDU. In particular, the process follows the steps here described. First, it optionally compresses the received data, according to a loss-less algorithm. Then, it protects the WTLS compressed structure, translating it into a WTLS Ciphertext. This step is performed by applying a Message Authentication Code (MAC) that is calculated through a keyed-hashing HMAC process [14];

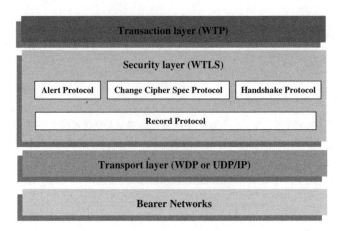

**Figure 9.8**   WTLS internal architecture.

the resulting entity (compressed data plus MAC) is encrypted with a symmetric ciphering algorithm, such as DES, triple DES, RC5 or IDEA. Finally, a WTLS header is added by the Record protocol (see [15] for details). At the receiving end, data is decrypted, verified and decompressed before being passed to higher levels.

Change Cipher Spec Protocol specifies, for the current transaction, the cryptographic algorithm and the hash algorithm used in HMAC. It is a simple protocol consisting of a single message (1 byte) of value 1. Either the client or the server can send it to notify the other peer that subsequent records will be ciphered under the newly negotiated ciphering algorithm and keys.

Alert Protocol signals alert messages; the alert level can assume one value among warning, critical and fatal. The reason for the alert message is signaled as well. In case of a fatal alert message, the secure connection is immediately terminated, while other connections on the secure session may continue; no new secure connections can be activated in the failed secure session.

Handshake Protocol is the most complex protocol of the WTLS internal architecture. It allows client and server to authenticate each other and to negotiate the secure attributes to be used for protecting data in the WTLS record. At the beginning of a communication, the client and the server agree on a protocol version and on a cryptographic algorithm, and they use public-key encryption techniques to generate a shared secret. They exchange a set of messages according to the following steps.

(1) A logical connection is established through the exchange of hello messages, in order to agree on the security parameters to be used.
(2) The server sends its certificate, if it has to be authenticated. A server key exchange message may be sent, if required, in order for some asymmetric algorithms to exchange the symmetric key. Then, the server may request a certificate from the client and finally it sends a hello done message. With this message (the only one that is mandatory in this phase) the server notifies the end of the initial phase and it waits for a client response. The messages here described are encapsulated in a lower layer message.
(3) Server certificate verification, client authentication (if requested) and key exchange take place in this step.
(4) A change cipher spec message is sent by the client, according to the change cipher spec protocol, and a finish message under the new ciphering attributes. The server responds with analogous change cipher spec and finished messages in order to complete the handshake procedure. From now on, client and server can exchange application layer data in a secure way.

### 9.2.5 Wireless Datagram Protocol

Wireless Datagram Protocol (WDP) is a protocol of the transport layer with some included functions of the network layer. WDP supports connectionless unreliable datagram service and bearer independence. WDP offers consistent services to the upper layer protocols of WAP and operates above the data capable bearer services supported by various air interfaces. Since WDP provides a common interface to upper-layer protocols and shields them from the bearer services provided by the network, security, session and application layers are able to operate independently of the underlying wireless network. At the mobile terminal, the WDP protocol consists of the common WDP elements plus an adaptation layer that is specific for the adopted air interface bearer. The WDP specification lists the bearers that are supported and the techniques used to allow WAP protocols to operate over each of them [16]. Figure 9.9 illustrates an example of WDP operating over a generic bearer (note that the bearer could be SMS, USSD, GPRS, etc.)

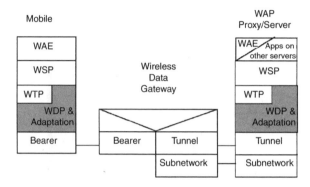

**Figure 9.9**  WDP architecture for a generic bearer.

The WDP protocol is based on UDP. UDP provides port-based addressing and Segmentation And Reassembly (SAR) in a connectionless datagram service. WDP must implement SAR functionality if it is not provided by the bearer service. The port number is related to the higher layer entity above WDP; this entity can be WTP or WSP or an application, such as e-mail.

When the IP protocol is available over the bearer service, the WDP datagram service offered for that bearer will be UDP and SAR service is provided by IP. WAP port numbers have been registered by the Internet Assigned Numbers Authority (IANA).

WDP is supported by Wireless Control Message Protocol (WCMP) for error handling functions [17], similarly to Internet Control Message Protocol (ICMP) adopted by IP.

## 9.3  WAP Languages and Design Tools

This section introduces the basics of WAP service creation, from languages and complementary technologies available to programmers, to the design principles of WAP services.

WAE provides a framework that includes a lightweight markup language (WML) and its scripting language (WMLScript), which are used to develop WAP applications. Using an analogy with the Web, WML plays the same role in WAP as HTML in the Web. Both languages are derived from XML.

The basic markup language in WAP 2.0, namely WML2 [18], is based on XHTML, as defined by W3C. By using the XHTML modularization approach, the WML2 language is very extensible, permitting additional language elements to be added as needed. WML2 is not intended for content authoring. The document type created according to the WML2 design guidelines consists of an XHTML core and WAP extensions. Classical WML contents can be delivered to WAP 2.0 clients in a transparent way for the user (backwards compatibility).

### 9.3.1 WML, WMLScript

Similarly to HTML, WML is a markup language that can be interpreted by specific software, the WAP micro-browser, which is optimized for lightweight use by limited capability devices. It specifies content and user interface, taking into account the constraints of small narrowband devices.

Since a mobile user cannot employ a QWERTY keyboard or a mouse, WML documents are structured into a set of well-defined units of user interactions called cards. Each card may contain instructions for gathering user input, information to be presented to the user, etc. A single collection of cards is called a deck, which is the unit of content transmission, identified by a Uniform Resource Locator (URL). After browsing a deck, the WAP-enabled phone displays the first card; then the user decides whether to proceed or not to the next card of the same deck (see Figure 9.10).

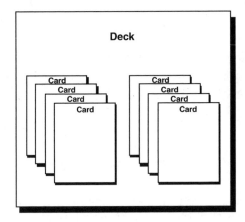

**Figure 9.10**   WML decks and cards.

WML decks can be stored in a static file in an origin server, or can be dynamically obtained by a content generator. WML content is scalable from a two-line text display on a basic device to a full graphic screen on the latest smart phones and PDAs. WML supports:

- text (bold, italics, underlined, line breaks, tables);
- black and white images (wireless bitmap format, WBMP);
- user input;
- variables;
- navigation and history stack;
- scripting (WMLScript) [19].

In particular, WML includes support both for managing the user agent state by means of variables and for tracking the history of the interaction. Moreover, WMLScripts are sent separated from decks and are used to enhance the client Man–Machine Interface (MMI) with sophisticated device and peripheral interactions. Figure 9.11 shows a WML document with various cards.

WMLScript is a procedural scripting language to be executed on the client device, derived from JavaScript and optimized for low-bandwidth communication and thin clients, like cellular phones and

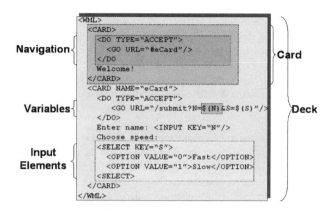

**Figure 9.11** WML document (deck) structured in cards.

pagers. It is part of WAE and it is used to complement WML, to overcome its limitations and to extend its capabilities by adding intelligence to the client and reducing the overall network traffic. WMLScript supports a set of locally installed standard libraries for string manipulation, mathematical operations, etc., supplying the programmer with the following capabilities:

- to check the validity of user input in order to prevent the transmission of invalid data to the server;
- to access the facilities and peripherals of the device, such as the phone book, SIM card, etc.;
- to enable dialogue locally with the user for displaying error and alert messages without involving the content server.

Unlike HTML, which can embed scripts, WMLScript is contained in separated files that are called from WML cards through specific links. WMLScript files are not delivered to the WAP client with WML files, but they are transferred separately, only when the client explicitly accesses a function contained in the WMLScript file.

### 9.3.2 Complementary Technologies

The Web standards specify many mechanisms to build a general-purpose application environment, such as the following.

- Standard naming model. All servers and contents on the Web are named with an Internet-standard URL [20].
- Content. All the contents on the Web have a specific MIME type, thus allowing Web browsers to process them correctly [21, 22].
- Standard content formats. All Web browsers support a set of standard content formats. These include HTML [23], scripting languages [24], and a large number of other formats.
- Standard networking protocols. Any Web browser is allowed to communicate with any Web server by means of these protocols. In particular, we can consider HTTP [25], operating on top of the TCP/IP protocol suite.

The WAP programming model is similar to the Web programming one. This fact provides several benefits to the application developer community, including a proven architecture and the ability to leverage existing tools (e.g., Web servers, XML tools, etc.). Optimizations and extensions have been made in order to match the characteristics of the wireless environment.

There are several WAP toolkits available for software developers to be used for WAP-based services, such as those of Ericsson, Nokia, Phone.com, etc. [26]. WAP allows customers to reply easily to incoming information on the phone by adopting new menus to access mobile services.

Existing mobile operators have added WAP support to their offerings, either by developing their own WAP interface or, more usually, partnering with one of the WAP gateway suppliers. WAP has also given new opportunities to allow the mobile distribution of existing information contents. For example, CNN and Nokia teamed up to offer CNN Mobile. Moreover, Reuters and Ericsson teamed up to provide Reuters Wireless Services.

One of the most important issues to be considered when designing WAP services is how to provide contents in a usable form to multiple disparate clients. XML and its associated technologies provide a solution to achieve such a goal by supporting different types of devices. Actually, XML defines a technology to create documents and to store information in a structured way and it allows one to access information using a variety of techniques in a platform-independent way. Unlike HTML, XML does not describe how to display data; it only defines how data are structured and organized. XML contains data and metadata, i.e., content description. It is designed to be readable by both operators and machines. WML is an XML language characterized by a Document Type Definition (DTD), where the DTD gives further indication of the XML document structure. The DTD contains the description of the possible tags that can be inside a document. A valid WML document follows the rules that are defined by the DTDs at the Web site with URL http://www.wapforum.org/DTD.

XML Stylesheet Language (XSL) specifies a mechanism to transform the structure or the contents of XML documents into other types of documents, such as WML to be delivered to a specific mobile device. In this way, an XML document can be the basic repository of structured contents, which can be extracted and transformed, through appropriate XSL stylesheets, in almost any output format, so as to adapt the same content to different models and types of mobile devices (e.g., PDAs and mobile phones with different characteristics in terms of display size, number of lines, supported colors, etc.).

As far as traditional Web browsers are concerned, XHTML defines an XML-compliant document format and it makes it easier to identify the specific features supported by a client. XHTML is indeed the reformulation of HTML into an XML-compliant format, allowing a traditional Web browser to process Web pages in the same way as XML documents.

Finally, the WAP protocol can be used for accessing value-added services by means of a built-in menu and a set of tools programmed on the SIM card by the operator. This is the SIM Application Toolkit (SAT) approach: a European Telecommunications Standard Institute (ETSI) standard for programming a special GSM SIM card to drive the GSM handset to access a service with an interactive exchange with the network [27]. SAT provides mechanisms that allow applications stored in the SIM to interact and to operate with the mobile equipment. SAT also provides a transport mechanism to download and to update applications. In the new 32K or 64K SIM cards, it is possible to have the capability to store different types of personal information such as social security ID, driver licence ID, and bank account numbers. Whenever an interactive service needs this sort of information, the customer can supply it by simply pressing few buttons on the handset.

Note that many languages for developing applications are based on XML, such as Simple Object Access Protocol (SOAP) and Synchronized Multimedia Integration Language (SMIL). In particular, SOAP is a lightweight protocol for the exchange of information in decentralized, distributed environments. It can be used to access service providers' databases. Finally, SMIL is employed to organize MMS presentations.

### 9.3.3 Conversion of Existing Webpages to WAP

The reason to convert existing HTML-based Websites to WAP relies on the impressive growth of mobile devices capable of accessing the Internet, even if, over the long term, optimal design methods for multi-client support will prevail, such as conversion of XML to WML/XHTML formats. Converters work on the fly, according to the following steps:

(1)  extraction of content from an HTML source page;
(2)  Reformatting of content and generation of the target markup language (WML).

The extraction process can be performed according to two different methods:

(1)  The fully automated approach, which consists of the extraction of all possible contents in a page, such as text, title, links and so on. Fully automated converters are dumb, in the sense that they apply the same built-in conversion rules to every page, without any attempt to customize individual pages.
(2)  The Configurable approach, which consists of the extraction of only specific parts of the page; in this case the developer should indicate which parts are to be converted and presented to the client.

Conversion can be also executed by modifying HTML documents. Appropriate 'helper' tags can be added in the HTML code in order to select the areas of interest for a converter when extracting data.

A final note to conversion of JavaScript. While the conversion of HTML to WML is viable, the automatic conversion of JavaScript to WMLScript is not possible. Unlike WMLScript, JavaScript is usually embedded within the HTML markup, therefore calls to JavaScript functions must be replaced by calls to prewritten WMLScript.

### 9.3.4 Dynamic Content Adaptation for WAP Pages Delivery

Besides the above conversion approach, this subsection deals with dynamic content generation and adaptation. In particular, we consider the provision of WAP pages starting from an original information, dynamically generated in XML format.

Content adaptations are related to information filtering (keep only relevant contents) and content variants (e.g., with or without images; images of different sizes and resolutions). Adaptation implies formatting the contents according to markup languages suitable for the mobile terminals, as described below.

Mobile phones have different characteristics in terms of the number of lines of the display. It therefore becomes necessary to adapt the presentation of a Webpage depending on the mobile terminal display capabilities. A dynamic adaptation process can be realized through a middleware [28] in the Web server (or content provider server), an intermediate layer based on XML technology. The header of the HTTP request received by a server from a WAP user can be used to extract important information for the automatic detection of the access device characteristics, such as:

- type of device
  personal computer, palmtop, mobile phone, etc.
- presentation / mark-up language
  (X)HTML (PC, Kiosk, Internet Point, WAP 2.0 mobile terminal), WML (WAP 1.X mobile terminal or PDA), SMS (mobile device), MMS (mobile device). See also Table 9.1;
- other important parameters for presentation
  number of colors, support for graphics, actual screen size.

**Table 9.1**  Languages for client devices

| Client | Markup language |
| --- | --- |
| PC-based browser | HTML, DHTML, XHTML |
| PDA | WML, XHTML |
| Mobile phone | WML, XHTML, VoiceXML |
| External server | XML languages |

The HTTP header fields used for this purpose are:

- HTTP_ACCEPT, containing the list of the MIME types accepted by the client;
- HTTP_USER_AGENT, which lists the most significant product information about the client (e.g., product/version).

Interesting WAP applications can be obtained with dynamic WAP pages by means of the following different technological (middleware) options:

- Microsoft ASP;
- Java and Servlets or Java Server Pages (JSPs) for generating an output dynamic page in XML format;
- the eXtensible Stylesheet Language Transformations[b] (XSLT) for generating WAP pages adapted for displays of different characteristics and sizes.

Let us consider a user requesting some 'personalized' information through a WAP service. The request reaches a server in the Internet that, after a query in distributed databases, can dynamically create an output information organized in the general XML format. Such information has to be converted into one of the formats in Table 9.1, depending on the client device characteristics. In particular, it is interesting to examine in detail how an XML document can be converted into a WML document. XSL includes an XML vocabulary for specifying formatting. XSLT describes how an XML document can be transformed into another XML document that uses a suitable formatting vocabulary.

A transformation expressed in XSLT describes rules for transforming a source tree into a result tree (a WML text or an HTML text, as an XML text, is a tree-structured document). The transformation is achieved by associating patterns with templates. A pattern is matched against elements in the source tree. A template is instantiated to create part of the result tree. The result tree is separate from the source tree. The structure of the result tree can be completely different from the structure of the source tree. In constructing the result tree, elements from the source tree can be filtered and reordered, and an arbitrary structure can be added.

A stylesheet contains a set of template rules; each device has a stylesheet in XSL. A template rule has two parts: a pattern, which is matched against nodes in the source tree, and a template that can be instantiated to form part of the result tree. This allows a stylesheet to be applicable to a wide class of documents that have similar source tree structures.

## 9.4 WAP Service Design Principles

Over the last few years we have seen the proliferation of powerful network applications with advanced graphical user interfaces, mainly targeted to Web users in front of desktop computers. The mobile Internet poses new problems related to porting these applications to different devices with limited capabilities and to users with different needs and requirements. Human Computer Interaction (HCI) deals with usability issues, where usability indicates the degree of user-friendliness of a system. A usable application should allow users to complete tasks in an easy way. Therefore, service developers should carefully take into account some important issues that are here introduced.

First, WAP devices have strong limitations in terms of display size and graphics, input capabilities, network bandwidth and processing power. In particular, the small screen of a typical WAP device does

---

[b] XSLT is a W3C language and the most important part of the XSL standards. It is the part of XSL that is used to transform an XML document into another XML document, or another type of document that is recognized by a browser, like HTML and XHTML.

not allow one to render fancy graphics or to show pages full of objects (text, images, etc.) in a clear way; consequently, user interfaces should be designed to be as small as possible in order to be displayed without unpredictable modifications on small terminals. Applications designed for big screens display badly on a small screen.

Secondly, since input techniques are mainly based on the small keypad of a handy phone, data entry should be minimized as much as possible and the input mode of the terminal should be compatible with the expected format for the data inserted by the user. Moreover, bandwidth limitations and the high costs of mobile services impose a limit on data transmissions in the radio channel, suggesting that WML decks be created with limited size, in order to realize effective services at an acceptable cost for the user.

Finally, WML pages are delivered to multiple disparate mobile devices, such as PDAs and phones, which can display the same page in very different ways. This makes the situation difficult and using multiple versions of an application, for example through XML and XSL transformations, can be a good solution.

The terminal is not the only element to be considered when designing WAP services; mobile users themselves are very different from Web users. A Web user can comfortably sit in front of his desktop PC, fully network-connected and fully equipped with a printer and a storage device to save navigation results. WAP users are usually on the move, at a congress, on a train, in their cars, and they need information quickly. Mobile subscribers are expected to behave in a more impulsive way than Web users. Mobile users are unlikely to 'surf the Web' as they might on a PC, because they need specific information quickly at their fingertips.

As Internet services have not been developed for mobile devices, simply shrinking a Web application to a WAP device can be really frustrating for a WAP user.

Taking into account these remarks, WAP developers should follow some general guidelines in order to realize efficient, effective and usable applications for WAP devices. Users are not programmers, hence the user interface should be intuitive and easy to learn without having to read a manual before use. A WAP user can also be unfamiliar with PCs and with the usage of traditional browsers.

Efficiency is another target to be pursued, because many users are impatient and they want an application to work as quickly as possible. Economy in using the communication channel can be achieved by minimizing interactions to complete a task.

Considering that users usually explore a small fraction of an entire application, sometimes they perform unusual operations. In such a case, it should be easy for them to repeat unusual operations with a small memory effort. Memorability describes this characteristic to be considered by developers in order to improve customer satisfaction.

Finally, a user should be 'forgiven' when entering a destructive action (e.g., canceling an ongoing transaction), thus, tolerance should be implemented, for example by asking the user for confirmation.

The features mentioned above can be achieved if the service design phase takes into account the following general rules.

- User activities should be clearly identified. When a Web application is ported to WAP, many functionalities should be discarded and only the main activities should be accessible for a mobile user, but in the fastest possible way. For example, a tourist could be interested in knowing which are the nearest restaurants, rather than getting a list of all the restaurants in the city, which could be a tedious and expensive operation with long waiting times for a mobile phone user.
- The most common activities should be performed through the shortest interaction path. This means that the application should be designed in a hierarchical way, where layers are sorted according to the usage probability of each activity. Hierarchy naturally supports backwards navigation, which is another mechanism to be implemented to allow the user to go back with just a click.
- Data entry should be kept to the minimum, because of the uncomfortable input mechanisms that are usually available in most WAP devices. When input is necessary, forms with select elements could be useful to avoid text entry and typing errors (user input validation can also be performed client-side through WMLScript).

- Usability can be improved with personalization. User data and preferences can be maintained by the service in order to adapt itself to the customer and to avoid repetitive inputs that could annoy the user. A possible mechanism for implementing personalization can be user registration and log-in to the service.
- Text should be short and clear and the possible actions a user can perform should be clearly identifiable.
- Consistency is very important: similar actions in the application should be performed consistently.
- Users should always be enabled to change their mind when exploring the application. This can be done by implementing a backward navigation functionality, which helps users to perceive that they are always able to reverse their actions.
- Push functionality can be very attracting for real-time services (weather forecasts, traffic info, etc.), especially when it is coupled with location-based services (note that push mechanism was not supported by the first WAP versions).

A final rule, but a very important one, is to test the application with typical users, in order to find out problems and to correct them according to an iterative design process.

The assessment of user-friendliness of an application can only be achieved if usability is carefully considered since the early stages of design. The design phase is crucial and a bad design that leaves out usability issues can compromise the accomplishment of the whole application.

## 9.5  Performance of WAP over 2G and 2.5G Technologies

The success of present (2G and 2.5G) and future (3G) mobile networks depends on the provision of attractive services for users. It is therefore important to plan network resources on the basis of appropriate models related to the traffic produced by an application. We focus our interests here on the model for the downlink traffic (i.e., from the network to the mobile user) generated by WAP browsing activities.

### 9.5.1  WAP Traffic Modeling Issues and Performance Evaluation

Some preliminary discussions on WAP traffic can be found in [29], where a comparison with the WWW traffic is shown. Here, we consider a downlink model, derived from [30]. In particular, we adopt a similar on–off model, where each source generating traffic alternates between an activity phase (when the user browses a deck) and an idle phase (when the user has a browsing pause), as shown in Figure 9.12.

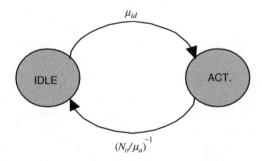

**Figure 9.12**    WAP browsing downlink traffic model related to a user.

The following description is valid for any version of WAP, even if the parameter values may also depend on the bearer used to convey WAP traffic. In this study we will refer to the GPRS packet-switched service of GSM as the bearer for WAP traffic.

Let us provide numerical details on the on–off downlink WAP traffic model corresponding to each user, shown in Figure 9.12 [31–33]. The idle phase length is assumed exponentially distributed with mean $1/\mu_{id} = 30$ sec. During an activity phase, the user browses a number of decks, which are assumed geometrically distributed with mean $N_d$ (typically, we will use $N_d = 2$, 3 or 4 decks, for a short browsing duration). Each requested deck is a datagram to be transmitted to the mobile user. The deck interarrival time in the activity phase is exponentially distributed with mean $1/\mu_a = 10$ sec. Hence, the activity factor for this source is $\Psi_w = (N_d/\mu_a)/(N_d/\mu_a + 1/\mu_{id})$ and the mean deck arrival rate is $\Psi_w/\mu_a$ decks/s. Note that $1/\Psi_w$ represents the burstiness degree of the downlink traffic related to a WAP user.

We have characterized the card length distribution in bytes by measuring the length of more than 1000 cards on different WAP servers. The resulting histogram is shown in Figure 9.13. The mean card

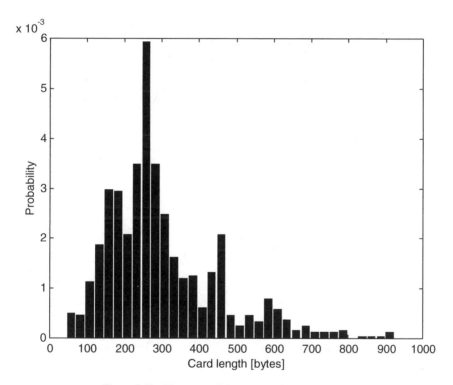

**Figure 9.13** Histogram of the number of bytes per card.

length $L_c$ was found to be about 300 bytes. The number of cards per deck $N_c$ is an important design parameter for WAP services. Of course, the distribution of the number of cards per deck depends on both the choice made by the designer of WAP pages and the service type. Considering many WAP sites in the Internet (see for example those in [32]), we have obtained the histogram of the number of cards per deck in Figure 9.14; in this graph we have also shown the fitting with the geometric distribution that has the same mean value (this distribution shows that the cases with a few cards per deck are more common). The results in Figure 9.14 show that there are on average $N_c = 2.5$ cards/deck and that the geometrical distribution allows an acceptable fitting. Hence, in what follows, we have assumed that the number of

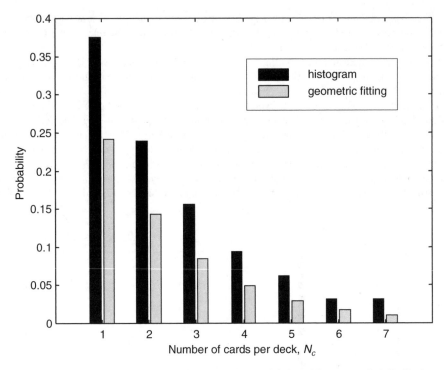

**Figure 9.14**   Histogram of the number of cards per deck and fitting with a geometrical distribution.

cards per deck is geometrically distributed with mean $N_c$ and possible values equal to 2, 3 and 4 (i.e., few cards per deck).

Finally, in our traffic model we have neglected the backlog effect due to the acknowledged nature of Web (and hence WAP) traffic. Therefore, new decks are transmitted during the activity phase without considering any special acknowledgment mechanism. Such assumption permits one to increase the traffic load and to evaluate the system performance under conservative conditions.

For the deck length, we have considered the distribution obtained by composing the card length distribution with the distribution of the number of cards per deck; moreover, we have assumed that the encoding process entails a deck length reduction of 80% [9]. The resulting encoded deck (here considered as a datagram) length $X$ has been modeled by a truncated Pareto (heavy-tailed) distribution that is well suited to account for the occurrence of exceptionally long decks. In particular, $X$ is a random variable with the following probability density function (pdf):

$$\text{pdf}(x) = \frac{\gamma k^\gamma}{x^{\gamma+1}} [u(x-k) - u(x-h)] + \omega^\gamma \delta(x-h) \qquad (9.1)$$

where $u(.)$ is the unitary step function, $\omega = (k/h)^\gamma$, $\delta(.)$ is the Dirac delta function.

Variable $X$ ranges from $k$ to $h$ in bytes. Note that parameter $k$ is related to a minimum content that is present in all the decks. We have considered that reasonable values are $k = 50$ bytes and $h = 5000$ bytes [31–33].

The $\gamma$ value in Equation (9.1) has been obtained by fitting the expected value of $X$, $E[X]$, with the expected deck length (i.e., $0.2N_cL_c$, where the factor 0.2 accounts for the deck length reduction due to the encoding process). According to [34], we have:

$$E[X] = \frac{\gamma k - h(k/h)^\gamma}{\gamma - 1}.$$ 
(9.2)

Hence, we impose the following fitting condition:

$$\frac{\gamma k - h(k/h)^\gamma}{\gamma - 1} = 0.2N_cL_c.$$ 
(9.3)

The obtained formula (9.3) is a transcendent equation in $\gamma$ (with parameters $N_c$, $L_c$, $k$, $h$) that has been numerically solved by means of the recursive Gauss–Newton method. We have obtained $\gamma = 1.68$, $1.27$, $1.08$, respectively for $N_c = 2, 3, 4$ cards/deck.

A Pareto distributed random variable, Z, with parameters $k$ and $\gamma$ can be obtained by means of the following transformation:

$$Z = \frac{k}{\sqrt[\gamma]{U}},$$ 
(9.4)

where $U$ is a random variable uniformly distributed from 0 to 1.

Finally, the encoded deck length in bytes, $X$, is obtained by means of truncation: $X = \min(Z, h)$; this length must be rounded to obtain an integer value.

Finally, the compiled deck must be packetized according to the layer 2 format. Let $E[L_w]$ denote the mean length of a datagram in packets. Let $T_s$ denote the packet transmission time. The traffic intensity contributed by the WAP traffic source in Figure 9.12 is

$$\rho = (\Psi_w/\mu_a)E[L_w]T_s.$$ 
(9.5)

It is easy to show that the above traffic source for a WAP browsing user produces a Markov Modulated Poisson arrival Process (MMPP) of decks [35].

Let us use the above WAP traffic model to study the performance of WAP browsing over the packet-switched GPRS service; we refer here to the case with $N_d = 4$ decks per activity phase and $\gamma = 1.08$, corresponding to $N_c = 4$ cards/deck.

In the GPRS network, mobile users access the Internet through the Gateway GPRS Support Node (GGSN). The physical GSM channels (i.e., slots) destined to GPRS are denoted as Packet Data CHannel (PDCH) and have a 52-frame period organization. The basic transmission unit on a PDCH is called a radio block: a slot in four consecutive frames is utilized to transmit a radio block. Every 13 frames, the slot of the PDCH is not used to transmit data, so that there are 12 radio blocks per multiframe. A radio block contains 456 bits, but the number of conveyed information bits depends on the coding scheme. GPRS supports four distinct coding schemes that are tailored to different channel conditions:

- CS-1 uses half-rate convolutional coding; it is the most robust scheme, used when there are bad channel conditions. CS-1 supports a data rate of 9.05 kbit/s per time slot in the frame.
- CS-2 allows a data rate of 13.4 kbit/s, but it is less robust than CS-1.
- CS-3 is a less reliable coding scheme than CS-2, but it supports data rates of 15.6 kbit/sec. CS-2 and CS-3 coding schemes are punctured versions of the CS-1 code.
- CS-4 uses no error correction; therefore, it can be only employed under good channel conditions. CS-4 supports data rates of 21.4 kbit/s per time slot.

The most typical coding schemes are CS-1 and CS-2. We refer here to the CS-1 case with a traffic capacity of 9.05 kbit/s per PDCH. Correspondingly, a block conveys about 22 information bytes.

Up to eight slots can be assigned to a multi-slot mobile terminal, but practical implementation aspects reduce this theoretical limit to four slots per frame. A Temporary Block Flow (TBF) is a physical temporary connection used to support unidirectional transfer on the air interface. In GPRS-Release'99, OSI layer 2 protocols at the base station are responsible for radio block allocation (on one or more PDCHs) on a TBF basis according to Quality of Service (QoS) profiles negotiated with the Serving GPRS Support Node (SGSN) through the activation of Packet Data Protocol (PDP) contexts.

The results shown in Figure 9.15 have been obtained according to the analytical approach validated in [35]. These results present the mean delay time to transmit an encoded WAP deck on the GPRS air

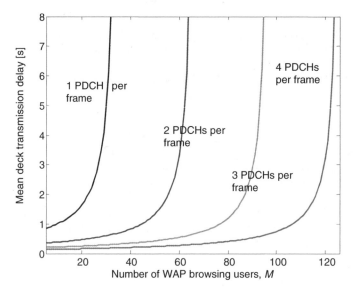

**Figure 9.15**    Mean deck transmission delay as a function of the number of WAP users that share a GPRS carrier with CS1 transmissions.

interface considering a single GPRS carrier, a variable number of users and a different number of slots (PDCHs) available for downlink transmissions (i.e., from 1 to 4). If, for instance, we consider that a mean deck transmission delay of 2 sec is acceptable, the obtained results highlight that about 20, 55, 85 and 105 users can be managed per carrier respectively for the cases with 1 PDCH, 2 PDCHs, 3 PDCHs and 4 PDCHs per frame used by WAP traffic over GPRS. This high capacity of users is achieved as a result of the selected requirement for the mean deck delay, the light traffic load produced by a WAP source, the multiplexing effect of different WAP sources and the availability of more GPRS resources per frame.

Referring to Figure 9.15, we note that the mean deck delays are on the order of seconds (and sometimes greater than 10 sec), so we may expect a backlog effect on the WAP traffic generator. As already explained, such an effect has been neglected for a conservative performance evaluation.

## 9.6  Examples of Experimented and Implemented WAP Services

Many European projects have investigated and experimented with the adoption of WAP in order to provide services to mobile users for different purposes (e.g., mobile commerce, tourists, citizens, businessmen, brokers, etc.). Below here we give a short survey of WAP-based information services that have been envisaged in some EU projects, both in the 'tourism area' and in the 'systems and services for the citizen area' [36].

### Tourism Area

CReation of User-friendly Mobile services PErsonalized for Tourism (CRUMPET), to allow for:

- personalizing services to the user's current location, personal interests and historical interaction with the system;
- adapting the representation of contents to changing technical environments;
- exploiting distributed tourism-related services and information.

Guide by Telematics to Enable Tourist Freedom at Sites (GUIDEFREE), to allow for

- integrating relevant tourist information from different Web sources;
- transforming the information into information that the tourist can easily understand;
- guiding tourists/travelers in using the new information;
- providing information that satisfies the personal preferences of the user;
- providing access both from home/work as well as on the move.

Personalized Tourist Services Using Geographic Information Systems via Internet (TOURSERV), to allow for

- providing online access to geographical and regional information (3-D maps of mountaineering, skiing and hiking regions given from a underlying GIS component) together with value-added services (i.e., bookings, etc.);
- visualizing information via a GIS-based user interface;
- personalizing the services and information provided;
- supporting all phases of a vacation (before, during, and after).

### Systems and Services for the Citizen Area

TELLMARIS, to provide for

- giving the user better tools for locating specific information;
- helping the user to orient his/her position in a geographical space;
- provide tourist information via a 2D or 3D map interface;
- develop new data structures for representing interactive 3D maps on mobile terminals.

The information must be available anywhere and any time, optimized for the available mobile device.

From the above list, we can conclude that a great deal of research activity has been carried out for WAP-based services to mobile users. We are now interested in providing more details on the WAP-based mobile information services developed in the Personalised Access to Local Information and services for tOurists (PALIO) European project [28] on the basis of a GSM and GPRS mobile communication system.

The PALIO project has designed an open communication and information architecture to support tourists in efficient navigation and interaction in a physical space. The PALIO information system is designed to provide information to tourists on the move.

Services are personalized and adapted at various levels according to:

- user preferences;
- context of use;
- access modalities (i.e., device type and transmission technology).

Adaptation and personalization are made possible by analyzing the following factors.

- User specific needs. The ability of the interface to manage personalization with options determined by the user profile and by their modality of service utilization.
- Device characteristics. The ability of the interface to provide an adequate utilization of contents through different enabling technologies (i.e., devices with different display size, processing power, network connectivity).

Adaptation to the device type is performed as described in Section 9.3.4.

Service personalization is achieved on the basis of a profile that the users define the first time that they accessed the service (registration phase); moreover, personalization rules are also dynamically updated to take into account the context of use and the history of interaction with the system. A context in the PALIO project comprises all 'circumstances' that may affect the interaction of the user, for instance the user geographical location, characteristics of the access device, characteristics of the network connection, date and time. The geographical localization is realized either through a GPS system (in an open area, with high precision) or by GSM (in urban areas with lower precision, related to cell size). The transition between the two methods is transparent to the user.

The WAP protocol allows mobile users to access contents stored in a suitable PALIO server at the regional level. The main WAP-based services that were used in the PALIO project were:

- real-time information on public transportation, traffic situation, road works and car parking;
- location-dependent information on museum, hotels, exhibitions, concerts, including navigation suggestions on the basis of user profiles.

Usually a tourist visits an urban area for just two or three days; the main appeal of a city being its culture and heritage. For example a tourist may want to visit Siena (Italy) and to go to a concert of 'Accademia Chigiana'. Therefore, we can conjecture that tourists arrive in Siena and go to hotels, where they ask for a city guide. The tourists find that it is possible to use their personal mobile phones as a city guide by means of the WAP technology. Now the tourists can easily look up a museum, a historical building, a restaurant or a concert in a very simple way.

Initially, the mobile phone provides a screen with a list of the principal services, i.e., a menu containing parking, restaurants, pubs, museums, monuments, churches, etc. When a tourist has choosen a service it is possible to specify how to do the search: the interface can return a complete list or a reduced list, referring only to an area near the tourist. For example, a user in 'Piazza del Campo' in Siena selects the 'monument service'. As a first level, the system returns the monuments closest to the user. Obviously the tourist can modify the first level and can choose to view the full list of monuments in Siena.

Every element of the service list can be selected in order to obtain detailed information, such as:

- location (address, telephone, etc.);
- local route;
- public transport (if the place is not reachable by foot), such as the nearest bus-stop;
- parking: address and route, free or not (some tourists use rental cars and parking in an unknown city can be a major problem);
- ticket office;
- info-services.

These characteristics are valid for every service offered by the PALIO project. If the user then requires a restaurant, the system will provide the previous information and additional information (e.g., type of cuisine for restaurants) for each element.

**Figure 9.16**   Interface for PDA for PALIO tourist services.

We now show some examples of the user interface on two devices: a PDA and WAP-enabled mobile phone. Figure 9.16 shows the service interface as it appears on a PDA. The characteristics of this configuration are:

- choice of presentation language: English or Italian;
- front-end tailored for 'small-screen', color- and graphics- capable terminals;
- recommendations (on tourist locations, hotels, restaurants, etc.) are in accordance with user preferences;
- recommendations are made at the city level or related to the user location (the accuracy of the location depends on whether GPS is used or not).

Finally, Figure 9.17 shows the WAP-based service interface as it appears on a mobile phone. The characteristics of this configuration are:

- choice of presentation language: English or Italian;
- front-end tailored for 'tiny-screen' terminals with no assumptions about color and graphics;
- because of the reduced size of the mobile phone display, only basic information is shown according to a filtering action operated by the middleware at the server level (see Section 9.3.4).
- recommendations are related to user neighborhoods.

**Figure 9.17**   Interface for a mobile phone for PALIO WAP-based tourist services.

## References

[1]   Open Mobile Alliance Web site: http://www.openmobilealliance.org/tech/affiliates/wap/wapindex.html

[2]   M. Mzyce, Wireless Application Protocol (WAP), *IEEE VTS News*, **48**(2), 7–12, May 2001.

[3]   World Wide Web Consortium, Extensible Markup Language (XML) 1.0, http://www.w3.org/TR/REC-xml/

[4]   Wireless Application Protocol Forum, Ltd., Architecture Specification', WAP-210-WAPArch-20010712.

[5]   World Wide Web Consortium, XHTML 1.1 - Module Based XHTML, http://www.w3.org/TR/xhtml11/

[6]   Wireless Application Protocol Forum, Ltd., WAP Public Key Infrastructure Definition, WAP-217-WPKI, April 24, 2001.

[7]   Wireless Application Protocol Forum, Ltd., User Agent Profile Specification, WAG UAPROF, November 10, 1999.

[8]   Wireless Application Protocol Forum, Ltd, WAP Provisioning Architecture Overview Specification, WAP-182-PROVARCH, 2001.

[9]   Wireless Application Protocol Forum, Ltd, Binary XML Content Format Specification, WAP-154 Version 1.2, November 4, 1999.

[10]  Wireless Application Protocol Forum, Ltd, Wireless Session Protocol Specification, WAP-230-WSP, July 5, 2001.

[11] Information Technology – Open Systems Interconnection – Basic Reference Model – Conventions for the Definition of OSI Services, ISO/IEC 10731, 1994.

[12] Wireless Application Protocol Forum, Ltd, Wireless Transaction Protocol, WAP-224-WTP, July 10, 2001.

[13] W. Stallings, *Cryptography and Network Security: Principle and Practice*, 2nd edn, Upper Saddle River, NJ, Prentice Hall, 1999.

[14] H. Krawczyk, M. Bellare and R. Canetti, HMAC: Keyed-Hashing for Message Authentication, IETF RFC 2104, February 1997, ftp://ftp.isi.edu/in-notes/rfc2104.txt.

[15] Wireless Application Protocol Forum, Ltd, Wireless Transport Layer Security, WAP-261, April 6, 2001.

[16] Wireless Application Protocol Forum, Ltd, Wireless Datagram Protocol, WAP-259, June 14, 2001.

[17] Wireless Application Protocol Forum, Ltd, Wireless Control Message Protocol, WAP-202, June 24, 2001.

[18] Wireless Application Protocol Forum, Ltd, Wireless Markup Language, Version 2.0, WAP-238, September 11, 2001.

[19] Wireless Application Protocol Forum, Ltd, WMLScript Specification, WAP-193, October 25, 2000.

[20] T. Berners-Lee, R. Fielding and L. Masinter, Uniform Resource Identifiers (URI): Generic Syntax, IETF RFC 2396, August 1998, http://www.rfc-editor.org/rfc/rfc2396.txt

[21] N. Freed and N. Borenstein, Multipurpose Internet Mail Extensions (MIME) Part One: Format of Internet Message Bodies, IETF RFC 2045, November 1996, http://www.rfc -editor.org/rfc/rfc2045.txt

[22] N. Freed, J. Kiensin and J. Postel, Multipurpose Internet Mail Extensions (MIME) Part Four: Registration Procedures, IETF RFC 2048, November 1996. URL: http://www.rfc-editor.org/rfc/rfc 2048.txt

[23] D. Raggett, A. Le hars and I. Jacobs, HTML 4.0 Specification, W3C Recommendation 18 December 1997, REC HTML40-971218, September 17, 1997, http://www.w3.org/TR/REC-html40

[24] David Flanagan, *JavaScript: The Definitive Guide*, OReilly & Associates, Inc., 1997

[25] R. Fielding, J. Gettys, J. Mogul, H. Frystyk, L. Masinter, P. Leach and T. Berners-Lee Hypertext Transfer Protocol – HTTP/1.1, June 1999, http://www.rfc-editor.org/rfc/rfc2616.txt

[26] WAP browsers: Nokia from http://www.nokia.com; Ericsson from http://www.ericsson.se/WAP; Phone.com from http://updev.phone.com; WinWAP from http://www.slobtrot.com; Motorola from http://www.motorola.com; Gelon.net from http://www.gelon.net; WAPman from http://palmsoftware.tucows.com

[27] Digital cellular telecommunications system (Phase 2+); Specification of the SIM Application Toolkit for the Subscriber Identity Module – Mobile Equipment (SIM - ME) interface (GSM 11.14), GSM 11.14, Version 5.2.0, December 1996.

[28] PALIO Project Web page with URL: http://www.palio.dii.unisi.it

[29] T. Kunz, T. Barry, J. P. Black and H. M. Mahoney, WAP Traffic: Description and Comparison to WWW traffic, *Proc. of the Third ACM International Workshop on Modeling, Analysis and Simulation of Wireless and Mobile Systems*, Boston, USA, pp. 11–19, August 2000.

[30] ETSI Selection Procedures for the Choice of Radio Transmission Technologies of the UMTS, UMTS 30.03 Version 3.1.0, ETSI, Sophia-Antipolis, Cedex, France, November 1997.

[31] A. Andreadis, G. Benelli, G. Giambene and B. Marzucchi, Performance analysis of the WAP protocol over GSM-SMS, *Proc. of IEEE ICC2001*, June 2001.

[32] A. Andreadis, G. Benelli, G. Giambene and B. Marzucchi, Analysis of the WAP protocol over SMS in GSM networks, *Journal on Wireless Communications and Mobile Computing*, John Wiley & Sons, **1**, 381–395, 2001.

[33] A. Andreadis, G. Benelli, G. Giambene and F. Petiti, Analysis of Downlink Scheduling for Web Traffics in 2G and 2.5G Mobile Networks, *Proc. of PIMRC 2002*, pp. 794–798, Lisbon, Portugal, September 2002.

[34] E. Brand and A. H. Aghvami, Multidimensional PRMA with Prioritized Bayesian Broadcast – A MAC Strategy for Multiservice Traffic over UMTS, *IEEE Trans. Veh. Tech.*, **47**(4), 1148–1161, November 1998.

[35] A. Andreadis, G. Benelli, G. Giambene and B. Marzucchi, A performance evaluation approach for GSM-based information services, *IEEE Trans. on Veh. Tech.*, **52**(2), 313–325, March 2003.

[36] WAP-related EU projects with URLs: http://www.ist-crumpet.org/, http://www.guidefree.gr/, http://www.tourserv.com/, http://www.tellmaris.com/

# 10

# Multimedia Messaging Service (MMS)

Alessandro Andreadis and Giovanni Giambene

## 10.1 Evolution from Short to Multimedia Message Services

Owing to the enormous success of SMS, new messaging types have been defined in order to enrich the exchanged contents. This section focuses on the evolution from SMS to Enhanced Messaging Service (EMS) to Multimedia Messaging Service (MMS).

The GSM Association on December 2001 reported that over 30 000 000 000 SMS were sent, with an exponential increase in time. Therefore, it was evident that the interest for messaging was high, and the principal operators worked on the possibility of sending messages with multimedia content (i.e., images, videos, etc.) using 2.5G and 3G mobile communication networks.

The EMS service is the first evolution of SMS and, like SMS, permits one to send easily several types of simple media content, like melodies, animations and icons (i.e., animated GIFs), but with a maximum size of about 640 characters, less than 1 kilobyte [1]. The delivery of the EMS is through the SMS path, without modifications in the GSM network; however, a greater memory capacity is needed on mobile terminals for storing images and melodies. EMS was commercially deployed in 2001, but it has not reached the popularity of SMS. Hence, it can be seen as an intermediate step from SMS to MMS.

MMS is an innovative messaging system, standardized by both the WAP Forum (now Open Mobile Alliance, OMA) and the Third-Generation Partnership Project (3GPP). It is similar to SMS and EMS, but allows one to insert multimedia contents in a message such as music, pictures, video clips, texts, etc. [2]. MMS is independent of the underlying mobile network (e.g., GSM, WCDMA, etc.) and allows content delivery to either mobile phones or e-mail addresses. On the basis of these features, it is estimated that this messaging service will increase telecommunication profits, addressing both professionals and personal user markets. The delivery of MMS to mobile phones through the air interface is based on the WAP protocol stack, due to its high efficiency. Note that here we use 'MMS' to denote both the service and the message itself; the distinction between the two above meanings will be evident from the context.

*Emerging Wireless Multimedia: Services and Technologies*   Edited by A. Salkintzis and N. Passas
© 2005 John Wiley & Sons, Ltd

It is possible to distinguish two scenarios for MMS messaging.

(1) *Person-to-person MMS*, that is messaging between people. This scenario is associated with the availability of multimedia accessories (i.e., cameras) that may be connected to the mobile phone. In this case, the user has the possibility of taking a snapshot of a scene and sending it to one or more recipients (i.e., Internet users, users of handsets, etc.).
(2) *Content (machine)-to-person MMS*, a Value-Added Service (VAS) may provide information about traffic, entertainment and museums through multimedia messages. In this context, the user(s) can activate several services and then receive information from content providers through MMS.

## 10.2 MMS Architecture and Standard

There are four key elements in the MMS architecture, defined in the MMS Environment (MMSE), as shown in Figure 10.1.

(1) *MMS Relay* provides access to different architectural elements for MMS provision and supports interactions with other messaging systems (it could be implemented together with the MMS server described below).
(2) *MMS Server* stores messages (this could include: an e-mail server, Short Message Service Center, fax).
(3) *MMS User Database* has databases with subscriber profile and information about user mobility.
(4) *MMS User Agent* is the MMS-enabled mobile phone.

These different elements interact in MMSE in order to provide functionalities across different systems. Note that MMS relay and MMS server can be integrated into a single platform, called the MMS Center (MMSC) [3].

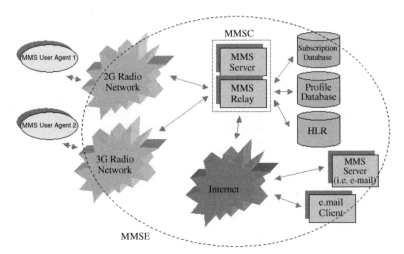

**Figure 10.1** Key elements of the MMS architecture.

MMS has been defined through a standardization process aimed at integrating different media (i.e., text, audio and video). Several standardization bodies have been involved in this process. In particular, 3GPP has defined high-level requirements: the MMS architecture, codecs and streaming protocols. As for the WAP Forum (now OMA), it has defined lower-layer aspects in order to connect the

MMS-enabled phone to the WAP environment. The MMS Interoperability Group (MMS-IOP), belonging to OMA, has focused on end-to-end operability and interoperability of MMS. Further standardization efforts are currently in progress in order to enrich the MMS specification.

The first 3GPP standard for MMS was the MMS Release'99 (for OMA, the standard is MMS 1.0), followed by MMS Release 4 (MMS 1.1) and, then, by MMS Release 5 (MMS 1.2). At present, the MMS Release 6 (MMS 1.3) is under preparation; some of its enhancements are described in subsection 10.2.2 [4].

### 10.2.1 Detailed Description of MMS Architecture Elements

In an MMS network it is possible to identify various system elements and interfaces, as detailed in Figure 10.2. In particular, one or more MMS servers can be present, depending on the specific service they provide. The different elements in Figure 10.2 are [5]:

- MMS Client, which is implemented as an application that interacts with the users on their mobile terminal;
- MMS Proxy-Relay and MMS Server, as already introduced;
- E-mail Server, for Internet e-mail services, supporting the Simple Mail Transfer Protocol (SMTP) protocol;
- Legacy Wireless Messaging Systems, representing various existing support systems for wireless messaging (i.e., paging or SMS systems).

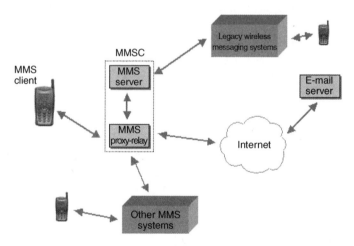

**Figure 10.2** MMS network representation.

Let us focus on the interfaces of the MMS reference architecture that are shown in Figure 10.3. In particular, there are eight different interfaces [5]:

(1) MM1: between the MMS user agent and the MMS relay/server;
(2) MM2: between the MMS relay and the MMS server;
(3) MM3: between the MMS relay/server and external messaging systems for e-mail, fax, Universal Mail Server (UMS)[a], etc.;
(4) MM4: between two MMS relay/servers in different MMSEs;
(5) MM5: between the MMS relay/server and the Home Location Register (HLR);

---

[a] UMS is a general-purpose POP3 and IMAP-compliant e-mail server.

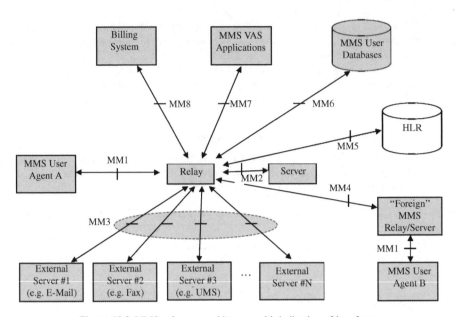

**Figure 10.3** MMS reference architecture with indication of interfaces.

(6) MM6: between the MMS relay/server and the MMS user databases;
(7) MM7: between the MMS relay/server and VAS applications;
(8) MM8: between the MMS relay/server and a billing system.

The protocol characteristics on the inter-connection between the mobile MMS client and the MMS relay/server (MM1 interface) are depicted in Figure 10.4.

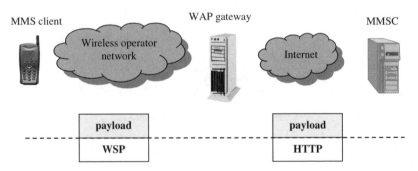

**Figure 10.4** MMS client-to-MMS relay inter-connection.

The MMS standard supports the use of e-mail addresses (i.e., an RFC 2822 routable address [6]) or Mobile Directory Numbers, MDN, (i.e., an E.164 number [7]) or both, to address the recipient of an MMS, as explained later.

We now provide some examples to describe the procedures to exchange MMS [3, 8].

Figure 10.5 deals with a content-to-person scenario, with user agent A and a VAS exchange MMSs. In particular, in the first part (i.e., procedures (1) and (2) we refer to the case where user agent A sends an MMS to the VAS; in the second part (i.e., procedures (3) to (6), a VAS sends a message to user agent A.

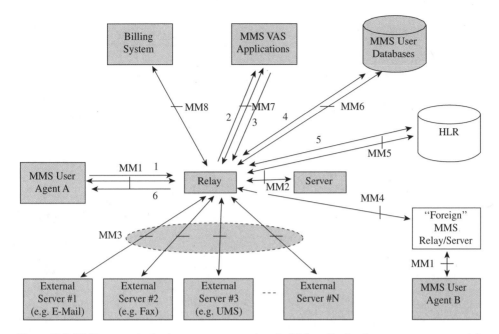

**Figure 10.5** MMS communication between user agent A and a VAS application (content-to-person scenario).

(1)  A multimedia message is sent by user agent A to the MMS relay/server according to an MMS-based subscribed service.
(2)  The MMS relay/server delivers the MMS to the VAS application.
(3)  The MMS-based VAS application sends an MMS to the MMS relay/server.
(4)  The MMS relay/server may control the MMS User Database in order to obtain information on the user profile (e.g., user enabled to receive messages from that VAS application).
(5)  Or it can make a query to the HLR in order to obtain user information (e.g., location of the user terminal in the network);
(6)  Finally, the MMS relay/server sends the MMS (received according to (3)) containing the requested information to user agent A.

The procedures where user agent A sends an MMS to user agent B who belongs to a network of a different operator are shown in Figure 10.6. We describe these procedures in two parts: the first phase (i.e., (1) to (3) below) concerns the operations in the home network of user A; the second phase (i.e., (4) to (8) below) is related to the operations in the network of user B (this second phase is, however, described still referring to the 'home network' of user A, by equivalently assuming that user B sends a message to user A).
  The different phases are detailed below.

(1)  User agent A sends an MMS that is received by its MMS relay/server.
(2)  This MMS relay/server interfaces with an external MMS relay/server that corresponds to the destination user agent B.
(3)  The destination MMS relay/server manages the procedures to send the MMS to user agent B.
    Let us now refer to the case where user agent B sends an MMS to A, to examine the procedures that occur in the network of A (note that this is the equivalent situation that occurs in the network of user B when the above MMS is routed to B after (3)).
(4)  An MMS is sent from user agent B and it is received by the related MMS relay/server.

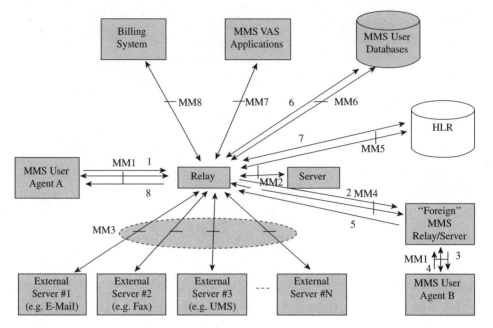

**Figure 10.6** MMS transmission and reception between two user agents in networks of different operators in a person-to-person scenario.

(5)  This MMS relay/server interconnects to the destination MMS relay/server.
(6)  The destination MMS relay/server can query the MMS User Database.
(7)  Or it can make a query to the HLR in order to receive information on the user.
(8)  Finally, the MMS is delivered to user agent A.

More details on the characteristics of the different interfaces and the procedures carried out at each of them are provided in the following sections.

### 10.2.2 Communications Interfaces

This section contains a description of all the reference points already shown (Figure 10.3) and a survey of the enhancements introduced by Release 6 of the MMS standard.

### MM1

This interface supports the communication between an MMS client (user agent) and the related MMSC (relay/server), providing a flexible protocol framework for the implementation of services. This communication protocol (based on the WAP protocol stack with binary encoding for MMS Protocol Data Units, PDUs) is used for submitting MMS, retrieving or pulling MMS, submitting push information (as part of a message notification) and exchanging delivery reports with the MMS relay/server. An MMS PDU sent through MM1 is subject to the WAP binary encoding for a more compact transmission on the air interface.

It is possible to have three types of commands that can be invoked by the MM1 interface through suitable PDUs:

(1)  Request: to request a service to be provided by another MMS entity (MMS client or MMSC);
(2)  Confirmation/response: to confirm the status of a previously requested service;
(3)  Indication: to notify another MMS entity of an event that has occurred.

The user agent provides MMS retrieval or delivery and terminal capability negotiation functionalities to the application layer. However, many vendor implementations of the user agent also support message creation, preview, submission and composition, encryption, decryption and digital signature.

MM1 enhancements provided by the MMS Release 6 include:

- the addition of information elements improving acknowledgments;
- handling of changes to the Multimedia Message Box (MMbox)[b] for storing, uploading, updating or deleting messages as well as for viewing information;
- a non-WAP MM1 implementation based on Mobile Station Application Execution Environment (MExE).

### MM2

This is the interface between the MMS server and the MMS relay. This reference point is required if the MMS server and the MMS relay are provided as two separated entities. Commercial implementations typically combine them into a single element (MMSC) and therefore an MM2 proprietary interface is used.

### MM3

This reference point is used by the MMS relay/server to send multimedia messages to (and to retrieve them from) servers of external messaging systems that are connected to the service provider and that support Message Exchange (Internet e-mail), Message Access (Voice Mail, Fax, etc.) and SMS Messages. The MMSC can periodically poll the external messaging server for incoming messages. MMS Release 6 also needs an MM3 interface to an external mail server to allow voicemail by means of VoiceXML.

MM3 could be used for routing and forwarding messages to the recipient using the appropriate Uniform Resources Identifier (URI) that can be obtained from the mobile terminal phone number on the basis of the E.164 standard representation.[c]

### MM4

This reference point is used to transfer messages between MMS relay/servers of different MMSEs. Transactions on MM4 can be implemented by using SMTP to transport Internet e-mail with Multi-purpose Internet Mail Extensions (MIME) attachments.

An MMS can be sent from an MMS user agent of one MMSE service provider and it can reach the MMS relay/server through MM1; then, it can be forwarded to an MMS relay/server of another MMSE service provider over MM4 by means of SMTP; finally, it is sent to the related MMS user agent (see also Figure 10.6 and related description).

MMS connectivity across different networks (i.e., MMSEs) is provided on the basis of Internet protocols. According to this approach, each MMSE should be assigned a unique domain name (e.g., mms.operator.net).

For sending a message to an external MMSE, the originator MMSC needs to resolve the recipient MMSC domain name to an IP address. If the recipient address is an e-mail address, then the originator MMSC can obtain the recipient MMSC IP address directly by interrogating a Domain Name Server (DNS) with the recipient address. If the recipient address is in the form of a phone number, then two methods have been identified by 3GPP.

(1) DNS-ENUM-based method. For those multimedia message recipients characterized by an MDN number (E.164 number) of an external MMSE, the originator MMS relay/server will resolve them

---

[b] The MMBox is employed to store the user MMS at the MMSC.
[c] E.164 is an ITU-T Recommendation that defines the international public telecommunication numbering plan used in the public switched telephone network as well as in the ISDN network. E.164 numbers can be translated by a DNS in order to obtain URI addresses (i.e., IP addresses).

to a routable RFC 2822 address (e-mail address). This method is known as DNS-ENUM. With this e-mail address, the originator MMSC determines the recipient MMSC IP address by the normal DNS method, according to what is described for the MM3 interface [6, 7].

(2) IMSI-based method. This address resolution method is based on a Mobile Application Part (MAP) query of the MMSC towards the Home Location Register (HLR) of the recipient MMSE in order to get the International Mobile Subscriber Identity (IMSI) associated with the recipient phone number (via the MM5 interface). From the IMSI, the originator MMSC can obtain the recipient MMSC Full Qualified Domain Name (FQDN)[d] and, then, by interrogating the DNS, the recipient MMSC IP address.

The GSM association has published recommendations for an efficient interworking over the MM4 interface. The realization of this interface that interconnects two mobile networks can be achieved over the public Internet (by a secure way) or over ATM or Frame Relay.

### MM5
The MM5 reference point may be used to provide information to the MMS relay/server about the subscriber by interrogating the HLR through MAP operations. The HLR can provide both information to route the message between two different MMSC (through MM4) and information on the user location. The MM5 reference point is not used in the case of notification through an SMS.

### MM6
This interface provides interworking between external user databases that have information on the MMS subscriber, including subscription details, user profiles and subscriber locations. Currently, this interface is undefined and further enhancements are expected.

### MM7
This is an interface between the MMS relay/server and various applications (i.e., VAS). Once a new VAS becomes available, users may subscribe to the service directly contacting the VAS provider and, thus, appearing in the distribution list maintained by the MMS relay/server.

The MM7 interface between a VAS provider and the MMS relay/server is implemented by using the World Wide Web Consortium (W3C) SOAP (Simple Object Access Protocol) language and HTTP (HyperText Transfer Protocol) as the protocol to transport SOAP messages (MM7 request/response are transferred in HTTP POST request/response messages). The VAS provider and MMS relay/server could play two roles of alternatively sending/receiving multimedia messages.

The implementation of the MM7 interface has the following characteristics.

- MM7 is based on SOAP 1.1 formatting language.
- SOAP messages consist of a SOAP envelope, SOAP header and SOAP body. The transaction ID is in the SOAP header; all the other information should be included in the SOAP body element.
- 3GPP as defined an XML schema for the MM7 interface to specify the structure of MMS PDUs embedded in SOAP messages [3].

Note that, the MMS relay/server and the Value-Added Service Provider (VASP) must be addressable by a unique URI-type address (placed in the host header field in the HTTP POST method).

Further innovations provided for MM7 by future MMS releases could be: defining Application Programming Interfaces (APIs), handling mailing lists and financial transactions, improved authorization/authentication, VASP interworking and support for enhancements of the interworking with VAS applications (i.e., based on the Open Service Architecture, OSA, extensions [9]).

---

[d] The FQDN address is a representation of an IP address for SMTP (e-mail) services. The FQDN address has a representation of the type: HostName.DNSDomainName.suffix

### *MM8*

This reference point is an interface that provides interworking with an external billing system. It is based on the OSA standard that specifies a number of interfaces (the set relevant for MMS deals with billing, charging and accounting) to allow systems to interact with each other in a well-defined way [10].

OSA allows implementing applications (i.e., services) that make use of network functionalities, offered in terms of a set of Service Capability Features (SCFs). OSA consists of: applications (implemented in one or more application server); framework (that provides applications to make use of the service capabilities in the network) and service capabilities servers (that provide the applications with SCFs that are abstractions from underlying network functionalities). OSA supports basic service mechanisms, such as Authentication, Authorization, etc.

In MMS Release 6, MM8 could be used to implement some prepaid services for MMS, charging enhancements (i.e., third party pays), advice of charge to user and multimedia message volume-based charging.

### *10.2.3 MMS Capabilities, Limitations and Usability Issues*

MMS represents a fundamental step for introducing multimedia services in mobile communication networks, involving several aspects, like network infrastructures, applications and terminals. MMS entails several advantages that can be detailed as follows.

- Benefits for users. The mobile messaging system is location independent, simple and personal. In addition to this, the new generation of mobile terminals is designed so as to enrich messaging with the possibility of integrating texts with images or sounds. MMS allows interesting features in the provision of multimedia contents, such as flexibility, scalability and compatibility with e-mail.
- Benefits for operators. MMS consists in a natural evolution of messaging, enlarging and differentiating this type of service. Operators can increase their revenues due to the support of MMS that add new value to classical messaging.
- Benefits for content providers. It is possible to individuate three factors that can add values to the information provided to the users: personalization, flexibility and mobility. These three aspects can enable content providers to create new attractive MMS-based services. Hence, MMS could be a starting point for the creation of new value-added services, providing existing contents through a new 'channel' or creating new specific contents based on MMS peculiarities with respect to SMS and EMS.

As explained before, every component of the market can reach benefits from the introduction of MMS. It is estimated that in 2006 MMS should reach 24% of incomes for operators and content providers. Finally, new enabling technologies are emerging for the support of this messaging system: picture messaging is followed by video messaging and it is possible to imagine even more sophisticated contents for videos and audio too.

Even if MMS have increased media potentialities with respect to SMS and EMS, we must consider that there are still some limitations to MMS-based services due to both handset capabilities (e.g., some displays are small, keypads on the mobile phone do not allow an easy input of text and data, not all the mobile phones have a memory card) and the fact that MMS does not provide a real-time service and delivery. Another problem concerns the Digital Rights Management (DRM) for the circulation of images. Other limiting aspects for MMS-based services are:

- the size of a 'full MMS provider', implementing all MMS services, is approximately 300 kbytes of object code for both client and server;
- the size of a 'full MMS server', implementing all MMS services, is more than 1 Mbyte of object code on a RISC machine.

As for the fruition of services in content-to-person scenarios, the MMS usability aspects are similar to those of WAP-based services, even if modern mobile phones have small joysticks or navigation tools that ease the exploration of contents. As for sending and receiving MMS in a person-to-person context, using it will be of a degree of complexity similar to that of SMS. Note that the current success of SMS relies exactly on the ease of use and management of SMS.

Moreover, usability refers to the easy use of the MMS technology by a customer without their knowing the complex technological processes that are behind this service. Hence, it is important to allow the personalization of the user interface to be as friendly as possible in order to permit the user to exploit completely the potentialities of MMS.

## 10.3 MMS Format

When an MMS is sent, it is forwarded as a Protocol Data Unit, PDU, as described in the subsection below. The structure of a PDU is standardized by the WAP Forum (now OMA) in the WAP-209-MMS encapsulation specification [11].

### 10.3.1 MMS PDU

MMS uses several types of PDUs to perform transactions that are described in Section 10.4. Here, the contents of the PDUs are described.

The PDU of an MMS consists of a header that contains specific information about message transactions, and a body. The message body may contain any content type such as text, image, audio and video. The message body is used only when an MMS is sent or retrieved. All other PDUs contain only the header.

In the MMS body, every part needs a further header with the related content type (e.g., image, text, etc.) and content location (e.g., picture1.jpg, text1.txt) [12]; such segmentation is called Multipart Message (MM), as shown in Figure 10.7.

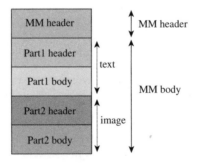

**Figure 10.7** MMS PDU format.

Below are described some different fields that may be present in the MM header [13]:

- *X-Mms-Message Type*, specifying the transaction type;
- *X-Mms-Transaction-ID*, identifying univocally the message;
- *X-Mms-Version*, specifying the MMS version used for this message;
- *To/From*, about the sender and the recipient of the message;
- *Subject* of the message;
- *Date* of the message transaction;
- *Content-Type*, defining the content of the MMS (i.e., image, sound or text).

In what follows, the header fields are described for the eight main PDUs. The name of an optional field is enclosed in square brackets; all other fields are mandatory. Note that the use of the PDUs below is described in Section 10.4.

### M-Send.req

The message body follows the headers. The transaction identifier is created and used by the sending client and it is unique for the transaction.

X-Mms-Message-Type
X-Mms-Transaction-ID
X-Mms-MMS-Version
[Date]
From
To, Cc, or Bcc
[Subject]
[X-Mms-Message-Class]
[X-Mms-Expiry]
[X-Mms-Delivery-Time]
[X-Mms-Priority]
[X-Mms-Sender-Visibility]
[X-Mms-Delivery-Report]
[X-Mms-Read-Reply]
Content-Type

### M-Send.conf

When the MMS relay has received the send request, it transmits back a response message to the mobile terminal indicating the status of the operation. The MMS relay must always assign an ID to the message when successfully received.

X-Mms-Message-Type
X-Mms-Transaction-ID
X-Mms-MMS-Version
X-Mms-Response-Status
[X-Mms-Response-Text]
[Message-ID]

### M-Notification.ind

MMS notifications inform the mobile terminal about the contents of a received message. The purpose of the notification is to allow the client to fetch automatically a message from the location indicated in the notification. The transaction ID is created by the MMS relay and it is unique up to the following M-Notify-Resp only. If the MMS client requests a deferred delivery with M-NotifyResp, the MMS relay may create a new transaction ID.

X-Mms-Message-Type
X-Mms-Transaction-ID
X-Mms-MMS-Version
[From]
[Subject]
X-Mms-Message-Class

X-Mms-Message-Size
X-Mms-Expiry
X-Mms-Content-Location

### M-NotifyResp.ind
The purpose of this message is to acknowledge the transaction to the MMS relay.

X-Mms-Message-Type
X-Mms-Transaction-ID
X-Mms-MMS-Version
X-Mms-Status
[X-Mms-Report-Allowed]

### M-Retrieve.conf
A client shall retrieve messages by sending a WSP get request to the MMS relay containing a URI to the received message. If successful, the response to the retrieve request will contain headers and the body of the incoming message.

X-Mms-Message-Type
[X-Mms-Transaction-ID]
X-Mms-MMS-Version
[Message-ID]
Date
[From]
[To]
[Cc]
[Subject]
[X-Mms-Message-Class]
[X-Mms-Priority]
[X-Mms-Delivery-Report]
[X-Mms-Read-Reply]
Content-Type

### M-Acknowledge.ind
An MMS acknowledge message confirms the delivery of the message from the receiving terminal to the. MMS relay.

X-Mms-Message-Type
X-Mms-Transaction-ID
X-Mms-MMS-Version
[X-Mms-Report-Allowed]

**M-Delivery.ind**   An MMS delivery report must be sent from the MMS Relay to the originating mobile terminal when the originator has requested a delivery report and the recipient has not explicitly requested for denial of the report. There will be a separate delivery report from each recipient. There is no response message to the delivery report.

X-Mms-Message-Type
X-Mms-MMS-Version
Message-ID
To
Date
X-Mms-Status

## Read Reporting

When the originating terminal requested the read-reply in the MMS, the recipient terminal may send a new MMS back to the originating terminal when the user has read the MMS. The content of the multimedia message is a terminal implementation issue. The read-reply MMS must have the Message-Class as Auto in the message. The MMS relay must deliver the read-reply message as an ordinary multimedia message. When the originating terminal receives the read-reply, it shall not create a delivery report or a read-reply message.

An example of a PDU header for MMS delivery (before binary encoding, according to the WAP specifications) is shown in Figure 10.8.

---

X-Mms-Message-Type : m-send-req

X-Mms-Transaction-ID : 13452917

X-Mms-Version : 1.0

To : +393475946381/TYPE=PLMN

From: +393389234175/TYPE=PLMN

Subject: Test MMS message

Date: 4 Oct 07:13:01 2003

Content-Type: application/vnd. wap.multipart.mixed

---

**Figure 10.8** Example of a MMS PDU header.

### 10.3.1.1 The SMIL Language

The Synchronized Multimedia Integration Language (SMIL, pronounced 'smile') is an XML-based language used to ease the organization of media objects contained in an MMS. It is the mandatory format for media synchronization and scene description of multimedia messaging.

The current SMIL release is 2.0, which has the following design goals:

- To define an XML-based language that allows authors to write interactive multimedia presentations. Using SMIL 2.0, an author can describe the temporal behavior of a multimedia presentation, associate hyperlinks with media objects and describe the layout of the presentation on the display.
- To allow reusing of SMIL 2.0 syntax and semantics in other XML-based languages, in particular those needing to represent timing and synchronization. For example, SMIL 2.0 components are used for integrating timing into XHTML.

SMIL 2.0 is defined as a set of markup modules that characterize the semantics and an XML syntax for certain areas of SMIL functionality.

The 3GPP MMS uses a subset of SMIL 2.0 as a format of the scene description [14]. MMS clients and servers shall support the 3GPP SMIL Language Profile. This profile is a subset of the SMIL 2.0 Language Profile, but a superset of the SMIL 2.0 Basic Language Profile. 3GPP TS 26.234 Recommendation has an informative Annex B that provides guidelines for SMIL content authors [15].

Additionally, 3GPP MMS should provide the following format: XHTML Mobile Profile. The 3GPP MMS uses a subset of XHTML 1.1 as a format for scene description. MMS clients and servers shall support XHTML Mobile Profile, defined by the WAP Forum. XHTML Mobile Profile is a subset of XHTML 1.1, but a superset of XHTML Basic.

SMIL permits the rendering of various media objects that can be organized dynamically on a predefined graphical layout to obtain a complete multimedia presentation. In particular, SMIL can be used to define a presentation as:

- to set-up the temporal behavior of the different involved objects;
- to describe the layout;
- to associate media objects to hyperlinks;
- to define conditional contents on the basis of several properties.

Designers of SMIL presentations can use a spatial description, where the rendering space is organized in regions that can be nested and can contain images, clips or text; the presentation is described in an XML-like mode by means of appropriate tags. SMIL also supports a temporal description of the different objects involved in a presentation; in particular, it is possible to synchronize the objects in sequence or in parallel (e.g., contemporary sounds or melodies). Moreover, different objects can be nested in more complex presentations.

With MMS SMIL, a multimedia presentation is composed of a slideshow, using slides with the same region configuration, organized (portrait or landscape with various sizes) to contain text, videos or images. Figure 10.9 shows a software tool for a Sony-Ericson mobile phone that helps in composing a slide show with images, sounds and a temporal relation between them.

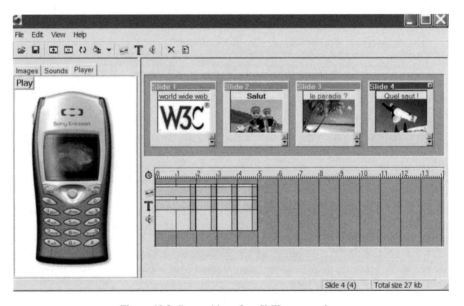

**Figure 10.9** Composition of an SMIL presentation.

The MMS SMIL profile was initially defined outside standardization in a document called MMS conformance document [16]. Such documents provide recommendations to ensure interoperability between MMS devices produced by different manufacturers. They identify SMIL 2.0 features that can be used for constructing the slideshow that should have a sequence of parallel containers that represent the definition of a slide. Each media object identified inside the parallel container represents one component of a slide. The MMS Conformance document also specifies that a slide may be composed of a text region only, of an image region only (this can contain either an image or a video clip), or of two regions. It also contains rules about region dimensions, timing information and maximum image dimensions.

## 10.4 Transaction Flows

When an MMS is sent, a transaction between the sending terminal, the MMS relay and the receiving terminal is originated using various transport schemes. The message exchange entails the following logically separate transactions (see Figure 10.10).

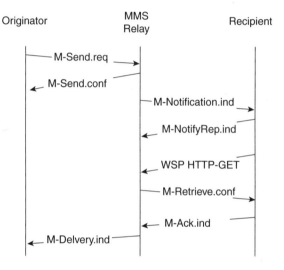

**Figure 10.10** MMS transaction flows between originator and recipient.

(1) The MMS client of the originator sends a message to MMS relay (M-Send.req, M-Send.conf).
(2) The MMS relay notifies the MMS client of the recipient about a new message (M-Notification.ind, M-NotifyResp.ind).
(3) This MMS client fetches a message from the MMS relay (M-Retrieve.conf).
(4) This MMS client sends a retrieval acknowledgement to the MMS relay (M Acknowledge.req).
(5) The MMS relay sends a delivery report about a sent message to the MMS client of the originator (M-Delivery.ind).

We now provide more details about the transaction flows shown in Figure 10.10. If a user agent creates an MMS, it sends an M-Send.req to the MMS relay (which replies with an M-Send.conf) by means of the WSP/HTTP-POST method. Then, the MMS relay sends, by means of WAP-PUSH, an M-Notification.ind (containing the MMS URI as data) to the receiving terminal for an incoming multimedia message, waiting for a reply (M-NotifyReport.ind). The MMS then is retrieved by using the URI through a WSP/HTTP GET connection [3]. Then, the user receives the MMS in the body of the PDU, named M-Retrieve.conf. The MMS client may further acknowledge the message retrieval by sending an M-Acknowledge.ind message to the MMS relay. The MMS relay, finally, replies to the message originator with an M-Delivery.ind message by means of WAP-PUSH.

   It is possible to analyze in detail the MMS transaction flows by referring both to the different interfaces for each kind of flow and to the relative transaction PDUs.

### MM1

The different operations and transaction flows that are possible over the MM1 interface are described below.

- Message Submission. This is the submission of an MMS from its originator (i.e., the client on the mobile phone) to the related MMSC. The originator needs a circuit-switched (i.e., GSM) or a packed-switched (i.e., GPRS) connection through a (usually dedicated) WAP gateway to reach the MMSC. For this transaction flow the message submission PDU (M-send.req) and the relative confirmation (M-send.conf) are generated, and the originator MMS client may provide date and time of the submission request, otherwise provided by the MMSC. At least one recipient shall be provided by the MMS client to the MMSC; if a Multimedia Message Box (MMBox) is not supported by the MMSC, the MMBox-related parameters are ignored. The client can request to pay for a reply message. If this functionality is not supported by the MMSC, it will generate an error message. Finally, the submission request can be accepted by the MMSC with a confirmation (using parameter Message-ID for unique identification) or rejected, specifying the reason (i.e., permanent or transient nature); if the submission is accepted, the originator MMSC parses the MMS identifying the recipient MMSCs[e] and then routes the message to them.
- Message Notification. This message is created by the recipient MMSC and sent to the recipient MMS client as part of a WAP-PUSH (encapsulated in several SMS) or as part of a data connection. The notification message PDU (M-notification.ind) delivered to the recipient MSS client is composed only of a header (essential information is contained in this PDU). The MMS client confirms the correct receipt of the notification with the response (M-notify-resp.ind), which is still only composed of a header. In this response the X-Mms-Status parameter is used to select different actions in order to defer the message retrieval or to do it immediately.
- Message Retrieval. This is the procedure for the delivery of a multimedia message from the recipient MMSC to the recipient MMS client and it occurs after the corresponding notification is received by the recipient MMS client. First, a retrieval request (WSP/HTTP GET.req) is issued; then, if the message retrieval is successful, the retrieval confirmation (M-retrieve.conf) is returned with the multimedia message contained in the PDU body. It is possible to perform an immediate retrieval; in this case, the client starts the retrieval procedure automatically after the receipt of the relative notification. Otherwise, in deferred retrieval, the MMS client informs the MMSC that the message corresponding to the notification may be retrieved later, not automatically.
- Delivery Report. The originator of the message can request the delivery report for every MMS recipient. Such a report specifies whether the message has been successfully retrieved or deleted or rejected or forwarded by the recipient MMS client; an indeterminate state is used if the message is sent to a message system where the delivery report is not supported. The forwarding of a delivery report to the MMSC is carried out over the MM4 interface, but the delivery from the MMSC to the originator MMS client is through the MM1 interface (with M-delivery.ind PDU).
- Read Report. Another possibility for the MMS originator during the message submission phase is to request a read report for the submitted message; such a report is made for every recipient and it notifies if the message was read or deleted without being read. A read report can be generated in two different ways if allowed by the MMS recipient:

(1) First method, allowed from Release 1.0. The recipient MMS client sends the read report in a PDU (with M-send.req PDU) to the message originator by using the MM1 interface and including the status of the corresponding multimedia message.

---

[e] The originator MMSC denotes the MMSC of the mobile phone generating the multimedia message; the recipient MMSC denotes the MMSC of the mobile phone that receives the multimedia message. These two MMSCs are different in the case that the MMS has to cross networks of different operators.

(2) Second method, allowed from Release 1.1. Two PDUs are used: the first (M-read-rec.ind) from the recipient MMS client to its MMSC (over MM1) and then to the originator MMSC (over MM4); the second (M-read-orig.ind) from the originator MMSC to the originator MMS client, containing the read report.

- Message Forward. Another capability of MMS clients (from MMS Release 1.1) is to support the forwarding of MMS (i.e., to an e-mail server). For this kind of request, the M-forward.req PDU is used with the related confirmation (M-forward.conf). Note that if the MMBox is not supported by the originator MMSC, the parameters of the forward request are ignored. However, if the forward request is accepted by the MMSC, it inserts the address of the forwarding MMS client in the 'from' field as well as a sequence number in the message to be forwarded (it can additionally insert date and time). The MMSC can forbid the forwarding of a message if it contains protected media objects.
- MMBox Interactions: If MMBox is supported by the corresponding MMSC (from MMS Release 1.2), the MMS client can:

employ the MMSC (M-Mbox-store.req) to store a message in the MMBox;
update the messages already stored in the MMBox;
upload a message to the MMBox, for messages created by the user or previously retrieved from MMSC (M-Mbox-upload.req);
delete a message (or more messages) from the MMBox in a single transaction (with M-Mbox-delete.req PDU);
view information on the contents of the MMBox: the MMS client can request information about message parameters or message contents (using M-Mbox-view.req PDU).

### MM2

The MM2 interface and its transaction flows are implemented in a proprietary manner because no technical realization of this reference point has been specified by standardization organizations.

### MM3

As seen in the above Section 10.2.2 on MMS interfaces, the MM3 reference point is used to permit the MMSC to exchange messages with external servers. There are various mechanisms for discovering incoming messages from external messaging servers: forwarding of the message to the MMSC; notifying the MMSC for a message waiting for retrieval; polling made by the MMSC to external messaging servers for incoming messages. It may be necessary for the MMSC to perform the conversion of the MMS in an appropriate format (supported by the external server), for example in a text-based MIME representation for the transfer through SMTP.

### MM4

At present, only one technical realization is available for this interface, standardized by 3GPP, using SMTP and based on the transaction flows described for the MM1 interface.

- Routing forward a message. This request is made by an originator MMSC to transfer an MMS to a recipient MMSC (MM4-forward.req). If it is possible, the recipient MMSC replies with a forward response (MM4-forward.res) composed by several parameters. If errors occur during the request process phase, an error code is generated.
- Routing forward a delivery report. This request allows the transfer of a delivery report between two MMSCs, that is the originator MMSC and the recipient MMSC. The recipient MMSC can request such a report (MM4-delivery_report.req) from the originator MMSC that thus sends its response (MM4-delivery_report.res). Several states can be reported: 'expired', 'retrieved', 'deferred', 'indeterminate', 'forwarded', 'unrecognized', and the value of the 'sender' field is the address of the recipient MMSC.

- Routing forward a read report. It is used for the transfer of a read report between two MMSCs. The request to forward a read report is made by the recipient MMSC (with MM4-read_reply_report.req); then, the originator MMSC acknowledges this request with the response (MM4-read_reply_report.res). Two read states are reported over MM4, that is 'read' or 'deleted without being read'.

### MM5
As seen previously, the transaction flows over this interface connect the MMSC with other network entities such as the HLR. Similarly to the MM2 reference point, at present no technical realization of the MM5 interface has been standardized.

### MM6
This interface supports flow interactions between the MMSC and user databases; this reference point has yet to be standardized.

### MM7
To enable interactions between VAS applications and MMSC, MM7 implements the following kinds of transactions:

- message submission;
- message delivery;
- message cancellation and replacement;
- delivery report;
- read report;
- message errors (MMSC error or VASP error).

A more detailed description of the above transaction flows over MM7 is provided in the Section 10.5, which deals with MMS-based VAS, since MM7 is actually the interface for VAS.

### MM8
Over this interface, network operators use proprietary transport protocols since at present a transaction flow is not standardized for sending charging data from the MMSC to the billing system.

## 10.5 MMS-based Value-added Services

In order to enable VASPs to support many multimedia contents (i.e., MMS) for multiple devices, the MM7 interface has been standardized by 3GPP, enabling communications between the MMSC of the MMS provider and the VASP (see Figure 10.11). This section contains a detailed description of the MM7 transaction flows.

The VAS server can perform several operations such as authentication, authorization, confidentiality and charging information. All these operations have to be supported by the standard interface and will be discussed in more detail below.

As for charging, the VAS server can indicate, in the case of message submission, which part is going to pay for message handling: the VASP, the recipient (one or more), both of them, with the cost shared between them or neither the recipient nor the VASP. Note that a VASP needs commercial agreements with the MMS provider to apply one of the previous billing options, and has to control how message contents are redistributed using appropriate DRM mechanism defined in the MMS 1.2 standard.

The use of common standard-based Web service interfaces will allow the possibility of supporting the quick and cost-effective deployment of revenue-generating mobile services by means of system integration. The mobile Web service interfaces are standardized by OMA, 3GPP and W3C. Such interfaces will be implemented on the basis of widely accepted standard technologies, like SOAP, XML,

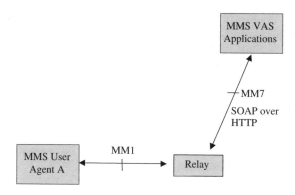

**Figure 10.11** VASP-relay-user interactions in the MMS reference architecture.

WSDL (Web Services Description Language) and HTTP. The first two technologies, SOAP and XML, are part of the OSA standard, defined by W3C. XML allows a general data representation, whereas SOAP is a simple communication protocol (HTTP is the underlying protocol); they allow open-source development of software and together provide a single and transport-neutral method of contents developed for different types of terminals like PCs, PDAs and, of course, mobile phones. In this environment, a VASP can use the HTTP POST method to exchange SOAP messages; the SOAP processing is simple, since it requires only HTTP and XML libraries for HTTP authentication mechanisms.

OSA supports network functionalities such as discovery and authentication. Moreover, it can enable new services without stopping already-in-progress applications.

It is important to note that there are other proprietary mechanisms that could be used to deliver VASP messages:

- by employing SMTP – the MMS content is attached to an e-mail sent to 'phonenumber@mms. domain';
- by using an HTTP POST with the MIME content type set as 'multipart/form-data';
- by means of an HTTP GET, if the MMS content is precompiled and resides on an external Web server.

Let us focus on the delivery of VAS over MM7. The MMSE allows VASs that may be supplied by the network operator or by third party VAS providers; hence, it is important to define how these VASPs interwork with the MMS relay/server.

### Message Submission from a VAS
A VAS application can send a submission request (MM7_ submit.req) for an MMS to the MMSC (directed to one or more recipients or a distribution list) over the MM7 interface. This request can be associated with several features:

- *Authorization*, with identification of both the service and the VASP;
- *Addressing*, where the recipient(s) is specified (the address of the originator may be present in the submission request);
- *MM7 version*, message type and transaction ID, for a quick link for the corresponding response;
- *Linked message ID*, for the association with a previous message that had an identification as part of the submission request;
- *Message subject*, that is a short text provided by the originator (like the subject of an e-mail);
- *Priority*, indicating the importance of the message;

- *Class*, specifying the category of the message content (i.e., advertisement, information, etc.);
- *Service code*, for billing purposes;
- *Time stamping*, with time and date of the request;
- *Time constraints*, for example a validity period;
- *Reply charging*, specifying if the message will be paid by the VASP and the related constraints (i.e., size);
- *Reporting*, because the VAS application can request the delivery and/or read–reply reports;
- *Restrictions* of content adaptation;
- *Content type* of the message.

If the submission request is successful, the MMSC sends an acknowledgment message (MM7_submit.res) with a response associated to the features described below:

- *MM7 version, message type and transaction ID*;
- *Message ID*, to the VAS application, if the submission request is accepted by the MMSC;
- *Request status* of the submission request (successful or failed due to an error).

### Message Cancellation and Replacement
If an MMS submission was successful (according to the previous point), the VAS application can cancel the sent message or the submitted message can be replaced by a new one (only if this message has not yet been forwarded or retrieved by the recipient).
   In the first case, the request to cancel the message (MM7_cancel.req) is associated with:

- *MM7 version, message type and transaction ID*;
- *Authorization*;
- *Addressing*, where the originator is specified;
- *Message ID*, to which the request applies, provided by the MMSC as part of the response.

The features of a replacement request (MM7_replace.req) are:

- *Content adaptation restrictions*, for adaptation or not;
- *Service code*, used by the MMS provider for billing purposes;
- *Reporting*;
- *Time stamping*;
- *Time constraints*;
- *Restrictions*;
- *Content type*.

The acknowledgments are MM7_cancel.res and MM7_replace.res, respectively, and are associated with:

- *MM7 version, message type and transaction ID*;
- *Request status* of the cancellation/replacement request (successful or failed due to an error).

### Delivery and Read–reply Reports
Delivery reports and read–reply ones can be requested by a VAS application as part of a message submission request. If these functionalities are allowed by the message recipient, the MMSC generates these reports and sends them to the VAS application. First, the MMSC has to send, respectively, the MM7_delivery_report.req and the MM7_read_reply_report.req requests to the VAS application, with the following features:

- *MM7 version, message type and transaction ID*;
- *Addressing*, where the originator is specified;

- *Message ID*, to which the report is related, is provided by the MMSC as part of the response;
- *Time stamping*, indicating time and date of the message handling;
- *Message status*, for instance, retrieved, deleted, etc.

Then, the VAS application acknowledges with MM7_delivery_report.res and MM7_read_reply_report.res respectively, associated with:

- *MM7 version, message type and transaction ID*;
- *Request status* of the report request (successful or failed due to an error).

### Message Delivery
The MMSC can request the delivery of a message to the VAS application (MM7_deliver.req). Such a request is associated with several features:

- *Addressing*, where both the originator and the recipients are specified;
- *Authentication*, provided by the MMSC as part of the request;
- *MM7 version, message type and transaction ID*;
- *Linked message ID*, used by the VAS application for handling a subsequent submission of a message related to a previously delivered message;
- *Message subject*;
- *Priority*;
- *Class*, specifying the category of the message contents;
- *Time Stamping*, containing time and date of message submission;
- *Reply charging*, specifying if the message will be paid for by the VASP and the related constraints (i.e., size);
- *Content type*.

When the delivery request arrives to the VAS application, it acknowledges with a response message (MM7_delivery.res) with the following features:

- *MM7 version, message type and transaction ID*;
- *Service code*, used by the MMS provider for billing purposes;
- *Request status*.

### Message Errors
If an error occurs during a request or a process, two generic error notifications can be used: MM7_RS_error.res (for MMSC to inform the VAS application about a request processing error) and MM7_VAS_error.res (for a VAS application to inform the MMSC about a request processing error) that are associated with the following features:

- *MM7 version, message type and transaction ID*;
- *Error status*, indicating the type of error that has occurred.

### 10.5.1 Survey of MMS-based Services

Many types of services can be provided by means of the MMS architecture, taking into account interactivity, entertainment and personalization. This section is devoted to giving an overview of different MMS-based services that can be attractive for users.

We can consider the following different typologies of multimedia messaging services:

- mobile personal communications;
- mobile dating;

- mobile marketing;
- mobile information services;
- mobile entertainment.

All the above categories can have an added value from localization that allows them to provide the user with specific information adapted to its location.

More details on all the above service categories are provided below.

### Mobile Personal Communications
Although MMS is used to communicate between two people, some audio features of an MMS allow it to be used as advanced voice mail. If the end-user is not available, a remote server can store the message in MMS format.

### Mobile Information Services
MMS provides a richer way to deliver information than SMS. MMS could be particularly suited for representing sporting events (i.e., sporting action), finance (i.e., some financial graphs) and news, in general. The key element for information services is that they be provided in a format that is compact and easy to view; this is exactly what MMS can do. For example, a Japanese application called 'Namidensetsu' provides sea, wave and weather information on surfing locations.

### Mobile Marketing
This type of service is well suited for MMS due to its graphical potentials; it can be a good complement to branding and advertising campaigns. For example, a user could opt to receive a free information service from a movie studio, which could increase interest in its products by distributing MMS messages (e.g., video clips).

### Mobile Dating
The convenience and the privacy of MMS mean that mobile dating can be a very popular service, better than the related Internet-based version. It is possible to have different types of messages: first text messages, then audio communication followed by pictures and video messages if both parties agree to maintain this type of contact (thus preventing a malicious use of this service).

### Mobile Entertainment
This kind of service could be divided basically into eight sub-classes (or even more). In particular, we can consider the following.

- Comics, jokes and icons. With MMS, comics and carrillon icons can be also delivered and additional traffic can be generated.
- Interactive music selection. Another feature of MMS is its ability to download music on demand (this could require the Digital Rights Management system).
- Interactive 'pick a path' video, which gives the user the ability to determine the outcome of a video film or story. At several points in the video, the viewer can be asked to determine in which direction the story should go by a simple selection or voting (i.e., for groups of friends). The technology to enable these types of services can be provided by VAS extensions.
- Collectable cards. MMS can deliver this service where users pay to receive a random set of cards (messages); in addition to the revenues generated by the delivery of each card, the service also generates traffic by allowing users to exchange cards between themselves.
- Celebration cards. These are prepackaged MMS, such as birthday cards, similar to the gift cards sold in shops.
- Adult services. This will play an important part in mobile data services, but these services are critical to be managed for many operators. Mobile terminals have small displays, they are personal, portable and can allow a discreet reception of such contents.

- Quizzes and competitions. MMS can allow service providers to deliver quizzes in a more compelling way than by means of simple text (SMS); these competitions can be sponsored by an advertiser or paid for by the users themselves.
- Horoscopes. MMS will make horoscopes look (and sound) much better than via SMS, but the basic functionality of the service could remain the same.

### Location-based Services

Other candidate services that are suitable for employing MMS are location-based services that use positioning technologies. With MMS, service operators can deliver services that might include:

- maps and routing;
- location of the nearest point of interest;
- information about locations.

Note that it is quite important for a VASP to be able to deliver location-based services using MMS rather than to require an application to be downloaded to support the service, which would severely limit the potential market for any service unless the application can be easily installed. In this regard, the PALIO (Personalised Access to Local Information and services for tOurists) IST project developed a service for providing location-aware information to tourists on traffic jams, museums, restaurants, hotels, etc. [17]. This experimental service has been based on simple textual contents (i.e., SMS), but it could be easily enriched by introducing MMS (providing museum photos, traffic maps or audio files), thus adding further value to the service itself.

## 10.6 MMS Development Tools

In order to facilitate the introduction of MMS in the market with a consistent set of services, many telecommunication companies provide tools to develop MMS software applications. There are several tools that permit the implementation of MMS-based applications in an easy way with several libraries. This section looks at some key examples.

Sony-Ericsson enables the download of the following tools from the developer on-line resources, known as Ericsson Mobility World [18]:

- MMS composer. This creates MMS messages containing a scene description by using a PC and, then, forwards these messages to a handset connected via serial cable.
- AMR (Adaptive MultiRate) and WAV (Waveform data) conversion tool. This has a graphical or DOS interface for the conversion of a WAV file in AMR format (and vice versa).

Nokia provides the following set of development tools and guidelines at the Nokia developer Forum [19]:

- Emulator for MMSC interface. This emulates the MM7 interface (at present, the Nokia MMSC is based on proprietary HTTP-based commands).
- MMS Java library. This includes message creation, encoding, decoding and sending to an MMSC emulator for the development of Java MMS-based applications.
- MMS developer's suite. This tool facilitates the creation of MMS with rich presentations (add-on for Adobe GoLive).
- Series 60 SDK Symbian OS and MMS extension. This is a PC emulator for creating and testing applications for mobile phones based on the 'Series 60' platform.
- Mobile Internet Toolkit. This tool allows the testing of multimedia messages and the creation of messages from many media objects.

Another tool for MMS development is MMS SDK by Openwave [20]. It operates at the MM7 interface, simplifies the development of MMS applications and can be used in combination with other Openwave software to view how MMS messages would appear on handsets. It includes the following components:

- MMS Library. This contains the tools necessary to develop MMS applications based on the MM7 protocol.
- MM7 Protocol. This provides a means of sending value-added service contents from third parties to subscribers;
- MMS Library API. This encapsulates the MM7 standard protocol in a Java API for the development of third party applications that can send multimedia messages to and from an MMSC.
- MMS Library Authentication and Security. This provides standalone applications supporting secure communication using HTTP client authentication schemes and the Secure Socket Layer (SSL) protocol.

Finally, the MMS Suite is a tool provided by one of the largest open-source software development Websites, SourceForge [21]. This Java-based software allows the creation of MMS documents with an unlimited number of pages that can be viewed with a mobile phone-like viewer. It is also possible to add images, sounds and text to a page (the following formats are supported: JPG, GIF and WAV) and it is possible to specify the layout of an MMS document. MMS Suite runs on Linux, Mac or Windows.

Note that for both the Sony-Ericsson and the Nokia tools, the access to resources for developers requires prior online registration.

## 10.7 MMS Evolution

Multimedia messaging as well as SMS may support the society by increasing interactions between people. Hence, by combining such novel services with location awareness, it is possible to provide the users with timely and location-dependent services. In addition to this, dynamic discussion groups and virtual meetings can be supported. Naturally, these innovative trends for MMS-based services also call for new rules in order to distinguish useful messaging from spamming. Personalization and adaptation play a fundamental role in addressing these problems. In this way, the new technologies can provide real support in everyday life for people such as citizens, workers, travelers and tourists.

MMS is attractive as it can provide the basis for novel services. For instance, a type of video-surveillance and remote control system that periodically sends (or in particular conditions) a user (or specialized security system) images of their home when they are away. This approach is also important for the remote control of plants, etc. Another significant example for the future evolution of MMS-based services can be guide services realized with maps sent through MMS. Finally, many other applications can be foreseen, for advertising products, etc.

As for the technological evolution of multimedia messaging, we can consider that different approaches are possible. In particular, we can refer here to MExE, a protocol that is able to provide a standardized environment to support different applications on mobile terminals. In fact, it can provide smart menus, personalized interfaces through the support of operators and service providers. Moreover, we can distinguish the following different MExE groups, depending on the potentialities of mobile terminals:

- Classmark 1, for WAP-enabled mobile phones with the possibility to access limited contents on the Web;
- Classmark 2, for mobile phones that support the JavaPhone Application platform. It is possible to support Java Applets on mobile phones with reduced processing capabilities for messaging, calendars, management of list of addresses, etc.;

- Classmark 3, for mobile smart-phones that support Java 2 Micro Edition (J2ME). In this case, more complex applications can be supported;
- Classmark 4, for enhanced mobile phones that employ Common Language Infrastructure (CLI) Compact Profile$^f$, providing an open development environment for applications written in different languages (e.g., Visual Basic). A minimal set of libraries and functionalities is specified in order to manage protocols such as HTTP, SOAP, etc.

Through the evolution from Classmark 1 to Classmark 4, an enhancement of MMS-based services is expected in the future years.

# References

[1] Simon Buckingham, *Next Messaging: From SMS to EMS to MMS*, Mobile Lifestreams, December, 2000.

[2] Jarkko Sevanto, Multimedia Messaging Service for GPRS and UMTS, *Wireless Communications and Networking Conference*, Vol. 3, pp. 1422–1426, 1999.

[3] 3GPP, Multimedia Messaging Service (MMS); Functional description; Stage 2; TS 23.140, Releases 1999, 4, 5 and 6.

[4] Website http://www.lebodic.net/mms_resources.htm

[5] Wireless Application Protocol Forum, Ltd, Multimedia Messaging Service, Architecture Overview Specification, WAP-205-MMSArchOverview-20010425-a, April 2001.

[6] P. Resnick, Internet Message Format, IETF RFC 2822, April 2001.

[7] P. Faltstrom, E.164 number and DNS, IETF RFC 2916, September 2000.

[8] Wireless Application Protocol Forum, Ltd, Multimedia Messaging Service, Client Transactions Specification, WAP-206-MMSCTR-20020115-a, January 2002.

[9] 3GPP, Open Service Access (OSA) Application Programming Interface (API); Part 1: Overview, TS 29.198-01, Version 6.0.1, February, 2004.

[10] Eurescom P920 project, Deliverable 'UMTS network aspects', January 2001.

[11] Wireless Application Protocol, Ltd, MMS Encapsulation Protocol, WAP-209-MMSEncapsulation-20020105-a, January 2002.

[12] Nokia, MMS Center Application Development Guide, Version 1.0, March 4, 2002 (available at the Website www.forum.nokia.com/).

[13] Nokia How To Create MMS Services, Version 3.1, August 28, 2002 (available at the Website www.forum.nokia.com/).

[14] 3GPP, Transparent end-to-end packet switched streaming service (PSS); 3GPP SMIL Language Profile, TS 26.246, V6.0.0 (2004-06).

[15] 3GPP, Transparent end-to-end Packet-switched Streaming Service (PSS); Protocols and codecs, TS 26.234 V6.1.0 (2004-09).

[16] Open Mobile Alliance, MMS Conformance Document 1.2, OMA-MMS-CONF-v1_2-20040727-C, July 2004.

[17] PALIO Project home page http://www.palio.dii.unisi.it/

[18] Ericsson Mobility Word Website http://www.ericsson.com/mobilityworld

[19] Nokia Developer Forum Website http://www.forum.nokia.com

[20] Openwave Website http://developer.openwave.com/dvl/tools_and_sdk/openwave_mobile_sdk/mms_sdk

[21] SourceForge Website http://sourceforge.net

---

$^f$ CLI Compact Profile specifies a minimal set of class libraries, supporting both common runtime library features and Web services infrastructure, including HTTP, TCP/IP, XML and SOAP.

# 11

# Instant Messaging and Presence Service (IMPS)

John Buford and Mahfuzur Rahman

## 11.1 Introduction

An instant messaging and presence service (IMPS) supports a community of users to exchange messages in real time and share their online status. This service is a linkage to other services such as file transfer, telephony and online games. The key technology directions include richer media, security, integration with location-based services and collaboration applications. This evolution is likely to occur as standards-based IMP systems become more prominent.

### 11.1.1 Basic Concepts

Following the rapid growth in the 1990s of consumer adoption of Internet services including e-mail and the web, instant messaging has become a mass market phenomena. Online international communities of millions of users have been drawn to instant messaging (IM) service providers such as AOL AIM/ICQ, Microsoft Messenger and Yahoo! Messenger. As IM popularity has grown, other uses in enterprise and consumer applications have emerged, including team collaboration, online gaming and news distribution.

Cellular service providers, which have provided two-way paging and SMS messaging services since the early 1990s, have more recently added instant messaging clients for the leading IM services. This is expected to fuel further growth in the use of IM. Awareness of the mobile user's location is also enabling new IMP applications.

The real-time character of instant messaging produces the experience of live conversation, albeit in text. Associated with this conversation is mutual awareness of the conversant's state, in IM terminology this awareness of other user's states is referred to as the user's presence. In typical IM systems, a user sets his presence state and makes the presence state available for other users by publishing it. Users who wish to receive the published changes subscribe to it and, if authorized by the publisher, receive updates when the state changes.

Although instant messaging and presence are closely associated, the publish/subscribe mechanism for presence has become central to other applications such as voice calls, online gaming and collaboration.

*Emerging Wireless Multimedia: Services and Technologies*   Edited by A. Salkintzis and N. Passas
© 2005 John Wiley & Sons, Ltd

A user has a live view of the presence of his online associates or buddies. When a buddy's presence changes, for example, indicating that the user is available, the user detecting the change can immediately initiate an invitation to the buddy to join an IM session, an online game or a multimedia call. In this way, a user's view of the active set of presence subscriptions is a launching point for a variety of multi-party interactions. However because of the close association of instant messaging and presence functions, it is convenient to refer to IMP with the understanding that the broader set of interactions is intended.

### 11.1.2 Brief History

Today's IMP systems are built on the legacy of early bulletin board systems (BBSs), which were followed in the late 1980s by distributed chat systems. The earliest Internet-based chat system, Internet Relay Chat (IRC) is still in use today with thousands of sites worldwide.

IRC, developed by Jarkko Oikarinen in 1988, is a multi-user chat system where participants convene on 'channels', usually with a topic of conversation, to talk in public or private groups (see Figure 11.1). It consists of various separate networks of IRC servers. The largest IRC networks are EFnet (the original IRC net), Undernet, IRCnet, DALnet and NewNet. Each IRC server hosts a set of channels and acts as a relay for clients that are connected to channels hosted by other servers [1]. Clients and servers communicate using the standard IRC protocol.

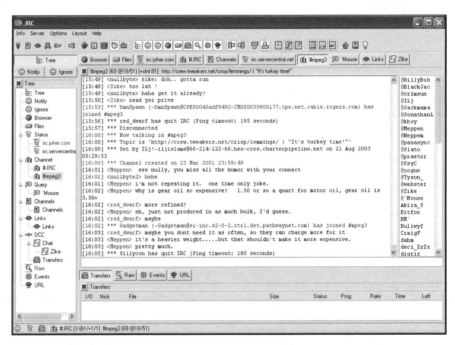

**Figure 11.1** IRC chat session (source: www.joher.com).

Mirabilis launched the ICQ ('I seek you'), the first instant messaging and presence service [2] in 1996. AOL's AIM, Yahoo! and MSN followed in 1997, 1998 and 1999, respectively.

In 1998 Jeremie Miller developed Jabber [3], an open-source IMP system that has grown into a large community. Jabber uses a standard protocol [4, 5] based on XML. SMS (Short Message Service) [6] and its successor MMS (Multimedia Messaging Service) [7] are distinguished from IM in that they provide store-and-forward messaging without presence.

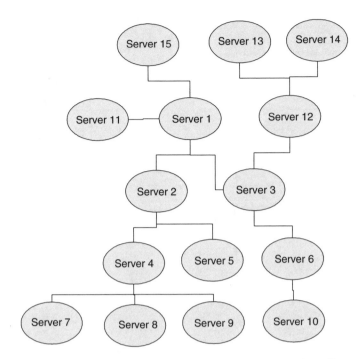

**Figure 11.2** IRC server network [1] is a spanning tree over which messages from clients are routed.

### 11.1.3 Standardization

Basic IMP systems design is well understood and multiple large-scale proprietary implementations have been operating for years. In the last few years, issues of interoperability, security, manageability and evolution have motivated several standardization efforts. Beginning in 1998, the IETF's IMPP (Instant Messaging and Presence Protocol) Working Group developed a model [8] and a set of protocol requirements [9].

Figure 11.3 shows the key IMP elements in the IMPP model. There are two services – Instant Message Service and Presence Service. The Instant Message Service has two clients. The Sender sends instant messages; the Instant Inbox receives instant messages. Messages may be filtered; delivery rules define the filters.

Presence Service has two clients. The Presentity (Presence Entity) publishes presence information. The Watcher receives presence information. The Watcher can be a Fetcher that receives presence information on a demand basis or a Subscriber that requests a subscription to a Presentity. The Subscriber receives presence information via notification when the Presentity publishes it. These entities interact with users (principals) through User Agents (UA), which represents the user interface function.

Presence Information is a sequence of tuples, each includes:

- Status (e.g., online/offline/busy/away/do not disturb)
  Open: IM accepted; Closed: IM not accepted;
- Communication address (contact address, contact type).

The Presence Service may make information about a Watcher available to other Watchers. For example, what is the list of Presentities that Tom's Watcher subscribes to or what are the Presentities that Sue's Watcher has received presence information for in the last eight hours? There are visibility rules that

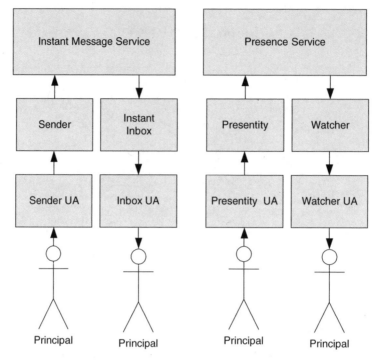

**Figure 11.3** Key IMP entities [8].

constrain how the Presence Service makes this available. The Watcher UA specifies the visibility rules for its associated Watcher entity.

The IMPP Protocol requirements [9] cover the following aspects:

- The relationship between presence service and instant message service;
- interoperability across domains;
- dimensions of scalability, including entities, domains and subscriptions;
- access control;
- end-to-end signaling and transport through proxies and firewalls;
- security, including message encryption and authentication.

The work on standard protocols then proceeded within IETF by two separate approaches. The SIP (Session Initiation Protocol) Working Group launched the SIMPLE Working Group, which is developing specifications for IMP that are compatible with the SIP specifications [10, 11]. IETF XMPP standardized the core XML protocol used by Jabber [4, 5].

Leading wireless handset manufacturers[a] formed the Wireless Village consortium in 2001 and produced specifications for a standard protocol. In 2002 the Wireless Village work was moved into the Open Mobile Alliance (OMA), which has published an updated set of Wireless Village specifications [12].

---

[a]Ericsson, Motorola and Nokia.

### 11.1.4 Issues

Key issues facing IMP evolution include interoperability, security and integration with other multi-user applications.

Interoperability addresses how users on different IMP services interact with one another and use other IMP services. There are two types of solutions that are discussed later in the chapter: gateways and multi-protocol clients.

A broad set of security issues includes:

- Spam, or unsolicited messages;
- messages or downloaded content containing viruses;
- denial of service attacks;
- impersonation or spoofing;
- stalking (using presence information to determine the location of a person);
- interception of messages.

Instant messaging can be integrated with other services to provide richer functionality. Such services include telephony, gaming and content delivery. Telephony enhanced with IMP can be part of a unified messaging service [13]. In multi-player games, IMP can be used by players to communicate with each other or with synthetic characters during the game.

### 11.1.5 Overview of Chapter

The remainder of the chapter surveys IMP clients, architecture principles, leading IMP protocols, security and future directions.

In this chapter, description and analysis is focused on the standard IMP protocols unless otherwise indicated.

## 11.2 Client

Each IMP service provider offers a PC-compatible client for using its IMP services. In addition, several clients that support a number of IMP services are available. There are also clients for mobile devices. This section presents one desktop and one mobile client.

### 11.2.1 Desktop

Figure 11.4 shows the contact list and messaging windows for Yahoo! Messenger. The contact list shows the list of buddies and their presence status. An IM session with one or more buddies can be launched from the contact list. The chat window shows message exchange between two buddies. New messages can be entered in the lower part of the window.

### 11.2.2 Mobile

Figure 11.5 shows the JustYak client for mobile devices that support J2ME applications. JustYak buddies are listed with their screen name and presence information in the 'Yakers' screen. A Yaker can be selected from the list and messages exchanged.

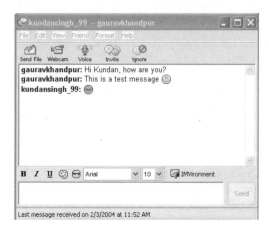

**Figure 11.4** Yahoo! Messenger 5.6.

**Figure 11.5** JustYak.

## 11.3 Design Considerations

Before describing specific systems, this section first reviews the basic functional elements of an IMP system. Some of the key systems design issues are then described. Finally, the basic end-to-end system architecture choices are discussed.

### 11.3.1 Basic Functional Elements

There are six functional elements:

(1) addressing and identification;
(2) address books and contact lists;
(3) messaging and content encoding;
(4) presence;
(5) groups;
(6) search.

**Table 11.1** Address formats

| Protocol | Example |
| --- | --- |
| SIP/SIMPLE | John Smith <sip:john.smith@somecompany.com:5060> |
| Wireless Village | wv:john.smith/friends@imps.com |
| XMPP | romeo@montague.net/orchard |

Entities that participate in IMP sessions must be identifiable and addressable. Addresses are used in the source and destination of messages, to route requests and responses, to name groups, to identify system entities, and to name arbitrary resources such as a mobile handset. In addition, systems may support other user identifiers such as user names, screen names and aliases. Table 11.1 shows example addresses for different protocols. The generic elements are:

$$\text{DisplayName Scheme}: \text{user@domain}:\text{port}/\text{resource}$$

Users maintain lists of addresses in address books or contact lists. Address books may be maintained at the client or at the server. The user selects entries in the address book to send invitations, messages and other requests.

Users exchange messages by first entering a dialog or session with one or more buddies. During the IM session, messages are transported over the network from the source endpoint to the destination endpoints and are displayed in arrival sequence order to the user. Some systems also support message storage at a server.

Messages may be encoded for various content types, although text messaging is the general medium. MIME type encoding [14] is used in several standard protocols.

User status and availability are collected in a set of state variables manipulated by the user at the client and referred to as the user's presence. The user publishes his presence whenever the presence attribute values are changed. Other users who subscribe to the publisher's presence receive the updates if they are online. Each IMP system defines a common content and format for presence attributes. Some systems provide extension mechanisms. Mechanisms for subscribing to and publishing presence information must also be defined.

Multi-party IM occurs in groups. A message sent to a group is sent to all users who have joined the group. Groups are addressible and typically have names. Additional operations on groups include:

- Create and Delete;
- Join and Leave;
- Moderate;
- Control attributes, such as public/private.

Finally, the ability to search for users, groups and resources is an important function in IMP.

### 11.3.2 Basic System Concepts

Assuming the availability of the necessary network infrastructure and devices, the basic system concepts for IMP include:

- end-to-end transport of messages, presence and control;
- distribution of function between edge and core network;
- resource consumption, particularly in wireless handsets and networks;
- systems architecture for scalability and reliability;
- security;

- gateways to other services;
- APIs and application integration interfaces.

In order to support a large collection of geographically distributed concurrent users, IMP systems are typically architected as a distributed set of servers that communicate with clients and other servers using predefined protocols. Generally the distribution of servers and the configuration of client connections to servers is a deployment decision that is invisible to users. Signaling may be initiated by a client or a server.

Communication between endpoints may be routed through the server network or may be routed between endpoints using peer-to-peer transport. Clients may be located in separate network domains where NAT (network address translation) is used.

The distribution of functionality and state between the edge (clients) and core network (servers) is a key architectural decision that affects performance and scalability. Thin client behavior, while desirable for resource-limited devices, may require more network traffic.

Scalability, the ability of a system to support incremental growth by adding incremental system capacity, is affected by the amount of new resources needed at a server per additional user, user session, network connection, routing tables and protocol message. IMP systems face tradeoffs similar to other Internet services in placing resource state at the server or client versus the protocol overhead. Other techniques include state life-time and frequency of state refreshing, and the ability to transmit partial updates.

Security, discussed later in the chapter, includes authentication of users and endpoints, protection of information, and privacy.

Gateways can be used to translate traffic from one domain to another. The translation includes protocol conversion and mapping of policies and identities.

Finally, in addition to protocols, APIs are needed for application developers to implement interoperable IMP clients and integrate IMP functionality into other applications.

### 11.3.3 Management Aspects

Both users and service providers benefit when the IMP infrastructure can be economically and efficiently administered. This requires system support for:

- configuration of client and server;
- management of a domain, its resources, and user accounts;
- establishing and administering security policy

Examples of management specifications include XCAP for SIP/SIMPLE [16] and management facilities of the Parlay platform [17].

### 11.3.4 Service Architectures: Client-Server and Peer-to-Peer

In a client-server system, clients communicate only with a server, and the server is responsible for routing requests and responses between clients. In a peer-to-peer system, clients communicate directly with each other.

Generally, peer-to-peer systems require more complex mechanisms for finding peers, discovering resources, and finding routes. Peer-to-peer avoids the single point of failure problems of the client-server and permits operation in ad hoc networks. Peer-to-peer systems may provide better scalability.

The majority of IMP systems are client-server; SIP/SIMPLE is a hybrid model. JXTA [18] is an experimental peer-to-peer system that includes an instant messaging client.

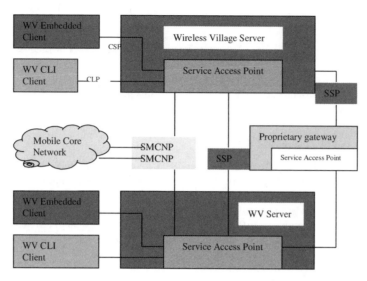

**Figure 11.6** Wireless Village service architecture [12].

## 11.4 Protocols

### 11.4.1 OMA Wireless Village

Wireless Village (WV) [12] is an open specification for mobile IMP services (see Figure 11.6). This system is designed for the mobile environment and is being standardized by the OMA (Open Mobile Alliance). The specification is primarily dedicated to 2G and 2.5G communication networks.

The goal of Wireless Village is to deliver symmetrical IMP communication between any mobile device and other wire line network IMP service. A WV system consists of a set of clients and servers. The WV Server is composed of Application Service Elements (ASE) and Service Access Points (SAP). An ASE includes:

- Presence Service Element. Manages presence information and contact lists.
- Instant Messaging Service Element. Manages the exchange of instant messages, including different types of message types such as text, video, image and sound.
- Group Service Element. Manages user groups including creation, deletion, and group properties.
- Content Service Element. Provides support for sharing images and documents between users.

The interface between a server and its environment is called the Service Access Point (SAP) and includes clients, other servers, the Mobile Core Network and gateways to non-WV servers. SAP supports the following features:

(1) Authentication and Authorization:
    - Application-Network Authentication/Authorization
    - User-Application Authentication/Authorization
    - Application-Application Authentication/Authorization
    - User-Network Authentication
(2) Service Discovery allows an application to identify the collection of service capability features that can be used

(3)  Service Agreement establishes service level agreement between clients
(4)  User Profile Management retrieves and updates user profile information.
(5)  Service Relay, using the Server-to-Server Protocol (SSP), routes all service request and responses between the servers.

A client can be either an embedded GUI client or a command line interface (CLI). Interoperability between servers and clients is achieved through the Wireless Village Protocol Suite.

### 11.4.1.1 Wireless Village Protocol Suite

The Wireless Village Protocol Suite consists of the Client-Server Protocol (CSP), Server-Server Protocol (SSP) and Command Line Protocol (CLP). The protocol stack is shown in Figure 11.7. The Client-Server Protocol (CSP) allows embedded mobile or desktop clients to communicate with a server. A collection of servers and gateways communicate using the Server-Server Protocol (SSP). A network of servers can interoperate with non-WV systems through gateways that support SSP. This facilitates the ability of clients to communicate with existing proprietary IMP networks. The SMCN (Server Mobile Core Network) Protocol allows Wireless Village server access to the mobile core network so that it can get presence and service capability information.

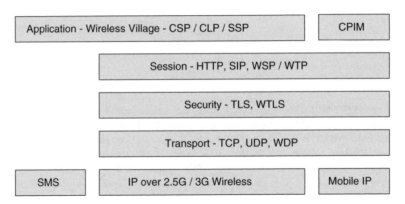

**Figure 11.7** Wireless Village protocol suite [12].

CLP is designed to provide the Wireless Village server and the CLI client with the means to communicate and interact with each other to support the IMP services in a legacy CLI client.

The Wireless Village Protocol Suite runs at the application level, and is compliant with IETF RFC 2778 [8], RFC 2779 [9] and the IMPP CPIM [22] model. The protocols may run over different transport layer and bearer protocols.

### 11.4.1.2 Presence and Instant Messaging Service

*Instant Messaging Service*
WV uses standard MIME (Multipurpose Internet Mail Extensions) encoding so that instant messages can carry a variety of content types. Out-of-band multimedia content can be shared through URI indirection. IM sessions can be one-to-one or one-to-many.

Figure 11.8 shows a stanza in an XML stream in the CSP protocol that is sending an instant message. Each request in CSP has an explicit response. In this case, a SendMessage request is given a SendMessage response (Figure 11.9). Each message has a unique transaction id. The response from

```
<SendMessage-Request>
        <DeliveryReport>T</DeliveryReport>
        <MessageInfo>
                <ContentType>text/plain</ContentType>
                <ContentEncoding>None</ContentEncoding>
                <ContentSize>38</ContentSize>
                <Recipient>
                        <User>
                                <UserID>wv:bob@somedoamin.com</UserID>
                        </User>
                        <Group>
                                <ScreenName>
                                        <SName>Bob</SName>
                                        <GroupID>wv:bob/group@somedoamin.com
                                        </GroupID>
                                </ScreenName>
                        </Group>
                        <ContactList>wv:bob/My_friends@smith.com
                        </ContactList>
                </Recipient>
                <Sender>
                        <User>
                                <UserID>wv:alice@anotherdomain.com</UserID>
                        </User>
                </Sender>
                <DateTime>20040725T1340Z</DateTime>
                <Validity>400</Validity>
        </MessageInfo>
        <ContentData>
                This is an example test message for WV
        </ContentData>
```

**Figure 11.8** SendMessageRequest for Wireless Village protocol [12].

```
<WV-CSP-Message xmlns="http://www.wireless -village.org/CSP1.1">
        <Session>
        <SessionDescriptor>
                <SessionType>Inband</SessionType>
                <SessionID>im.user.com#48815@server.com</SessionID>
        </SessionDescriptor>
        <Transaction>
                <TransactionDescriptor>
                        <TransactionMode>Response</TransactionMode>
                        <TransactionID>appl#12345@ex100</TransactionID>
                        <Poll>F</Poll>
                </TransactionDescriptor>
                <TransactionContent xmlns="http://www.wireless -
                village.org/TRC1.1">
                        <SendMessage-Response>
                                <Result>
                                        <Code>200</Code>
                                        <Description>Successfully
                                        completed.</Description>
                                </Result>
                                <MessageID>0x0000f132</MessageID>
                        </SendMessage-Response>
                </TransactionContent>
        </Transaction>
        </Session>
</WV-CSP-Message>
```

**Figure 11.9** Response sent by the server in response to a send message [12].

the Wireless Village server indicates that the message has been properly delivered to the recipient. It also shows the session and transaction description tags associated with the instant message.

CSP also enables multimedia content to be uploaded to a third-party content sharing server (e.g. Yahoo! Photo Album). The user can inform other users of the shared content through the one-to-one or one-to-many IM that contains the URI of the shared multimedia content. Or the user can inform other users through an invitation that carries the URI of the shared multimedia content. The capability negotiation feature enables the server and client to recognize the capabilities and supported MIME types of each terminal in order to facilitate interoperability in multimedia content sharing.

### Presence Service

The presence service in Wireless Village follows the subscribe-notify paradigm. In order to get presence information of a presentity, a user subscribes to the presence information of that presentitity, and in response notification messages are generated by the server. The example, Figure 11.10 shows a

```
<WV-CSP-Message xmlns="http://www.wireless-village.org/CSP1.1">
      <Session>
      <SessionDescriptor>
            <SessionType>Inband</SessionType>
            <SessionID>im.user.com#100@server.com</SessionID>
      </SessionDescriptor>
      <Transaction>
            <TransactionDescriptor>
                  <TransactionMode>Request</TransactionMode>
                  <TransactionID>appl#100@exam100</TransactionID>
            </TransactionDescriptor>
            <TransactionContent xmlns="http://www.wireless -
            village.org/TRC1.1">
                  <SubscribePresence-Request>
                        <ContactList>wv:alice/ContactList-5@alice.com
                        </ContactList>
                        <PresenceSubList
                        xmlns="http://www.wirelessvillage.
                        org/PA1.1">
                              <OnlineStatus/>
                              <Registration/>
                              <ClientInfo/>
                              <TimeZone/>
                              <GeoLocation/>
                              <Address/>
                              <FreeTextLocation/>
                              <UserAvailability/>
                              <PreferredContacts/>
                              <PreferredLanguage/>
                              <StatusText/>
                              <StatusMood/>
                              <Alias/>
                              <StatusContent/>
                              <ContactInfo/>
                        </PresenceSubList>
                  </SubscribePresence-Request>
            </TransactionContent>
      </Transaction>
      </Session>
</WV-CSP-Message>
```

**Figure 11.10** Subscription for presence information in Wireless Village protocol [12].

subscribe request for presence information such as on line status, availability, geographical location, address and status of user Alice. This subscription request is identified by the transaction id. When a response to this subscription request is received, it is matched against this transaction id.

### 11.4.2 IETF SIP/SIMPLE

SIP [10] is a standard IETF signaling protocol used for setting up, controlling and tearing down 'interactive communication sessions' with two or more participants. SIP sessions include, but are not limited to, multimedia sessions and telephone calls. On the other hand, SIMPLE [11] is a set of extensions to SIP made specifically to support instant messaging and presence. SIP is an application-layer text-based protocol modeled after HTTP/SMTP protocols. In other words, each SIP request is an attempt to invoke some methods on the called party similar to HTTP.

Figure 11.11 shows a SIP-based network architecture. The main entities in the architecture are the SIP User Agent (or SIP Client) and the SIP Proxy/Redirect Server.

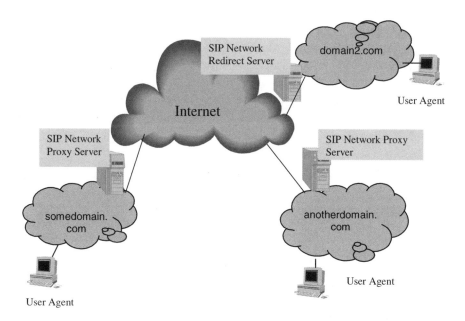

**Figure 11.11** General architecture of SIP-based network.

- SIP Client or User Agent (UA). An end system that acts on behalf of someone who wants to participate in a communication session. It comprises a User Agent Client (UAC) that initiates a SIP request and a User Agent Server (UAS) that answers the request. The presence of UAC and UAS enables peer-to-peer communication.
- SIP Proxy/Redirect Server. After receiving a SIP request, a proxy server determines the next hop server on the way to the destination and then forwards the request to the next hop server. A redirect server, instead of forwarding the request to the next hop server, notifies the sender to contact the next hop server by sending a redirect response to the sender.

Apart from the methods defined by SIP, SIMPLE provides extensions to SIP in order to support instant messaging and presence functionality such as

- SUBSCRIBE, which supports subscription of presence information, the address (URI) of the presentity of the presentitiy whose presence information is sought is included in the SUBSCRIBE request;
- NOTIFY, which generates a message that conveys presence information;
- MESSAGE, which is used when a user agent wishes to send an instant message. The body of the MESSAGE method contains the instant message itself. The headers of MESSAGE method contain sender address and recipient address in addition to other SIP related headers.

The SUBSCRIBE and NOTIFY methods are primarily intended for asynchronous notification of events due to changes in presence information. A SUBSCRIBE request for presence information usually contains an EXPIRES header that specifies the duration of the subscription of presence information. In order to keep subscriptions in effect beyond the duration communicated by the EXPIRES header, subscribers need to refresh the subscriptions on a periodic basis by reissuing the SUBSCRIBE request. NOTIFY messages are sent to inform the subscribers of changes in the state of presence information. Usually a single SUBSCRIBE request can generate multiple notification messages because the presence state may change multiple times during the expiration period of the SUBSCRIBE request.

Some of the important components of SIMPLE are as follows.

### Presence User Agent (PUA) [15]

This is a logical entity that manipulates the presence information for a presentity. Each presentity can have several PUAs' allowing a user to have multiple devices such a cell phone or PDA. Each of these devices independently can generate a component of the overall presence information for the presentity. The PUA is responsible for pushing or publishing presence information to the presence system. The PUA is not responsible for receiving SUBSCRIBE messages and also does not send NOTIFY messages.

### Presence Agent (PA) [15]

This entity is a SIP user agent and has the following functions:

(1) receives SUBSCRIBE messages;
(2) responds to SUBSCRIBE messages;
(3) generates notifications when there is a change in presence state.

As the PA has the capability to receive SUBCRIBE requests and to send out NOTIFY messages, the PA needs to access the presence information generated by the PUA. There are several ways in which this can be achieved:

(1) by co-locating the PA with the Proxy/Registrar;
(2) by placing the PA at the PUA at the presentity.

However, these are not the only locations where a PA can be located. The SIMPLE specification does not mandate any specific location for PA. A PA is addressable by a SIP URI and there can be multiple PAs associated with a particular presentity each handling a subset of currently active subscriptions for the presentity.

SIMPLE provides two services: Presence Service and Instant Message Service, which are described below.

### 11.4.2.1 Presence Service

The Presence Service receives, stores and distributes presence information. The clients are either presentities or watchers. The role of the presentities is to provide presence information. The role of the watchers is to receive presence information. Figure 11.12 shows a Presence Service system. The watcher

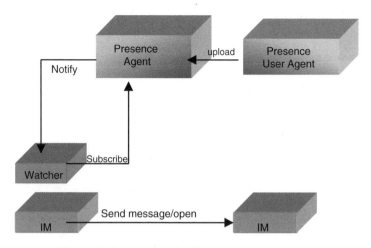

**Figure 11.12** Interaction of different SIMPLE components.

subscribes to presence information of a presentity via the Presence Agent (PA). The PA notifies the watcher of any changes to the presence information. The Presence User Agent (PUA) publishes presence information on behalf of a presentity to the PA. In this example the PUA is a logical entity that provides support to a presentity.

Figure 11.13 shows an example SUBSCRIBE message where Alice is subscribing to Bob's presence information. In response to a successful subscription, notification messages will be sent to Alice containing the state of Bob's presence information. For simplicity, this example only shows the communication between the two end parties. We have omitted any intermediaries such as SIP proxy/ server between the two end parties. Figure 11.14 shows a typical NOTIFY message.

```
SUBSCRIBE sip:bob@example.com SIP/2.0
To: <sip:bob@example.com>
From: <sip:alice@example.com>;tag=78923
Call-Id: 1349882@alice-phone.example.com
CSeq: 1 SUBSCRIBE
Contact: <sip:alice-phone.example.com>
Event: presence
Expires:600
Accept: application/cpim-pidf+xml
Content-Length: 0
```

**Figure 11.13** SUBSCRIBE to presence information of Alice's camera [11].

### 11.4.2.2 Instant Message Service

SIP provides two modes of instant messaging: page-mode and session-mode. Multimedia content can be transmitted using either of these two modes. In page-mode, a User Agent sends and receives instant messages to and from other User Agents without establishing a messaging session. The User Agent uses the MESSAGE method to send such messages. There are no explicit associations between messages. Each instant message stands alone. Page-mode messaging is only suggested for use for short messages

```
NOTIFY sip:alice@alice-phone.example.com SIP/2.0
To: <sip:alice@example.com>;tag=78923
From: <sip:bob@example.com>;tag=4442
Call-Id: 1349882@alice-phone.example.com
CSeq: 20 NOTIFY
Contact: <sip:bob.example.com>
Event:  device-info
Subscription-State: active;
Content-Type: application/cpim-pidf+xml
Content-Length: ..

  <Presence Information Data Format(PIDF)Document>
```

**Figure 11.14** NOTIFY message [11].

as it uses the regular signaling channel. Every MESSAGE method is acknowledged by the recipient with a SIP 200 OK response.

Figure 11.15 shows how Alice can send an instant message to Bob using the SIP MESSAGE method. A SIP 200 OK acknowledgment will be sent once Bob receives the message (Figure 11.16).

```
MESSAGE sip:bob@domain.com SIP/2.0
Via: SIP/2.0/TCP alicepc.domain.com;branch=z9hG4bK776sgdkse
Max-Forwards: 70
From: sip:alice@domain.com;tag=49583
To: sip:bob@domain.com
Call-ID: asd88asd77a@1.2.3.4
CSeq: 1 MESSAGE
Content-Type: text/plain
Content-Length: 18

Hi Bob. This is an example message.
```

**Figure 11.15** SIP MESSAGE method [11].

```
SIP/2.0 200 OK
Via: SIP/2.0/TCP user1pc.domain.com;branch=z9hG4bK776sgdkse;
                                        received=1.2.3.4
From: sip:Alice@domain.com;;tag=49394
To: sip:Bob@domain.com;tag=ab8asdasd9
Call-ID: asd88asd77a@1.2.3.4
CSeq: 1 MESSAGE
Content-Length: 0
```

**Figure 11.16** 200 OK response of received message [11].

There are many applications where it is useful for a sequence of instant messages to be associated together in some way. Session-mode messaging is intended for where a sequence of messages is expected and the message content may be large, such as multimedia data, or there is some need to be able to manage the resources associated with the message exchange. A new protocol for session mode messaging, MSRP (Message Session Relay Protocol) [19], is under development. MSRP is a text-based transaction-oriented protocol. It provides a relay mechanism for NAT traversal.

MSRP introduces relays on the outside edge of the firewall/NAT. For every MSRP connection between two clients MSRP has two TCP connections. One is between the end-client and the relay, and the other is between two relays or between a relay and the remote end-client.

MSRP defines two methods: SEND and REPORT. The SEND method is used to deliver partial or complete messages and the REPORT method conveys the status of an earlier SEND request. Every SEND request has a transaction id to identify the instant message contained in the SEND request.

The example in Figure 11.17 shows MSRP session establishment and an exchange of instant messages. The MSRP session is arranged using the SIP INVITE message. Alice's SIP user agent sends an INVITE message to Bob indicating its intention to establish an MSRP session. The INVITE message contains an MSRP URL in the session description indicating the location where Alice would be expecting MSRP requests from Bob. Once MSRP parameters are negotiated using INVITE, the two parties can start sending instant messages using the MSRP SEND request. Every SEND request received by Bob will be acknowledged by MSRP 200 OK message. The transaction id in the 200 OK message will identify the message being acknowledged. The termination of an MSRP session is done using a SIP BYE message.

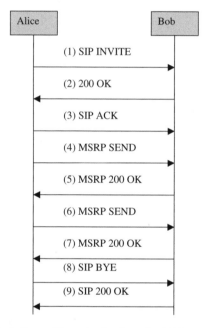

**Figure 11.17** Basic Instant Messaging Session using MSRP without relay [19].

### 11.4.3 XMPP

XMPP (Extensible Messaging and Presence Protocol) is the XML-based protocol underlying the Jabber system. Whereas Jabber evolution and specifications are managed by the Jabber Foundation [20], XMPP is standardized by the IETF. The XMPP core protocol [4] is a request-response protocol in which two endpoints exchange XML stanzas in two parallel XML streams. In XML terminology, an XMPP stream has a root element denoted by the <stream> tag. XML fragments within the stream are stanzas and represent a balanced sequence of XML tags such as shown in Figure 11.18.

Figure 11.19 shows the architecture for XMPP. Clients connect to and are authenticated by servers, which in turn route to other servers if the destination client is not locally connected. Messages to clients on foreign IMP systems go through a gateway.

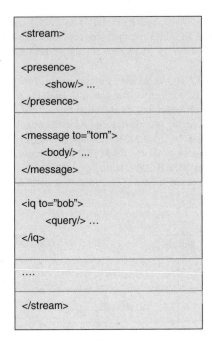

**Figure 11.18** An XMPP stream with a sequence of stanzas [4].

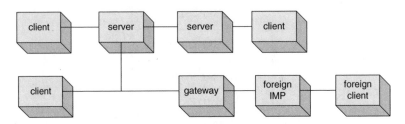

**Figure 11.19** XMPP architecture [4].

### 11.4.3.1 Instant Message Service

An example XMPP stream exchange is shown in Figure 11.20. Both client and server transmit <stream> elements to each other. Once the initial stream is successfully negotiated, XMPP stanzas can be exchanged. In this example the element <message> is used to envelop the message parameters, and the <body> tag marks the content of the message. Unlike SIP/SIMPLE and Wireless Village, XMPP does not use MIME-encoding of content types.

### 11.4.3.2 Presence Service

The <presence> stanza is used to publish user presence to subscribers. Normally notifications are sent from a client to its server and then broadcast by the server to those entities that are subscribed to the presence of the sender. After establishing a session with a server, the client sends its initial presence. The

```
C: <?xml version='1.0'?>
   <stream:stream
       to='example.com'
       xmlns='jabber:client'
       xmlns:stream='http://etherx.jabber.org/streams'
       version='1.0'>
S: <?xml version='1.0'?>
   <stream:stream
       from='example.com'
       id='someid'
       xmlns='jabber:client'
       xmlns:stream='http://etherx.jabber.org/streams'
       version='1.0'>
... encryption, authentication, and resource binding ...
C: <message from='juliet@example.com'
            to='romeo@example.net'
            xml:lang='en'>
C:     <body>Art thou not Romeo, and a Montague?</body>
C: </message>
S: <message from='romeo@example.net'
            to='juliet@example.com'
            xml:lang='en'>
S:     <body>Neither, fair saint, if either thee dislike.</body>
S: </message>
C: </stream:stream>
S: </stream:stream>
```

**Figure 11.20** Example XMPP stream with IM messaging [4].

server in turn publishes the presence stanza to all subscribers. Figure 11.21 shows an example presence with the availability set to 'dnd' (do not disturb). Other possible availability settings include: away, chat and xa (extended away).

```
<presence xml:lang='en'>
  <show>dnd</show>
  <status>Wooing Juliet</status>
  <status xml:lang='cz'>Ja dvo&#x0159;&#x00ED;m Juliet</status>
</presence>
```

**Figure 11.21** Example XMPP presence stanza [5].

### 11.4.4 Functional Comparison

Table 11.2 compares SIP/SIMPLE, Wireless Village and XMPP based on current standard specifications. All three IMP systems are being actively extended in their respective standards groups. It is likely that these systems will converge functionally over the next several years.

Both Wireless Village and XMPP use a client-server architecture whereas SIP/SIMPLE uses client-server for some functions and allows peer-to-peer communication after a session is established. Wireless Village explicitly separates the client-server protocol from the server-server protocol, which makes it possible to do client-specific optimization of the client-server protocol.

Tables 11.3 and 11.4 show presence attributes available in these three systems. Table 11.3 shows user presence attributes such as availability and identifies whether the protocol supports the attribute or not (Y/N) and provides details about the scope of the attribute in some cases. Table 11.4 shows client device attributes and similarly identifies whether the protocol supports the attribute or not.

**Table 11.2**  Functional comparison of three standard IMPS protocols

|                          | SIP/SIMPLE | Wireless Village | XMPP |
|--------------------------|:----------:|:----------------:|:----:|
| Resource-based addressing |           | ×                | ×    |
| Contact list             |            | ×                |      |
| Privacy list             |            |                  | ×    |
| Rich presence            |            | ×                |      |
| Watcher subscription     | ×          |                  |      |
| Message storage          |            | ×                |      |
| Search                   |            | ×                |      |
| Groups                   |            | ×                |      |

**Table 11.3**  Comparison of user presence attributes in three standard IMP protocols

| Attribute         | User meaning                 | SIP/ SIMPLE         | Wireless Village | XMPP              |
|-------------------|------------------------------|---------------------|------------------|-------------------|
| UserAvailability  | Availability for communication | open/closed       | Y                | away, chat, xa, dnd |
| PreferredContacts | Contact preferences          | <contact>URI        | Y                | N                 |
| PreferredLanguage | Language preference          | N                   | Y                | N                 |
|                   |                              | N                   | Y                | N                 |
| StatusText:       | User specified status text   | N                   | Y                | <status>          |
| StatusMood        | Mood                         | N                   | Y                | N                 |
| Alias             | Alias name                   | N                   | Y                | N                 |
| StatusContent     | Media info for user status   | N                   | Y                | N                 |
| Contactinfo       |                              | N                   | Y                | N                 |
| Timestamp         |                              | Y                   | N                | N                 |
| Priority          | contact priority             | Y                   | N                | −128 to 127       |

**Table 11.4**  Comparison of client device presence attributes in three standard IMP protocols

| Attribute            | Client device meaning                        | SIP/ SIMPLE | Wireless Village                                                                                        | XMPP |
|----------------------|----------------------------------------------|:-----------:|--------------------------------------------------------------------------------------------------------|:----:|
| OnlineStatus         | Logged in a server                           | N           | Y                                                                                                      | N    |
| Registration         | Registered in mobile network                 | N           | Y                                                                                                      | N    |
| ClientInfo           | General device info                          | N           | ClientType (MOBILE_PHONE, COMPUTER, PDA, CLI, OTHER); DeviceManufacturer; ClientProducer; Model; Language | N    |
| TimeZone             | Local time zone                              | N           | Y                                                                                                      | N    |
| GeoLocation          | Geographical location                        | N           | Longitude Lattitude Altitude Accuracy                                                                   | N    |
| Address:             | Network address                              | N           | Country, City, Street, Crossing1, Crossing2, Building, NamedArea, Accuracy                              | N    |
| Free Text Location   | Free text description of the location of the user | N      | Y                                                                                                      | N    |
| PLMN                 | PLMN code of the network                     | N           | Y                                                                                                      | N    |
| Comm. Capabilities   | Communication capabilities of the client     | N           | Y                                                                                                      | N    |

### 11.4.5 Gateways

While the use of gateways seems like the most likely solution for producing interoperability, limited progress has been made on defining interoperable mappings between the leading protocols. The main approach has been to define mappings [24, 25] of specific protocols to CPIM [22], which leads to a common denominator mapping.

### 11.4.6 Standard APIs

A variety of efforts have approached the interoperability problem by developing protocol-independent APIs. A notable effort is [26, 27] which provides protocol-independent application control of standard instant messaging and presence protocols. Applications using the core features of this API are intended to be portable across implementations written to different IMP protocols including IETF SIP/SIMPLE, OMA Wireless Village, and IETF XMPP.

The key application level entities available are:

– Group: join or manage a group to participate in a chat room style communication
– IMClient: send instant messages to buddies or contacts
– InstantMessage: create and access instant message content
– PresencePublisher: publish user's presence attributes and authorize access to user's presence information
– Search: search for users, groups, or other addressable resources
– PresenceWatcher: obtain access to another user's presence information

## 11.5 Security and Protocols

This section addresses the following topics related to IMPS security and privacy:

• Spam or unsolicited messages;
• messages or download content containing viruses;
• denial of service attacks;
• impersonation or spoofing;
• stalking (using presence information to determine the location of a person);
• interception of messages;
• privacy.

### 11.5.1 Authentication, Confidentiality/Privacy and Integrity

In general, IMP communication involves a set of clients communicating through a network of servers (Figure 11.22). These interconnections can be intercepted by a third party, either inadvertantly or intentionally. Such interceptions can be misused for various purposes including:

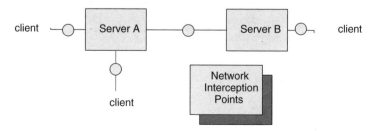

**Figure 11.22** Interception points in an IMP network.

- eavesdropping or interception of private messages by a third party;
- masquerading as a user in order to use that user's resources;
- other man-in-the-middle attacks.

There are various techniques to secure these connections against interception, typically involving encryption of connections. It is important to note that hop-by-hop signaling typically needs to be exposed at each routing point, while message content should not be exposed until it reaches the endpoint. So a combination of end-to-end content encryption with hop-by-hop mutual authentication might be used.

Hosts are also subject to attack and compromise, either at a client or server. Mutual authentication protocols are one technique to avoid propagation of a security penetration at a client or server.

Finally, a user typically wishes to protect the privacy of his presence information. Most IMP systems associate grant/block lists with access to privacy information, so that a user can explicitly determine who can and who cannot access the presence information. Further granularity of access control can be provided by defining access control filters so that a user can grant/block access to a specified subset of presence attributes.

Additionally, a user can subscribe to the Watcher list, to be notified of each subscription to his presence information. Each time a subscriber is added or removed from a user's presence information, the user receives a notification.

These techniques provide control over direct access of presence information to trusted parties, but do not protect second-hand leakage of presence information to untrusted parties.

### 11.5.2 Spam in IM

The vulnerability of IM systems to spam is similar to email (Figure 11.23). A source of spam (or spammer) must either obtain an account from an IM service, masquerade as another user, or insert a new server into an IMP service. Then the spammer creates a message addressed to either a group or a user whose address has been previously captured.

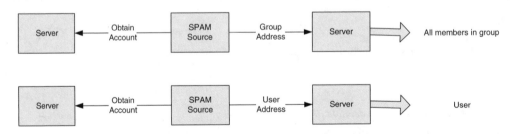

**Figure 11.23** Spam insertion into IM.

However, unlike Internet e-mail, IM systems have additional facilities to protect the IM from spam. First, an IM recipient can create grant and block lists that explicitly permit or deny other users the ability to send an IM to the recipient. If the spammer isn't on the grant list, or is on the block list, then the IM is blocked at the server and never reaches the recipient. Second, unlike internet e-mail, the 'From' field in IM cannot easily be spoofed or manipulated, meaning that the account that is sending the spam can be easily identified. Because accounts can be traced, service administrators can quickly disable accounts used by spammers.

### 11.5.3 Gateways and End-to-End Services

Each IMP service domain may connect to other domains using a gateway. If the domains support separate protocols with different security mechanisms, incompatibilities between the domain's security

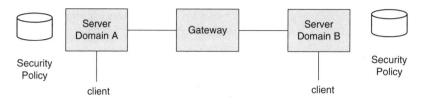

**Figure 11.24** Security policy mapping between two domains.

mechanisms may lead to exposures. As a general principle, the domain with the weaker security model will become the focal point of attacks. Figure 11.24 shows an example gateway configuration between two domains with different protocols and security policies. There are a number of issues with this configuration.

- There may not be a corresponding entity in each domain, e.g., user accounts, groups, etc., or the entity may not have the same access rights in each domain.
- Encryption mechanisms may not be the same, leading to the gateway having to de-crypt/re-encrypt.
- Authentication and authorization mechanisms may differ or may be non-existent in one of the domains.

It is a policy decision within each domain as to whether to permit weaker security mechanisms to be mapped to or from the given domain's mechanism.

### 11.5.4 Denial of Service

A denial-of-service attack is based on forcing a server or client to consume a large amount of resources to servicing requests that are not intended for normal use. For example, Figure 11.25 shows scenarios where either a single client or a collection of clients might create a huge volume of requests at a server. A single client running at full-speed might generate a large volume of messages, subscription requests, registrations, send messages of extremely large size, etc. This could cause the server to become overloaded, and prevent it from serving normal requests. In a distributed denial-of-service attack a client in one network causes multiple network clients to send a large volume of data to a target server. A server could guard against this by rejecting protocol messages that exceed a certain rate, or blocking connections from a known source if the volume exceeds a threshold. A number of denial-of-service attacks can be prevented if registration requests to an IMP server are properly authenticated and authorized by registrars.

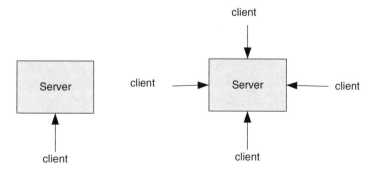

**Figure 11.25** Denial of service scenarios against an IMP server.

A similar attack could be performed by creating a large set of clients that simultaneously generate requests from a server. This type of attack is more difficult to defend against.

### 11.5.5 Security Features of Specific Standards

#### 11.5.5.1 Wireless Village

The Wireless Village interoperability framework and protocol suite addresses user privacy at several levels. The session management authenticates the user whether or not the user is allowed to use the service, and authorizes the user to a certain service level through service and capability negotiation.

Upon successful authentication and authorization, the user can control his or her own presence information so that it is only made available according to the user's wishes. The access control of presence information is achieved through proactive presence authorization and reactive presence authorization. In proactive presence authorization, the user can define several levels of access privilege with certain presence attributes, and select others in advance to grant them one level of access to the presence information. In reactive presence authorization, the user can decide on-the-fly whether or not to grant the access when someone is trying to subscribe to the presence information. The user can block and unblock the IM and invitation from others through a block list and a grant list. A decision tree is shown in Figure 11.26 The user can also reject others from his group through a reject list, and cancel the content sharing by canceling the invitation to content sharing.

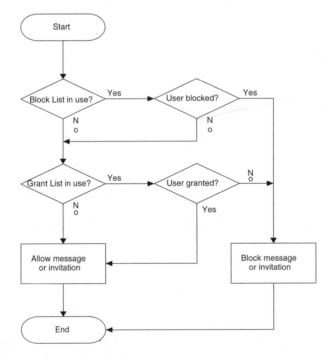

**Figure 11.26** Access control algorithm [12].

#### 11.5.5.2 SIP/SIMPLE

SIP derives it security features from existing security mechanisms used in HTTP and SMTP. SIP defines mechanism for the following security features:

- confidentiality and integrity of message content;
- authentication and privacy of participating parties.

Confidentiality and integrity can be achieved in the SIP/SIMPLE framework at the transport and network layer, which encrypts signaling traffic and hence provides message integrity and confidentiality. IPSec and TLS are the two popular transport and network layer protocols that could be used to provide confidentiality and message integrity.

IPSec is a network layer security protocol. In most circumstances IPSec does not require integration of SIP/SIMPLE applications and hence is suitable for deployments where adding security features to a host is troublesome. TLS, on the other hand, is a transport layer connection oriented security protocol, which works on top of TCP. TLS needs to be tightly coupled with a SIP/SIMPLE application. However as transport mechanism in SIP is specified in hop-by-hop basis, there is no assurance that TLS can be used end-to-end.

SIP also defines a mechanism to use the SIPS URI scheme. The use of the SIPS URI scheme signifies that each hop that a SIP message is forwarded over must be secured with TLS until the message arrives at the entity that is responsible for the domain portion of the message request URI.

SIP also suggests the S/MIME [12] mechanism to provide the integrity of the MIME contents carried in the body of the SIP message. Using S/MIME it is possible to encrypt only the body part of a SIP message. The encryption of an entire SIP message end-to-end for the purpose of confidentiality is not appropriate as the network intermediaries such as proxies need to see the headers of the SIP message. S/MIME can provide end-to-end confidentiality and integrity of message bodies as well as mutual authentication. S/MIME can also be used to provide a form of integrity and confidentiality for SIP headers fields by using SIP message tunneling.

SIP provides an extension of the HTTP digest authentication mechanism for authenticating the messages sent between user agents and also between user agent and SIP proxy. Any time that a proxy server or UA receives a request, it can challenge the initiator of the request to provide the identity information. Upon authentication, the recipient can ascertain whether or not this user is authorized for a particular service. This mechanism can be used for authenticating the subscription of presence information of a user.

### 11.5.5.3 XMPP

The fundamental security features of authentication, confidentiality and message integrity in XMPP can be achieved in following manner.

Authentication. XMPP protocol mandates the use of SASL (Simple Authentication and Security Layer) [21] for the authentication mechanism. SASL is used for identifying and authenticating a user to an XMPP server and for optionally negotiating protection for subsequent protocol interactions.

TLS. The TLS protocol is used for encrypting XML streams in order to ensure the confidentiality and integrity of data exchanged between two entities.

TLS plus SASL EXTERNAL. Authentication and Confidentiality

XMPP mandates the use of both TLS and SASL for server-to-server and client-to-server communication.

Figure 11.27 shows the order of layers in which protocols must be stacked in XMPP implementation.

## 11.6 Evolution, Direction and Challenges

### 11.6.1 Rich Presence

There is significant interest in extending the basic presence model to provide additional attributes about the user, such as [28]:

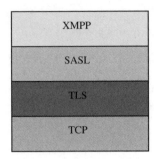

**Figure 11.27** Security and Protocol Layers for XMPP.

- the user's geo location;
- future availability (based on calendar entries);
- recency of user's activity on the device;
- further details about the user's activity, e.g., meeting, in-transit, meal, etc.

### 11.6.2 Home Appliance Control

There has been wide interest in using IM for controlling home appliances [29, 30]. Appliances might have integrated support for sending and receiving instant messages carrying control and status information. Alternately, a home gateway could have an agent that translates instant messages to a device control protocol such as X10.

Figure 11.28 shows a home network configuration in which a SIP proxy resides on a gateway with a firewall and/or NAT. SIP messages from anywhere on the Internet could be routed to the proxy for any endpoint in the home network.

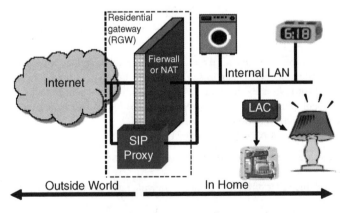

**Figure 11.28** Device control using IM [30].

Figure 11.29 shows an example SIP message carrying an XML-encoded device command. The SIP method type is 'DO', an experimental method that is not part of the SIP standard. Alternately, SIMPLE messages could be used to carry the XML-encoded command.

```
DO sip:[slp://d=lamp,r=bedroom,u=stanm]@home,net
From: sip:stan@co.com
To: sip:[slip://d=lamp,r=bedroom,u=stanm]@home.net
Via: SIP/2.0/UDP anypc.com
Content-function: render
Content-type: application/dmp
<?xml version="1.0"?>
<DMPAction>
<Device>lamp_device_id</Device>
<Control>
<Action>Power On</Action>
</Control>
</DMPAction>
```

**Figure 11.29** SIP control messaging to turn on a light [30].

The advantages of using a protocol such as SIP/SIMPLE are:

- features for security and session mobility are automatically available;
- existing IM applications can be used;
- the appliance can send information to the user regardless of where the user is.

Most likely, additional mechanisms would be needed to protect the home appliances from malicious or accidental command scenarios.

### 11.6.3 Context-Aware Instant Messaging

A powerful change in the IM paradigm is to permit IM and presence operations based not only on users' identity but also on their role and context. For example, in a study using IMP in a hospital [31], staff can send IMs to 'the attending physician in room 212 between 8 am and noon'. Thus physicians can communicate with their counterpart in a later work shift, regardless of who the actual physician is in the later shift.

Figure 11.30 shows the system architecture of the context-aware IM system. Figure 11.31 shows the PDA client with location maps to facilitate place/time IM targets.

### 11.6.4 Virtual Presence at Websites

LLuna [32] is an innovative virtual presence application that enables users to encounter other users who are visiting the same websites. Users are depicted with animated avatars, and can exchange instant messages depicted as bubbles (Figure 11.32). The underlying IMP platform for Lluna is Jabber.

### 11.6.5 IMP as Application Middleware

The evolution of IMP services as a component of application platforms is a result of the trend towards integrating communication services in applications to enable collaboration. In a document sharing collaboration, presence attributes visible to the collaborators might include which sections are currently being edited and recent change history. This gives each participant a view into the work process as well as specific tasks.

An example of middleware being developed for the Java platform are four specifications being standardized for SIP/SIMPLE IMP [33, 34] and protocol agnostic applications [26, 27]. These APIs are representative of a number of efforts towards extending software platforms with richer personal communication features.

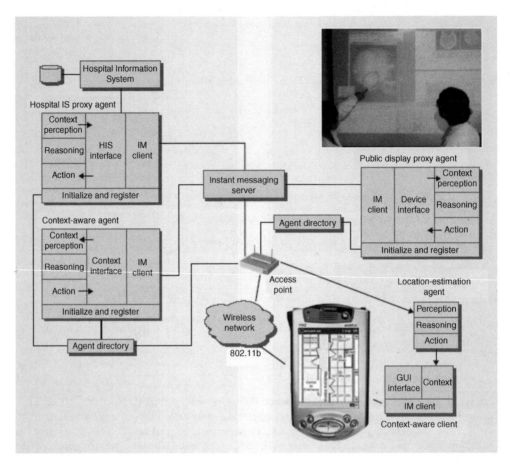

**Figure 11.30** Context-aware IM in a hospital [31].

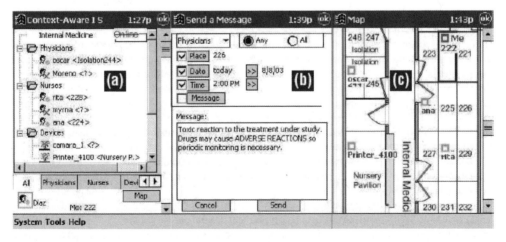

**Figure 11.31** Mobile client interface [31].

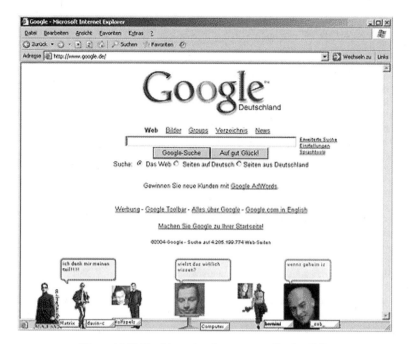

**Figure 11.32** The Lluna virtual presence application [32].

## 11.7 Summary

An instant messaging and presence service supports a community of users to exchange messages in real-time and share their online status. This service is a linkage to other services such as file transfer, telephony and online games. The key technology directions include richer media, security, integration with location-based services and collaboration applications. This evolution is likely to occur as standards-based IMP systems become more prominent.

Three approaches to IMP have been described: SIP/SIMPLE, Wireless Village and XMPP. SIMPLE leverages the infrastructure being developed to support SIP clients for packet telephony. Wireless Village was initiated by leading mobile phone vendors. XMPP standardizes the core protocol for Jabber, an open-source XML-based IMP system. Combined with the plethora of proprietary IMP systems that are widely adopted, the segmented IMP environment is likely to continue for some time.

There are a number of interesting emerging uses of IMP, including remote device control and context-aware applications. Integration of IMP functions with other desktop applications is an important trend consistent with growing emphasis on collaboration enablers in the enterprise.

## References

[1]  J. Oikarinen, Internet Relay Chat Protocol. IETF RFC 1459, May 1993.
[2]  Y. Goldfinger, S. Vigiser, A. Vardi and A. Ammon, Communication System. US Patent 6,449,344. September 10, 2002.
[3]  J. Hildebrand, Nine IM accounts and counting, *Queue* November 2003, 44–50.
[4]  P. Saint-Andre, Extensible Messaging and Presence Protocol (XMPP): Core. IETF RFC 3920. October 4, 2004.
[5]  P. Saint-Andre, Extensible Messaging and Presence Protocol (XMPP): Instant Messaging and Presence, IETF RFC 3921, October 4, 2004.

[6]  G. Le Bodic, Mobile Messaging, SMS, EMS, and MMS, John Wiley & Sons, November 2002.

[7]  3GPP2. 3GPP2 Multimedia Messaging System, MMS Specification Overview, X.S0016-000-A, May 2003.

[8]  M. Day, J. Rosenberg and H. Sugano, A Model for Presence and Instant Messaging, IETF RFC 2778, February 2000.

[9]  M. Day, S. Aggarwal, G. Mohr and J. Vincent, Instant Messaging / Presence Protocol Requirements, IETF RFC 2779, February 2000.

[10] J. Rosenberg, H. Schulzrinne, G. Camarillo, A. Johnston, J. Peterson, R. Sparks, M. Handley, E. Schooler, SIP: Session Initiation Protocol, IETF RFC 3261, June 2002.

[11] B. Campbell, J. Rosenberg, H. Schulzrinne, C. Huitema, D. Gurle, Session Initiation Protocol (SIP) Extension for Instant Messaging, IETF RFC 3428, December 2002.

[12] The Wireless Village IMPS Standard v1.2, Open Mobile Alliance, March 2003.

[13] B. Raman, R. Katz and A. Joseph, Universal inbox: providing extensible personal mobility and service mobility in an integrated communication network, *Workshop on Mobile Computing Systems and Applications* (*WMSCA '00*), December 2000.

[14] N. Freed and N. Borenstein, Multipurpose Internet Mail Extensions (MIME) Part One: Format of Internet Message Bodies, IETF RFC 2045, November 1996.

[15] J. Rosenberg, A Presence Event Package for the Session Initiation Protocol (SIP), IETF RFC 3856, August 2004.

[16] J. Rosenberg, An XML Configuration Access Protocol (XCAP), draft-ietf-simple-xcap-06, Feb 2005, work in progress.

[17] The Parlay Group. Parlay 4.1 Specification. http://www.parlay.org

[18] Bernard Traversat, Ahkil Arora, Mohamed Abdelaziz, Mike Duigou, Carl Haywood, Jean-Christophe Hugly, Eric Pouyoul, Bill Yeager, Project JXTA 2.0 Super-Peer Virtual Network, www.jxta.org. May 2003.

[19] B. Campbell, R. Mahy and C. Jennings, The Message Session Relay Protocol, IETF draft-ietf-simple-message-sessions-08.txt. February 2005, work in progress.

[20] Jabber, http://www.jabber.org

[21] J. Myers, Simple Authentication and Security Layer (SASL), IETF RFC 2222, October 1997.

[22] J. Peterson, A Common Profile for Instant Messaging (CPIM), IETF Draft <draft-ietf-impp-cpim-01>, November 2000, work in progress.

[23] B. Ramsdell. S/MIME Version 3 Message Specification, IETF RFC 2633, June 1999.

[24] P. Saint-Andre, Mapping the Extensible Messaging and Presence Protocol (XMPP) to Common Presence and Instant Messaging (CPIM), IETF RFC 3922, October 4 2004.

[25] B. Campbell and J. Rosenberg. CPIM Mapping of SIMPLE Presence and Instant Messsaging. IETF draft-ietf-simple-cpim-mapping-01, June 26 2002, work in progress.

[26] Java Specification Request 186, Presence, December 2004.

[27] Java Specification Request 187, Instant Messaging, December 2004.

[28] H. Schulzrinne, RPIDS-Rich Presence Information Data Format for Presence Based on the Session Initiation Protocol (SIP), February 2003.

[29] S. Tsang, Computing everywhere: harnessing the Internet for networked appliances, *Pervasive Computing 2001*, May 2001.

[30] S. Khurana, A. Dutta, P. Gurung and H. Schulzrinne. XML Based Wide Area Communication with Networked Appliances, *2004 IEEE Sarnoff Symposium on Advances in Wired and Wireless Communications*, April 26–27 2004, Princeton, NJ.

[31] M. Munoz, M. Rodriguez, J. Favela, A. Martinez-Garcia and V. Gonzalez, Mobile communications in hospitals, *IEEE Computer*, September 2003, 38–46.

[32] Lluna, http://www.lluna.de

[33] Java Specification Request 164. SIMPLE Presence, April 2005.

[34] Java Specification Request 165. SIMPLE Instant Messaging, April 2005.

# 12

# Instant Messaging Enabled Mobile Payments

Stamatis Karnouskos, Tadaaki Arimura, Shigetoshi Yokoyama
and Balázs Csik

## 12.1 Introduction

### 12.1.1 Mobile Payments

According to an old telecom saying, no service can be considered as such, unless you can charge for it. Mobile payment (MP) is a term used to describe any payment where a mobile device is used in order to initiate, activate and/or confirm this payment. Contrary to popular belief, mobile payments do not restrict themselves to payments via mobile phone but can be made by virtually any mobile device such as Tablet PCs, PDAs, Smartphones, or even merchant-operated mobile terminals. Several approaches have been developed [1–4], but up to now none of them has managed to reach the critical mass and establish itself as a global mobile payment service. The availability of such a service will be a driving force for the development of new mobile applications, will accelerate the growth of mobile commerce, will generate new business opportunities for the mobile operators and as such will contribute to the overall economic growth [5]. Several institutions predict that, in the next few years, payment by mobile phone will become common. [1, 6]. Some go even further, naming MP as the future killer application for mobile commerce in a 2.5 G and beyond infrastructure. Mobile Payment is expected to boost both m- and e-commerce as users will be able to pay for e/m content, in vending machines, use m-ticketing, reload their prepaid cards Over the Air (OTA), do m-shopping and pay in real and virtual Points of Sale (POS).

With today's mobile and wireless networking technologies such as GPRS, UMTS and WiFi, the Internet and its services are available to almost any kind of end-device. It is therefore an evolutionary step that many applications undertake, appearing in a mobile version that takes into account the limitations set by the end-devices such as memory, speed, capacity and connectivity. Most of the mobile payment approaches try to use widely available phone features in order to address a wide customer base. Therefore, most of them are based on SMS, but some use a combination of SMS and IVR (Interactive Voice Response) for user authentication. Some more advanced approaches take into account the last technical developments in mobile devices and make use of protocols such as IrDA and Bluetooth, while others even use existing execution environments such as JAVA and data connections such as GPRS, EDGE, etc. The debut of UMTS, wireless LAN, WiMAX and other 3G and beyond technologies will provide new capabilities [7] that will free MP from some of its limitations and allow more sophisticated approaches to be developed. In such an infrastructure, data services will become more important and provide the real revenues for the Telecom operators and their partners. Even the voice, the killer application for existing mobile phones, is going to be data based and replaced by Voice-over-IP (VoIP).

Moving towards all-IP networks means that applications and services over mobile devices will increase in number and importance, and the demand for a method of paying for them will be evident. Strict telecom billing (e.g. via the mobile phone bill or as a prepaid amount) is only one of the approaches that fail within the mobile payment domain. Mobile network operators (MNO) can handle micro-payments (usually amounts under $2) and mini payments (usually amounts ranging from $2 up to $20). Although this can provide some flexibility, we need to cover also a wider spectrum on payment amounts and, for that, the help of banks and third party financial service providers (e.g. credit card organizations) is needed. They could successfully handle mini payments, but also cover macro payments (typically any amount above $20). So, as we can see, there is a need to develop approaches that will provide a global mobile payment service that has the right business model and simultaneously takes advantage of the infrastructure and capabilities that will be common within the next years.

In the mobile payment domain, many standardization organizations and consortia [1] are working towards the goal of finding the right approach. In general, we can distinguish the following categories in existing consortia.

- MNO driven. Simpay (www.simpay.com), Starmap Mobile Alliance, GSM Association (www.gsmworld.com), European Telecommunications Standards Institute (ETSI – www.etsi.org), Universal Mobile Telecommunications System forum (UMTS – www.umts-forum.org).
- Bank driven. Mobey Forum (www.mobeyforum.org).
- Cross industry driven. Mobile Payment Forum (MPF – www.mobilepaymentforum.org), Mobile Payment Association (MPA – mpa.ami.cz), Paycircle (www.paycircle.org).
- Device manufacturer driven. Mobile Electronic Transactions (MeT – www.mobiletransaction.org).
- Technology driven. Open Mobile Alliance (OMA – www.openmobilealliance.org), Infrared Data Association (IrDA – www.irda.org).
- Identity driven. Radicchio (www.radicchio.org), Liberty Alliance (www.projectliberty.org).

We can clearly see that there is a lot of interest in mobile payments and, although there are some successful and promising approaches, there is still a long way to go before we can realize a global mobile payment service that will empower future mobile and electronic applications.

### 12.1.2 Instant Messaging

Instant messaging (IM) is a widely used service in fixed Internet infrastructure. Successful examples include ICQ (www.icq.com), Microsoft MSN Instant Messenger (messenger.msn.com), Yahoo! Instant Messenger (messenger.yahoo.com) and AOL Instant Messenger (www.aol.com/aim). Standardization fora and consortia have been working on interoperable instant messaging and presence protocols such as the SIMPLE [8], XMPP [9] and the IMPP [10] of IETF (which has become more of a standard that encompasses SIMPLE and XMPP). IM enables online users to check the status of people in their contact list and send messages in real time to each other. Therefore, any IM approach deals with synchronicity and presence awareness. The infrastructure follows the client-server architecture where the IM application is stored on the client and connects to a server in order to request the presence status of specific users. By making such a service available to mobile users, the 'anytime, anywhere' flexibility of mobility could make the already highly popular IM even more widely used, and new kind of applications can be developed that will rely on IM as an underlying message carrier. It is predicted [7] that by 2005 the revenue from mobile instant messaging in Europe could be as high as 760 million euros. Existing efforts support near real time message distribution in one-to-one or one-to-many connections. Mobile IM is one of the first presence enabled applications and, although it is basically used for transmitting text messages, it can be used for transmitting images and support multi-user applications such as shared content, white-board, conferencing, etc. Instant messaging should not be seen as a standalone service. In any future mobile scenario, context awareness sets an important new paradigm [11]. The majority of context aware applications nowadays focus on location awareness, therefore instant messaging should also be seen in that context, and especially coupled with the concepts of presence and location. This integrated approach is expected to empower future personalized mobile Internet applications that will adapt themselves to the current user's context. By adding mobile payment, it will be possible to enrich the polymorphism of such services but also their attractiveness since a personalized payment function is there. Open standards for Mobile instant messaging have been defined by the Wireless Village initiative and Open Mobile Alliance (OMA) [12] and mobile phones with integrated IM clients are already on the market.

### 12.1.3 Instant Messaging Enabled Mobile Payments (IMMP)

As we have mentioned, both instant messaging and mobile payment are promising approaches in their respective domains. Combining both of them would create a powerful duo that we consider needs to be further researched. Existing mobile payment approaches use SMS, IrDA and Bluetooth for communication. SMS has been proven to be not only insecure and unreliable, but also expensive. Therefore it may suit any archaic efforts on MP, but definitely cannot be used neither for macro payments (for security reasons) or for mini/micro-payments (due to its high cost). IrDA and Bluetooth are two protocols that require either a line of sight between the transacting devices or a limited distance between them, therefore they demand that both transacting partners in a payment scenario are more or less in the same physical space. It can be clearly seen that IM could easily slip into the role of any of these protocols. It can support security (that can be embedded on the application) and can be cost effective, since there is data communication. Furthermore both transacting parties do not necessarily have to be in the same physical space. Therefore, IM can generally replace all the aforementioned protocols in any mobile payment scenario.

However, taking a closer look at it we can see that (i) IM is a suitable medium for real-time communication, and (ii) it can be personalized based on our current context. Therefore it makes sense to use IM in mobile payments, especially within the context of person-to-person (P2P) mobile payments where both parties are known to each other (e.g. belong to the 'buddy' list). In this chapter we take this as a use-case and explore how such a service can be designed and implemented.

## 12.2 Instant Messaging Mobile Payment Scenario

It is already 2008, the technology world has survived the. com crash and investments on technology related areas have started increasing again. Internet based services are flourishing, however they are not alone this time. Mobile services are also gaining momentum and, due to their nature (anywhere, any time, in any form), have far outrun their Internet siblings in some sections. People no longer have to go home and log into a terminal to do their job; the mobile city vision has successfully made its first steps and a wide variety of people, ranging from youngsters who simply want to try the latest mobile games to business professionals who travel around the globe and want to know in real-time their portfolio performance at the stock exchange, constitute a large diverse clientele for the mobile service market. In such an era where the mobile services are starting to become integrated into the tasks of everyday life, payment for such services is a must. The mobile services and the user have set demanding requirements such as real-time payment processing and interaction with a wide variety of virtual and real points of sale around the world in virtually any currency. The good news is that this trend has been recognized early enough and such global payment services exist. In 2008 mobile payments are not only possible, but form a generic service that other more intelligent services in e- and m-commerce can easily integrate and with which they can interact.

Evelyn is a child of this era and, although she is only twelve, she has been a mobile phone owner for many years and really cannot imagine how people had managed their daily life before mobile phones. Having finished her school day, she is on her way back home, when she passes through the shopping center. Her phone beeps; a new notification has arrived from her favorite toy store (which thanks to location based services has noticed her presence) that just for today the doll that Evelyn wanted is reduced in price by thirty percent, a special discount for her as a loyal customer and, of course, as gift for her upcoming birthday. Evelyn cannot really believe her luck. She immediately looks at her instant messaging tool to see if her father or mother are online, and ask them if she can buy it. Yes, her father who is currently on a business meeting abroad is online. She quickly drafts an instant message to him, informing him of the discount and adding a photo of the doll, just to encourage his approval. Some seconds later her father replies, 'OK, go ahead sweetheart. I was planning to get this tomorrow, but you can have it today if you want'. Evelyn lets the store personnel pack the doll and is ready to pay. She can't pay with real cash as the doll costs much more than the money she carries on her or the money stored in her phone, and she is too young to have a credit card. However, this is not a problem today as her father can authorize the payment sent from the store, via his mobile from anywhere in the world although he is not physically with her. Evelyn's father receives a signed instant message from the merchant (directly or duly forwarded from Evelyn) containing an invoice for the purchase his daughter is making with her mobile phone. Subsequently he authorizes the payment and a real time receipt arrives not only at his mobile phone, but also at his daughter's and with the merchant. The transaction is complete, the doll is paid for and Evelyn can now leave the store with her birthday present. Evelyn's father knows that since he subscribed to this service he has made his family life easier by being able to handle similar situations while being mobile. The happy smile on the face of his loved ones and the easie and flexibility that this has brought to his everyday life is the reason why he keeps subscribing and, frankly, he also considers it strange that once such a service only a vision.

## 12.3 The Generic MP and IM Platforms of IMMP

Developing an IMMP approach from scratch would be like reinventing the wheel, especially when there are already prototypes available that can be brought together. Therefore we have concentrated on sticking together two existing approaches by creating the necessary APIs and the message exchange via which the two systems could cooperate. We have therefore chosen the Secure Mobile Payment Service (SEMOPS [13]), an innovative mobile payment service prototype, and the Air Series, a mobile IM service [14]. The authors of this chapter were active participants in the development of these two prototypes, so our work in trying to define the common APIs and integrate the systems was eased. In order to better facilitate our dilemmas in the design of the IMMP, we give a short introduction to these services and the way that they operate. The same operation and functions are also available on the integrated version of these two, namely the IMMP.

### 12.3.1 The Secure Mobile Payment Service

SEMOPS was initiated with the aim of effectively addressing most of the challenges bundled with a mobile payment service, and developing an open, cross-border secure approach [15]. The service is built on the credit push concept and is based on the cooperation between banks and mobile network operators (MNO). An innovative business [16] model that allows revenue sharing is combined with state of the art mobile technology with the goal of developing a real time, user-friendly mobile payment service, for virtual and real points of sale (POS), as well as for person-to-person transactions. The solution establishes new ways of interaction between the mobile commerce players, thereby relying on the already established traditional trust relationships between customers/merchants and their existing home bank or MNO.

The aim is to combine the new payment solution with various forms of proven and state-of-the-art mobile and wireless technology in order to achieve a high level of security, availability, user friendliness and interoperability. We have therefore taken into account existing approaches and, after evaluating them, an architecture that overcomes the already identified problems was designed. As in every payment system, the aim is to transfer the funds from the customer side to the merchant side. This usually happens via a financial service provider such as a bank. Figure 12.1 shows a simplistic view of what a global MP is targeting. The developed service [17] wants to provide a secure global payment service that will accommodate a wide range of sophisticated functions and basically will compete with the use of cash payments as we know today. The merchant and the customer exchange transaction data and then the fund transfer is made via the trusted payment processor (in Figure 12.1 it is the bank). The DataCenter simply routes the information flow between the actors.

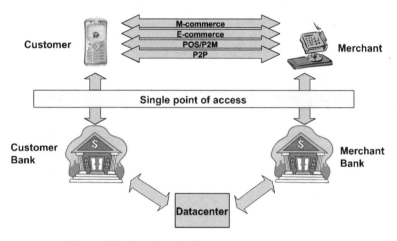

**Figure 12.1**   Overview of MP concept (bank-based model).

The proposed mobile payment service is based on the structured interaction of individual modules as can be seen in Figure 12.2. There are different transaction and channel specific front-end modules developed to reflect the underlying dependencies in heterogeneous environments and to provide a user-friendly interaction. In the mobile environment the customer modules are tailored to the specifics and technical quality of the handsets, as the design contains not only the SIM toolkit based applications but also the more modern Java and OS based modules. The payment service developed is novel in the sense that it establishes a process flow that allows cooperation between banks and mobile operators or, in general any other third service providers that can slip into these roles, e.g. a financial service provider that can assume the tasks of a bank.

In Figure 12.2 one can distinguish the main players and components in a mobile payment scenario. Each user (customer or merchant) connects with his home bank/MNO only. The banks can exchange messages between them via the Data Center (DC). We should mention that the legacy systems of the bank and the merchant are integrated in the proposed infrastructure and are used as usual. In order to give an idea of the interworking of the approach we describe one of the possible scenarios.

(1) The merchant (in general any realPOS/virtualPOS) provides the customer with the necessary transaction details. This is generally done via SMS, IrDA or Bluetooth.
(2) The customer receives the transaction data and subsequently initiates the payment request, authorizes it and forwards it to his payment processor (at the customer's bank or MNO).

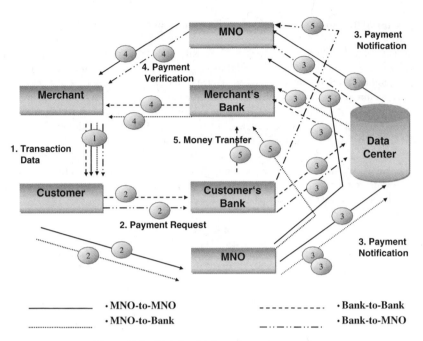

**MNO-to-MNO**
**MNO-to-Bank**
**Bank-to-Bank**
**Bank-to-MNO**

**Figure 12.2**   High-level information and money flow.

(3) The payment processor identifies the customer, verifies the legitimacy of the payment request and forwards this request to the merchant's payment processor via the DC.
(4) The merchant's bank receives the payment notice, identifies the merchant and asks him to confirm or reject the transaction.
(5) Once the merchant side confirmation comes, the funds transfer is made and all parties are notified of the successful payment.

The description above refers to a prompt payment, and is a fraction of the area covered by the MP service. However, the service is more versatile and also allows deferred, value date and recurring transactions. The service also has a refund feature and, in case of cross border transactions, automatic conversion between currencies is also possible. Further information on the architecture, design and economics of the approach can be found in the authors, previous work [17].

### 12.3.2 Air Series Wireless Instant Messaging

The Air Series wireless instant messaging [14] is a platform developed for deployment as an added value service for mobile users, and to aid the later introduction of more sophisticated context aware services. As expected, it can offer all of the capabilities of an IM platform and is purely Java based on the server and client side.

Three different parts have been developed, namely the Air Messenger client software, the AirBridge (a gateway for communication protocol conversion), and the AirBOT (an agent like application development framework), all of which compose a fully mobile messaging solution. Taking advantage of instant messaging technology, it is possible to go beyond today's Web services for mobile terminals, and develop more interactive and real-time services, one of which could be mobile payments. Figure 12.3 depicts the overall architecture of the Air Series IM solution. One can clearly see the technologies and modules of the architecture, the major ones are described below.

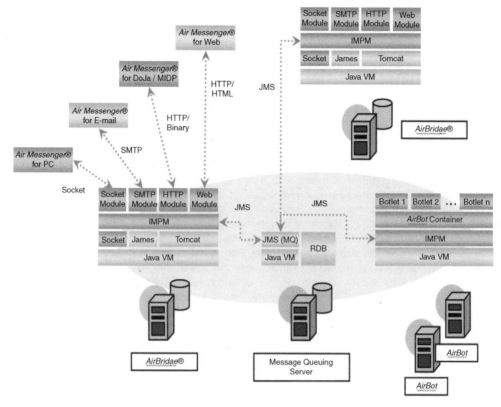

**Figure 12.3** The IM platform architecture.

- The AirBridge is a module that provides a robust and reliable framework for message and presence exchange, and seamlessly connects PCs and mobile terminals over unstable wireless connections. As depicted in Figure 12.3, the communication modules are located on the front-end of the server and handle protocol translation. For instance, when a sender sends an SMTP-based message to a receiver using HTTP-based AirMessenger, the message is forwarded to the HTTP module which compresses it to binary formats (which may be device or application dependant) for transmission efficiency. In addition to this, each module is also responsible for delivery or read-reply report management. The IMPM is a core API library to provide IM services such as messaging, presence management, and contact list management. IMPM also controls sessions between the Message Queuing Server and communication modules. Each communication module utilizes these APIs and communicates with others via the Message Queuing Server.
- The Message Queuing Server provides message queues of each user's messages. Because wireless/ mobile connections are periodically unstable, messages are often lost or resent. In order to address this unreliability problem, AirBridge uses the message queuing function of the server to reliably send messages to clients.
- The AirBOT is an application development framework for IM based real-time, agent-enabled service called BOTlet. With AirBOT, developers can easily build BOTlets on their framework and extend IM functionality from a simple messaging function to an advanced application. On the AirMessenger, entries of the AirBOT agent are listed along with 'buddies' in the contact list and users can talk with them just as they do with their friends. Answering questionnaires or information retrieval are examples of AirBOT enabled services.

In general, any HTTP/Socket/Mail based client application communicates through AirBridge with the
IM server which is based on the Java Message Service (JMS) [18]. The AirBridge development has
taken into account the work done in standardization fora such as the Wireless Village and relevant
technologies such as SIP/SIMPLE [8] and PAM4.0 [19]. However, in the prototype a proprietary
protocol is also used, in order to enhance the functionality of the IM platform and the AirBot.

## 12.4 Design of an IM-enabled MP System

We have designed and implemented a prototype of IMMP for wireless/mobile devices. IM in this
context is used as an additional channel in order to allow the transacting parties in a mobile payment
scenario to initiate contact and exchange data that will lead to the realization of the payment process.
Although many existing communication aspects within the existing flow [17] can be performed via IM,
we consider IM as most appropriate for the front-end communication, i.e. that of customer and merchant
between themselves (peer-to-peer) and their respective banks. The payment process starts with the
merchant side providing the transaction data to the customer, which is bundled into a transaction data
message according to the specifications [13]. This initial step can be done via IM, especially in cases
where the customer and the merchant are not in the same physical space, e.g. Internet purchases.

Basically, our goal was to integrate two different components, one that has been developed to handle
the mobile payment transaction (semops) and the other that will provide the add-on functionality of
instant messaging. This integration can be done in different ways according to the location and degree of
integration between the two components. Each approach has its own pros and cons. In general, we
considered the following possibilities:

- the server-based approach (Figure 12.4);
- the 'cooperating clients' approach (Figure 12.5);
- the integrated module approach (Figures 12.6 and 12.7).

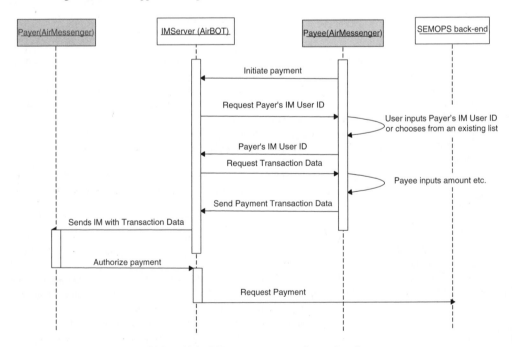

**Figure 12.4** P2P payment process (server-based).

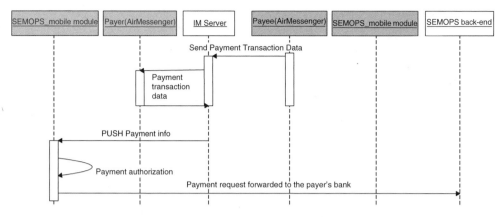

**Figure 12.5** P2P payment with 'cooperating clients'.

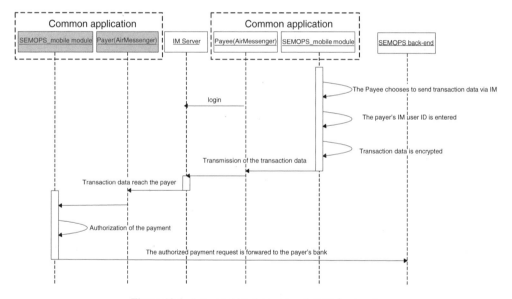

**Figure 12.6** Integrated Module approach (MP front-end).

The grey-shaded boxes in Figures 12.4–7 reside on the mobile device of the customer while the dotted-shaded ones reside on the mobile device of the merchant. The remaining components are hosted on the service provider side and are expected to be connected via the Internet. The following sections provide an insight into some possible implementation approaches when one has two independent components (in the future these can be considered as commercial of the self – COTS) that need to be integrated in order to provide a new service. However, the examples are not exhaustive, as each one of the proposed scenarios can be extended by changing the message flow among the main components or the level of dependability on the IM channel.

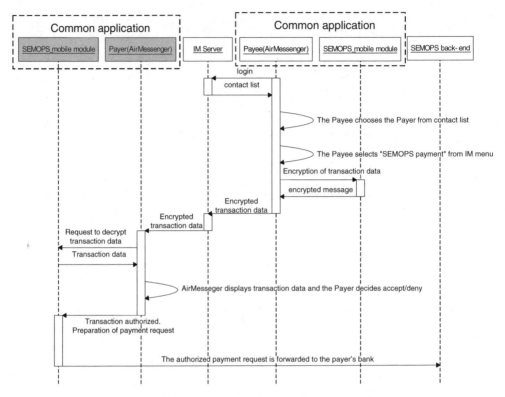

**Figure 12.7**   Integrated module approach (IM front-end).

### 12.4.1 Server-based Approach

Figure 12.4 depicts a server-based IMMP. This approach generally assumes that the core service runs on a server and the interactions are proxied via a thin client such as an application on the mobile device. In such a framework the payee interacts with the AirBOT, which is the presence/messaging handling application located on the fixed line. The AirBOT provides the main SEMOPS module functionality, which is then proxied to the mobile device and presented to the user via the AirMessenger. The AirBOT is assumed to integrate the IM server. The payee initiates the payment by providing the IM user ID (which is unique) and the transaction data, which are then processed internally by the AirBOT, and at the end the user authorization is requested. Once the user authorizes the transaction, the AirBOT sends the payment request to the SEMOPS backend, which is installed in the user's payment processor, e.g. his bank. This approach implements a wallet like server-based approach. The whole payment functionality is wrapped at the server side by the AirBOT, and IM is used to send the transaction data and request the real-time user authorization. This provides several advantages as the core application is always under the control of the service provider who can enhance it with new functionality, without having to update the client side in parallel. This can lead to incremental updates and maintenance of a wide variety of clients at end-devices (which could be tailored to the device's capabilities) without affecting the whole IMMP functionality since the core application and logic is on the server side. However, this approach also implies that some personal sensitive data are stored on the service provider side, something that limits the control that the user has on them. Other security and privacy issues include interaction between the operator's gateway and the server components in the Internet. The approach places a direct trust in the service provider that it will manage all personal data of the user. It could eventually intercept the user's communication with other parties, as the service provider is the middle man in any communication scenario and builds user profiles with private information. The proposed MP design [17] allows totally anonymous payments to be made, something that comes into conflict with this implementation approach. Furthermore [17], in general, it does not provide a wallet-like server based application as the approach here implies, but rather considers that the payment application is under the control of the user in a device that he trusts, i.e. his mobile phone. This is one possible implementation, however it does not satisfy all privacy/security concerns of the proposed MP approach, therefore we will take a look at some alternative solutions.

### 12.4.2 'Cooperating Clients' Approach

The engineering design of this approach is powered by the fact that we have two different applications, i.e. one of IM and the other of MP, which may evolve independently. In order to do this, we need to define only the abstract communication model between these two applications that will lead to cooperation between them, while both run on the mobile device of the user. Therefore this approach constitutes a thick client; one with minimal support from the server side, e.g. IM tasks. One of the technical problems that arose is that, since both applications will be running as MIDlets, it is impossible (at least for the moment) to have the two independent modules on the mobile device in a cooperation status, e.g. one controlling the other. Of course, a possible implementation approach would be to make this cooperation with external help from the server-side (as depicted in Figure 12.5), which would bridge the two MIDlets with some server-side support, however this would complicate our approach unnecessarily and possibly decrease overall performance.

This approach does not interfere with the parallel evolvement of the MP and IM modules on the mobile device and is therefore compatible with all possible MP scenarios envisaged. As already mentioned, the communication between the MP module and the IM is not done on the device, but by using the server side, which gets the information from the IM and pushes back to the MP module and vice versa. The server bridges the communication between two applications (IM and MP) on the handset, i.e. we have the payee's AirMessegner application transmitting the data to be given to the payer's MP module. The data are received from the payer's AirMessenger and, with the agreement of the user, are forwarded to the IM server, who then pushes then back to the handset, but now to the MP module, which is listening on a specific port. A PUSH like functionality (if not already supported) can be implemented by constantly polling the IM server. However, this approach requires two applications in mobile terminals and it means that the user (payee or payer) has to launch both applications before initiating a transaction. This is a step that we would like to avoid, as leaving too much responsibility on the user side is not desirable, since if one of the applications is not started the whole payment procedure fails. Therefore, we abandon this scenario (although it is feasible) for a future where multi-threaded applications and inter-MIDlet communication and management become a reality.

### 12.4.3 Integrated Module Approach

As we have seen in the previous sections, it is possible to have two different applications either cooperating on the mobile device with external help, or the IM module proxying a server based MP module, but it may add complexity or introduce unwanted behavior into our system. Therefore, the next logical step is to try to integrate both applications into one 'common application' that will have the functionality of both modules (the IM and the MP). The approach does not allow independent evolvement of its distinct parts, as the integration is now at source level.

Again, we face two possibilities with regard to the front-end interface via which the user will initiate the mobile payment. Basically we have:

- the MP front-end (realized by SEMOPS) with hidden IM functionality;
- the IM front-end with hidden mobile payment functionality.

Each of them, of course, assures an easier and friendlier usage to their respective users. Therefore existing users of SEMOPS would prefer the first case, while those familiar with the IM would probably prefer the second.

The interaction for the first case is depicted in Figure 12.6. The merchant (payee) starts the MP application on his mobile device and chooses to send the payment transaction data via IM. The payer is selected by simply typing his IM UserID. Subsequently the transaction data are sent to the IM server, who then forwards it to the payer's application via IM. The payer accepts the payment and the authorized payment request is then forwarded to his bank. In this scenario, we use IM only to get the transaction data and forward it to the bank once the payer has authorized the payment. This scenario assumes minimal IM intervention and simply proves that IM can act as an alternative to SMS, Bluetooth, IrDA and alike protocols. Furthermore, the user has very limited interaction with the whole IM process; therefore the learning curve is kept to a minimum. Usability research points out that this would mostly help existing SEMOPS users who are already accustomed to the standard process of payments and who would not like any drastic changes to the whole procedure (such as learning how to interact with an IM system).

The second case would have the IM application as the front-end for realizing mobile payments according to the proposed approach [17]. This interaction is depicted in Figure 12.7. The user is assumed to be accustomed to the IM system and its usage for communication with his 'buddies'. Therefore we want to provide added value by integrating a payment service into what he already uses. It is assumed that the payer and the payee have an initial contact over IM, and one wants to send money to the other. In this case, the payee selects the payer from his contact list (or searches through the AirBOT, which features a system-wide search service). The transaction data (M1 message) is encrypted and sent to the payer via IM. At the payer side, the transaction data is decrypted, and the user authorization is requested. Once the user confirms the payment, the payer's MP module prepares and sends a payment request directly to his payment processor, e.g. his bank.

Both Figures 12.6 and 12.7 show a fraction of the whole spectrum of possible mobile payment processes that the proposed architecture [17] can manage. In addition, the diagrams indicate that the payer always accepts the payment and that the payment request is forwarded to the selected payment processor. Of course, in case of a failure, e.g. if the payer rejects the payment, the appropriate notifications are distributed to all parties according to the MP specifications. Both scenarios can be further extended so that the IM is also used among the payment processors as well as the DataCenter. However, in our initial prototype we have focused only on the user mobile front-end scenarios.

## 12.5 Implementation

A prototype has been developed, implementing the 'integrated module' approach with the IM application as a front-end, proving that the concept is viable. Figure 12.8 shows various screenshots of the Graphical User Interface (GUI). As can be seen, the payee selects IM in his mobile device and logs into the IM server. Subsequently he selects the payer from his contact list and chooses the menu allowing him to fill in the payment info (which will create an M1 message containing the transaction data). After the payer receives the transaction data and accepts the payment, the integrated module forwards the information to the payment processor to realize the actual transaction.

Figure 12.9 shows the internal sequence realized by the integrated IMMP module. The iExternalApp interface allows access to the internal SEMOPS methods from external application. This scenario focuses on handling payments among friends (names existing in the IM 'buddies' list) but not among strangers. However it is desirable that the user can also send messages to persons not existing in his predefined 'buddies' list but also on the fly communicate with new users or be able to search for them in a user database (DB). The latter would enable the user to make payments to total strangers without having to add them to his contact list. This could be typical for one-time transaction scenarios, e.g. when a taxi driver wants to send a payment request to his one time customer. In order to handle this, he either

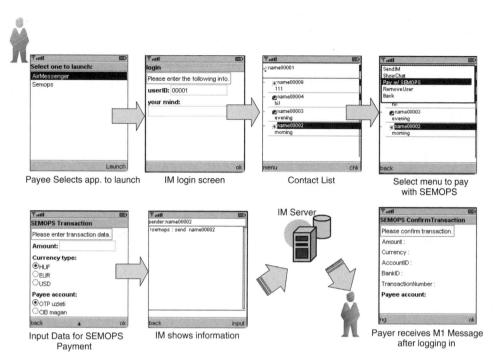

**Figure 12.8**   IMMP transaction realization.

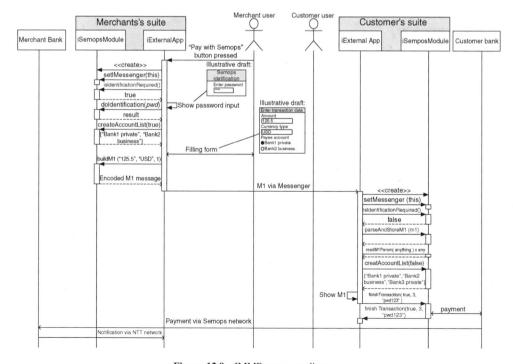

**Figure 12.9**   IMMP sequence diagram.

has to type the client's IM ID or search for it. Therefore, we developed an additional search functionality (like a white page directory) using the AirBOT. This implies a text-based search with given parameters. The AirBOT can locate a user's IM ID based on several criteria, e.g. search only online users. In the future and in order to protect the user privacy this will be more user controlled (e.g. the user will specify if such a search functionality should list him, and how many of his profile properties should be available to other parties), and even presence features may be added (e.g. the taxi driver is allowed to search only for users within 10 meters of his current location).

Figure 12.10 shows the user interface of our prototype implementation with the search feature as described above. The user only has to choose 'Search User' from his menu or just select the AirBOT which is displayed as a user on his contact list. He can interact with the AirBOT simply by choosing or typing in keywords for the temporary user. These keywords could be associated with a user name, an alias, his mobile phone number, his online or offline status, his current location or other criteria.

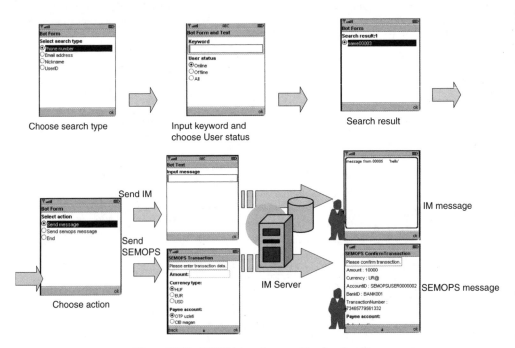

**Figure 12.10**   IMMP interactive searching functionality.

With regard to the technology side, the implementation was done in Java and is based on MIDP 1.0 specifications without any vendor-specific APIs. The JAR application is about 100 kbytes, which today limits the choice of mobile devices on which this application can run. However, we do not expect this to be a problem, as the latest generation mobile phones on the market do not have these limitations. The current implementation does not use HTTPS for secure transfer of data; instead, the payment related data are encrypted/decrypted as specified in the SEMOPS specifications. We have here to mention that we have tried to adhere to specific requirements [17] and (i) develop an application mainly targeting limited-capability mobile devices, (ii) use open interfaces in order to address a large number of end-devices, and (iii) provide an easy-to-maintain viable approach.

In the future, we plan to migrate to MIDP2.0 and take advantage of the PUSH-like functionality as well as the HTTPS support for specific communication parts. We are also interested in experimenting with mobile Public Key Infrastructure (mPKI) and attribute certificates in order to support fine-grained user, client and server authentication schemes, as well as policy-based management. Making the approach more user-friendly, secure, personalized and robust, and including cooperation schemes among different IM servers in various administrative domains to enhance the search functionality also provides future challenges.

## 12.6 Security and Privacy in IMMP

Security and privacy are essential elements for the success of mobile commerce and applications. They are business enablers and not just add-on features. This is because both elements are critical in fostering users' trust towards any mobile services and applications. Security, privacy and trust have been identified as critical enablers for the success of mBusiness by many European Union funded roadmap projects such as PAMPAS (Pioneering Advanced Mobile Privacy and Security [20]) and MB-net (A Network of Excellence on mBusiness Applications & Services [21]) as well as others such as Accenture and CERIAS [22]. IMMP is a combination of a specific MP approach [17] and IM. Because payment procedure is already specified and used by channels complementary to IM in a uniform way, it is clear that integrating IM support must fulfill the security, trust and privacy requirements set by the specifications.

The MP security framework was built with real operational payment processor environments in mind, which put some constraints on the overall approach, as outlined below.

- Banks do not allow encrypted information into the Intranet; therefore, decryption must be done in the Demilitarized Zone (DMZ).
- Banks usually have their own authentication systems, therefore any new MP approach must co-operate with existing infrastructures.
- A global MP approach should be extensible and use heterogeneous channels, including 'strange' ones, like USSD; therefore secure protocols such as SSL/TLS cannot always be used to encrypt the transmission channel.
- Regulations in different countries prohibit the usage of the same keys for encryption and signing; therefore, a new MP approach must have multiple key pairs if encryption and digital signatures are to be used.

Based on these limitations, our MP approach [17] (and therefore also IMMP) gives the security model depicted in Figure 12.11. The termination of the physical channels and the decryption of the messages takes place in the DMZ. The decrypted information reaches the bank module (residing on the Intranet of the bank) through the bank's standard authentication system, which is already used for applications such as home banking. We currently use 1024 bit RSA encrypted XML with 3DES message keys, and 1024 bit RSA digital signatures on the messages, but with a different key pair. The hardware security modules execute all the cryptographic operations in the system, resulting in the split security operations depicted in Figure 12.11.

Strong end-to-end encryption for the transferred data is provided, and different authentication techniques embedded into this encryption can be realized. Purely IM related activities such as logging in an IM server, etc. can be done over secure channels in order to minimize the risk. This seems a viable solution, but in live environments it must be adapted to the usual practices of banks, which insist on not allowing anybody else to authenticate their users, as this task has to remain within the banks' legacy procedures (at least until the policy/technology framework in the bank changes). The user data that reside on the payment processor are protected with state of the art technology solutions such as encryption, limited availability, key management, etc. The same is true for the data relying on the mobile phone, to which the user has access via an IMMP-specific authentication, such as entering a password. Furthermore, all MP interactions that contain user private data are exchanged securely with a single user-trusted entity, namely his financial service provider. The IDs used in the MP approach masquerade the specific transaction, which is untraceable from only that known information. Since the

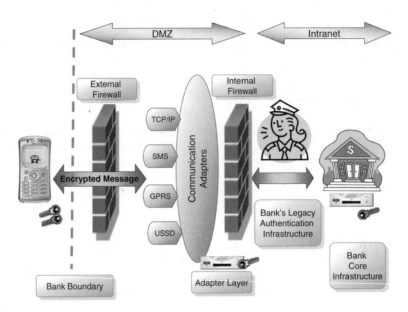

**Figure 12.11**   Split security operations at payment processor's side.

business model of our MP approach [16] also tackles the privacy, there is no immediate need at the moment to address it further via add-on solutions such as anonymity/pseudonymity services, etc. Add-on services, such as the 'search' function that we introduced, are at the moment in the hands of the service provider. However in the future it is expected that the mobile user will be able to selectively control not only who can access his presence information and when they can access it but also what portion of it would be available for a specific task. This user-controlled privacy coupled with the capabilities of out MP model can make anonymous payment possible, and provide a real alternative to today's cash dominated market.

In the future the IMMP will follow a more sophisticated approach. MobilePKI, mobile digital signatures, encryption, and biometric authentication are expected to be widely available in the future mobile devices (for instance Fujitsu's F900iC 3G handset features a fingerprint scanner). Therefore, it needs to be examined how these methods can be integrated in the system for providing strong security and privacy whenever it is required, and always balancing other requirements such as usability and performance. Furthermore, Identity Management efforts are ongoing for the Internet community and several standardization consortia such as Radicchio (www.radicchio.org) and Liberty Alliance (www.projectliberty.org) are working towards federated identity in the virtual world. If such efforts are successful, they will have a catalytic effect on MP domain, as they will provide a homogeneous identity framework capable of universally bridging the real and virtual worlds. Therefore, efforts, like the newly announced (March 2004) cooperation of NAC, OMA, OSE, PayCircle, SIMalliance and WLAN Smart Card consortium with the Liberty Alliance to demonstrate that federated identity is, among others, a key enabler in mobile payments are steps in this direction. Once this is available it will quickly be integrated to IMMP-like solutions.

## 12.7  Conclusions

Mobile payments are expected to be of critical importance for the e-/m-commerce domain within the next few years. It is expected that there will be different business models and different technology approaches in the quest for a successful service launch that will reach a critical mass and establish itself as a global payment service. Within this context we believe that IM coupled with context awareness can be successfully combined with mobile payments. In order to prove this, we designed and implemented such an IMMP service based on a global MP service [17] and an IM platform [14]. Instant messaging is a very interesting approach to realize almost real-time message exchange between the parties in a mobile payment scenario. We have presented the basic technologies that were used, as well as the design dilemmas that arose while trying to realize such a system. We have commented on the pros and cons of the different scenarios and finally implemented one of these as a proof of concept. The prototype described in this chapter has been successfully demonstrated by NTT DATA and the SEMOPS consortium to the general public at the premier information technology event CeBIT 2004 [23].

The authors believe that IM is a promising approach and, once it is extended with presence management capabilities, it can be used as a generic service in future mobile services such as mobile gaming, mobile digital rights management (mDRM) scenarios, etc. The IMMP approach depends on IM and therefore faces the same problems as do IM systems. In future IMMP users may use different IM systems, which brings to the surface the interoperability problem as well as the scalability one for services such as 'search' which we generically introduced. However, there is ongoing work [24] towards developing standardized IM applications, and new approaches [25] that may arise can be easily integrated. Mobile payments and especially person-to-person payments are the most interesting ones, without of course neglecting the IM interaction with a virtual POS. Almost in all the current scenarios of IM, the end-user is assumed to be a human entity or a machine that interacts via a pre-defined process. However, in the future such interaction could be at a more flexible level with the introduction of intelligent agents or expert systems that slip into the customer/merchant roles. Finally, wireless/mobile instant messaging may do for the 2.5 G and beyond infrastructures what e-mail did for the Internet, i.e. open a new technology to the masses. Finally it may become the de facto standard for 3G content services and other applications that may require a real-time presence-enabled framework. Mobile payments within that vision are seen as one successful IM-powered financial service, standalone or as part of more sophisticated applications.

# References

[1] Stamatis Karnouskos, Mobile payment: a journey through existing procedures and standardization initiatives, *IEEE Communications Surveys and Tutorials, Vol. 6, No. 4, 4th* Quarter 2004.

[2] Norman Sadeh, *M Commerce: Technologies, Services, and Business Models*, John Wiley & Sons, 2002.

[3] J. Henkel, Mobile payment: the German and European perspective. In *Mobile Commerce*, Gabler Publishing, Wiesbaden, Germany, 2001, http://www.inno-tec.de/forschung/henkel/M-Payment%20Henkel%20e.pdf

[4] David McKitterick and Jim Dowling, State of the Art Review of Mobile Payment Technology, Department of Computer Science Trinity College Dublin, Technical Report, http://www.cs.tcd.ie/publications/tech-reports/reports.03/TCD-CS-2003-24.pdf

[5] EU Blueprint on Mobile Payments, Accelerating the Deployment of Mobile Payments throughout the Union, Working Document, Version 1.1 (draft) – 12-July-2003.

[6] Mobey Forum White Paper on Mobile Financial Services, June 2003, http://www.mobeyforum.org/public/material

[7] Durlacher Research, UMTS Report – an Investment Perspective, London, Bonn, 2001, http://www.dad.be/library/pdf/durlacher3.pdf

[8] SIP for instant messaging and Presence Leveraging Extensions (simple), http://www.ietf.org/html.charters/simple-charter.html

[9] Extensible Messaging and Presence Protocol (XMPP) of the Internet Engineering Task Force (IETF), http://www.ietf.org/html.charters/xmpp-charter.html

[10] Instant messaging and Presence Protocol (IMPP) of the Internet Engineering Task Force (IETF), www.imppwg.org

[11] *The Book of Visions 2001 – Visions of the Wireless World*, Wireless World Research Forum, version 1.0, http://www.wireless-world-research.org, 2001.

[12] Open Mobile Alliance, Wireless Village Initiative, http://www.openmobilealliance.org/WirelessVillage

[13]  Secure Mobile Payment Service (SEMOPS), www.semops.com

[14]  I. Tanaka, T. Arimura and S. Yokohama, Wireless Instant Messenger Development in the Japanese Market, *Second International Conference on Mobile Business*, 23–24 June 2003, Vienna, Austria. www.mbusiness2003.org

[15]  A. Vilmos and S. Karnouskos, SEMOPS: Design of a new Payment Service, International Workshop on Mobile Commerce Technologies and Applications (MCTA 2003). In *Proceedings 14th International Conference (DEXA 2003)*, IEEE Computer Society Press, September 1–5 2003, Prague, Czech Republic pp. 865–869.

[16]  S. Karnouskos, A. Vilmos, P. Hoepner, A. Ramfos and N. Venetakis, Secure Mobile Payment – Architecture and Business Model of SEMOPS, EURESCOM summit 2003, Evolution of Broadband Service, Satisfying User and Market Needs, 29 September – 1 October 2003, Heidelberg, Germany.

[17]  S. Karnouskos, A. Vilmos, A. Ramfos, B. Csik and P. Hoepner, SeMoPS: a global secure mobile payment service. In Wen-Chen Hu, Chung-Wei Lee and Weidong Kou (eds), *Advances in Security and Payment Methods for Mobile Commerce*, IDEA Group Inc., Nov. 2004.

[18]  The Java Message Service (JMS), http://java.sun.com/products/jms

[19]  Presence & Availability Management (PAM) Working Group, http://www.parlay.org/about/pam

[20]  Deliverable D04: Final Roadmap (Extended Version), Pioneering Advanced Mobile Privacy and Security (PAMPAS – www.pampas-eu.org).

[21]  G. Giaglis, P. Ingerfeld, S. Karnouskos, P. Lee, A. Pitsillides, N. Robinson, M. Stylianou and L. Valeri, mBusiness Applications and Services Research Challenges, White Paper, 24th November 2003, MB-net Project (IST-2001-39164).

[22]  Roadmap to a Safer Wireless World, Security Report, Accenture and CERIAS, Oct 2002. http://www.cerias.purdue.edu/news_and_events/events/securitytrends/

[23]  CeBIT – Center for Office and Information Technology, http://www.cebit.de

[24]  Michael McClea, David C. Yen and Albert Huang, An analytical study towards the development of a standardized IM application, *Computer Standards and Interfaces Journal*, **26**(4), 343–355, August 2004.

[25]  A. C. M. Fong, S. C. Hui and C. T. Lau, Towards an open protocol for secure online presence notification, *Computer Standards and Interfaces Journal*, **23**(4), 311–324, September 2001.

# 13

# Push-to-Talk: A First Step to a Unified Instant Communication Future

Johanna Wild, Michael Sasuta and Mark Shaughnessy

'Push-to-Talk', or PTT, is a familiar term for anyone who has played with two-way radios, walkie-talkies, or even intercom systems. It simply refers to how one controls a transmit button to transmit one's voice to others over a half-duplex communication path: push [the button] to talk (PTT); release [the button] to listen.

'Push-to-Talk' also defines an instant communication application that has long been used in the private two-way radio domains, providing efficient communication service, from enhancing the performance of on patrol Public Safety officers through instantaneous communications with other Public Safety members, to allowing family members to stay in touch with each other via their 'walkie-talkies' while out hiking the wilderness. The Push-to-Talk application is now being integrated into public cellular systems.

Enhancements in communications technology have been the keystone of new and improved service offerings. Push-to-Talk over Cellular is such a keystone through its use of IP-based technology and packet data services. Now a user can activate the 'Push-to' button, send their voice over an IP channel, and talk to one or more participants of the call. In the near future this will be any media, not just voice, which is shared with the call audience. With the heart of the PTT service being IP based, it is strategically situated to leverage multimedia support for video, multimedia messaging, virtual reality applications (e.g., gaming), etc. Push-to-Talk over Cellular is a first step to Push-to-'Anything' over Cellular, a suite of 'Push-to' services that enable sending any media, from anywhere to anywhere, in a unified instant communication service set.

The following sections provide a view into what Push-to-Talk is, how it works in the cellular domain, and how this may evolve as an instant communication offering in a multimedia world, fulfilling the vision of a Push-to-Anything future.

*Emerging Wireless Multimedia: Services and Technologies*   Edited by A. Salkintzis and N. Passas
© 2005 John Wiley & Sons, Ltd

## 13.1  Short History of PTT

Early in the 20th century, radio communication was adapted for use by the public safety community to provide timely broadcast information via radio receivers in police squad cars, improving the efficiency of the police force out on patrol. This replaced the need for the mobile police officer to interrupt their normal patrol periodically to telephone or visit the Police Station to get updated information before they resumed their patrol. A radio dispatcher at the Police Station could now transmit bulletins and information to the police forces in the field (e.g., on patrol) to address current needs via the radio channel. This was still a half duplex broadcast paradigm requiring the police officers responding to the radio message to seek out a telephone box and call back to the dispatch position via wireline telephone services to update the dispatcher and receive any further information.

Late in the 1930s this paradigm was changed to allow more real-time communication. The police squad cars were equipped with mobile transceivers. Now the officer in the car could not only hear the bulletin broadcast from the dispatcher, but could immediately respond to the dispatcher message, and receive any necessary additional information at that time; no longer did the addressed officer need to delay a response until a means to make a telephone call back to the station was available. The officer could now simply depress a button on the microphone, the Push to Talk button, and talk to the dispatcher via radio. During WWII this technology was brought to the battlefield in the form of portable radio transceivers (e.g. Handie-Talkies) to allow troops to have instant communications with headquarters without the need to lay telephone lines between the field and the headquarters locations.

These two-way radio systems allowed for general analog voice communication among a group of people monitoring a common radio channel. This was used extensively by the Public Safety communities (e.g., Police, Fire, Emergency Medical) to enable real-time communication between people in the field and a central dispatch position, often located at the Public Safety office. Typically these radio systems would incorporate 'repeaters' to allow the coverage area of the radio system to be vastly expanded over the somewhat limited direct range between mobile transceivers. The repeater would act as the intermediary receiving a transmission from a mobile device, and retransmit or repeat the information in a new strong radio transmission for the other mobile devices to receive. Here the dispatcher could communicate in a real two-way conversation with one or more people in the field. This was used predominantly for simple voice communications, but could also be used to send limited bursts of slow speed data to the radio user devices.

A more simple and inexpensive two-way radio voice communication was also available to the general population, known as the popular 'walkie-talkie' that allowed people to communicate directly between two or more suitable, typically handheld, radio transceivers over short distances. These were used for such purposes as: 'family' communications, construction site communication, warehouse communication, etc.

The two-way radio system working on a radio channel shared by many users required channel access management to be exercised by the individual users. Prior to transmitting on the channel one needed to listen to the channel to determine if anyone else was actively talking; only when the channel was unused should one start one's transmission. Even so, there was still the potential for collisions of courteous users, or the intentional collision of rude users who did not care to respect the transmission rights of others on the channel, in general corrupting the transmissions. The determination of who could talk on the channel was essentially based on common perception of 'politeness' (or who had the strongest transmitter or loudest voice).

As the use of two-way radio grew, the term PTT evolved to describe not only the operation to allow transmission of voice, but also a communication service or application. The PTT communication service can be defined as an instant, half-duplex communication service that allows callers to connect rapidly with each other and enjoy short, bursty conversations that get the important parts of the message across in less time than a phone call. A key aspect of PTT service is one-to-many group conversations where the calling parties can instantly and simultaneously communicate with all the members in a group. This was well positioned to meet the needs of activities involving groups and instantaneous communications.

Public Safety markets, as well as industrial and small business concerns, could incorporate the PTT application to their day-to-day communications needs.

The next improvement for handling PTT activity was to introduce 'trunking' techniques in the 1970s to automate the channel resource management. While there can be a large number of people normally associated with a channel, generally there is a small number of these people who want to make a transmission at any given time. This is the same scheme used by the telephone company to eliminate the need for having dedicated lines (channels) for each user to every other user. Instead it 'trunks' a smaller number of channels among all the users, with the number of channels being adequate to handle the rate of normal, expected call service requests without running out of resources but with far less than a channel per user. In the two-way radio domain this scheme employs a request to send (RTS) condition from a user wanting to make a PTT session. The RTS is recovered at a computerized controller for the radio system, which determines what channel resource is available for the transmit session. The controller then sends a channel grant assignment to the originator of the RTS to inform him that he may use the assigned channel for the transmission (more precisely, to his radio unit which automatically switches to the assigned channel). Simultaneously, the idle users on the radio system are informed that there will soon be a transmission by the originator, targeting a particular audience of users. These users' radio units can now go to the assigned channel to hear the transmission from the originator. The channel resource is essentially dedicated to this call for the duration of this transmission and will not be shared with any other call during this time. Once this transmission has terminated, this channel resource is made available to be assigned to another user's call needs. The transmitting unit returns to an idle condition and awaits indication of a new transmission that it should monitor. Other members of the group in the call may make an RTS to start their own transmission (perhaps a response to the previous transmission) and a communication message of multiple individual transmissions can be built. There is no need for the requesting user to monitor the channel to see if it is clear to transmit, since the radio system controller now assumes this functionality and will only assign idle channel resources to the requesting user. The PTT communication system controller is now exercising talker control and arbitration over those users requesting to transmit. Schemes based on request received time, user priority, call identity, requestor identity, etc., can be employed to administer transmit permission for the PTT call.

In early trunked PTT systems, the channels supported analog voice communications and their quality was still directly affected by the RF channel environment. However, the signaling directing the usage of the channels was being accomplished digitally which allowed for more faithful recovery of the necessary signaling information to allow the radio user units to perform the proper radio system function. The next improvement in the PTT evolution was to apply digital technology to the voice payload too, such that the analog voice signal is converted to a digital representation as a packet of digital bits, and the digital packets are sent over the RF channel. This digital representation allowed for error handling (error detection/correction) of the voice information to allow for a more faithful recovery of the voice even when the RF channel environment was less than perfect.

Trunked PTT communication systems are well-behaved networks that generally incorporate proprietary schemes for signaling and call control. This affords the ability to deliver high performance wireless networks that are tailored to the PTT experience. Typically these communication systems can achieve sub-second initial channel access times in response to a transmission request, while supporting thousands upon thousands of active users. There are ongoing activities that are standardizing the digital trunking radio system environment to eliminate the proprietary aspects of the basic service and allow easier interoperability between multiple vendors' product offerings.

The ability to handle telephone calls across radio carriers was also being developed as another communication choice in the 1970s and has culminated in today's cellular and personal communication system offerings. Here the end user enjoys a communication experience similar to that of the typical wireline telephone call. This is generally a full duplex communication lasting for a relatively long duration (typically 120 seconds or more) in which the end parties simultaneously talk to each other over radio resources dedicated to their use for this communication.

With the advent of IP technologies in the wired world, a packetized voice session could be treated as another data session over an IP network, e.g., Voice over IP (VoIP). As the voice in the wireless domain is now routinely being converted to digital packets, the voice service needs to be handled over wireless channels as just another data application running over an IP network.

There is now interest in uniting cellular services with the PTT application. At least one system exists that combines a standard cellular network with a proprietary PTT solution to provide high quality telephony and PTT service to the end users. Another approach is to provide proprietary signaling to support PTT service, with the voice portion of the service being handled as standard cellular telephony-like voice. This approach reuses the existing voice scheme of cellular, but does not directly capitalize on the half duplex nature of PTT, nor is it open to easy multimedia support. Another approach is to provide signaling and voice as part of an integrated packet data service over the cellular network. This approach separates the PTT voice service from the standard telephony voice service by supporting it as VoIP in the packet data domain of the cellular network. This latter approach is the direction of the PTT over Cellular initiative to define a standards compliant scheme for supporting the PTT service over current cellular networks. This approach can easily leverage the PTT feature and support multimedia needs in the common packet data domain. (The packet data based approach is the focus for the following descriptions of PTT service.)

## 13.2 Service Description

Push-to-talk has evolved to become a suite of real-time voice and multi-media services. PTT services are of great value and interest in the consumer, industrial and enterprise markets. Consumers find the ability to communicate quickly and easily with family and friends with the push of a button to be a real convenience. Teens enjoy being able to 'chat' with all of their friends as they utilize the group call features of PTT. Businesses have discovered the benefits of being able to communicate quickly with individuals or groups of colleagues. Commercial users, such as taxi companies and delivery firms, also find increased value in the service since PTT offered on the cellular network provides many more features than the basic Walkie-Talkie service of years past. Some vendors provide a personal computer client for PTT, and this enables commercial users to communicate with and coordinate the activities of their field staff from the convenience of their desk. All of these market segments benefit from the versatility and 'quick delivery' aspects of the PTT communication service.

### 13.2.1 Features

The list of PTT features offered by each vendor varies, but most offer at least the following basic set.

### 13.2.1.1 Call Start Techniques

The following call start techniques are applicable to most PTT call types.

#### Forced Calls (Automatic Answer)
The default for PTT calls is Automatic Answer mode, also called 'barge' or 'forced audio'. This means that when the PTT call starts the initiator is given permission to begin talking immediately. The target handset will begin to play out the speech automatically as soon as it receives it without the target user needing to take any action to answer or accept the call. This is the normal way in which things work on a walkie-talkie and is useful in environments where fast conversations are desired.

*Invited Calls (Manual Answer)*

Optionally, PTT calls can be set to allow the target party(s) to be 'invited' to the PTT call, and the target user must accept the call rather than just receiving 'forced' audio. In the invited mode, the handset will alert the target user (e.g. ring or vibrate) to indicate that a PTT call is pending. The user can then decide to accept or reject the call based on the caller's ID, which is displayed on the handset display. This type of call start procedure is useful in environments where users may be disturbed by audio that suddenly starts playing out of the handset speaker. However, it has the disadvantage of much slower call setup since the target user must answer the call before audio can be delivered.

## 13.2.1.2 Call Types

*Private Call, or One-to-One Call*

A Private Call is a call from one subscriber to one other subscriber. It allows a subscriber to place a PTT call to any other PTT subscriber by either choosing a buddy target user from the PTT contact list (or handset phone book), explicitly dialing the target's address, or by selecting the target using the recent calls menu. The initiator is usually granted permission to send voice or other media (e.g., given the floor) while the other party listens after the call is set up. The initiator releases the floor when he releases the PTT button. Either party can then request the floor by pressing and holding their PTT button. The Private Call is terminated if either subscriber presses the call end button, or when no one has made a transmission for a pre-defined period of time (usually a few seconds).

*Group Call*

This allows an initiator to place a one-to-many call. This is useful when the calling party wants to include a number of people in a conversation all at once, such as when a supervisor wants to give instructions to her team, or when a group of friends want to make some plans.

- Static or pre-defined groups. Many vendors offer at least a couple of different ways to select the target participants for a Group Call. The simplest method is known as 'static' or 'pre-defined' groups. In this case, the group membership list is set up in advance and stored in either the handset or the network. The list can be created on the handset and sent into the network, or the user, through a personal provisioning web page, may create it. A group name is usually stored with the list. The Group Call is started when the initiating subscriber selects the group name from their PTT phonebook or contact list. The initiating subscriber then presses and holds the PTT button to set up the call. An advantage of Pre-Defined Groups is that their definition remains available and the group name is valid after the call is over, so it is relatively easy to start another Group Call to the same participants as often as desired. Also, some vendors offer the ability for multiple subscribers to access the same group list definition(s), so, for example in a commercial environment, the groups only need to be defined once and many subscribers can then make calls to them.
- Dynamic or ad hoc groups. A more flexible way to select the target participants for a Group Call is to define them 'on-the-fly', creating a dynamic or ad hoc group, just before the call is desired to be started. In this case, the initiator selects individual members from their PTT phonebook or contact list by highlighting them in some way, then immediately presses and holds the PTT button to set up the call. An advantage of Dynamic Groups is that the initiator can select the specific participants needed for the call as communication needs dictate. However, once the call has ended, the group definition is deleted from the network. If the initiator or any participant then wants to start a Group Call again with the same participants, they have to go through the process of selecting the members individually to recreate the ad hoc group before pressing the PTT button.

For either type of participant selection method, Static or Dynamic, once the initiator presses the PTT button to start the call, the system will initiate a call to all of the members of the group. Typically, as soon as the first member joins the call, the voice path will be opened and the initiator can begin talking with the first responding participant. As other participants join, they will be added to the call. All others on the call can hear whoever is talking. Any participant can talk once they have been granted the permission to transmit.

The group call ends when there is only one participant left on the call, or if there are no transmit requests. Some vendors opt for an automatic release of the call over a pre-defined period of time (usually a few seconds) of no transmit activity. Some implementations also allow the call to be ended manually, either by the originator, or by other designated user(s), by pressing the call end button.

### 13.2.1.3 Talker Arbitration / Floor Control

Since the PTT call is typically half duplex and often involves a group of participants, there is a need to control explicitly who is allowed to transmit on the call at any given time. The talk permission, or floor, is granted to one member of the call at a time, usually under the control of the PTT server, with the other members of the call listening to the transmission of this granted member. In this case, each time a user wants to speak, an explicit floor (transmit) request is sent to the server. The server can then decide if that user should be allowed to speak at that time. The server can incorporate a number of factors into that decision such as the requesting user's priority, who else is talking, etc.

Alternatively a handset contention floor control method can be employed for the talker arbitration. In this case, the handset clients implement a back-off mechanism so that if more than one person attempts to start talking at the same time, known as a 'glare' condition, each of those handset clients will detect contention and stop transmitting, giving a failed indication to the user attempting to transmit. The users must then manually re-press the PTT button to attempt to obtain the floor again. Usually the timing works out where one user will be first and successfully get the floor by locking out the other user(s) on the call while he is speaking. As can be seen, however, this method is not robust and tends to cause user frustration, especially as the number of parties on the call gets large.

### 13.2.1.4 Late Join

Users may join an ongoing Pre-Defined Group Call late if, for example, they missed the start of the call while busy in another activity, or while the handset was turned off. In this case, if they select the group and press the PTT button, the server will add them to the group call currently in progress. This user will begin to hear the current transmitted speech for the ongoing call, if there is someone currently transmitting. The late joiner is normally not automatically granted the floor (e.g., allowed to transmit) and must vie for the floor with the other members of the PTT call in the normal PTT talker arbitration fashion.

### 13.2.1.5 Chat Groups

Some PTT solutions provide a Chat Group service. Chat Groups differ from normal groups in that they are assumed to be always active, and users can enter or leave the Chat Group as they wish. Subscribers can create their own private Chat Groups, or the Operator can create public Chat Groups. In either case, as long as a subscriber knows the name of a Chat Group and is authorized to be a member of it, he can select it from his contact list and join any conversation currently ongoing in the Chat Group. Most vendors do not give the floor to the newly entering subscriber because there may be a conversation ongoing for the Chat Group. The members of the Chat Group must vie for the floor with the other members of the Chat Group in the normal PTT talker arbitration fashion.

### 13.2.1.6 Call Alert

A Call Alert allows the initiator to signal to a target that he wishes to contact the target, but not necessarily start a call at that moment. This allows for a more 'polite' method of starting communications. This is just a signaling message, and as such is a less intrusive way of letting the target know that the initiator wants to talk. Some implementations include a small set of short text fields in the Call Alert that can be displayed on the target handset, such as 'Please call me', or 'I'll talk to you later'. The initiator sends the alert after selecting a target, selecting the Call Alert option, then pressing and releasing the PTT button. The target subscriber receives an indication that there is an incoming Call Alert and has the option to respond or ignore it (reject it). To respond to the Call Alert, the target subscriber presses the 'accept' soft-key or simply presses and holds their PTT button. (The PTT action typically starts a Private Call back to the Call Alert initiator.) To ignore the alert, the target subscriber presses the 'ignore' soft-key, or simply does nothing and the Call Alert notification times out on his handset. The initiator will receive an indication that the target has accepted, rejected, or was unavailable for the Call Alert.

### 13.2.1.7 Presence

Many PTT vendors include Presence functionality bundled as part of their PTT service. When a subscriber enters the PTT phonebook, the handset client shows the list of subscriber's contacts and groups. Next to each entry in the phonebook is a presence indicator to show PTT call availability for a user or group. Groups are assumed to be available if at least one member of the group is available. Presence status is updated to and from the subscriber's handset on a periodic basis. Usually this update does not occur exactly as the presence state changes, but is held or aggregated with other user status information until a periodic interval time expires, in order to conserve use of wireless resources. This presence update interval is typically configurable by the operator. Alternatively, the presence status can be dynamically updated as changes occur to provide a more timely representation of the current conditions, but typically at the cost of increased network and client/server processing resources.

***Presence State***
In addition to the basic available/not-available presence state, which is usually determined automatically by the network (based on registration), many vendors allow users to set their presence state. For example, many handset clients allow the user to provide a state of: 'busy', 'do-not-disturb', 'in meeting', 'away', 'available', etc. A special presence state is the 'do-not-disturb' state and will result in the PTT server blocking calls to this particular user. The presence state for a user is available to other users that have subscribed to it via special icons at their handset.

***User Specified Presence Indication Blocking***
This feature allows a subscriber to hide his/her presence to all users through a setting in the handset client or in the self-provisioning web interface.

### 13.2.1.8 Multiple Simultaneous Sessions

PTT over cellular's use of a protocol designed for multi-media allows for the potential of actually supporting multiple service sessions simultaneously, not just switching between them. Clearly the handset client platform and network need to be able to support the extra signaling and media traffic that results from this, but if they do, the user can benefit from using two or more simultaneous services. Perhaps one is a voice PTT session with a friend, while another is a stream of the latest video from a favorite music group. This type of multi-media and multi-session flexibility is expected to increase in popularity as the network technology is scaled to support it. Even today, multi-session capability can be supported on wireless networks such as 802.11b. Future wireless networks will allow for even higher bandwidth across wider geographic areas.

### 13.2.1.9 Caller ID and Current Talker ID

PTT signaling messages contain information about who originated the call as well as who is currently talking. Most vendors display this information to the user in the handset client.

- Caller ID blocking. Many PTT solutions allow the user to block all other subscribers from seeing their phone number, in a similar fashion to caller ID blocking on standard cellular telephony.

### 13.2.2 Service Management Functions

#### 13.2.2.1 Add Member

The PTT service provides the ability to add a new party to an ongoing call. While in a PTT Call, any authorized participant can select a new member to be added to the call and signal that to the server. The PTT server then initiates a call to the new target user and connects the media path to them once they have joined.

#### 13.2.2.2 Group Participant Lists

During a group call, it is often helpful to know who is actually present and participating in the call. Particularly if the call is a Dynamic Group Call, the targets of the call won't know who else is participating. Many of the PTT solutions include signaling while in a group call to convey the IDs of all of the parties that have joined. This is useful in Chat Group calls as well.

#### 13.2.2.3 Contact and Group List Management

A key benefit of the current generation of PTT on Cellular solutions is they take advantage of both packet data to the handset and the World Wide Web to allow for flexible service provisioning options. Operators, as well as end users, can typically provision contact lists, group lists, and many other user preferences through a web browser or directly through the handset.

#### 13.2.2.4 White/Black List

The White/Black list allows a user to decide who will be allowed to call them or who will be allowed to see their presence state. Some vendors use a common white/black list for calls and presence, while others allow for separate lists for each. The white list includes all of the users who are allowed to call or view presence, while the black list includes all of the users who are blocked. These lists are typically stored on the server and synchronized with the handset periodically.

#### 13.2.2.5 User Notification Prior to Being Added to a Contact/Group List

If enabled by the subscriber, a message is sent to the subscriber whenever someone attempts to add them to a contact or group list. The target user indicates whether to accept or reject this request to be added to another user's contact list or group. The user can then decide if they would like to add this user to their white or black list to enable or disable calls from this subscriber, as well as enable or disable the delivery of presence status to this subscriber. A user can set the options for this feature via the handset or the user self-provisioning web server.

#### 13.2.2.6 Charging

In a standard cellular telephony call, the charging for the resources utilized to support the service by the Operator are assigned to the originator or target of the call, or sometimes both. With the introduction of the PTT Group call, there is now the potential for many individual PTT users to be active in a single PTT Group call. There are resources being used in the network to support each of the PTT users active in the call. The Operator may charge for all the resources being used in the PTT call in a number of ways, from charging only the call originator, to sharing the charges for the call across all the participants of the call. The shared charges can further be defined to address the resources being used by each of the participants in sending and receiving PTT Group call related information.

In the future, a PTT group call may include one or more PTT participants who are PTT subscribers of several different operators. Each operator charges post and pre-pay PTT subscribers independently from any other operator's charging policy. The charging of the participant can be based on, for example, the PTT call type call time, or a per transmission basis per participant. Specific actions initiated by a participant in a PTT call might be charged to the initiator of the action, but also possibly to other participants in the call since the action may incur increased resource utilization for all users. Users need to understand that actions like inviting a new participant into the PTT call, subscribing to the participant information in a Group call, as well as the amount of participant information sent to the requesting PTT call participant fall into this category. Restrictions might be imposed by the operators and/or subscribers to such actions in order to control charging. In the future, for PTT calls using different bearer streams, the network infrastructure may capture QoS parameters negotiated by each participant to allow for additional QoS based charging.

## 13.3 Architecture

The Push-to-Talk over Cellular (PoC) service is being defined in the standards arenas to promote a solution that will enable interoperability of multiple vendor PoC products. The architecture depicted here is the general architecture for a typical PTT system offering.

### 13.3.1 PoC System Diagram and Key Elements

The architecture diagram in Figure 13.1 shows major components of a system providing PTT Service to wireless devices. Several of the key components belong to existing wireless networks connected by an IP backhaul. The most common key elements and interfaces will be introduced in the next sections.

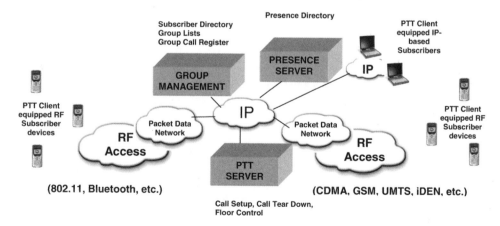

**Figure 13.1** PoC system architecture.

### 13.3.1.1 PTT Client

The PTT Client contains the application software running on a mobile wireless device, computer, PDA or other user device. The PTT Client provides for registration of the user to the service, authentication of the user to the network, multimedia session control, sent and received media encoding and decoding, floor control, and PTT related configuration data management. Group management and presence functionality are also supported in order to provide the expected PTT user experience. The PTT Client is essentially unaware of the underlying communication environment that is transporting the PTT signaling.

### 13.3.1.2 PTT Server

PTT Server hosts functionality required in the network to provide for the PTT service. Typically for a PTT call, the PTT server needs to support PTT session handling, access control, policy enforcement, media distribution, floor control and floor control negotiation, management of participant information, and collection of charging data. The PTT Server handles user related functionality including but not limited to: signaling and media relay on behalf of the user, media adaptation, policy enforcement, floor control and floor control negotiation, collection of media quality information, collection of charging related data, etc. Options like parallel simultaneous sessions for a particular user can also be handled by the PTT Server.

### 13.3.1.3 Group Management Server

The Group Management Server hosts functionality pertaining to the notion of a group audience or function to support the expected PTT user experience for pre-defined groups. This functionality comprises management of user defined groups and lists (e.g. contact and access lists) that are needed for the PTT service. The management operations include creation, modification, retrieval, and deletion of groups and lists for authorized users. The Group Management Server holds the master data of user defined groups and lists and provides upon request notifications of changes to subscribers.

A valid group consists of three or more users who are logically linked to a common group name. PTT activity can be addressed to this common group name, and all active members of the group will be automatically included in the PTT activity. A user can be associated with as many groups as desired/ allowed, but generally will only participate in activity for a single group at any time. The groups can generally be composed of individual PTT users, other groups of PTT users, or a combination of both.

### 13.3.1.4 Presence Server

The Presence Server performs functions that can enhance PTT Service such as publication, maintenance and retrieval of presence status for PTT users, and authorization of who can view PTT user presence information. With the PTT presence information, PTT users can be aware of availability of other PTT users prior to attempting initiation of PTT calls to the users, allowing for a more efficient use of system resources and user time.

### 13.3.1.5 RF Access Network

Access to PTT is available over a cellular private or public network (e.g., GSM or CDMA) as well as fixed wireless networks (e.g., 802.11 or Bluetooth). The PTT service makes use of the packet data service support of the underlying network to facilitate IP messaging.

### 13.3.1.6 Packet Data Core Network

The Packet Data Core Network supporting the PTT service is assumed to be SIP/IP enabled. It includes a number of SIP proxies and SIP registrars and is expected to fulfill functions like PTT user registration and maintenance of registration state, discovery and address resolution services, authentication and authorization of the PTT user, routing of the SIP signaling between the PTT Client and the PTT Server, and SIP compression.

*13.3.2 Interfaces*

Implementation of a PTT service uses a wide range of wireless technologies supported by the PTT enabled subscriber devices, the packet based networks, and the attached PTT application servers. The interfaces described below (see Figure 13.2) will address separately the over-the-air interface and the intra-network interfaces essential for the support of the service. The inter-network interface essential for enabling roaming is also introduced.

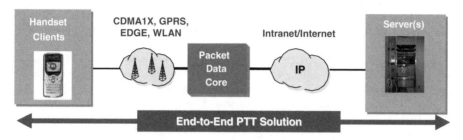

**Figure 13.2**   Overview of interfaces.

### 13.3.2.1 Subscriber Device to Network Infrastructure

This is the over-the air interface that carries both control and user data between the PTT enabled Subscriber Device and the Network Elements of the infrastructure. The control messages or signaling messages have to provide for:

- RF access;
- Packet Data Network;
- PTT application control.

RF access signaling is dependent on the air interfaces supported by the subscriber device and the accessed network. It includes, but is not limited to, GSM, EDGE, UMTS, CDMA2000, 802.1x and Bluetooth.

Packet Data Network (PDN) access is standardized for GPRS (GSN) and CDMA2000 (PDSN) access. Deployments of non-standard PDN access can be also found. These protocols are often complemented by AAA protocols in support of authentication, authorization and accounting.

Standard solutions for the PTT service are IP based, with IPv4 deployments available today and IPv6 deployments planned. These solutions subscribe to using a multitude of standard protocols from the IP and packet data arena to control various aspects of the PTT service. Some of the major impacts and associated protocols are highlighted below.

- Session Initiation Protocol (SIP) and the associated Session Description Protocol (SDP) are the primary control protocols.
- SIP, Real-time Transport Protocol (RTP) and Real-time Transport Control Protocol (RTCP) are some of the protocols supported to accomplish the task of talker arbitration.
- Control of PTT configuration data and management in support of the PTT Group/user information is accomplished typically by a transaction protocol using an Extensible Markup Language (XML) encoding of the data. The use of XML Configuration Access Protocol (XCAP) is anticipated when it becomes available.

- Recommended presence related signaling in support of PTT is SIP for Instant Messaging and Presence Leveraging Extensions (SIMPLE). Some non-standard implementations are currently already deployed.
- RTP/RTCP is the preferred protocol typically used to transport traffic or user data over the air.

### 13.3.2.2 Intra Network Interfaces

Intra Network Interfaces are interfaces to shared network functions, and other services and applications. PTT Service Entities will have interfaces to the Presence Service and the Group Management Service. The interface of PTT to the Presence Service is based on SIP/SIMPLE. Existing deployments of PTT have integrated non-standard presence functionality or other interfaces (e.g. Wireless Village) to pre-existing presence services. The interface to Group Management will be based on IETF defined XCAP. Existing deployments of PTT have integrated Group Management functionality or other interfaces (e.g. based on XML) to pre-existing group and list management services.

PTT Service Entities have interfaces to other shared network functions, like: Charging, Service Discovery, Domain Name Server (DNS), Security (including Authentication and Authorization).

These interfaces depend on the underlying network technology. The standardization of the PTT Service is based on the assumption that the underlying technology is an IP Multimedia Subsystem (IMS) / Multi-Media Domain (MMD) enabled network. Implementing the PTT Service using other underlying technology should be possible. The assumption is in all cases that there is equivalent functionality to that which is provided by an IMS/MMD enabled core network.

### 13.3.2.3 Inter Network Interfaces

In order to enable roaming of PTT subscribers between different operators domains, the Network-to-Network Interface has to be standardized as well. The areas of standardization for this interface are:

- the Standard Control Plane;
- the Standard User Plane.

It is expected that these interfaces for the PTT Service will use SIP/SDP and RTP/RTCP protocols for the control and user plane respectively ensuring interoperability between different PTT enabled networks.

## 13.4 Standardization

Push to Talk Standardization for cellular networks started in 2003 in the Open Mobile Alliance (OMA) with the creation of the Push to Talk over Cellular (PoC) work group. As planned for all OMA Service Enablers, the PoC Standard should be independent of the technologies of the underlying packet networks used.

In an effort to speed up the time to market of PoC, a consortium of several manufacturers supported by a series of operators interested in PTT functionality was formed to develop a series of PoC specifications targeted to the standardization process in OMA. The specifications, focused initially on the User-to-Network Interface (UNI), were expanded to allow for interoperability on the Network-to-Network Interface (NNI) as well, and provide some basic presence features in conjunction with the PoC Service. Interoperability tests based on this set of so-called Industry Consortium (IC) Specifications are planned for the second half of 2004.

The OMA PoC standard, targeted for completion by the end of 2004, uses a reference implementation of an IP Multi-media Subsystem (IMS) or Multi-Media Domain (MMD) enabled network, as specified by 3GPP and 3GPP2. Networks providing equivalent functionality with IMS/MMD enabled networks can be used as well. IMS/MMD enabled networks follow 3GPP/3GPP2 standards, which define a system architecture leveraging IETF protocols whenever possible.

The following sections describe the main activities and their interdependencies related to the PTT service within the standards organizations listed above.

### 13.4.1 OMA

The Open Mobile Alliance (OMA) was formed with the goal of helping create interoperable services that work across countries, operators, and mobile terminals and are tailored for consumers' needs. Companies participating in the OMA work toward the wide adoption of a variety of new enhanced services for mobile information, communication and entertainment. OMA includes all key elements of the wireless value chain, and contributes to the timely and efficient introduction of services and applications.

The PoC Work Group within OMA is chartered to develop specifications to permit the deployment of interoperable PoC services. Owing to critical performance objectives for this service, support for the lower-level network capabilities required for this set of application enablers requires some direct engagement with the network defining groups (e.g. 3GPP and 3GPP2). This close engagement may be a key differentiator with other OMA work groups that develop similar application enablers. The expectation is that this PoC Work Group would further develop the PoC Service and address extensions of this in the future.

The OMA specifications are contribution driven. PoC Specifications are mostly driven by contributions made by manufacturers and operator members of the Industry Consortium noted above which produced a set of PoC Service specifications in an effort to speed up standardization work.

The architecture of the OMA standard PoC Service is part of the overall OMA Services Architecture, leveraging synergies between OMA defined service enablers and usage of common functions as provided by the supporting packet data networks. The PoC Service specifications are independent of the prospective underlying technologies to be used. Related Presence and Group Management enablers are specified in parallel by the OMA Presence and Availability (PAG) Work Group. The interoperability tests for the PoC Service Enabler are planned to be made in conjunction with the required Presence and Group Management Service Enabler Functionality.

### 13.4.2 3GPP and 3GPP2

3GPP has a work item on the 3GPP Enablers for services like PoC. This work item was approved in September 2003. The objective of this work item is to provide possible enhancements of service requirements and architecture necessary in the 3GPP system to support PoC service using IMS. The PoC service enabler requirements are being defined within OMA. The necessary enhancements of the service requirements and the architecture of the 3GPP system will use the OMA PoC requirement specifications as their basis. 3GPP2 is also writing a requirement document to identify possible enhancements needed to support PoC.

### 13.4.3 IETF

IETF defines all IP related protocols and standards, including SIP, RTP, RTCP, etc. In recent years 3GPP and 3GPP2 have adopted a large number of IETF standards to be used in their respective specifications. Many of the key building block protocols for PoC specifications will be from IETF. However, several of them are still at draft stage. IETF and OMA are working on the rules of engagement for collaboration. For now, individual contributions are the basis of OMA related work in IETF.

At present, there is no work ongoing in IETF that is specific for PoC. However, many PoC specifications directly rely on protocols or draft protocols from IETF. Of particular relevance are the SIP protocol suite and its extensions (e.g., SIMPLE, SIPPING, XCAP), SIGCOMP, RTP/RTCP, XCON, HTTP, etc.

Some examples of what must be standardized for PoC are as follows.
Session Initiation Protocol (SIP):

- the selection and format of the fields used within the SIP messages;
- the recommended SIP compression;
- the PoC specific application tags.

Session Description Protocol (SDP)

- negotiation of the media characteristics during session invitation (including address endpoints for the bearer path);
- formats for identifying and negotiating vocoders and vocoder modes such as AMR, EVRC, SMV, G711, voice packing options, jitter buffer size, etc.;
- negotiation of the talk burst control protocol to be used in the session.

Floor Control or Talker Arbitration is an area where proprietary and semi-proprietary protocols are currently used. Examples of usage of existing protocols to transfer floor control signaling are SIP messages, RTP header extensions, and RTCP: APP. RTCP: APP messages have been identified by OMA to implement the floor control for the initial release of the standardized PoC Service Enabler. It is planned that the Binary Floor Control Protocol (BFCP) should also be added when it becomes available.

With regard to bearer traffic, RTP is used to carry the bearer traffic in the form of multiple vocoder frames bundled in an RTP payload. The RTP payload format is subject to OMA PoC standardization. Separate RTP sessions are planned for simultaneous transmitted voice and video streams.

## 13.5 Service Access

As shown on the architecture diagram, the PTT service is realized as an application client running on a wireless subscriber device communicating with an application server over a packet switched network. The OMA standardized PoC Service is using an IMS/MMD enabled network or a SIP/IP network providing similar capabilities. A sequence of procedures needed for a successful PoC communication establishment between a PoC User A and PoC User B is shown in Figure 13.3. Each subscriber device has to be authenticated and registered in its own packet access network as well as SIP registered for the use of the PoC Service via, for example, the IMS enabled core network prior to initiating or participating in any PTT service activity. During the SIP registration process the validity of the PoC subscription is verified as well. Triggered by PoC User A, the PoC session establishment takes place using the SIP protocol. Associated with the bearer, the floor control of the PoC Session is handled using RTCP:APP based floor control procedures. Media starts flowing using the SDP negotiated RTP session.

In the standardized architecture, access to the PoC Service is allowed for subscribers who are authenticated and registered with the IMS/MMD enabled core network or equivalent SIP/IP network. During the registration process the network infrastructure checks the validity of the subscription for the invoked PoC Service. In an IMS/MMD environment the subscription data is part of the Home Subscriber System (HSS), whereas in other systems it might be part of an application specific Subscriber Database.

The following sections describe the session control (establishment and termination), floor control approaches, and transfer of media based on the OMA PoC Standard approach applied to an IMS enabled 3GPP system.

### 13.5.1 Service Control

In the proposed OMA PoC Standard two mechanisms for session establishment are described. In both cases, the session is first established between the PoC user client and the controlling PoC application

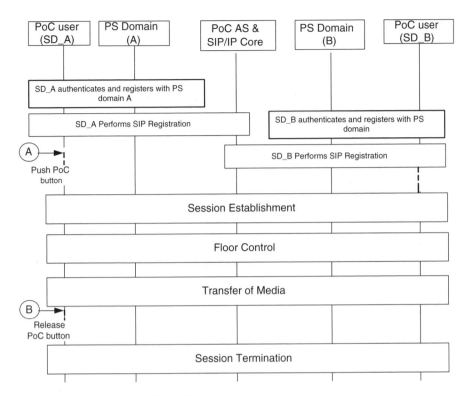

**Figure 13.3**  Service access procedures.

server that is serving that user, and then the other party session members are invited to the session. These session control schemes both deal with negotiating the media parameters such as IP address, ports and codecs that are used for sending the media and floor control packets between the PoC Client and the home PoC Server when the user wants to initiate a PoC communication with other PoC users. These mechanisms for session setup are referred to as the On-Demand Session and the Pre-established Session and are described below.

### *On-Demand Session*
The On-Demand Session provides a mechanism to negotiate media parameters using the full session establishment procedure each time the user wants to establish/receive/join a PoC session.

### *Pre-established Session*
The Pre-established Session provides a mechanism to negotiate media parameters prior to initiating a PoC session. This mechanism allows the PoC Client to invite other PoC clients or receive PoC sessions without renegotiating the media parameters. After the pre-established session has been set up (once the PoC user has registered), the PoC Client is able to activate the media bearer either immediately after the general PoC session pre-establishment procedure or when the actual SIP signaling for the PoC media session establishment is initiated. The pre-established session is an option in the OMA standard allowing for a simplified setup of the PoC session. This option will not be described here.

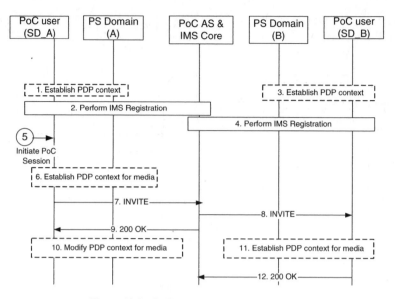

**Figure 13.4**   On-Demand Session establishment.

A simplified flow for the On-Demand Session in an IMS enabled network is shown in Figure 13.4, as an example.

Each subscriber device is attached to its packet domain.

(a)   Subscriber Device of user A (SD_A) has established a PDP context and has registered with the IMS (Steps 1 and 2).

(b)   Subscriber Device of user B (SD_B) has established a PDP context and has registered with the IMS (Steps 3 and 4). Note that the subscriber devices may go through the listed procedures at different times.

(c)   User A chooses the address of user B and presses the push-to-talk button on the subscriber device A to indicate that he wishes to communicate with the user at terminal B (Step 5).

(d)   If SD_A does not have a PDP Context active, it establishes a new PDP Context (Step 6), and initiates a SIP session for the PoC communication by sending the SIP INVITE into the IMS (Step 7). The INVITE request contains the information about the invoked service; the IMS identifies the service and routes the session establishment request to the PoC Application Server.

(e)   The PoC Application Server (after determining that the PoC communication should be completed), together with the IMS, forwards the invite to the terminating terminal for the PoC communication (Step 8).

(f)   After determining that the PoC communication should be completed, the PoC Application Server and the IMS send a SIP 200OK message to the SD_A (Step 9). The terminal might decide to modify the PDP context for media (Step 10).

(g)   Receiving the INVITE (Step 8), SD_B accepts the session by returning 200 OK (Step 12) and establishes the PDP context for the media (Step 11). Depending on the terminal setting (automatic answer mode or manual answer mode), the PDP context may be established in a different order.

### 13.5.2 Floor Control

Floor Control provides a means for a floor control server to grant or deny requests from users' transmit media in a call. Floor control as described by PoC standards includes the following generic actions.

- Floor Request. The action provides the capability for a participant in a talk session to ask for permission to talk.
- Floor Release. The action taken by a granted user to indicate release of their permission to talk.
- Floor Grant. An action from the network to inform a requesting participant that the floor has been granted.
- Floor Idle indication. An action from the network to inform participants that the floor is available.
- Floor Deny. An action from the network to inform the requesting participant that the floor request is denied.
- Floor Taken. An action from the network to inform all participants that the floor has been granted to the indicated user.
- Floor Revoke. The action from the network to remove the permission to talk from a user who has previously been granted the floor.

Additional actions may be needed in support of options (e.g. queueing floor requests).

The OMA PoC Work Group has chosen to use RTCP:APP messages to implement the Floor Control protocol. It also specifies a Talk Burst Control Protocol negotiation that allows interworking with existing non-proprietary solutions and the introduction of new protocols (e.g. XCON BFCP) as they become available.

Figure 13.5 shows an example message exchange between the PoC Server in an IMS enabled network and the PoC enabled subscriber devices after a one-to-one PoC session establishment.

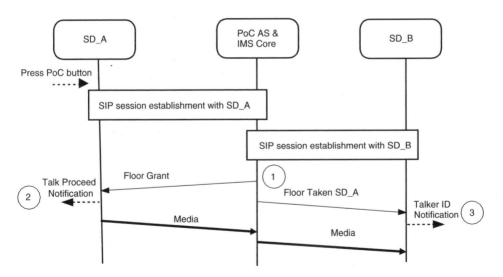

**Figure 13.5**   Floor Control during PoC communication establishment.

(1) The PoC server determines that originating user (SD_A) may have the floor and issues a Floor Grant message to the requesting subscriber device and a Floor Taken message to all other subscriber devices currently in the talk session. The Floor Taken message contains the talker identification of the user granted the resource (in this example SD_A).

(2) When SD_A has received the Floor Grant message it may provide the user with a Talk Proceed Notification, and the user may begin speaking.
(3) When SD_B has received the Floor Taken message, it may display the identity of the granted user (SD_A) to the user.

### 13.5.3 Media

In simplest terms, the PTT service is a framework for sending content or 'media' to other parties upon the press of the PTT button. Voice is the traditional media one thinks of, and it is in fact the most common media transferred in PTT calls today. However, a great deal of media flexibility is offered with IP based PTT through the use of the SIP and SDP. This flexibility allows for many choices about the media to be negotiated both at the start of a call and, to a lesser extent, within a call. Attributes that can be negotiated include not only the type of media to be sent (voice, pictures, video, etc.) but also many details about each type of media such as encoding, vocoder, bit rate, framing, etc. This makes PTT a service that can be relatively easily enhanced over time.

The SDP also contains information about the endpoint IP addresses that should be used for the media. The server uses SDP to tell the handset client to which address to send media, and the client tells the server the address at which it wants to receive media.

### 13.5.3.1 Media Transport

With PoC being an IP based service, media is sent by encapsulating it in the Real-Time Transport Protocol (RTP). Most implementations of PTT carry voice media in RTP on UDP/IP since voice is fairly resilient to an occasional lost or late packet, however, some carry RTP in TCP/IP. TCP/IP is typically not used for real-time media in wireless environments because it is difficult to get good performance with TCP's retransmission algorithms.

As an example, let's look at how voice is typically transported for PTT over GPRS, the packet data subsystem of GSM. First, a vocoder type and bit rate needs to be selected to optimize the use of GPRS air-interface resources. Since this version of PTT client needs to run in GSM handsets, a vocoder that is already available in many GSM handset implementations is ideal. In this case, the Adaptive Multi-Rate (AMR), 3GPP reference vocoder was selected. AMR supports many bit rates. Ideally, one of the higher bit rate versions would be selected as default since higher bit rates generally mean better voice quality. However, there is a channel capacity tradeoff. It turns out that, after adding the overhead of IP, a relatively low rate version of AMR is needed in order for the resulting VoIP packets to effectively fit in GPRS channels. Most implementations use standard AMR frames at a rate of 4.75 or 5.15 kbps. AMR framing for VoIP is specified in RFC3267 from the IETF, and most vendors follow that specification.

### 13.5.3.2 Speech Packet Framing

Without going into great detail about GPRS channel coding and capacity here, it was desirable to keep the total bit rate of voice media for PTT on GPRS, once the IP and framing overhead was added, to under 8 kbps. Compared with the AMR vocoder payload, the RTP/UDP/IP overhead is relatively large. To maximize the payload packing efficiency, multiple AMR frames are bundled into each RTP packet. For maximum efficiency, as many AMR frames as possible should be bundled into an RTP packet. This improves the ratio of payload to IP overhead. However this leads to another tradeoff, bundling too many voice frames into an IP packet leads to two problems. First, more frames in the packet results in longer audio delay. This is because, since this is a real-time media, the vocoder is producing frames at a fixed output rate. The RTP packet can't be sent until all of the AMR frames for that packet have been collected from the vocoder. This directly impacts the amount of time between when the user speaks a sound and when the RTP packet containing that sound is completed and sent into the network. The second problem is related to the impact of errors on audio quality. Since the RTP is carried on UDP/IP, any lost or delayed packets in the network are ignored at the receiver. If the lost RTP packet contained a large number of AMR vocoder frames, an objectionable audio hole may be produced at the vocoder playback end. For these reasons, most vendors have opted to bundle eight AMR4.75 frames into each RTP packet.

Figure 13.6 demonstrates this typical framing and the resulting bit rate of the VoIP stream.

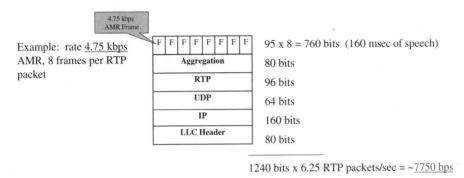

Figure 13.6   Example vocoder frame packing for GPRS.

### 13.5.3.3  Voice Clarity

Some of the issues related to voice quality were introduced in the previous section. Probably the most fundamental contributor to voice quality is the clarity or Mean Opinion Score (MOS) of the vocoder itself, when sent in error free conditions. The typical voice coders used for PTT are EVRC (for CDMA 1X implementations), AMR (for GPRS), and AMR or G.711 for WLAN and Internet clients. Voice quality can never get better than that achievable by the base vocoder; errors will only make it worse.

The voice quality is also directly related to the available channel data rate, which sets the maximum rate at which the vocoded information can be sent. For CDMA 1X, two fundamental channel (FCH) rates are allowed by the standards: 9.6 kbps and 14.4 kbps. When using the 9.6 kbps channel rate, the EVRC vocoder for PTT must be restricted to the so-called 'half-rate max' at 4.0 kbps in order to still fit in the channel once IP overhead is added. With the 14.4 kbps channel rate, EVRC can be allowed to go to full rate at about 8.0 kbps. As expected, the Mean Opinion Score (MOS) for full rate EVRC is significantly better than for 'half-rate max' EVRC.

For GPRS, the AMR vocoder at 4.75 or 5.15 kbps results in a fairly good MOS. For WLAN and Internet clients, higher rates of AMR or G.711 are typically used for PTT since those networks are not so severely bandwidth limited. These higher vocoder rates result in correspondingly higher clarity.

### 13.5.3.4  Vocoder Negotiation and Transcoding

With all these vocoder possibilities for PTT, one begins to wonder how PTT clients on different networks could ever communicate with each other. The answer is by using vocoder negotiation, or transcoding if negotiation is not possible.

#### Negotiation
Some vendors are recommending in PTT standards that the client devices support a common interoperable vocoder as default, and that the SIP session establishment support a vocoder negotiation

procedure to select among a common set of vocoders if the common one is not implemented or otherwise available in the client for some reason.

A proposal for the high level flow of the negotiation process is as follows, and is the typical offer/answer model for service negotiations specified by SIP. The vocoder offer and answer information will be carried in the SDP, which is part of the SIP INVITE, OK and ACK messages. The call initiator includes a vocoder offer in the initial SIP INVITE message that is sent to the server. Ideally, this offer will include at least the standard default vocoder, but it may also include others that the initiating client can support. The server will send an INVITE message to each participating client as part of normal call setup. In a slight variant to the simple SIP negotiation model, the INVITE sent to the clients will not include any vocoder information.

Each participating client will reply with a SIP OK (offer) that will include a list of vocoders that the client can support. The server will evaluate the initiator's SIP INVITE and the participant's SIP OK to determine the best vocoder to use for this call. The server will then instruct the call initiator to use the selected vocoder by sending the vocoder selected in the SIP OK message (answer). The server will tell each participant what the vocoder selected is by including the vocoder in the SIP ACK message (answer).

### *Transcoding*

Vocoder negotiation as part of normal SIP call setup is the preferred method for supporting multiple vocoders. However, negotiation may not be possible in the early phases of PTT deployments, since many vendor's handsets do not have sufficient hardware/software to support multiple vocoders (for example, they don't have enough DSP resource to support both AMR and EVRC). Another solution that could be employed before the handsets are capable of multiple vocoders is to insert a transcoding resource in the network. Some PTT vendors have proposed deploying large-scale transcoding elements in the network to convert the media codecs as necessary among the participants of a call. In this case, the server could accept differing offers of codecs from each participant and route their media to a transcoder device that would convert one codec stream to another.

Transcoding seems at first to be a straightforward solution, however it suffers from a couple of major problems. First, the media quality is already somewhat marginal given the low-rate vocoders needed to fit VoIP in current wireless channels. Transcoding between low-rate vocoders often results in significant degradation in voice quality, sometimes to the point of loss of intelligibility. A second problem is that transcoding elements are quite expensive. Hardware costs can be quite high because a significant amount of DSP processing power is needed to transcode each stream. Scaling a network transcoding solution to support millions of PTT users is problematic. For those reasons, negotiation among the PTT clients using SIP is the preferred long-term solution to codec interoperability.

## 13.6 Performance

PTT over Cellular is intended to be a service that allows real-time conversations between users. To enable this, the underlying performance of the transport network needs to be quite good.

The PTT solution as described in this chapter is a packet data service on IP. Until recently, the performance of wireless IP packet data subsystems was insufficient to support adequate performance of PTT, but the advent of CDMA2000 1x and GSM – GPRS have changed that. While these current generation packet data networks provide what most would consider 'acceptable' performance for PTT, they are still not quite able to support a conversation at the rate at which people are accustomed to speaking. Most PTT users must be 'trained' in a sense to wait a brief time before speaking, and to be patient as the other party(s) responds. As we move into the future, with even higher performance wireless packet data networks such as CDMA2000 EV-DO and EV-DV, UMTS, HSDPA, 802.16, etc., the performance of VoIP services like PTT will continue to improve. Already the performance of PTT on wired and wireless LANs is truly conversational. Increased performance will not only improve the user's perception of a natural conversation, but it will also allow for more complex media and simultaneous transfer of multiple streams.

Some of the key performance characteristics of the PTT service from a user's perspective are the following:

- Voice Delay;
- Call Setup Delay;
- In Call Talker Arbitration or 'Subsequent PTT' Delay.

The network operator is also concerned with some additional performance characteristics, particularly in the area of capacity impacts on the network:

- Air-Interface Capacity Impact;
- Network Capacity Impact.

A short description of each characteristic and an example of current performance is given in each of the following subsections. For these examples, the GPRS cellular packet data technology is given as a representative case.

### 13.6.1 Voice Delay on GPRS

Voice delay is measured from the time User A speaks a sound into his microphone to when that same sound is played out of User B's speaker. It includes the entire audio path through User A's microphone, speech encoder, IP stack, air-interface, network transport and possible packet duplication (for group calls), air-interface on User B's side, IP stack, speech decoder, and playout through user B's speaker.

The one-way audio delay for PTT is typically between 1 and 2 seconds on average when modest loads on GPRS channels are present. This would be much too long for a full-duplex telephone call, but it is marginally acceptable for half-duplex PTT. For telephone calls, voice delays longer than about 300 msec make conversation very difficult. Anyone who has made an overseas telephone call through a telecommunications satellite can attest to this. However, since PTT is a half-duplex voice service, the longer voice delay essentially appears to the talker as a longer response time by the listener. The listening party may actually respond quickly, but since the speech took a while to get through the network, it appears that he is slow to respond.

As the system packet data load goes up, voice delays can grow to 2 or 3 seconds or more. A problem with high call loads on GPRS is that, between speech bursts, the channel may be difficult to acquire again if many users are contending for it. Therefore, the perceived user experience will degrade if the system becomes heavily loaded without increased capacity added or QoS mechanisms in place.

### 13.6.2 Packet Arrival Jitter

Since the media for PTT is sent as small packets through a wireless data network, the transit delay of each individual packet is constantly changing. Delays in acquiring the channel, retransmissions over the air due to errors (if used), and network element queues all contribute to packets arriving at varying times at the listening client. This effect is common to most packet data systems and is generally referred to as 'jitter'. A jitter buffer (also called smoothing buffer) is used at the receiving client to accumulate a few packets before starting to play them through the decoder. This allows the receiving client to ride through or 'smooth out' any short gaps between receipt of packets. It's similar to a water pitcher that has a plug at the bottom and is being filled by a source from the top that has a varying rate. The pitcher is allowed to fill a little before the plug is removed. As long as the water empties from the hole at about the same rate as the pitcher is being filled (on average), it won't run dry. Thus, jitter buffers help to ensure there are no gaps in playback at the receiver. The disadvantage of a jitter buffer is it adds directly to voice delay since the receiver must 'fill the buffer' a little before playback can begin. Typical jitter buffers for PTT on GPRS may range from 500 msec to a second or more.

### 13.6.3 Call Setup Delay on GPRS

Initial call setup time for PTT is measured from the time User A pushes the PTT button until the talk proceed tone is played at his client to indicate the target is ready to receive the voice. It is desirable that this time should also be as short as possible, since once a user has decided to say something, he would really like to just press the button and start talking. The traditional way in which this time is measured includes generating and then sending the call request at User A's client, processing the request at the server, forwarding the request to User B (which often includes paging User B), replying back to the server from User B, and finally delivering the response back to User A, plus all of the network transit and channel acquisition times for each of those messages. This call setup method has been called 'late media' because User A isn't given the go-ahead to speak (i.e. send media) until after the response from User B's client is received. Other call setup methods that use a couple of different forms of 'voice spooling' are described below.

Initial call setup time for PTT on GPRS is typically between 4 and 5 seconds. It turns out, in cellular packet data systems, that most of the time is spent acquiring channels, locating and paging the called party, and transit over the air. Relatively little time is spent in the client and server as they process the messaging. This points out how end-to-end performance of PTT is highly dependent on the underlying performance of the network. Figure 13.7 shows the main components of this time. The blue numbers in this figure represent processing, buffering, and transmission times in milliseconds at the corresponding nodes.

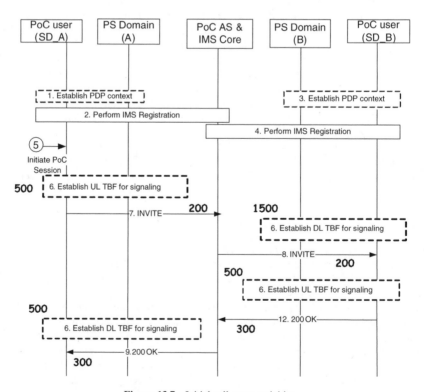

**Figure 13.7**   Initial call setup activities.

*13.6.4 Call Setup Optimizations*

It is desirable to make the initial call setup time as low as possible. However, achieving a setup time much below 3.5 seconds with currently fielded GPRS RANs is believed to be impossible without implementing some form of voice spooling technique. These spooling techniques, which have been implemented by many PTT vendors, can help reduce perceived call setup delay. They don't actually reduce the setup delay, but the initiator is given the impression that it is faster. The possible call setup methods are defined as follows.

- No Spooling (also called Late Media). This is the classical model that requires a definitive response from the target client before the talk proceed is provided to the call initiator.
- Server Spooling (also called Early Media). This method allows the talk proceed indication to be given as soon as the initiator has exchanged signaling with the server. The server will spool (store) any audio coming from the originator handset, while in parallel the target is being paged and put on the traffic channel. When the target is available, the audio is played out from the server to the target.
- Handset Spooling. This method allows the handset client to immediately provide a talk proceed indication upon the PTT push, and spool the audio on the handset while the handset sends the INVITE and the target is paged and put on channel. Then the originating handset will play out the audio to the target handset.

These techniques all have benefits and drawbacks. While the talk permit time from the originator's perspective may be improved, the initial Round-Trip Response time may be significantly increased, and must be considered when picking a particular approach. Round-Trip response time is defined as the time from when the first talker finishes speaking until the earliest possible time that he could hear a response from a second speaker. It takes into account the talker arbitration signaling delays, plus voice transmission delays. Figure 13.8 demonstrates the expected call setup and Round-Trip Response times for the various methods. (These are typical times on fielded GPRS networks supplied by one of the major PoC vendors.)

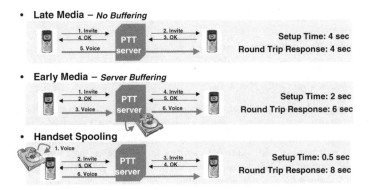

**Figure 13.8**   Three call setup methods.

One concern with spooling techniques is that all of them have issues with 'false positives', which allow media to be sent from the initiator without having a definitive confirmation that the target is available. This may lead to a negative user experience. Some vendors provide presence indication, which helps by showing when the target user is at least 'expected' to be reachable, thus reducing the number of false positives to just those cases where the target is already engaged in another call, or has briefly lost coverage.

As discussed above, spooling moves or 'hides' the start-up delay, moving it into the Round Trip Response Time delays, but only for the initial call setup. Subsequent PTT actions within a call do not use spooling techniques. In the non-spooling model (also called Late Media), the setup delay and round trip response time are both approximately 4 seconds. Server spooling (Early Media) is a compromise between the two other models, providing relatively fast setup followed by a longer response time, on the order of about 6 seconds. In the last case, Handset Spooling, the talk proceed tone can be given nearly instantaneously, but the true setup delays are pushed into the round-trip response time, which includes the time for both the originating and target handsets to get on channel and process the requests. This can grow to as long as 7 or 8 seconds without including the time for the listener to think about a response. If 2 or 3 seconds are included for the listener to think, the initiator may be waiting 9–12 seconds for a response, and may have terminated/exited the PTT session, having perceived it as unsuccessful. This is a risk associated with handset spooling.

Most vendors offer at least one of these spooling techniques, often as an administrator configurable option. This allows the network operator to experiment with them and weigh the relative tradeoffs of faster perceived setup time against 'false positives' and longer round-trip response times.

### 13.6.5 Talker Arbitration Delay on GPRS

Talker Arbitration Delay (also known as Subsequent PTT delay) is how long it takes a new talker to be given the 'floor' to speak, within an ongoing call. It is measured from the time a second user presses the PTT button until the talk permit indication is given. In most implementations, it includes generating and sending the floor request to the server, followed by the accept or reject response from the server, plus channel acquisition and transit delays for these two messages.

Talker Arbitration Delay is typically around 1.0 to 1.5 seconds today on GPRS. This is longer than most users are comfortable with. It directly impacts the smoothness of conversation. Again, the ideal situation would be to just press the button and start talking. There are a number of ways that vendors are currently trying to optimize this time. One way includes taking the server out of the path and using a contention based mechanism among the parties on the call (whoever starts talking first wins), but this leads to user frustration when more than one person tries to start talking at the same time. The bulk of the time for Talker Arbitration delay is again taken up by establishing the path through the packet data network, so another way in which vendors are trying to remove some of the delays is to reduce the channel acquisition time by adjusting air-interface related timers.

### 13.6.6 Capacity Impacts on GPRS Networks

Of concern to network operators is the additional capacity impact on their network due to deployment of the PTT service. Ultimately the network provider wants to be sure that revenues generated from users using the PTT service exceed the expense of network equipment needed to support it. In some cases network operators will need to add additional capacity to the wireless base site equipment, transport equipment, and IP routing network equipment to support the increased packet data traffic demands caused by the PTT service (or any other interactive packet data service for that matter). This section describes the two PTT capacity impacts that have the highest cost to the operator:

(1)  air-interface capacity impact;
(2)  network capacity impact.

Before discussing the capacity impacts of the PTT service, let's review the basic attributes of PTT traffic.

### Typical PTT Call Attributes
A 'typical' PTT call on GPRS has the following characteristics:

- speaker pushes (and holds) button, waits until Talk Permit Tone, then speaks;
- mean call duration: $\sim 40$ seconds;
- five half-duplex talk bursts (e.g., transmissions);
- talk spurt – average 4 sec – then release button;
- thinking time – average 1–2 sec;
- thus, average voice activity per user per call is about 12 sec or 30%;
- vocoder rate: 4.75 kbps AMR, which is bundled in RTP/UDP/IP.

### Typical Data Usage per PTT Voice Call
By knowing the size of the messages, the number of talker changes, and the size of the vocoded voice packets used in a PTT call, it is possible to calculate the total number of bytes transmitted on average per PTT voice call.

An example using one of the popular vendor's PTT solutions shows an average per user load of about 11.4 kbytes on the uplink (from handset to network), and 11.9 kbytes on the downlink (from network to handset) for the average PTT call. To be more conservative, a group call factor may be added to the downlink, which results in about 13.1 kbytes per user per call.

### 13.6.7 PTT Capacity Relative to Cellular Voice Calls

Numerical and simulation analyses demonstrate that, from a capacity utilization perspective, PTT minutes of use on a cellular packet data network like GPRS cost less than about 50% of the cost of a conventional GSM telephony voice call, even counting the packet overhead needed for VoIP. This savings is due primarily to the fact that PTT is a half-duplex service, and is coupled with the typical vocoder rate in the range of 4.0 to 4.75 kbps. Some cellular packet data solutions that do not have independent assignment of uplink and downlink (that is, both are assigned together in more of a 'circuit' mode) do not get this benefit, therefore their minutes of use for PTT are about equivalent to a telephony voice call.

One of the key conclusions from this analysis is that the loading of PTT users on packet data networks (those which do support independent uplink and downlink assignments) can be higher than is typical for telephony users on voice telephony channels. Depending on a network operator's charging model for PTT, this means that there is a good potential for gaining increased revenue for minutes of use compared with an equivalent amount of resource usage for telephony.

## 13.7 Architecture Migration

Most participants in the PoC standardization process believe that PoC should be evolved to become a 3GPP IP IMS based application, or equivalently, a 3GPP2 MMD based application. The IMS (per 3GPP) and the MMD (per 3GPP2) are multimedia IP-based architectures for mobile environments that are intended to support a broad range of applications and services. PTT is a perfect candidate service for the IMS/MMD.

For the remainder of this chapter, the term IMS is used to broadly describe the IP Multi-media Subsystem, whether applied to GSM/UMTS, CDMA, or other future cellular networks. This is not intended to slight the MMD work being done in 3GPP2, but it is simply a useful shorthand for the purposes of this chapter since the IMS and MMD architectures are essentially the same from the perspective of applications using them.

At the time of writing, all vendors who currently offer a SIP based PTT solution on cellular networks provide their solution as a handset client plus a packet domain server connected above the mobility anchor point in the cellular network. In GSM/UMTS networks, that anchor point is called the Gateway GPRS Support Node (GGSN). In CDMA 1x networks, it is called the Packet Data Serving Node (PDSN).

A process for how the current PoC architecture may be migrated to full integration within an IMS is as follows. This is just an example and individual PTT vendors may not follow this migration path.

Initial implementations of PTT deploy it as a Packet Switched domain application. The PTT server, consisting of Call Control, SIP Registrar, Subscriber Database, and Packet Duplication functions are deployed as IP servers in the network operator's intranet. Some vendors bundle the call control and packet duplication functions together in one server platform, and the SIP registrar and subscriber database in a separate server platform, while other vendors bundle all of those functions together in a single server platform. PTT client applications in the handsets interoperate with the PTT server via the GPRS or CDMA2000 packet data network. The IMS elements are shown shaded [grey] in Figure 13.9 since they do not yet exist in commercial deployments. In this phase, client interoperability requires agreements among vendors since the PoC standards are not yet ratified.

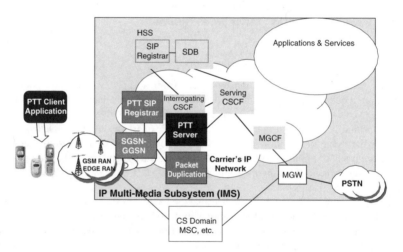

**Figure 13.9**    Standards evolving.

As standards work progresses and standards documents are released in draft form, many of the vendors that have currently deployed PTT will upgrade the PTT server functions to be compliant with the ISC (SIP-based) interface of the Call State Control Function (CSCF) in the IMS. The PTT server functions will move to their proper place in the Applications and Services domain. New standardized client functionality may be introduced into handsets at this time.

Once standards are finalized (Figure 13.10), it will be possible to have the HSS provide SIP Registrar functions for all SIP services, including those for PTT. The subscriber database will also be harmonized so that subscriber records in the HSS will include PTT profile information. Client standards will be complete, ensuring interoperability of handsets with other vendor's servers. Inter-network standards will also be complete, ensuring interoperability of PTT between cellular operators and clients in the Internet. The packet duplication function may be moved to the Media Resource Function (MRF) at this time, depending on vendor's desire to do so, or that may occur later.

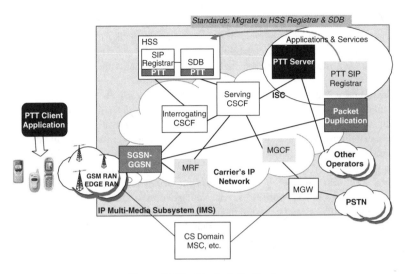

**Figure 13.10**   Standards finalized.

## 13.8  Possible Future, or PTT Evolving to PTX

The PTT application has been basically focused on 'voice' media, with the ability to provide source control for the media in a group environment as a foundational function. As an IP data domain offering, the PTT service can gracefully evolve to have other media and multiple medias simultaneously included in a session. Based upon standard IP schemes (e.g., RTCP, TCP/UDP, SIP) different media streams can be controlled in a PTT session to augment or supplement the typical voice session. Each of these media streams can be handled separately regarding source control to allow multiple points of entry into the session to facilitate a more flexible flow of information during the communication session. These streams would need to be synchronized across the communication and would need to take into account any limitations of the communication environment or the terminating device capabilities to handle particular media streams. This is now not just a single service, which is tied to voice media (e.g., PTT), but is a suite of instant communication services (e.g., PTX) that can handle variegated media. Below are some potential future applications of the 'push to' scheme.

### Push to Text: Text to Speech/Speech to Text Automatic Conversion
This service allows an information session to be automatically converted between speech and text presentation mode to satisfy the needs and capabilities of the end users for this session without the need for the sender to be aware of the receiving end user's mode of presentation. For example if a user is in a meeting or another situation in which voice messaging would be considered inconvenient (or rude), a voice message received by this user's device could be converted to text for display at the handset. Similarly, a response could be entered via a text input device, which will be converted to a voice message for the other users that can process voice messaging. In this way multiple modes of input/ output can be automatically facilitated in a common PTX session.

### Push to View: Send an Image (e.g., Camera Photo) to Members of the PTX Call
This service allows a user to send an image instantly (e.g., a stored image file, a photograph from a handset camera) to one or more other users, not as an e-mail attachment function, but as a real-time transaction. This service allows voice or other streams to accompany the sending of the image to add to the value of the transmitted image (e.g., you can talk about what is being shown in the image to the

receiving parties, or you can send along an audio file to provide 'bird call' for an image of the rare species you have sighted). The image can be formatted appropriately to account for such things as: current channel environment, end user capabilities, cost of service, etc. This can also be used as part of a Caller ID scheme by providing a User selected image to accompany the call.

### Push to Video: Send a Video Session to Members of the PTX Call
This service allows the user to send not only an image, but a true video stream to other select users. This service would support video sessions of varying levels of quality from stop frame, to streaming, to live action. There can be multiple media streams (e.g., voice commentary) to accompany this video media to add value to the video session.

### Push to Game: Real-time Sharing of Information in a Group Gaming Environment
The instantaneous nature of the 'push-to' service along with the ability to provide source arbitration across a group of participants can be used in interactive gaming services. Additionally this could be used as an introduction to a gaming environment in which the end user could simultaneously play the game and converse (via voice, text messaging, etc.) with other people to get help in operating the game.

### Push to Meet: Manage Online Meetings Using Presence Information
By using the presence information to determine who/when select users can meet to communicate via a PTX session, a more efficient online meeting environment can be created. This can also be mated with location information for each of the potential call members; potentially a distance attribute could be included in the selection criteria to determine suitable members for the call (e.g., members who are currently within 2 miles of location x). This could be used to schedule communications or gain necessary information more efficiently and on a timely basis from select subject matter experts. Based upon the current attributes of pertinent users, calls can be dynamically scheduled and call membership can be dynamically adjusted as other pertinent users become available for the session. Each of the members of the PTX session could provide and receive session communication based upon particular characteristics of the user's PTX end device and the particulars of the PTX communication itself.

### Push to Share: Virtual Reality Shared Environment
This leverages virtual reality multimedia offerings to provide the environment of a user to other select users to allow for more efficient/realistic sharing of information or an experience. In this mode the end user would literally be able to experience a version of the sensed environment of the sender and could share/interact in the current experience.

The perspective of the PTT service evolving beyond simply 'talk' to encompass any type of media and service represents a range of bright opportunities for the Push-to-'Anything' future.

# 14

# Location Based Services

Ioannis Priggouris, Stathes Hadjiefthymiades and Giannis Marias

## 14.1 Introduction

Location Based Services (LBS) can be considered as the most rapidly expanding field in the mobile communications market. Their first appearance, in a much more primitive form than that known today, is traced back in the middle of 1990s, propelled by the advent of the GPS positioning system. However, only a few years later the proliferation of the mobile/wireless Internet, the constantly increasing use of handheld, mobile devices and position tracking technologies and the emergence of mobile computing prepared the grounds for the introduction of this new type of services, with an impressively large number of applications for the general public.

According to recent research results, the LBS market will generate over US$5 billion for European operators alone, by the end of 2005 [1]. Such estimates justify the increased interest of all the key players in the LBS chain (e.g., telecom operators, content providers, service providers, etc.), in the development of advanced solutions that can boost the market.

So what exactly is a Location Based Service? In the popular context, Location Based Services have come to mean solutions that leverage location information to deliver consumer applications on a mobile device. Application opportunities can be grouped into the following categories:

- navigation and real-time traffic monitoring;
- location-based information retrieval;
- emergency assistance;
- concierge and travel services;
- location-based marketing and advertising;
- location-based billing.

These categories target to the wider portion of the LBS market and will be accessible to millions of users by large players in the telecommunication, automotive, and media industries. Apart from these services, however, an additional set of applications, focused on specialized target groups (e.g., the corporate sector, the health sector) will be developed. These include:

- dispatch and delivery route optimization;
- monitoring person location, which includes data for health care, emergency calls and prisoner tagging;

*Emerging Wireless Multimedia: Services and Technologies*   Edited by A. Salkintzis and N. Passas
© 2005 John Wiley & Sons, Ltd

- third party tracking services for Enterprise Resource Planning (ERP) and the corporate and consumer markets (e.g., fleet management, asset or individual tracking);
- security and theft control;
- people finding.

Apparently, Location Based Services cover the whole range of user needs from emergencies (such as the FCC E911 mandate for mobile emergency calls) to amusement (e.g. friend finder, POIs), as well as a large range of business needs (e.g. fleet management). For the development and provision of LBS services, a synergy of different, yet complementary, technologies and architectures is required. An overview of these technologies, along with other critical technical issues, is given in the sections below, in an attempt to define the requirements and the architectural aspects of an LBS system; that is a system focused on the delivery of location based services.

## 14.2 Requirements

Before stating the requirements for delivering Location Based Services to end users, we will establish a common nomenclature and identify the essential components that make up such a process. There are two main entities involved in the LBS provisioning model:

(1) The 'LBS server', which is responsible for providing the location sensitive information to the 'LBS client'. Providing such answers may also require invocation or queries to other network entities.
(2) The 'LBS client', who asks the 'LBS server' for location sensitive information and is the recipient of the response produced by the corresponding service; an LBS client may range from a notepad computer to a mobile phone or any other handheld mobile device.

Hereinafter, we will refer to the combination of these two entities as the 'LBS system'. It is evident that the LBS provision model greatly resembles the standard client-server model (Figure 14.1). There is always an entity that asks for the information and another who should provide it. Although, this is true from a logical perspective, sometimes it is not very easy to physically separate the server and the client entities, as they may both reside in the same physical device.

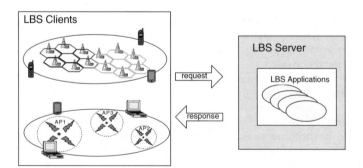

**Figure 14.1**   LBS client-server architecture.

This was the usual case until a few years ago, when the concept of an LBS system would normally bring to mind an electronic device, equipped with a GPS receiver and a display, where the user could see his current position, possibly on a map. However such proprietary LBS solutions were characterized by certain drawbacks. First the consumer found them expensive; the cost of the electronic device was usually high. Moreover, the device was capable of providing only a certain range of LBS services and no enrichment was possible due to the lack of open interfaces.

Currently, such solutions, although not completely abandoned, are targeted to a specific class of users, and have been replaced in solutions that apply to the classical Client-Server model, where the entities involved are physically apart. A key factor in this development was the growth in the capabilities of mobile and handheld devices, as well as the growing maturity of the wireless infrastructure in terms of positioning technology. Moreover, the new model completely separates the service logic from the client side, thus allowing new services to be developed and be accessed using the same terminal equipment. Having made this brief analysis, we can now proceed to some of the basic requirements that an LBS system should meet. Prime requirements include the following.

- Cost-effectiveness. Service delivery should not require the end-user to buy costly equipment.
- Openness. Service provisioning should support a variety of access protocols so that the service is available through different networks and different client equipment.
- Reusability. The LBS system should be capable of hosting a number of different services, with different requirements and functionality. The introduction of new services should not require changes to either the LBS Server or the Client and potential changes to the Server should maintain downward compatibility (i.e, should not affect the execution of existing services).
- Security and privacy. Security in the interfaces with external entities is essential so that secure communication and privacy is achieved. This is a fundamental requirement as the end user would not normally be willing to have his personal information (e.g., location) revealed to a third party.
- Scalability. The system should be able to host a large number of services, each capable of serving numerous concurrent requests.
- Extensibility. The system should be extensible and capable of accommodating new technologies. In order to achieve this, there must be independence from underlying technologies. Therefore the system should not be bound to any specific positioning or GIS technology. This will allow it to easily adopt newly evolved technologies from both sectors, thus increasing its life expectancy.

Additional requirements, which are optional, but may greatly enhance the potential of the LBS system include the following.

- Support for many operation paradigms. This means that apart from the classic request/response functionality, the platform should support services using the push model as well as event scheduling, which can be based on time or location events (such as notifying a user using SMS when he enters a shop about the day's offers).
- Roaming across different infrastructures. Both indoors and outdoors environments should be supported.
- Support for flexible service creation processes.
- Support for service deployment and operation, which should be provided through automatic procedures.
- Portability. Independence of operating systems and hardware platforms is a characteristic of prime importance as it guarantees integration with every infrastructure.

Having established the mandatory and desirable requirements as listed above, in the following section, we will proceed with the definition of its architecture and a comprehensive analysis of the heterogeneous technologies that such a system can merge.

## 14.3 LBS System

As already mentioned, an LBS system is composed of two elements; the LBS Server and the LBS Client. In this chapter we focus mainly on the Server side, while a brief discussion on LBS clients, their capabilities, and special configurations that may be needed is provided at the end of the chapter.

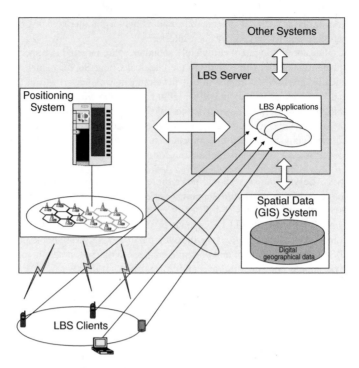

**Figure 14.2**   Generic LBS provisioning model.

A generic LBS provisioning model is shown in Figure 14.2. In the figure the main entities involved in the LBS provisioning scheme are depicted. Hence, apart from the core LBS System, as was defined earlier the model contains also three additional types of systems:

- the Positioning System;
- the Spatial Data (GIS) System;
- Supplementary Systems.

The first two systems are essential to the LBS provisioning chain, as the information they provide to the LBS Server is mandatory for executing the LBS application. The 'Supplementary Systems' category includes auxiliary or optional entities (e.g., billing systems), which although not needed for the basic LBS provisioning process, can greatly enhance it and allow the implementation of advanced concepts such as service personalization, QoS differentiation and different provisioning policies. Further details of all these systems are given below.

An important issue is that although each system in the generic model is depicted as topologically distinct, this is not mandatory. The model in Figure 14.2 shows the logical separation between the different components of an LBS System. The LBS Server, for example, may integrate the GIS system as well as the Positioning system in the same physical node.

### 14.3.1 LBS Server

The LBS server is the actual middleware that is used for the provisioning of LBS services. Running location-based applications requires the integration of many different technologies from both the Information Technology and Telecommunications (ITC) sector. Given that ITC technology is rapidly evolving, designing an extensible and open LBS middleware infrastructure is essential, in order to interact with heterogeneous systems that facilitate inter-operation with these evolving technologies. Moreover, according to the generic LBS model (Figure 14.2) these systems may belong to different vendors, each having its own communication protocols and APIs. Despite the fact that standardization institutes and fora (e.g., ETSI, 3GPP, OMA, OGC, etc.), together with key business players in the communications and GIS sector, have specified standard protocols and APIs for implementing such interaction, there is a large portion of the market that still uses proprietary interfaces. In order to cope with such peculiarities a generic LBS server should provide a framework that is capable of adapting to all them, with the minimum possible changes.

Application server architectures provide such extensible frameworks. An application server allows one to develop and shelter the business logic that is needed for executing an LBS service and at the same time saves the need to re-engineer your system if a component in the architecture needs to be changed.

There are many reasons to approach a location server infrastructure as a series of logically discrete components integrated through business logic stored in an application server [3]. It allows you to create infrastructure services based on industry standards for the various specialized components required. Your positioning interface might be based on the specification recommendations provided by the Open Mobile Alliance (OMA) and your spatial data server interface might be based on the Geography Markup Language (GML) proposed by the Open GIS consortium (OGC). However, the major advantage of this logical separation is that any one piece of your infrastructure is insulated from problems in another component. It allows you to substitute components that do not deliver acceptable results without affecting the rest of the system, and it also allows you to enhance components' functionality easily. Hence, interfacing with any proprietary system is possible through replacing standard infrastructure services with new ones that will comply with the peculiarities of the desired protocol.

Apart from making the LBS server extensible, the use of an application server as a basis for the LBS server provides automatic fulfillment of two other basic requirements for service provisioning systems that are reusability and scalability.

The basic functional components of the generic LBS server are shown in Figure 14.3. Four main functional areas have been identified, namely:

(1)  the LBS Application layer;
(2)  the Infrastructures Services layer;
(3)  the Presentation layer;
(4)  the Management layer.

We will further elaborate on the functionality of each layer in the following subsections.

### 14.3.1.1 LBS Application Layer

We will start our analysis with the LBS Application layer, which is the hosting system for all LBS applications running on the server. Each layer contains the business logic of a corresponding LBS service. Upon being called, by the LBS client, each LBS application collects the requested information and returns it to the end-user. In order to achieve the required functionality, it has to perform a number of actions: parse the incoming request, consult a number of services from the Infrastructure Services layer, create a response and post it to back to the client. We will further examine the overall process flow that takes place inside the LBS Server after we review in detail the internal architecture of the LBS system.

This layer also features an API, which allows communication with external service creation systems. This interface permits service deployment from such entities and could allow automatic service deployment tools to be built and integrated with the LBS server. In the same perspective, advanced service creation platforms can be connected through the same interface.

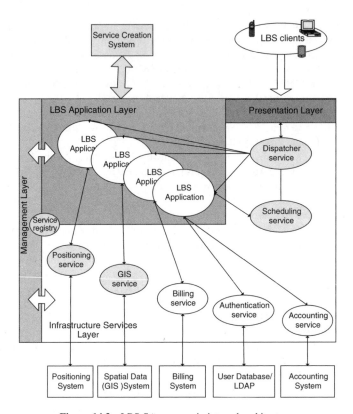

**Figure 14.3** LBS Server generic internal architecture.

### 14.3.1.2 Infrastructure Services Layer

The most important part of the LBS server is the Infrastructure Services layer. This layer contains the business logic, which is usually used by the LBS applications and which facilitates the interaction with external systems. In the absence of this layer, each location service would need to incorporate the corresponding logic in itself. Such an approach is far from optimal for a number of reasons, which will become obvious from the following example.

Assume that the LBS server is connected to a positioning system using a well-established protocol such as the Open Service Access (OSA) protocol [12]. The OSA specification defines an asynchronous communication protocol, where both the client and the server side need to implement predefined interfaces. Whenever, the client side, which in this case is the LBS server, needs to retrieve the position of a device, it initiates an interaction with the OSA compliant positioning system. A couple of messages are exchanged between the two entities, before the result of the initial query, is returned to the client (Figure 14.4).

In the presence of the Infrastructure Services layer, a positioning service takes care of all the aforementioned interaction. The service hides all the complexity of the OSA communication protocol, its particularities, asynchronous nature and specific parameter types. The LBS application only needs to perform a single call towards the positioning service and wait for the answer. This approach not only removes the programming burden from the developer of the LBS application, but also promotes the re-usability principle, which is considered a prime requirement for the LBS system engineering.

Now, let us review the case where no infrastructure services are present. Given that no positioning service exists, communication with the positioning system is handled internally by the LBS application. LBS application developers have to incorporate identical business logic in their applications

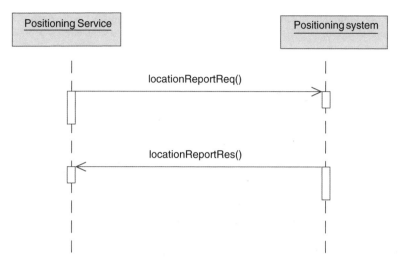

**Figure 14.4**   OSA-specific position retrieval sequence diagram.

every time they want to acquire location information. However, most important of all: if for some reason the interface with the positioning system has to be changed, the same should apply to all installed LBS applications. All of them will need to be rewritten to conform to the new communication protocol. Note that in the first case, the same change would impact only the infrastructure positioning service, which would have to be rewritten according to the new protocol.

This paradigm showed that the Infrastructure Services layer is the actual middleware, which is placed between the LBS applications and the various external systems and provides, to the former, a means to retrieve information efficiently from the latter. It strongly resembles what the JDBC middleware does for Java applications, in terms of interfacing with an RDBMS. Basic infrastructure services, pertaining to an LBS system, include:

- the Positioning service, which implements communication with positioning systems (a detailed analysis on available positioning systems will follow later in the chapter);
- the GIS service, which creates a communication link with the spatial data system (e.g. the GIS server), in order to perform spatial requests and acquire the needed data.

Additional infrastructure services that are not required for providing the core LBS functionality include the following.

- The Authentication service, used for interfacing with repositories containing user-specific info. The LBS application developer may utilize this service for restricting access to certain users either to the LBS system, or to specific applications in it. Moreover, the information acquired during the authentication process may be used for customizing the LBS application according to the user's preferences (i.e., personalisation).
- The Billing/pricing service. This service facilitates communication with billing/pricing gateways. It provides the application developer with the means to create billing accounts, checking pre-paid accounts in order to determine whether the user can use the service and it allows the implementation of advanced billing and pricing schemes from the side of the service operator.
- The Accounting service. This provides a convenient way to interface with external entities (e.g., databases) and store accounting information, pertaining to the usage of the LBS server. This information may be used for a variety of purposes, such as monitoring performance, tracking access to the server, tracking errors, etc.

Finally, there are some infrastructure services that are not meant to interact with external systems, but whose presence is essential for delivering the LBS application. We have identified two such services.

(1) The Dispatcher service. The task of this service is to intercept incoming requests from client applications and dispatch them to the corresponding requested service. It acts as the central entry-point for all incoming requests and provides a hidden discovery-like service for them independent of the used communication protocol. If, for example, a communication between the LBS client and the server uses the HTTP protocol, the Dispatcher service may be a simple servlet application which will perform a lookup in an internal service repository for determining the availability of the requested LBS application, and invoking it. If another protocol is used, then the servlet is replaced by another application, featuring the same interface and identical behavior and everything is as it should be. More than one client-server communication protocols (HTTP, WAP, RMI, TCP/IP, CORBA, SMS, etc.) could be supported simultaneously, by using multiple dispatcher services, one for each supported protocol.

(2) The Scheduler service. A scheduler service is essential both for supporting the delivery of applications to the end-user as well as for scheduling internal tasks that may be required by the LBS server. An example of the former includes the triggering of the execution of certain LBS applications, in certain time intervals. For instance, a user may register for receiving weather forecast SMSs, once per day, or a father may register for being informed on the location of his underage son every 2 hours. Several paradigms of such time-scheduled service provision can be devised. Regarding the LBS server itself and its internal operation, a scheduling service could be used for implementing operations like clearing cached data, restructuring internal repositories after a certain amount of time, freeing memory and many others which are typical for such enterprise systems.

An archetypal sequence diagram, depicting the internal flow of data inside the LBS system from the time a request is posted by the client until the corresponding response is received, is presented in Figure 14.5.

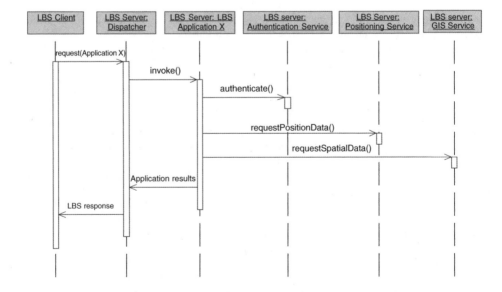

**Figure 14.5**  Typical processing of an LBS request.

### 14.3.1.3  Presentation Layer

Presenting the answer produced by an LBS application to the end-user is a fundamental part of the LBS provisioning system. As we have stated in previous paragraphs, the LBS server can potentially support a variety of protocols for communicating with the client devices. In a typical two-tier architecture the presentation logic would normally be embedded with the business logic. For an LBS server, this means that each service has to check the protocol of the request and format the response accordingly. It is sensible to assume that such an approach is not optimal as there are too many communication protocols available today, each with it own peculiarities and characteristics. Handling all of them inside the service logic would not only be overwhelming for the service creator, but it would also require the whole service to be rewritten in order to support additional protocols.

For the three-tier architecture model, which is followed by all modern software systems, the proposed generic LBS architecture features a presentation layer, whose purpose is to provide the necessary modules and tools that will format the results, produced by the underlying layers, in an attractive and comprehensive way by the end terminal. Inserting this layer between the LBS server and the client we automatically gain a significant advantage, as now the service can adopt a single format for its answer, which will then be transformed in the presentation layer to that understood by the client.

The XML specification is the most appropriate format for the initial response produced by the LBS application, due to its simplicity and the plethora of tools and technologies that are available today (e.g., XSLT, DOM, SAX) for its processing and transformation. XML allows the development of a single interface to the LBS server that uses different style sheets (e.g., XSLT) to customize output for the format the client is expecting (Figure 14.6). HTML, HDML, cHTML, WML and VXML are all based on the XML specification. If the presentation layer is based on XML, supporting a new client (or protocol) requires nothing more than creating a new style sheet. It does not require a new interface to be engineered.

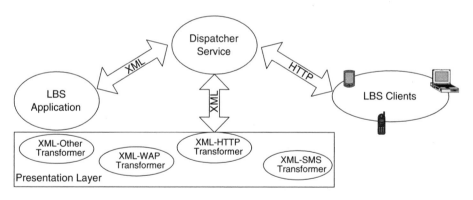

**Figure 14.6**   Use of XML transformers for presenting results to end users.

### 14.3.1.4  Management Layer

A flexible management interface is required in order to administer the LBS server effectively. The management interface should have methods that will, at least, allow the following operations to be performed.

- Starting an LBS application. Each LBS application upon being deployed on the Server should not be eligible for access by the end users. Enablement of the application should be done explicitly through the LBS Server's management console.

- Stopping an LBS application. The administrator of the platform should be allowed to terminate the availability of a specific service at any time. This is required for security (e.g., to prevent unauthorized access) or other reasons (e.g., freeing resources, etc.). Stopping an application should be allowed either massively (i.e., for all users or a group of users) or individually (i.e., for a specific user). Moreover, at least two stopping types must be supported:
  - Immediate or Hard stop, which will force termination of all active instances of the LBS application. Users already accessing the platform may receive a message of unavailability and the service will cease execution until it is explicitly restarted.
  - Soft stop, where users already accessing the location service will continue to be served until their requests are completed, but no further admittance of additional users will be permitted. When all pending requests are served the service will stop completely.
- Configuring an LBS application. LBS applications deployed on the server may require some configuration prior to becoming eligible to end-users. This is because the service provider and the mobile operator are usually different actors in the LBS chain (see Figure 14.12). The service creator develops applications that can generally be applied to any network infrastructure. The easiest way to do this is to embed in the service configuration options, which could effectively be changed in order to adapt the service to any potential infrastructure. Configuration may come before the deployment phase, from the service creator. However what if, later on, the mobile operator changes its infrastructure? Will there be a need to re-deploy the service from scratch? In order to avoid this and instead allow applications to operate in a dynamic environment, the management layer should provide the means to configure these services, by providing an infrastructure for dynamically passing parameters to them.

Support for all the abovementioned operations is essential, not only for the LBS applications, but especially for the infrastructure services that run on the LBS server. The infrastructure services are the actual mediators between the LBS applications and the network equipment, so the need for supporting dynamic on-the-fly configuration is much greater in their case. Moreover starting and stopping an infrastructure service is extremely crucial not only for security and resource management issues, but also in order to avoid unexpected behavior during their configuration. For example, imagine re-configuring the connection to the positioning system while positioning requests are being served.

Operation of each hosted service can be modeled by the three-state machine shown in Figure 14.7. Maintenance of states could be done either by the service itself, through the use of a status flag, or by

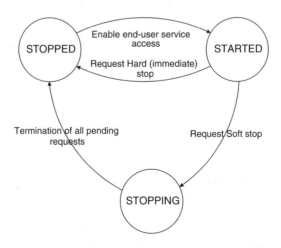

**Figure 14.7** State machine for LBS services hosted by the LBS server.

use of a Service Registry, which keeps track of all services states at any specific time. In the latter case, invocation of each service/application is performed through consulting the registry.

### 14.3.2 Positioning Systems

By the term 'positioning system', we refer to the network entity that is responsible for determining the location of a client device. Positioning technology [4] is a key point within the LBS context. There are two key components that form a positioning system. The first is the technology used to determine the user's location. The second is the method that an application uses to get access to this information. Positioning systems can be grouped in two major categories:

(1)  outdoor positioning systems, to provide location estimates for outdoor environments;
(2)  indoor positioning systems, tailored to the specific needs of indoor environments (e.g., buildings).

As seen in Figure 14.8, each of the above is further divided into additional categories, with reference to the technology used for location estimation. We will review separately indoor and outdoor systems and their sub-categories in the sections that follow. For each system both positioning methods and position retrieval protocols will be reviewed in detail.

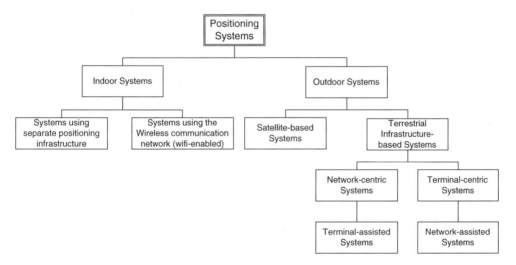

**Figure 14.8**  General categorization of positioning systems.

Another categorization of location systems is related to the semantics of their output (Figure 14.9). A location system can provide two types of information: physical and symbolic. An example of a physical location is the global coordinates (i.e., the longitude and the latitude of a point). Symbolic Location, on the other hand, encompasses abstract ideas of where something is (e.g., 'room32' or 'Corridor'). Physical location systems may further separated to those producing absolute location and those that return relative location information. The former include systems using a shared reference grid for all located objects. For example, information like latitude, longitude and altitude or their equivalents, such as Universal Transverse Mercator (UTM) coordinates always use the same reference point.

In relative systems, each object can have its own frame of reference. For example, the main entrance of a building can be used as the beginning of the grid, and thereafter all location information is provided relative to it. An absolute location can be transformed into a relative location i.e., relative to a second

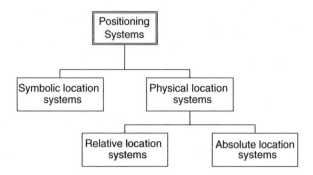

**Figure 14.9**   Categorization of location systems, with reference to their produced output.

reference point. However, a second absolute location is not always available. Conversely, there are also techniques (e.g., triangulation) for determining an absolute position from multiple relative readings if we know the absolute position of the reference points.

Additionally, positioning systems might support different types of location requests, such as Request Response (RR), Event-driven Request (ER) or Periodic Request (PR) [40]. The first type of request is materialized through a solicited approach, where the LBS Server asks from the positioning system the retrieval of the mobile terminal's coordinates. Requests of the ER type contain certain attributes and result in position reporting based on certain location-related events occurring in the underlying network (e.g., a user enters or exits a specific geographical area). In PRs, the LBS server explicitly defines the mean time (period) between the resulting location reports.

### 14.3.2.1 Outdoor Positioning Systems

Outdoor positioning systems usually produce absolute physical location, either in UTM coordinates or longitude and latitude.

***Positioning Methods***
The positioning method identifies the technology and the procedures used for determining the exact location of a mobile device. Recent research results are impressive. A plethora of solutions that provide an estimate of the user's location have emerged, each with distinct characteristics and capabilities, and hence applicable in different circumstances. Two main categories of outdoor positioning methods are widely used today:

(1)   satellite-based;
(2)   terrestrial infrastructure-based.

The Global Positioning System (GPS) [5] is the most noticeable example of the first category. GPS consists of a constellation of satellites that transmit information 24 hours per day, thus enabling a GPS receiver to determine its own position. In GPS, location determination uses the timed difference of arrival of satellite signals and is performed either entirely within the mobile unit or within the network. The method provides 10–40 meter accuracy.

Other more accurate positioning methods have been developed that involve post-processing, network assistance, and relative or differential positioning. These methods are known as Differential GPS (D-GPS) and Assisted GPS (A-GPS) [3]. A-GPS uses a GPS reference network (with fixed coordinates) to accelerate and improve the accuracy of positioning. The reference network provides the GPS receivers on the terminal with data that they would ordinarily have to download from the GPS satellites, thus resulting in faster and more accurate responses. D-GPS is practically the same technique; D-GPS [7] describes the situation where the reference infrastructure is equipped with fixed GPS receivers that provide relative positions for correcting the position estimates received by the corresponding receiver of the device. Both aforementioned methods can potentially provide location accuracy at the meter and sub-meter level by using infrastructure-based assistance. A certain drawback of the traditional GPS solution is its inability to support positioning in indoor environments. However A-GPS can use additional equipment, known as correlators, in order to provide results in indoor environments. The corresponding technique is known as massive parallel correlation.

Advanced research has been performed in recent years in the area of satellite-based position methods and already two new systems are emerging. The European Space Agency (ESA) works on Galileo [8], a system consisting of 30 satellites, planned to be commercially available after 2008. Galileo, according to its specification, will deliver highly accurate (4–10 m) positioning services. On the other hand, the Russian Federation Government in coordination with the Russian Space Forces has launched GLONASS [9], which aims to provide significant benefits to the civil community through a variety of applications, but with significantly lower performance (accuracy in the range 57–70 meters). One significant problem with satellite-based solutions is that their application requires expensive add-ons to the terminal device.

Terrestrial Infrastructure-Based positioning methods, on the other hand, rely on cellular networks equipment in order to determine the user's location. They don't normally require expensive add-ons for the terminal equipment, but this advantage is balanced by the fact that they are less accurate than their satellite-based counterparts. Two options have been proposed for location determination using terrestrial infrastructure: the 'terminal-centric' approach, where the terminal itself determines its position and the 'network-centric' approach, where the position of the terminal is determined exclusively by the network infrastructure. Sub-options of both solutions exist. Hence, the 'network-centric' approach might require additional information from the handset resulting in the 'terminal-assisted' case, while in the 'terminal-centric' approach there might be some interest in getting corrections from the network – the 'network-assisted' case. The most popular terrestrial-based methods are as follows [3].

- The Global Cell-ID (GCI) method. This is the simplest terrestrial, infrastructure-based, method. This network-centric method is based on the capability that is inherent in every cellular network, as part of their mobility specification, to identify the cell where a specific mobile terminal is located. Although easy to use, GCI provides a rather rough estimate of the user's actual location, because the radius of radio cells can vary significantly. The GCI's accuracy can be improved with the Timing Advance (TA) technique. TA is a special parameter of cellular networks, especially of the GSM, and provides the base station to mobile terminal roundtrip delay.
- The Time of Arrival (TOA). This network-centric method, referred to as uplink time of arrival (UL-TOA), is based on the time of arrival of a known signal sent from the mobile device and received by three or more base stations. All measurements are performed at the BTSs and forwarded to the Mobile Location Centre for generating a position fix, through triangulation.
- Angle of Arrival (AOA). This is another network-centric method. It does not require a mobile device upgrade to operate. AOA is based on the assumption that multiple base stations simultaneously receive the signal produced by a mobile device. The base stations have additional equipment that determines the compass direction from which the user's signal is arriving. The information from each base station is sent to the Mobile Location Center, where it is analyzed and used to generate an approximate latitude and longitude for the mobile device.
- The Enhanced-Observed Time Difference (E-OTD) method. This is similar to time of arrival (TOA), but is a terminal-centric positioning solution. E-OTD takes the data received from the surrounding base stations to measure the difference in time it takes for the data to reach the terminal. The time difference is used to calculate where the mobile device is in relation to the base stations. For this to work, the location of the base stations must be known and the data sent from the different base stations must be synchronized. All measurements in this method take place at the terminal and are

used to generate the position fix, or they are forwarded to the Mobile Location Center for position fix generation. Accuracy of E-OTD is expected to be as good as 50 meters using GSM and even greater with 3G networks.

### Position Retrieval Protocols

The 3rd Generation Partnership Project (3GPP) has specified a standard configuration of location services entities in GSM/GPRS and UMTS Public Land Mobile Networks (PLMNs). To support LBS the 3GPP initiative has defined the Gateway Mobile Location Center (GMLC) [41]. The GMLC is the first node that an external location services client accesses in a 2G/3G PLMN. The architecture for the provision of LBS as defined by 3GPP is shown in Figure 14.10. Although the architecture seems a little complicated, the part we focus upon is the communication interface between the Location Service (LCS) client, and the Gateway Mobile Location Centre (GMLC). This interface (Le), is used by the former in order to get access to the location information describing the position of the target device. Several attempts have been made for standardizing this interface, which have resulted in the emergence of two protocol specifications: the Mobile Location Protocol (MLP) by the Location Interoperability Forum and the Open Service Access (OSA) by the ETSI-Parlay-3GPP bodies. This section presents an overview of these activities as well as some others from individual vendors, such as the Mobile Positioning Protocol by Ericsson.

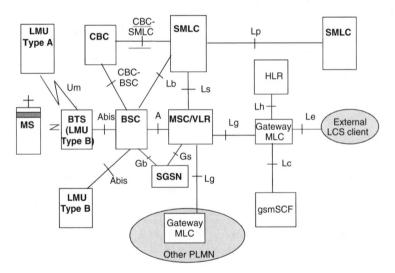

**Figure 14.10**   3GPP LCS network reference model (taken from [41]).

### Mobile Location Protocol (MLP)

The Mobile Location Protocol (MLP) [15] was proposed and standardized, by the Location Interoperability Forum (LIF), an open industry organization dedicated to promoting the development of location-based services. In 2003 LIF was consolidated into the Open Mobile Alliance (OMA), and it no longer exists as an independent organization. However, the work originated in the LIF continues within OMA's Location Working Group (LOC).

   MLP enables location applications to interoperate with the wireless networks irrespective of their underlying air interfaces and positioning methods. MLP defines a common interface that facilitates the exchange of location information between location-based applications and the wireless networks represented by location servers. MLP supports privacy and authentication to ensure that users' whereabouts are protected, and are only provided to those who are authorized to know them.

MLP is an application-level protocol for querying the position of mobile stations independent of underlying protocol technology. It serves as the interface between a Location Server and a location-based application. Possible realizations of a Location server are the Gateway Mobile Location Center (GMLC), which is the location server defined in GSM and UMTS and the Mobile Position Center (MPC), which is defined in ANSI standards. Since the location server should be seen as a logical entity, other implementations are possible.

In most scenarios, the client entity initiates the dialogue by sending a query to the location server and the server responds to the query. MLP can be implemented using various transport mechanisms like HTPP, HTTP/SOAP, etc. MLP is currently at version 3.0.

### Open Service Access (OSA)

In order to be able to implement future applications/end-user services that are not yet known today, a highly flexible Framework for Services was required. The Open Service Access (OSA) specification [10], proposed by the Parlay group jointly with ETSI and 3GPP, enables third-party applications to make use of network functionality in an easy and protocol agnostic manner. The Parlay Group is a multi-vendor consortium formed in 1998 in order to develop open, technology-independent application programming interfaces (APIs) that enable the development of applications that operate across multiple, networking-platform environments. Members of the Parlay Group include: industry leading IT companies, Internet service vendors (ISVs), software developers, network device vendors and providers, service bureaus, application service providers (ASPs), application suppliers and large and small enterprises.

In OSA, network functionality offered to applications is defined in terms of sets of Service Capability Features (SCFs). The aim of OSA is to provide a standardized, extensible and scalable interface that allows for inclusion of new functionality in the network in future releases, with a minimum impact on the applications using it. Specifically, the OSA specification consists of three parts.

(1) The Applications (e.g. VPN, conferencing, location based applications). These applications are implemented in one or more Application Servers (Figure 14.11).
(2) The Framework [11]. This provides applications with basic mechanisms that enable them to discover and make use of the service capabilities in the network.
(3) The Service Capability Servers (SCSs). These provide the applications with Service Capability Features (SCFs), which are abstractions of the underlying network functionality.

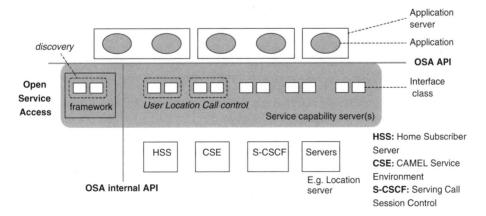

**Figure 14.11**   Open Service Access reference model.

SCFs provide access to the network capabilities for service designers; they are open and independent of any programming language, thus enabling easy development of new applications or enhancements of already existing ones. Core network elements (e.g., gateways, routers, servers, etc.), on the other hand, are free to use their specific underlying protocols, in order to provide the expected functionality. Their design and implementation details do not affect applications using them, as the latter do not interface directly with them but through the corresponding SCSs. In other words, a SCS serves as a gateway between the network entities and the applications. Moreover, in order to abstract from operating systems and programming languages for both the client (application) and the server (SCS) side, the OSA specification proposes either CORBA or Web Services as the most fitting candidates for providing SCFs. OSA SCFs are specified in terms of APIs, which should be implemented by both the application and the network element (e.g., location server) [12]. It should also be noted that the scope of OSA's SCFs are very wide covering, apart from positioning capabilities, issues such as messaging, charging, accounting, user interaction and generally almost every aspect of network capabilities.

### Mobile Positioning Protocol (MPP)
The Mobile Positioning Protocol (MPP), proposed by Ericsson, is an Internet-based protocol that enables location-dependent applications to communicate with a Mobile Positioning Server (i.e., at the Le interface of the 3GPP LCS network model) and request the position of mobile terminals. MPP is a proprietary protocol, that was introduced in order to cover Ericsson's need for a position gateway protocol, until LIF's MLP be finalized. However Ericsson continues to support it, in its Mobile Positioning System (MPS). The Gateway Mobile Positioning Center (GMPC) is the main entity of the MPS system; it comprises the mediator between the mobile network and the location-dependent application, thus, playing the role of the LCS Server as defined in the LCS Reference Model. MPP offers a carefully designed generic interface towards the GMPC, thus making the applications independent of the underlying physical positioning technology that is used when retrieving the position of a mobile terminal. MPP is purely XML-based and makes use of the HTTP protocol, thus making the GMPC available from any platform with TCP/IP capabilities, (e.g., a servlet running on an application server). MPP is currently at version 5.0 [16].

### 14.3.2.2 Indoor Positioning Systems

Indoor positioning systems are specifically targeted at indoor environments and usually produce responses that are specifically tailored to their application domain. Some of them return physical location, much like their outdoor counterparts. However, because indoor environments are local and isolated, most indoor positioning systems adopt the relative location approach, using a predefined point inside the building as their reference point. Finally, there are numerous systems that produce symbolic information as output.

### Positioning Methods
There are three main methods that are used for location estimation in indoor environments: triangulation, scene analysis and proximity. These techniques are used either on their own or jointly. The latter case can further enhance the accuracy and precision of the positioning method.

(1) Triangulation is a technique that uses the geometric properties of triangles to compute objects location. There are two triangulation approaches:
   (i) Lateration, which measures the distance from a number of multiple reference points. Measurements are through the following.
      • Time of flight, which measures the time that it takes for an emitted signal to be reflected by the located object.
      • Attenuation, which measures the decrease in the emitted signal as the object's distance from the transmitter increases.
   (ii) Angulation, which uses angular measurements to determine the position of an object.

(2) Scene analysis uses features of a scene observed by a reference point in order to draw conclusions about the location of the observer or of objects in the scene. It usually requires a database of signal measurements, which is used from the positioning system for location estimation.

(3) Proximity, which determines when an object is 'near' a known location. The object's presence is sensed using a physical phenomenon with limited range. Proximity sensing can be done through:

- physical contact through pressure sensors, touch sensors and capacitive field detectors;
- monitoring wireless access points for determining when an object is in their range;
- observing automatic ID systems, through the proximity of systems like credit card point-of-sales terminals, computer login histories, landline telephone records, etc., which can be used to infer the location of a mobile object.

From these techniques, some require specialized positioning infrastructure to be installed (e.g., sensors, RF-tags), while others rely on the infrastructure of the wireless communication network (e.g., IEEE 802.11), thus resulting in two primary classifications. Systems of the first category are:

- Active Badge [17, 18, 19], a proximity system that uses infrared emissions emitted by small infrared badges, carried by objects of interest. A centralized server receives the emitted signals and provides the location information.
- Active Bat [17, 18, 19] system resembles the Active Badge system but uses an ultrasound time-of-flight lateration technique for higher accuracy.
- MIT's Cricket system [20] relies on beacons that transmit a RF signal and an ultrasound wave, and on receivers attached to the objects. A receiver estimates its position by listening to the emissions of the beacons and finding the nearest one.
- SpotON [21] is a location technology based on measuring RF signal strength emitted by RF tags on the objects of interest and perceived by RF Base Stations.
- Pseudolites [22] are devices that emulate the operation of the GPS satellites constellation, but are positioned inside buildings.
- Pin Point 3D-iD [23] is a commercial system that uses the time-of-flight lateration technique for RF signals emitted and received by proprietary hardware.
- MSR Easy Living [18, 19] uses computer vision to recognize and locate the objects in a 3D environment. The 3D cameras that are used provide a stereoscopic image of an indoor environment and, with the assistance of an image recognition algorithm, locate the objects.
- MotionStar Magnetic Tracker [18, 19] incorporates electromagnetic sensing techniques to provide position-tracking.
- Smart Floor utilizes pressure sensors to capture footfalls in order to track the position of pedestrians.

Systems belonging in the second category include, among others:

- MSR RADAR system, which uses both scene analysis and triangulation based on the received signal's attenuation, of the underlying IEEE 802.11 network [24, 25].
- UCLA's Nibble, which uses scene analysis to estimate the location of the user and provides him/her with symbolic and absolute positioning information [26, 27]. It operates on IEEE 802.11b WLANs. Location estimates are based on measurements of the signal quality of nearby access points and determination of the location uses a Bayesian filter, in order to distinguish the location from other locations with different signal quality characteristics. Nibble does not generate location coordinates, but a symbolic name corresponding to the current area of the terminal device.
- Ekahau's Positioning Engine (EPE) is a commercial product that combines Bayesian networks, stochastic complexity and on-line competitive learning, to provide positioning information [28] through a central location server. It operates on IEEE 802.11 WLANs. EPE produces relative location coordinates and its accuracy is estimated to 1–3 meters.

The main advantage of the methods used by the first category is the high accuracy with which they can estimate the position of an object. However, the disadvantage is that additional equipment is required to be carried by the located object, which in most cases is small and economic, but it does not help in the user-friendliness envisaged for these systems. Moreover, a main drawback is the deployment costs and the operation-maintenance of a second location-specific infrastructure that runs in parallel with the wireless data communication infrastructure.

### Position Retrieval Protocols

Although there is significant activity is the area of indoor positioning, no established specification exists regarding the communication interface with external LBS applications. Almost all existing solutions incorporate proprietary interfaces for disseminating the estimations calculated by the underlying positioning technique towards the LBS applications. Despite the fact that no standardization activity exists for indoor position retrieval, it is evident that for the systems that produce absolute location estimates (e.g., global coordinates such as longitude and latitude), the OSA APIs can be adopted for returning location results to an LBS client.

For the rest of the systems a unified interface that will abstract from the specific attributes and parameters used by each method should be devised and standardized. Towards this direction, several WLANs and indoor positioning architectures have been proposed. WhereMoPS [29] provides a layered system model for indoor geolocation systems that includes data collection, location computation, location normalization, and location provisioning components. The Location Stack [30] model, focused on location context, consists of a seven-layer stack that includes sensors, measurements, fusion, arrangements, contextual fusion, activities and intentions. Finally, the Gateway WLAN Location Center (GWLC), hides the heterogeneous functions of the indoor positioning architectures, providing a unified interface towards LBS applications for retrieving location data of users and objects, using capability brokering, authentication, security and location cashing methods [31]. This interface greatly resembles the OSA mobility SCF API.

### 14.3.3 Spatial Data (GIS) Systems

Location based services require a Spatial DataBase (SDB) server for providing effective and efficient retrieval and management of geo-spatial data [36]. Spatial database systems serve various spatial data (e.g., digital road maps) and non-spatial information (e.g., navigation instructions) on request to the client. SDB servers provide efficient geo-spatial query-processing capabilities such as find the nearest neighbor (e.g., petrol station) to a given location and find the shortest path to the destination. In the OGC location services reference model, the SDB (Location Content database according to OGC [32]) server acts as a back-end server to the GeoMobility server. The GeoMobility server comprises abstract data types (ADTs) and core services through which a service provider can provide location application services and content. Core services include:

- location utilities services;
- directory service;
- presentation services;
- gateway services, and
- route determination services.

The OpenLS standard [32] aims at the development of interface specifications that facilitate the use of location and other forms of spatial information in the wireless Internet. The exact objective of the OpenLS initiative is the production of open specifications for interoperable LBS that will integrate spatial data and processing resources into the telecommunications and Internet services infrastructure.

The GeoMobility server is a core component in the OpenLS framework. It comprises a series of core services as listed above.

- The Location Utilities Service (LUS) includes the geo-coding and reverse geo-coding functionality. Additionally, the LUS contains an abstract data type (ADT) known as the address. The geocoding service is a network accessible service that transforms a description of a location into a normalized description of the location with point geometry. The reverse geocoding service maps a given position into a normalized description of a feature location. Geocoding is the process of assigning an $(x, y)$ coordinate pair to a given address. Address interpolation is a well-known geocoding technique. Given a street segment with start and end coordinates and an associated address range we can interpolate the (approximate) location of any given address that falls within the given range by simply dividing the length of the road segment by the number of houses. Reverse geocoding finds an address given an $(x, y)$ coordinate pair. Owing to the low precision of the information provided by systems like the GPS, the most likely segment of the road network needs to be identified given the estimated location. This technique is also known as map matching. Map matching can be geometric, probabilistic or fuzzy.

- The Directory Service (DS) provides a search capability for one or more points of interest (e.g., a place, product or service with a given position) or area of interest (e.g., a sports complex or a bounding box). An example query is 'where is the nearest restaurant to the university campus?'. The types of queries addressed to the DS can be: (i) attribute queries based on non-spatial attributes, and (ii) proximity queries based on spatial attributes. Proximity queries are divided into three categories: point queries (given a query point find all spatial objects that contain that point), range queries (given a query polygon find all spatial objects that interest the polygon) and nearest neighbor queries (given a query point find the spatial objects with the smallest distance to the point).

- The Presentation Service (PS) deals with the visualization of the spatial information as a map, route or textual information. Routes, points of interest, object locations and textual information such as navigation instructions are overlaid on road maps through the PS. Currently, most presentation services are based on a visual interface but in the future the situation is expected to move towards voice-based interfaces (especially in vehicle navigation systems).

- The Gateway Service (GS) enables the retrieval of the present position of a mobile terminal from the network infrastructure. Specifically, the GS is the interface between an Open Location Services Platform and Positioning Systems through which the platform obtains real-time position data for mobile terminals. Real-time position information is acquired by means of a satellite network such as GPS/Galileo or through terrestrial radio based positioning systems.

- The route determination service provides the ability to find a best route (and navigation instructions) between two locations. User constraints are also taken into account. Route determination should support two services: the first handles the determination of a new route (optimum path between two endpoints), the second deals with the determination of alternative paths. Algorithms incorporated in the route determination service include the established Dijkstra algorithm, the A* algorithm, the IDA* (iterative deepening search algorithm) algorithm, and others.

As previously mentioned, a Spatial Database Server aims at the effective and efficient management of data related to a space in the physical world. An SDB provides conceptual, logical and physical data modeling facilities to build and manage spatial databases. Spatial information contained in the SDB is in the form of digital road maps and can be modeled and managed as discussed below.

- Digital road maps. Location based services rely on digital road maps, postal addresses and point of interest data sets. These maps are indispensable for any location-based utility that involves position- or route-based queries. Current road navigation systems use digital road maps available on CDs or DVDs. Many other applications require updateable backend spatial databases. Typically, digital road maps are available from governmental organizations (e.g., Departments of Transport). Nowadays, this situation has changed as many private companies also offer on-line digital road maps enriched with point-of-interest data sets. Data quality is one very important feature of such sources of information. Data quality refers to the accuracy and precision of the GIS database and the involved spatial data. Specifically, data quality is reflected in the following components: lineage, positional accuracy, attribute accuracy and completeness.

- Data model for digital road maps. In terms of the conceptual data model used to denote the application entities and their relationships, in the LBS world the most popular technique is the Pictogram-Enhanced Entity-Relationship (ER) model (PEER). PEER covers spatial semantics through geometric attributes and spatial relationships. In terms of the logical data model, a hybrid object-relational model is gaining popularity nowadays for the implementation of SDBs. The OGC has also provided a series of SQL extension specifications for handling spatial information (e.g., functions on geometry datatypes, topological relationships) in terms of the physical implementation (indexing, memory management, etc.) of the SDB. Established spatial indexing structures are the R-trees.

Commercial examples of SDB are the Oracle Spatial [37], the DB2 Spatial Extender [38] and ESRI's Spatial Database Engine [39].

### 14.3.4 Supplementary Systems

#### 14.3.4.1 Service Creation System

The LBS server is a housing system for the LBS services that are delivered through it. The main task of the server is to handle incoming requests, dispatch them to the correct services and then collect the results produced by the service and deliver them to the corresponding end-user, who performed the initial request. A detailed elaboration on how all these are handled was given above. However, another important aspect of service provisioning is the process of creating a new service.

Creating a new service requires first writing the service, and then deploying it on the server. Depending on the type of the server, the service can be written directly in some general-purpose high-level programming language (e.g. Java, C++) that is understood by the server, or in other proprietary languages that are server-specific. The latter are, usually, XML based scripting languages (e.g., CPL-like or other) that are tailored to the specific LBS system that they support. At the deployment phase, service scripts are translated, compiled and installed on the platform in an executable form (e.g., Java bytecode or executable files).

Modern LBS systems integrate in their architectures tools that take care of both service creation and service deployment. We refer to these tools as the service creation system. The tools could range from a simple text editor for writing the service and some command line tools for deploying it, to integrated development environments (IDEs). Such environments can greatly support service creators and ease their tasks. Usually, they feature advanced editing perspectives (i.e., structured around the service specification language) and tools that allow for the automatic installation of new services on the platform. Many of these service creation systems can operate remotely (i.e., as stand-alone systems) and can be linked to the server only for deploying the necessary services.

The potential of such stand-alone service creation systems is great as they clearly separate service creation and deployment from service provisioning, thus, allowing those two processes to be managed by different entities. Consequently, a variety of business models can be applied in the LBS provisioning chain (see Section 14.3.4.2, Billing systems).

A crucial issue in remote service deployment is that of security. Secure communication between the service creator and the platform host is essential for applying such a model. Although different security protocols can be used for securing the aforementioned transaction, a straightforward solution involves the use of Web services and their implementation over HTTP/SOAP using the SSL protocol. Such a solution is both lightweight – as it does not require the significant overhead both in terms of administration and data exchange that is imposed by other protocols (e.g., IPsec) – and easy to implement, considering the plethora of Web services software kits available on the market today. Moreover, a significant advantage of the proposed security infrastructure is that it does not make any assumptions about the type of the client (i.e., the service creation system) and the server (i.e., the LBS server). Hence, heterogeneity of the two systems is not an issue, providing that the client respects the Web-based protocol exposed by the server.

### 14.3.4.2 Billing Systems

Before analyzing the proposed architecture of the billing system, we must first identify the different business entities involved in the provisioning of LBS applications, as well as the potential revenue flows between them. Five different actors have been identified in the LBS provisioning chain:

- Users that are the final consumers of LBS;
- Service Providers developing value added services using the LBS provision platform;
- ASPs owning the LBS Server;
- Network Operators providing location data to the LBS server through mobile networks;
- Content provides (e.g. GIS providers), managing and handling spatial data.

Figure 14.12 shows the businesses involved in LBS provision and the potential revenue flows between them.

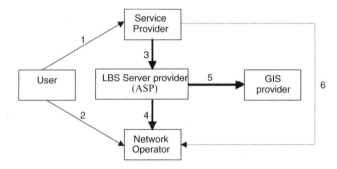

**Figure 14.12**   Businesses involved in LBS provision and expected revenue flows.

Additionally, the following revenue flows have been identified.

(1) The Service Provider may directly charge the end-user for service use, especially in the case of transactions performed while using the service. Note, however, that the most used business model does not presuppose this revenue flow, but rather adopts the revenue flow (6) (see below).
(2) The Network Operator charges the user for the regular mobile user subscription and issuing the SIM card, as well as airtime/volume of data downloaded to the terminal.
(3) The LBS Server provider may charge the Service Provider for various functions:
- for service creation and deployment (this should normally be a one-time fee);
- for service execution, mainly in the case that a Service Provider application has particular scalability and latency requirements (e.g., it needs to support a large number of user requests per second). Fees could be different for various classes of deployment, resulting in different scalability/performance levels;
- For the positioning/GIS queries incurred by the service. Prices could depend on the number/QoS of requests to the Network Operator or GIS provider.

(4) The Network Operator may charge the LBS Server provider for the location information that is retrieved from his network (revenue flow (4)). The Network Operator-LBS Server provider agreement could be limited to a flat fee, could include guarantees for QoS (e.g., min/max requests per second, latency, etc.) or could even specify the calculation of charges per number of requests (combined with the corresponding QoS received). This latter model may be the least possible.

(5) The LBS Server provider (ASP) may be charged by the GIS provider (if the latter is an entity different from the LBS platform provider) for the retrieval of the GIS data.

(6) Network Operator and Service Provider may establish an agreement for sharing the revenue from service usage. This, however, does not involve or impose any requirements on the LBS Server provider. Another case would be for the network operator to also provide the service. In this case revenue flows (1) and (2) overlap, while flow (6) becomes obsolete.

Note that although a business-level relationship between the end-user and the LBS platform provider could be identified, such a relationship is not required (the user should be charged only by the Network Operator and also probably by the Service Provider). Generally, LBS provisioning systems should be as 'transparent' as possible to the mobile user.

An advanced billing system, discussed above, considers all possible revenue flows and is capable of supporting most of them. Flows (3), (4) and (5) are directly related to the LBS Server provider, and if the service provider also operates the LBS server revenue, flow (1) should also be considered. The billing system should provide the means for each entity involved in LBS provisioning to acquire specific information that will be used for creating billing accounts.

A generic billing architecture is depicted in Figure 14.13. A billing system is an application framework that consists of three main components: an RDBMS, which is used to store information that will be used later for billing the user, a billing model and a charging module. The billing model

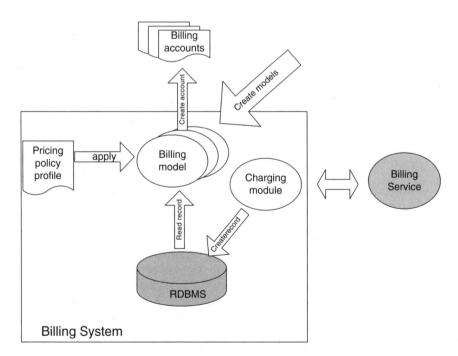

**Figure 14.13**  Generic billing architecture.

contains the business logic that generates the final billing accounts, based on a certain pricing policy profile. The pricing policy profile contains only price-specific information (e.g., the cost in euros of each request, or the cost of each byte transferred) and may be considered part of the billing model as well. However, separating the two makes the overall framework much more dynamic and scalable as a new billing model does not need to be created each time a different pricing policy is applied. The existing model can be used and the new policy profile applied to it. An interface for creating new business models is also available, so that new business models that meet the various requirements imposed by each business entity, can be imported. Finally the charging module provides the business logic for logging all necessary information needed for billing the user in the RDBMS. This may include creation of CDRs, checking of pre-paid profiles, etc.

Most of the existing billing models perform billing based on subscription, the time the service is used or the amount of transferred data. Logging the performed activity in the RDBMS, thus creating CDR-like records that could later be used for creating billing accounts, could provide the desired functionality in this 'post-paid' scheme.

However, there are also pre-paid schemes, which require enhanced functionality. Implementation of such schemes requires a bi-directional communication interface between the billing system, and the peer billing service. The billing system not only needs to accept requests for logging charging data, but must also provide answers to requests querying the balance of the user's account, etc.

Billing models for LBS applications may also consider charging based on the delivered QoS level. QoS in LBS systems is mainly valued in terms of the accuracy provided by the location estimation method, the response time and the time-to-first fix. A network operator that supports different positioning technologies each with, for example, a different precision level, may want to incorporate QoS-specific information in his charging records, and use a pricing policy that will consider it, instead of adopting a flat pricing model.

Most billing systems adopt proprietary interfaces for communication with external entities (e.g. billing services), however some standardization activity in this area is also available. Specifically, 3GPP has published a specific part of the OSA API that defines the interaction between a charging application (e.g., our LBS billing service) and the network operator [14].

### 14.3.4.3 Security Systems

Computer security is often thought of in terms of secrecy or confidentiality. Equally important to computer security are data integrity, authentication, authorization and systems availability. Securing access to a computer system is a common requirement, which explains also why security services are always integrated in such systems. In this section some specific requirements that should be considered when designing a security service for an LBS system are analyzed. In location-based systems, security is focused on two main areas:

(1)  controlling access;
(2)  anonymity and confidentiality.

Controlling access to the LBS server is essential in order for the service provider to support different operation paradigms. Services delivered by the LBS server may vary in terms of functionality, content and capabilities. It is evident that a service provider that aims to cash in on the usage of the services offered by his platform will require a mechanism that will control access to them. User authentication techniques are a common practice for achieving this functionality. As displayed in Figure 14.3, an authentication service, which communicates with an external repository containing the credentials of all users, allowed to access the Server is necessary. Such repositories are usually implemented in Directory/ LDAP servers, while large-scale implementations may involve RDBM systems. Moreover user repositories may be part of the LBS server or reside externally, acting as authentication gateways.

An example of such an external authentication gateway is the HLR of the GSM Authentication subsystem. Information pertaining to eligible services per user and their access rights can be incorporated in it and further used for authorizing the user's entry into the system. Part 3 of the OSA protocol [11], describing the framework, provides, among others, an open and standard mechanism (in the form of an API) for implementing communication with such authentication gateways. If the repository is inside the LBS server, proprietary APIs can be used for retrieving the information needed for authorizing the user's access.

However, building an authentication service for an LBS system is not straightforward, as ethical and legal issues need to be considered. Specifically, the anonymity and confidentiality of crucial personal data should be maintained. To elaborate further, data flow in LBS platforms is usually complex and, as already explained, it normally requires the contacting of other systems in order to retrieve certain information (e.g., the GMLC or the GIS server, etc.). These systems are also considered to be provisioning platforms and require some kind of authentication before accessing their services (e.g., position retrieval in the case of GMLC or map retrieval from the GIS server). Two different authentication procedures can be followed. First the LBS system can authenticate itself to these external gateways and then proceed with retrieving the requested data. This method may be sufficient for accessing the GIS server as no personal data are requested, but it is insufficient for the Positioning server (e.g. the GMLC). This insufficiency derives from the need to maintain the confidentiality and privacy of the user's location. The positioning server comprises part of the PLMN of the mobile operator and, although the latter should track the location of its subscribers (according to EU-112 and US-911 directive), this information should be revealed only in the case of an emergency. Considering the above, the authenticated LBS server as a third-part entity mediating between the mobile operator and the end user is not authorized to have access to such information.

An alternate authentication procedure would be to have the user authenticate to the Positioning Server, for retrieving his own location. However, even this type of authentication may not be sufficient as the authentication request passes through the LBS server and therefore the latter should be a trusted entity. The usual way of solving such security issues is to move the LBS system inside the mobile operator's PLMN, in which case the privacy and confidentiality of the requested information is always maintained. Towards this direction, IETF formed the geographic location privacy (geopriv) workgroup [34], whose main task is to assess the authorization, integrity and privacy requirements that must be met in order to transfer such information, or authorize the release or representation of such information through an agent. Further issues covered by the geopriv initiative are the way that the authorization of requestors and responders is handled as well as the authorization of proxies. The ultimate goal of geopriv is the specification of a service that is capable of transferring geographic location information in a private and secures fashion (including the option of denying transfer). The geopriv activities could significantly contribute to the maintenance of the privacy and confidentiality of location data transferred between the positioning servers and the LBS server; however currently the geopriv protocol is still in the development phase[a].

Another service that is common to security infrastructures is the accounting service. The accounting service maintains a per-user account of performed actions (e.g., authentication requests, accessed services, data requested, etc.), with the purpose of providing a means of determining possible attacks or unauthorized admissions to the system. Since LBS systems do not impose specific requirements regarding this service, we will not elaborate further on it. Accounts are usually stored in RDBMSs, which can be internal or external to the LBS server. For communication with external accounting systems the OSA specification contains a certain Part [13], which defines the way in which such a communication is implemented in an open and standard way.

---

[a] A series of Internet drafts and RFCs [33, 35] have been published.

### 14.3.5 LBS Clients

The location services client is the interface through which end-users interact with the LBS server. It is the client's device where all technologies are brought together and, in most cases, its usability and performance also determines the user's opinion of the delivered service. Mainstream consumers will judge products on issues such as the aesthetics of the form factor and ease of use, in addition to application performance and delivery.

The proliferation of the Internet and the World Wide Web technology has impacted significantly on the way client/server software is developed. Creating massive installed bases has never been easier than today, owing to the advancements of many versatile technologies such as distributed environments, object-oriented computing, connectionless application protocols, and text-based markup languages.

The incorporation of business logic completely on the server side allowed the development of thin client systems with improved capabilities and low power consumptions, which are ideal for the mobile world, while at the same time eliminated the headaches of software upgrades and multi-version support. Application functionality is downloaded on the fly as needed. This section elaborates on the type of available LBS clients and discusses common platforms that currently exist or are planned for release in the near future.

In terms of functionality LBS clients can be divided into two large categories:

(1)  dedicated LBS clients;
(2)  non-dedicated LBS clients.

The first category includes devices that are exclusively oriented to delivering this type of service. Typical examples of such devices are the navigation and pilot systems that are integrated into vehicle dashboards or sold as standalone units. Most of these devices feature an integrated GPS receiver for acquiring the coordinates of the supporting mobile unit and their operation is offline. Maps or other location-specific information are retrieved from CD-ROMS or DVD-ROMS integrated with the device. Today, many vendors provide many such products (Figure 14.14). Products range from in-dash navigation systems (Blaupunkt's TravelPilot-DX series [http://www.blaupunkt.co.uk], Harman Kardon's TrafficPro series [http://www.hktrafficpro.com], Pioneer AVIC series [http://www.pioneer.co.uk], KENWOOD's KNA-DV+KVT/DDX series [http://www.kenwoodeurope.com]) to remote-mount (Garmin's StreetPilot and GPSMAP series [http://www.garmin.com], Alpine's INA-N333RS and NVE-N099P units [http://www.alpine.com], Magelan's ROADMATE series [http://www.magellangps.-com], AvMap's GeoSat2 [http://www.avmap.it]) and hand-held ones (Thales MobileMapper and ProMark series [http://www.thalesnavigation.com], Garmin's eTrex, Rino, Geko, GPS, Foretrex/forunner and GPSMAP series [http://www.garmin.com], Magelan's Meridian and SporTrack series [http://www.magellangps.com]).

Many of the abovementioned devices, and especially those aimed for inside-vehicle use, also offer supplementary services (e.g., radio receiver capabilities, etc.); however as their prime usage is the delivery of navigation information or some other type of LBS, they are considered as belonging in this category. A major shortcoming of dedicated LBS clients is their high cost, but this cost does also include the delivery of the services. Moreover, the fact that they usually operate off-line makes them difficult to adapt to changes of data, and the type service offered is usually pre-defined and cannot easily be enriched. An advantage is the precision and the quality of the results.

Non-dedicated LBS clients can be used for different purposes, including their use in the LBS domain. This category includes devices whose primary functionality is not the pure delivery of location-based services but they also feature LBS-specific capabilities (e.g., GPS receivers, etc.). A further classification of the devices in this category is necessary. On the one hand we have non-dedicated LBS clients, which have integrated LBS capabilities, while on the other there are similar devices that do not integrate such functionality in their architecture but which can achieve it through additional hardware and

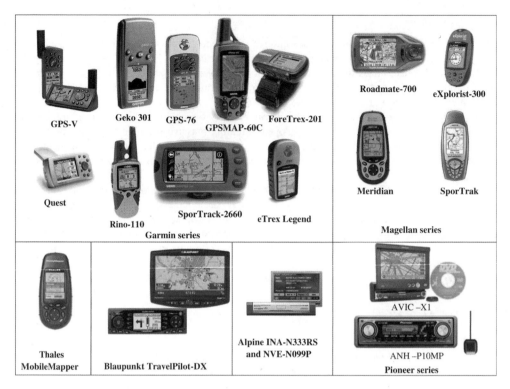

**Figure 14.14**   Dedicated LBS client devices.

software upgrades. The simplest example of a device that belongs to this second group is that of ordinary notebooks or PDAs that, with the addition of a PCMCIA card and the appropriate software, can become powerful LBS clients. Even without the addition of new hardware, these devices can use their network connection for accessing the location-based services offered by an LBS server (e.g., a standard GPRS or UMTS-enabled phone). The first group includes mobile phones, PDAs or other similar devices, that integrate LBS-specific circuits. Several organizations are providing such devices) (Garmin[http://www.garmin.com], Motorola [http://www.motorola.com], Hitachi [http://www.hitachi.com], Sanyo [http://www.sanyo.com], etc.). Some of these devices are shown in Figure 14.15.

Owing to their limited size and storage capabilities, all these devices are considered thin clients and require server-side support (i.e., the LBS server discussed in Section 14.3.1) for its proper operation in the LBS domain.

Non-dedicated LBS-clients have certain advantages over the dedicated ones. The most significant is their on-line operation. LBS-specific data is usually located on the server sides, where it can be updated on a frequent basis, easily and without cost to the end-user. They are used for a variety of purposes, and probably many of the consumers already own one. Their cost is usually lower than that of dedicated clients, but the user is also has the additional cost of buying access to the service. Although some of these devices (e.g., laptops and PDAs) can be really expensive, considering their multi-purpose functionality, the end-user would probably be willing to pay the cost.

Non-dedicated clients do not usually possess the precision and quality of the LBS services that are available to the dedicated clients. However, this varies, depending on the type of the device and the additional hardware that is used. Some of them can easily receive hardware and software upgrades (e.g., possibly enabling new positioning techniques). These upgrades can potentially improve the quality and precision of the delivered service.

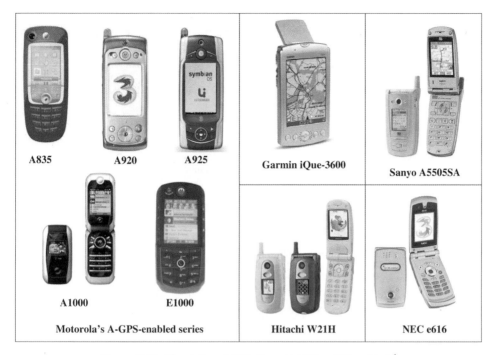

**Figure 14.15**   Non-dedicated LBS clients (A-GPS or GPS-enabled).[b]

## 14.4  Available LBS Systems

Many software vendors, prompted by both the boom of the mobile Internet and the lack of ready LBS provisioning solutions, have developed software tools and middleware platforms for handling both the creation and delivery of such services. Most of these platforms offer a robust environment that, to some extent, can be used for developing LBS services. The vast majority of LBS platforms are built using Java enterprise technologies, and most of them comply with industry standards through open interfaces. A list of the most popular LBS platforms that are available today is given in Table 14.1.

Table 14.1 illustrates that only a few solutions cover all the phases of the LBS provisioning chain, that is from the specification of a new service to its actual deployment and delivery to end users. Most of the available platforms focus on the core functionality of an LBS server, which is the delivery of the service to the end user; they neglect other issues such as service creation and deployment. Others do consider these latter issues, but restrict their capabilities to a limited subset of the full LBS spectrum. Indoor and outdoor environments are sufficiently covered by the variety of existing platforms but only a few unified solutions, covering both cases, really exist. Finally, open interfaces and compliance with widely acceptable protocols are a characteristic common to most of them.

Platforms, such as Celltick's Interactive Broadcast that supports only the SMS and WAP interfaces for GSM/ GPRS, or Ericsson's MPS are enhanced position-tracking solutions for second or third generation mobile networks. Kivera's Location Engine does not implement any interface with network operators

---

[b] Many of the pictures of the devices, along with information about the capabilities of the mobile phones can be found at http://www.3gtoday.com/devices

**Table 14.1**   LBS systems and their characteristics

| Location platform | Provided by | Key characteristics | |
|---|---|---|---|
| ArcLocation Solutions | ESRI | • WAP/SMS/HTTP connectivity<br>• GMLC connectivity using MLP | http://www.esri.com |
| Autodesk Location Services | AutoDesk | • Service deployment through Java or web services APIs | http://www.autodesk.com |
| Canvas Location-Enabling Server | Telenity | • SCE provided, Services specified in SCML<br>• OSA/Parlay support | http://www.telenity.com |
| Interactive Broadcast Platform | Celltick Technologies | • Supports GSM and GPRS networks using SMS or WAP | http://www.celltick.com |
| Location Engine | Kivera Inc. | • No interface to positioning infrastructure | http://www.kivera.com |
| LocationAgent | Mapflow | • Service deployment over 2G/3G networks | http://www.mapflow.com |
| LocationNet | LocationNet | • LBS middleware with integrated GIS engine.<br>• GMLC connectivity using LIF's MLP.<br>• Support for the HTTP, SMS/MMS and WAP protocols | http://www.locationet.com/ |
| MapInfo MapXtreme Java Edition | MapInfo | • Java middleware for LBS but without positioning interface | http://www.mapinfo.com |
| Mobile Positioning System | Ericsson | LBS for 2G/3G networks | http://www.ericsson.com |
| PanGo Proximity Platform | PanGo Networks | • Proximity services for WLAN environments | http://www.pangonetworks.com |
| Webraska Products | Webraska Mobile Technologies | • GMLC positioning interface<br><br>• SOAP HTTP/XML APIs for service development and deployment | http://www.webraska.com |
| The Cellpoint MLS/MLB architecture | cellpoint | • 2G/3G networks support | http://www.cellpoint.com |
| Location Studio | Openwave | • Tools for creating LBS applications<br>• Web-service compliant interfaces towards GIS systems and positioning components | http://www.openwave.com |
| Appear Server | Appear Networks | • Supports WLAN 802.11 and Bluetooth environments<br>• Application downloaded and executed on the device | http://wwwappearnetworks.com |
| PoLoS platform | IST research project | • HTTP, WAP, SMS protocols supported<br>• Open web-services enabled interfaces towards external entities<br>• OSA/Parlay support for communication with positioning systems<br>• Indoor and Outdoor support<br>• SCE provided, Services written in SCL | http://www.polos.org |

for acquiring positioning, and thus can be considered a spatial data provider platform and not an autonomous LBS platform.

The most advanced platforms available today cater for all standard interfaces used by mobile phones (SMS, Multimedia Messaging Service (MMS), WAP) as well as HyperText Markup Language (HTML) for 2.5G and 3G HTML-enabled mobile phones and PDAs. In terms of positioning technologies, those supported are mostly those interfacing with GMLC-enabled 2G and 3G networks. LocatioNet and Cellpoint's MLS/MLB platforms cover all the aforementioned features together with a set of off-the-shelf applications, while the former comes with a high-performance GIS engine. Webraska and ESRI's ArcLocation platforms add support for GPS-enabled handheld devices. Telenity's Canvas Location-Enabling Server, apart from supporting standardized XML interfaces with positioning and GIS servers, facilitates service creation through a Service Creation Environment (SCE). The result is a service in the service-creation markup language (SCML) format that is deployed and executed on the platform. Deployment of LBSs through Web services or Java Application Programming Interface (API)s is supported by Autodesk's LocationLogic, whilst Openwave with Location Studio acts more like a mediator than an integrated platform hosting application allowing the development of LBS applications, with Web-services compliant interfaces towards GIS and positioning components. In indoor environments and specifically in IEEE 802.11 enabled infrastructures, PanGo provides the Proximity Platform for deploying and delivering LBS, while Appear Networks follows a different approach with its Provisioning Server that focuses on fast and effortless delivery of all types of mobile applications (including location-aware applications) for local execution on the wireless device.

Finally, the PoLoS platform [42, 43], which is a prototypical LBS system developed in the context of the IST framework, provides an integrated solution for creating and delivering LBS services in both indoor and outdoor environments. The platform comes with a service creation subsystem that allows the easy creation of services as well as their remote deployment and administration. Services are specified using the Service Creation Language (SCL) [44], which is based on the XML specifications. During the deployment phase, the SCL scripts are transferred to the server through a secure web-services interface, where they are translated in Java, compiled, and installed on the platform. The PoLoS middleware offers its hosted services access to positioning and GIS systems as well as to other supportive entities (e.g., billing components, authentication servers, other external data repositories, etc.). For communication with PLMN components (e.g., positioning or SMS gateways), the widely acceptable Parlay/OSA protocol has been adopted. PoLoS provides a complete provisioning infrastructure that allows the provision of services either on-request, through a wide range of supported protocols (HTTP, WAP, SMS), or their scheduling on a time or event (e.g., when the user enters a specific area) basis.

## Acknowledgement

This work is supported by the PYTHAGORAS program of the Greek Ministry of National Education and Religious Affairs (University of Athens Research Project No. 70/3/7411).

## References

[1] C. Pawsey, J. Green, R. Dineen and M. Munoz Mendez-Villamil, Ovum Forecasts: Global Wireless Markets 2002–2006 – An Ovum Report, February 2002.
[2] J. Spinney, A brief history of LBS and how OpenLS fits into the new value chain, ESRI, 2003.
[3] A. Jagoe, *Mobile Location Services: The Definitive Guide*, Prentice Hall, 2002.
[4] J. Hightower and G. Borriello, Location systems for ubiquitous computing, *IEEE Computer*, **34**(8), 2001.
[5] A. Leick, *GPS Satellite Surveying*, 2nd edn, John Wiley & Sons, 1995.
[6] A. El-Rabbany, *Introduction to GPS: The Global Positioning System*, Mobile Communication Series, Artech House, 2002.

[7]  B. W. Parkinson and P. K. Enge, *Differential GPS, Global Positioning System: Theory and Applications*, Vol. II, ed. B. W. Parkinson and J. J. Spilker Jr., American Institute of Aeronautics and Astronautics, Inc., Washington DC, pp. 3–50, 1996.

[8]  J. Benedicto, S. E. Dinwiddy, G. Gatti, R. Lucas and M. Lugert, *GALILEO: Satellite System Design and Technology Developments*, European Space Agency, November 2000.

[9]  http://www.rssi.ru/SFCSIC

[10]  3GPP TS 29.198-1, Open Service Access (OSA); Application Programming Interface (API); Part 1; Overview, September 2004.

[11]  3GPP TS 29.198-3, Open Service Access (OSA); Application Programming Interface (API); Part 3; Framework, September 2004.

[12]  3GPP TS 29.198-6, Open Service Access (OSA); Application Programming Interface (API); Part 6: Mobility, September 2004.

[13]  3GPP TS 29.198-11, Open Service Access (OSA); Application Programming Interface (API); Part 11; Account Management, September 2004.

[14]  3GPP TS 29.198-12, Open Service Access (OSA); Application Programming Interface (API); Part 12: Charging, September 2004.

[15]  LIF TS 101 Specification, Mobile Location Protocol specification, Version 3.0.0, June 2002.

[16]  Mobile Positioning Protocol Specification, Version 5.0, Ericsson 2003.

[17]  J. Hightower and G. Borriello, 'A Survey and taxonomy of location systems for ubiquitous computing, University of Washington, Computer Science and Engineering, Technical Report UW-CSE 01-08-03, August 2001.

[18]  J. Hightower and G. Borriello, 'Location systems for ubiquitous computing, University of Washington, Computer Science and Engineering, August 2001.

[19]  J. Hightower and G. Borriello, Real-Time error in location modeling for ubiquitous computing, University of Washington, Computer Science and Engineering, October 2001.

[20]  N. B. Priyantha, A. Chakraborty and H. Balakrishnan, The cricket location-support system, *6th ACM MOBICOM*, August 2000.

[21]  J. Hightower, G. Borriello and R. Want, SpotON: an indoor 3D location sensing technology based on RF signal strength, University of Washington, Computer Science and Engineering, Xerox Palo Alto Research Center, UW CSE Technical Report #2000-02-02, February 2000.

[22]  C. Kee, D. Yun, H. Jun, B. Parkinson, S. Pullen and T. Lagenstein, Centimeter-accuracy indoor navigation using GPS-like pseudolites, November 2001.

[23]  J. Werb and C. Lanzl, Designing a positioning system for finding things and people indoor, PinPoint Corp., September 1998.

[24]  P. Bahl and V. N. Padmanabhan, RADAR: An in-building RF-based user location and tracking system, Microsoft Research, INFOCOM 2000.

[25]  P. Bahl and V. N. Padmanabhan, A. Balachandran University of California at San Diego, 'Enhancements to the RADAR user location and tracking system, February 2000, Technical Report, MSR-TR-2000-12, Microsoft Research.

[26]  P. Castro, P. Chiu, Ted K. and R. Muntz, A Probabilistic room location service for wireless networked environments, In *Proc. UBICOMP 2001*.

[27]  The Nibble Location System, http://mmsl.cs.ucla.edu/nibble

[28]  Ekahau Positioning Engine™, http://www.ekahau.com/products/positioningengine

[29]  M. Wallbaum, WhereMoPS: an indoor geolocation system, *13th IEEE International Symposium on Personal, Indoor, and Mobile Radio Communications*, Lisbon, Portugal, September 5–18, 2002.

[30]  J. Hightower, B. Brumitt, and G. Borriello, the location stack: a layered model for location in ubiquitous computing, *Proc. 4th IEEE Workshop on Mobile Computing Systems and Applications* (WMCSA 2002), NY, June 2002.

[31]  G. Papazafeiropoulos, N. Prigouris, I. Marias, S. Hadjiefthymiades and L. Merakos, Retrieving position from indoor WLANs through GWLC, *Ist Mobile and Wireless Communications Summit*, Portugal, June 2003.

[32]  OGC Implementation Specification document No. 03-006r3, OpenGIS Location Services (OpenLS): Core Services [Part 1–5], version 1.0, Open GIS Consortium Inc., 16 January 2004, http://www.opengeospatial.org, accessed September 2004.

[33]  J. Cuellar, J. Morris, D. Mulligan, J. Peterson and J. Polk, Geopriv Requirements, RFC 3693, IETF February 2004.

[34] Geographic Location Privacy (geopriv), http://www.ietf.org/html.charters/geopriv-charter.html, accessed September 2004.

[35] M. Danley, D. Mulligan, J. Morris and J. Peterson, Threat Analysis of the Geopriv Protocol, RFC 3694, IETF February 2004.

[36] Shashi Shekhar, Ranga Raju Vatsavai, Xiaobin Ma and Jin Soung Yoo, Navigation systems: a spatial database perspective, in *Location-Based Services*, eds J.Schiller and A.Voisard, Morgan Kaufmann, Elsevier, 2004.

[37] Oracle spatial user's guide and reference, 10g Release 1 (10.1), Oracle Co., December 2003.

[38] IBM DB2 Spatial Extender – User's guide and reference, Version 8, IBM Corp., 2002.

[39] http://www.esri.com/software/arcgis/arcsde

[40] G. F. Marias, N. Prigouris, G. Papazafeiropoulos, S. Hadjiefthymiades and L. Merakos, Brokering positioning data from heterogeneous infrastructures, *Wireless Personal Communications Journal*, September 2004.

[41] 3GPP TS 03.71 V8.8.0 (2004-03), Technical Specification Group Services and System Aspects; Digital cellular telecommunications system (Phase 2+); Location Services (LCS); (Functional description) - Stage 2 (Release 1999).

[42] A. Ioannidis, I. Priggouris, I. Marias, S. Hadjiefthymiades, C. Faist-Kassapoglou, J. Hernandez and L. Merakos, PoLoS: integrated platform for location-based services, *Proc. IST Mobile and Wireless Communications Summit*, Portugal, June 2003.

[43] M. Spanoudakis, A. Batistakis, I. Priggouris, A. Ioannidis, S. Hadjiefthymiades, and L. Merakos Extensible platform for location based services provisioning, in *Proc. 3rd International Workshop on Web and Wireless Geographical Information Systems* (*W2GIS 2003*), Rome, December 2003.

[44] A. Ioannidis, M. Spanoudakis, P. Sianas, I. Priggouris, S. Hadjiefthymiades and L. Merakos Using XML and related standards to support location based services, in *Proc. SAC'2004*, 19th *ACM Symposium on Applied Computing, Web Technologies and Applications Special Track*, Nicosia, Cyprus, March 2004.

# Index

Abstract Data Type, 412, 413
accounting, see service accounting
Acknowledged mode (AM), 80
active badge, 411
Active bat, 411
Adaptive Differential Pulse Code Modulation
    (ADPCM), 26
adaptive encoding, 56
Adaptive Multi-Rate (AMR), 249
Adaptive Multi-Rate Wideband (AMR-WB), 35
Adaptive Resource Reservation Over Wireless
    (ARROW), 207–211
    performance, 211–213
Additive Increase Multiplicative Decrease, 61
ADPCM, see Adaptive Differential Pulse Code
    Modulation
ADT, see Abstract Data Type
Advanced Mobile Phone Service (AMPS), 237–238
Aggregate Control
    in RTSP, 99–101
AIM, 320
Algorithm
    A*, 413
    Dijkstra, 413
    Iterative Deepening A* (IDA*), 413
AMR, 78
AMR-WB, 78
Angle of Arrival, 407
angulation, 410
AOA, see Angle of Arrival
API, see Application Programming Interface
Application Programming Interface, 399, 409, 410, 412,
    417, 418, 422, 423
Application Server (AS), 243, 248
application server, 399, 409, 410
Application Service Provider, 409, 415, 416
Application-defined RTCP Packet, 69
ARROW, see Adaptive Resource Reservation Over
    Wireless
ASP, see Application service Provider
Asynchronous Transfer Mode (ATM), 241
attenuation, 410, 411
authentication, see service authentication

Background QoS Class, 253
Base Station Controller (BSC), 241
Base Station Subsystem (BSS), 241
Base Transceiver Station (BTS), 236–237, 241, 251
Bearer Service, 239, 252
billing
    model, 416, 417
    service, 400, 416, 417
    system, 398, 400, 414, 415, 416, 417
Bluetooth security, 179
Bluetooth, 172
bluetooth, 422
Breakout Gateway Control Function (BGCF),
    246, 249
Broadcast / Multicast Service Center (BM-SC), 249–
    251, 254

Call Session Control Functions (CSCF), 112
Card, 276
CBR encoding, 52
CDR, 417
CelB, 75
Cell Broadcast Center (CBC), 244
Cell Broadcast Service (CBS), 244, 250
cell, 236
CGw/CCF, 220
Circuit Switched (CS) domain, 240–241, 243, 245–246,
    249, 252
Code Division Multiple Access (CDMA), 238
Code Excited Linear Prediction Coding (CELP), 34
color Representation, 37
Common Open Policy Service (COPS), 254, 256
Constant Bit Rate, 79
continuous media (CM), 50
Control Plane
    IETF protocols for, 88
    ITU-T protocols for, 85–86
controlled distortion, 20–21
conversational applications, 55
conversational audio, 53
Conversational QoS Class, 253
COPS for Policy Provisioning (COPS-PR), 254
COPS, 112

---

Core Network (CN), 239, 241, 245, 249, 252
Cricket, 411

Data Plane
  IETF protocols for, 87
  ITU-T protocols for, 85–86
Deck, 276
decoder, 51
de-jitter buffer, 51
delay jitter, 58
delay-constrained retransmission, 60
Dialog SIP, 108
DIAMETER, 112
Differential Pulse Code Modulation (DPCM), 25–26
Digital Advanced Mobile Phone Service (D-AMPS),
  238
discrete media, 50
Domain Name System (DNS), 247
DPCM, see Differential Pulse Code Modulation
DS, see service Directory
DVI4, 74
dynamic RSVP (dRSVP), 214
    flow rejection, 214–215
    performance, 216–217

Ekahau's Positioning Engine, 411
end-systems, 63
Enhanced Data Rates for GSM Evolution (EDGE), 239,
  242
Enhanced Messaging Service (EMS), 293
Enhanced Observed Time Difference, 407, 408
entropy, 19–20
    coding 20
    reduction 20
E-OTD, see Enhanced Observed Time Difference
EPE, see Ekahau's Positioning Engine
ESA, see European Space Agency
ETSI, 399, 408, 409
European Space Agency, 407
European Telecommunications Standards Institute
  (ETSI), 239
extensibility, 397
eXtensible HyperText Markup Language (XHTML),
  264, 267, 268, 275, 279, 305
eXtensible Markup Language (XML), 264
eXtensible Stylesheet Language Transformations
  (XSLT), 280

First Generation (1G), 236–239
fleet management, see management fleet
Forward Error Correction (FEC), 58
Frequency Division Duplexing (FDD), 237, 242
Frequency Division Multiple Access (FDMA),
  237, 242
functionality additions
  RTSP, 98

G722, 74
G723, 74
G726–40, 74
G729, 75
Galileo, 407, 413
Gatekeeper
    Discovery of, 86
Gateway GPRS Support Node (GGSN), 249–252,
    254–256
Gateway Mobile Location Center, 408, 409, 418, 422, 423
Gateway Mobile Positioning Center, 410
Gateway Mobile services Switching Center (GMSC),
    240–241
GCI, see Global Cell Id
General Packet Radio Service (GPRS), 238–243,
    246–247
General Packet Radio Service (GPRS), 285
Geography Markup Language, 399
GIS
    server, 401, 418, 422
    service, 400, 401, 402
    system, 398, 412, 422, 423
Global Positioning System, 406
Global System for Mobile Communications (GSM),
    237–239, 241–242
Glocal Cell Id, 407
GLONASS, 407
GML, see Geography Markup Lanaguage
GMLC, see Gateway Mobile Location Center
GMPC, see Gateway Mobile Positoning Center
Goodbye RTCP Packet, 69
GPRS Attach, 241, 247
GPRS Support Node (GSN), 241
GPRS, 408, 420, 421, 422
GPS, 395, 396, 406, 407, 411, 413, 419, 423
    Assisted (A-GPS), 406, 407, 421
    Differential (D-GPS), 406, 407
Group of Blocks (GOB), 79
GS, see service Gateway
GSM EDGE Radio Access Network (GERAN), 239,
    241–242, 249, 252
GSM, 407, 408, 409, 418, 421, 422
GSM, 75
GSM-EFR, 75

H.261, 42
H.263 video codec, 78
H.263, 42
H.264, 45
H261, 77
H263, 77
H263–1998, 77
H323
    Comparison with SIP, 89
handover, 236, 242
Headers in SIP messages, 105

Hierarchical MRSVP (HMRSVP), 201
High Speed Circuit Switched Data (HSCSD), 238
HiperLAN/1, 125
HiperLAN/2, 121, 127
    comparison with 802.11e, 155
    convergence layer, 132
    DLC, 128
    error control, 130
    MAC, 128
    performance, 132–133
    physical, 123
Home Subscriber Server (HSS), 112
Home Subscriber Server (HSS), 240–241, 246–248
HyperText Markup Language (HTML), 264, 267, 275, 278

indoor
    environment, 397, 405, 407, 410, 411, 421, 423
    positioning, see positioning indoor
    systems, 405
IAPP, see Inter-Access Point Protocol (IAPP)
IAPP+, 204–206
ICQ, 320
IEEE 802.11, 411, 422, 423
IEEE 802.15, 169
IMMP, 349, 351, 356, 363
IMS Session, 242
Instant Messaging and Presence Service, 319
   Addressing, 325
   API, 339
   architecture, 326
   context awareness, 345
   groups, 325
   home control, 344
   Instant Message Service, 321
   interoperability, 339
   management, 326
   presence service, 321
   presentity, 321
   rich presence, 343–344
   security, 23, 339–343
     denial of service, 341
     end-to-end security, 340–341
     privacy, 340
     spam, 340
   Subscriber, 321
   User Agent, 321
   virtual presence, 345
   Watcher, 32
instant messaging, 350, 354
Integrated Services Digital Network (ISDN), 235, 237, 240, 246, 249, 252
Inter-Access Point Protocol (IAPP), 203
Interactive QoS Class, 253
interleaving, 58, 60, 71
inter-media synchronization, 50, 70

International Mobile Telecommunications 2000 (IMT-2000), 238
International Telecommunications Union (ITU), 238
Internet Group Management Protocol (IGMP), 250–251
Internet Relay Chat, 320
Internet Service Vendor, 409
Interrogating Call State Control Function (I-CSCF), 246–248
InterWorking Function (IWF), 237
intra-media synchronization, 50, 70
IP Multicasting, 244, 251–252
IP Multimedia Subsystem (IMS), 239, 242–243, 245–249, 252–253, 256
IP telephony, 55
IP version 4 (IPv4), 248, 250
IP version 6 (IPv6), 235, 246–248, 250
ISV, see Internet Service Vendor

Jabber, 320
Java library, 315
jitter, 70
JPEG, 75
JustYak, 323–324

location
    physical, 405, 406, 410
    symbolic, 405, 406
    UTM, 405, 406
L16, 75
L8, 75
lateration, 410, 411
layered encoding, 75
LBS, 395–406, 408, 412, 414–423
LCS, see Location Client Service
LIF, see Location Interoperability Forum
Line ending
    in RTSP, 96
    in SIP, 105
linear prediction, 23
Lluna, 345, 347
Location Based Service, 395, 396, 412, 413, 419, 420
Location Client Service, 408, 410
Location Interoperability Forum, 408, 410, 422
Location service, see service location
Location Stack, 412
Location Utililities Service, 413
location-dependent information services, 269, 282, 288
LPC, 75
LUS, see Location Utilities Service

Management
    fleet, 396
    interface, 403
    layer, 399, 400, 403, 404
MBMS Bearer Context, 251
MBMS Session, 249

MBMS UE Context, 251
MBOA, 193
Media description
  DESCRIBE in RTSP, 98
  field in SDP, 92
Media GateWay (MGW), 241, 246, 249
Media Gateway Control Funcion (MGCF), 112
Media Gateway Control Function (MGCF), 246, 249
Media Gateways (MGW), 112
media on Demand (MoD), 55
media on demand delivery, 53
media-Independent FEC, 59, 71
media-specific FEC, 59, 61, 71
MEGACO, 112
Midcall Mobility, 110
middleware, 398, 399, 401, 421, 422, 423
MMBox, 299, 308
MMS PDU, 302
MMS Relay, 294, 295
MMS transaction flows, 307
MMS-based services, 310
mobile commerce, 349
Mobile Ip with SIP, 109
mobile payment, 349, 351
Mobile Positoning Center, 409
Mobile Positoning Protocol, 410
Mobile RSVP (MRSVP), 201
Mobile services Switching Center (MSC), 240
Mobile Station (MS), 236–237
Mobile Station Application Execution Environment
  (MEXE), 299
Mobile Terminal (MT), 240, 252
model-based rate control, 57
MotionStar Magnetic Tracker, 411
MP2T, 77
MP3, 27–29
MPA, 75
MPC, see Mobile Positoning Center
MPEG-1, 43
MPEG-2, 44
MPEG-4 AAC, 78
MPEG-4 video codec, 78
MPEG-4, 44
MPP, see Mobile Positioning Protocol
MPV, 77
MSR Easy Living, 411
MSR-Radar, 411
Multicast Listener Discovery (MLD), 250–251
Multimedia Broadcast / Multicast Service (MBMS),
  239, 242, 244–245, 249–254
Multimedia Messaging Service (MMS), 293
Multimedia Resource FUnction (MRF), 112
Multimedia Resource Function Controller (MRFC), 246,
  248
Multimedia Resource Function Processor (MRFP), 246,
  248

multimedia session, 63
multimedia traffic, 170
Multipoint Control Unit (MCU), 56
Multi-Pulse Excitation Coding (MPE), 31

navigation, 395, 412, 413, 419
Nibble, 411
non-adaptive Encoding, 56
non-real-time (non-RT), 50
NTP
  use in SDP, 91

OGC, see Open Gis Consortium
OMA, see Open Mobile Alliance
one-way streaming, 53
Open Gis Consortium, 399, 412, 424
Open Mobile Alliance (OMA), 263, 293
Open Mobile Alliance, 399, 408
Open Service Access, 401, 408, 409, 410, 412, 417, 418,
  422, 423
Open Service Architecture (OSA), 300, 301
openLS, 412
OPTIONS
  RTSP method, 98
OSA, see Open Service Access
outdoor
  environments, 397, 405, 421, 423
  positioning, see positioning outdoor
  systems, 405, 406

Packet Data Protocol (PDP) Context, 241, 247–248,
  251, 253, 255–256
packet loss rate, 70
Packet Switched (PS) Domain, 240–241, 245, 249
parity-based FEC, 59
Parlay (Group), 408, 409, 422, 423
Parlay, 326
PAUSE
  RTSP method, 101
PCMA, 75
PDP Context, 113
PEER, see Pictogram-Enhanced Entity Relationship
piconet, 172, 184
Pictogram-Enhanced Entity Relationship, 414
Pin Point 3D-iD, 411
PLAY
  RTSP method, 101
PLMN, see Public Land Mobile Network
Policy Based QoS, 254
Policy Control Function (PDF), 254–256
Policy Enforcement Function (PEF), 254–256
Policy Information Base (PIB), 254
positioning
  indoor, 405, 410, 412
  method, 405, 406, 407, 408, 410
  outdoor, 405, 406

Server, 410, 418
system, 395, 398, 400, 401, 404, 405, 406, 410,
    411, 413, 422
technology, 397, 405, 410
power control, 242
priority-based retransmission, 60
privacy, 397, 408, 418
probe-based rate control, 57
proximity, 410, 411, 413, 422, 423
Proxy Call Session Control Function (P-CSCF),
    246–248, 254–256
PS, see service Presentation
Pseudolites, 411
PTT architecture, 375
    Migration, 391
PTT description, 370
PTT standardization, 378
    Open Mobile Alliance (OMA), 379
    3GPP and 3GPP2, 379
    IETF, 379
Public Land Mobile Network, 408, 418, 423
Public Switched Telephone Network (PSTN), 235, 237,
    240, 246, 249, 252
Pulse Code Modulation (PCM), 249
Pulse Code Modulation (PCM), 25
Push modality, 264, 266, 269, 298
push model, 397

QCELP, 75
QoS requirements, 55
QoS, 398, 415, 416, 417
Quality of Service (QoS), 109, 112
Quality of Service (QoS), 242, 243, 247–248,
    250–256
qualizers, 22

Radio Access Network (RAN), 239–242, 249–252
Radio Network Controller (RNC), 241–242
Radio Network Subsystem (RNS), 242
Range Header
    use in RTSP, 101
rate adaptation control, 57, 71
Real Time Control Protocol (RTCP), 62
Real Time Protocol (RTP), 246, 249
Real Time Protocol (RTP), 62
real-time (RT) multimedia transmission, 51
Real-time Transport Protocol (RTP), 62, 77
Receiver Report RTCP Packet, 67
receiver-based rate control, 58
RED, 75
REDIRECT
    method en RTSP, 103
redundancy reduction, 21–22, 25
Reed-Solomon (RS) codes, 59
Regular Pulse Excitation Coding (RPE), 33
Rendering Time, 101

retransmission 58, 71
reusability, 397, 399
RF-tags, 411
roaming, 397
Round Trip Time (RTT), 70
routable address, 296,
Route Determination Service, see service Route
    Determination
RSVP mobility proxy, 202
RSVP over wireless, 213
RTP/RTCP
    use in SDP, 92
RTSP
    overview, 94–95

service
    accounting, 400, 401, 418
    authentication, 400, 401, 402, 417, 418
    creation, 397, 399, 400, 414, 415, 421, 423
    creation environment, 423
    creation system, 400, 414, 415, 423
    deployment, 397, 399, 414, 422
    Directory, 413
    Gateway, 413
    location, 400, 404, 408, 412, 413, 419, 422
    logic, 397, 403
    Presentation, 413
    route determination, 413
    scheduler, 402
    specification language, 414
scalability, 397, 399, 415
scalable encoding, 57
SCE, see service creation environment
scene analysis, 410, 411
SCF, see Service Capabiltiy Feature
Scheduler, see service scheduler
SCS, see Service Capabiltiy Server
SDB, see Spatial Database
SDP
    Extended syntax in 3GPP, 116
    Session Establishement, 88
    Syntax of Messages, 90
seamless service continuity, 219
Second Generation (2G), 237–240
security, 396, 397, 404, 412, 414, 417, 418
SEMOPS, 352
Sender Report RTCP Packet, 66
sensors, 411, 412
service adaptation, 279, 280
Service Capability Feature, 408, 409, 410, 412
Service Capability Server, 409, 410
service capability, 240
Service Set Identifiers (SSIDs), 222
service usability, 280, 301
Serving Call State Control Function (S-CSCF),
    246–249

Serving GPRS Support Node (SGSN), 241, 244–246, 249–251
Session Initiation Protocol (SIP), 247–249, 254–256
SETUP
  RTSP method, 102
  3GPP adaptaion, 117
Short Message Service (SMS), 264
Signaling GateWay (SGW), 246, 249
SIM Application Toolkit (SAT), 278
Simple Object Access Protocol (SOAP), 278, 300
SIMPLE, 322, 331–335, 342–343
  API, 345
  architecture, 331–332
  instant messaging service, 333–334
  Message Session Relay Protocol, 334–335
  page-mode messaging, 333–334
  presence agent, 332
  presence service, 332–333
  presence user agent, 332
  security, 343
  session mode messaging, 334–335
SIP
  Comparison with H323, 89
  headers, 105
  methods, 105
SIPS URI Scheme, 105
Smart Floor, 411
SMS, 397, 402, 403, 421, 422, 423
SNR, 182
software tools for mobile service development, 275, 315
Source Description RTCP Packet, 68
source-based rate control, 57
Spatial
    data server, 399
    data system, 398, 400, 401, 412, 414, 415, 423
    database, 412, 413, 414
    information, 412, 413, 414
speech compressors, 25
SpotOn, 411
Stateful and Stateless SIP proxy, 104
streaming QoS Class, 253
Subscription Locator Function (SLF), 246–247
supplementary service, 240
Synchronized Multimedia Integration Language (SMIL), 278, 305

teleservice, 240
Terminal Equipment (TE), 240, 252
Terminating a session
  in RTSP, 102
  int SIP, 108
Third Generation (3G), 236–240
Three way handshake in SIP, 107
Time Division Duplexing (TDD), 242
Time Division Multiple Access (TDMA), 237–238, 242
Time Of Arrival, 407

time of flight, 410, 411
Timing in RTSP, 101
TOA, see Time Of Arrival
tracking, 395, 396, 411, 421
traffic model, 282
Transmission Control Protocol (TCP), 61
triangulation, 406, 407, 410, 411
TTL
  use in RTSP multicast/unicast, 92–93

UL-TOA, see Time Of Arrival
Ultra Wideband (UWB), 191
UMTS, 408, 409, 420
Unacknowledged Mode (UM), 80
Uniform Resource Locator (URL), 264, 276
Universal Mobile Telecommunications System (UMTS), 239–242, 252–254
Universal Terrestrial Radio Access Network (UTRAN), 239, 241–242, 249, 252
Universal Transverse Mercator , see location UTM
User Agent Client (UAC), 103
User Agent Server (UAS), 103
User Datagram Protocol (UDP), 62
User Equipment (UE), 239–240, 246–251, 252–256
UTM, see location UTM

Variable Bit Rate, 79
VBR encoding schemes 52
VDVI, 75
videoconferencing (videophone), 55
Visitor Location Register (VLR), 240–241
Visitor Mobile services Switching Center (VMSC), 240

WAP bearers, 263
WAP Forum, 263
WAP gateway, 265
WAP proxy, 265
WAP-based services, 286
Wf interface, 220
WhereMoPS, 412
Wideband Code Division Multiple Access (W-CDMA), 238–239, 242
Wi-Fi, 181
WiMedia, 169
Wireless Application Environment (WAE), 267
Wireless Datagram Protocol (WDP), 275
Wireless Markup Language (WML), 264, 276
Wireless Personal Area Network (WPAN), 169
Wireless Session Protocol (WSP), 269
Wireless Transaction Protocol (WTP), 271
Wireless Transport Layer Security (WTLS), 273
Wireless Village, 322, 327–331, 342
  access control, 342
  application service elements, 327
  instant messaging service, 328–330

presence service, 330–331
protocol suite, 328
WMLScript, 264, 268, 276
Wo, interface, 221
Wr/Wb interface, 220
Wx, interface, 221

XMPP, 322, 335–337, 343
architecture, 335–336
instant message service, 336
presence service, 336–337
security, 343

Yahoo! Messenger, 323–324

ZigBee, 194

3G AAA Proxy, 220
3G, 77
3GPP, see 3rd Generation Partnership Project
3G/WLAN, 217
architecture, 218
performance, 224–231
QoS, 222
3rd Generation Partnership Project (3GPP), 238–239,
252
3rd Generation Partnership Project 2 (3GPP2), 238
3rd Generation Partnership Project, 399, 408, 409,
410, 417

802.11, 123
DCF, see Distributed Coordination Function

Distributed Coordination Function, 133
MAC, 133
PCF, see Point Coordination Function
performance, 146
Point Coordination Function, 135
standardization, 136–137
802.11, 181
802.11, see IEEE 802.11
802.11b, see IEEE 802.11
802.11e, 137
ACK policies, 144
Admission control, 143
comparison with HiperLAN/2, 155
EDCA, see Enhanced Distributed Channel
Access
Enhanced Distributed Channel Access,
137
HCCA, see HCF Coordination Channel
Access
HCF Coordination Channel Access, 139
HCF, see Hybrid Coordination Function
Hybrid Coordination Function, 139
power management, 143
Simple scheduler, 142
traffic scheduling, 206
Video over 802.11e, 152
VoIP over 802.11e, 150
802.15.1, 172
802.15.2, 180
802.15.3, 184
802.15.3a, 191
802.15.4, 194